EPA ENVIRONMENTAL ENGINEERING SOURCEBOOK

Edited by

J. Russell Boulding

CRC Press
Taylor & Francis Group
Boca Raton London New York

CRC Press is an imprint of the
Taylor & Francis Group, an **informa** business

CRC Press
Taylor & Francis Group
6000 Broken Sound Parkway NW, Suite 300
Boca Raton, FL 33487-2742

© 1996 by Taylor & Francis Group, LLC
CRC Press is an imprint of Taylor & Francis Group, an Informa business

First issued in paperback 2019

No claim to original U.S. Government works

ISBN-13: 978-0-367-44865-3 (pbk)
ISBN-13: 978-1-57504-002-8 (hbk)

Visit the Taylor & Francis Web site at
http://www.taylorandfrancis.com

and the CRC Press Web site at
http://www.crcpress.com

ABOUT THE EDITOR

J. Russell Boulding first began working in the environmental field in 1973 when he helped set up the Environmental Defense Fund's Denver Office, and has been a freelance environmental consultant since 1977 when he established Boulding Soil-Water Consulting in Bloomington, Indiana. He has a B.A. in Geology (1970) from Antioch College, Yellow Springs, Ohio, and an M.S. in Water Resources Management (1975) from the University of Wisconsin/Madison. From 1975 to 1977 he was a soil scientist with the Indiana Department of Natural Resources and mapped soils in southern Indiana on a cooperative program with the U.S. Soil Conservation Service. Since 1984 he has been Senior Environmental Scientist with Eastern Research Group, Inc. in Lexington, Massachusetts.

Mr. Boulding is the author of more than 120 books, chapters, articles and consultant reports in the areas of soil and ground-water contamination assessment, geochemical fate assessment of hazardous wastes, mined land reclamation, and natural resource management and regulatory policy. From 1978 to 1980 he served as a member of the Environmental Subcommittee of the Committee on Surface Mining and Reclamation (COSMAR) of the National Academy of Sciences (NAS) and as a consultant to the NAS Committee on Soil as a Resource in Relation to Surface Mining for Coal. Mr. Boulding is an ARCPACS Certified Professional Soil Classifier and National Ground Water Association Certified Ground Water Professional. His professional memberships include the Soil Science Society of America, International Society of Soil Science, Association of Ground Water Scientists and Engineers, and the International Association of Hydrogeologists. Since 1992 he has been a member of the American Society for Testing and Material's Committee D18 (Soil and Rock) and active in subcommittees D18.01 (Surface and Subsurface Characterization), D18.07 (Identification and Classification of Soils) and D18.21 (Ground Water and Vadose Zone Investigations). In 1993 he became chair of D18.01's Section on Site Characterization for Environmental Purposes.

The U.S. Environmental Protection Agency (U.S. EPA), through a number of its national research laboratories, publishes several series of relatively short documents that provide up-to-date information on the current state of knowledge about environmental site assessment and remediation of contaminated soil and ground water. Many of these papers have become classics, and widely cited in the literature on environmental site assessment and remediation. Other equally valuable papers have not received the attention they deserve. For several years I have been collecting these documents from EPA's R.S. Kerr Environmental Research Laboratory (Ada, OK), Environmental Monitoring Systems Laboratory (Las Vegas, NV), and Risk Reduction Engineering Laboratory (Cincinnati, OH)[1], and it occurred to me that it would be nice to make them available in a convenient form for more widespread use. This volume, *EPA Environmental Engineering Sourcebook*, includes papers and bulletins that focus on remediation of contaminated soils and ground water. The companion volume, *EPA Environmental Assessment Sourcebook*, includes papers that focus on contaminant behavior and transport processes and modeling and environmental site characterization and monitoring.

Three types of EPA documents are included in this volume:

1. *Ground Water Issue Papers* (5 chapters). These papers have been developed by the Regional Superfund Ground Water Forum, a group of ground-water scientists, representing EPA's Regional Superfund Offices, organized to exchange up-to-date information related to remediation of Superfund sites.
2. *Engineering Issue Papers* (2 chapters). These papers have been developed by the Regional Superfund Engineering Forum, a group of EPA professionals representing EPA's Regional Superfund Offices organized to identify and resolve engineering issues impacting the remediation of Superfund sites.
3. *Engineering Bulletins* (21 chapters). Engineering Bulletins are a series of documents that summarize the latest information available on selected treatment and site remediation technologies and related issues.

Editing of the chapters in this volume include minor editorial corrections, reorganization for consistency, current affiliation of authors (if changed), updating of references originally cited as in press, and additional relevant EPA Superfund Innovative Technology Evaluation (SITE) program publications. Where possible, the National Technical Information Service (NTIS) acquisition number for other EPA documents has been added, and some additional annotations may be added, such as the availability of more recent editions of books cited in the original paper. EPA Engineering Bulletins do not list authors at the beginning, as ground-water issue papers do, but authors are identified in this volume, if specific authors could be identified.

Many of the EPA contacts listed in the original bulletins have retired or been reassigned to other EPA programs. I would like to thank Stephen Schmelling, RSKERL-Ada, and Trish Ericson, RREL-Cincinnati, for taking the time to help me check and update the current affiliations of authors of the

---

[1] Recently all of EPA's laboratories were reorganized and renamed. RSKERL and RREL are now part of the National Risk Management Research Laboratory and EMSL-Las Vegas is now the National Exposure Research Laboratory. This volume retains the old names and abbreviations because they are firmly embedded in all the documents included in the volume and because they more clearly differentiate the various sources.

documents included in this volume, and to make sure that the names of EPA contacts and their phone numbers are up-to-date.

EPA CONTACT ADDRESSES

The location and telephone number of one or more EPA contacts is included at the end of each chapter. The mailing addresses for RREL-Cincinnati and RSKERL-Ada are as follows:

U.S. EPA Risk Reduction Engineering Laboratory (RREL)
26 West Martin Luther King Drive
Cincinnati, Ohio 45268

U.S. EPA Robert S. Kerr Environmental Research Laboratory (RSKERL)
P.O. Box 1198
Ada, OK 74820

CONTENTS

Part I: Containment, Pump-and-Treat, and In Situ Treatment

Part II: Ex Situ Treatment Methods for Contaminated Soils, Ground Water and Hazardous Waste

ABBREVIATIONS AND ACRONYMS

APA	Air pathway analysis
APC	Air pollution control
ARAR	Applicable or relevant and appropriate requirements
ASTM	American Society for Testing and Materials
BDAT	Best demonstrated available technology
BTEX	Benzene, toluene, ethylene, xylene
BTX	Benzene, toluene, xylene
CB	Cement bentonite (slurry mixture)
CERCLA	Comprehensive Environmental Response, Compensation, and Liability Act (Superfund)
COD	Chemical oxygen demand
CST	Critical solution temperature
DCE	Dichloroethylene
DE	Destruction efficiency
DNAPL	Dense nonaqueous phase liquid
DOC	Dissolved organic carbon
DRE	Destruction removal efficiency
EFS	Electrode feed system
EM	Electromagnetic
EMSL-LV	U.S. EPA Environmental Monitoring Systems Laboratory, Las Vegas, NV (recently renamed Technical Support Center for Monitoring and Site Characterization)
EPA	Environmental Protection Agency
GAC	Granular activated carbon
HC	Hydrocarbon
HDPE	High density polyethylene
HTAS	High temperature air stripping
ISV	In situ vitrification
LDR	Land disposal restrictions
LNAPL	Light nonaqueous phase liquid
MCL	Maximum contaminant level
MeCl	Methylene chloride
MEK	Methyl ethyl ketone
NAPL	Nonaqueous phase liquid
NPDES	National pollution discharge elimination system
NPL	Superfund National Priority List
O&G	Oil and grease
O&M	Operation and maintenance
ORD	U.S. EPA Office of Research and Development
ORNL	Oak Ridge National Laboratory
OSC	Superfund On-Site Coordinator
OSWER	U.S. EPA Office of Solid Waste and Emergency Response
PAH	Polyaromatic hydrocarbon
PCB	Polychlorinated biphenyl
PCDD	Polychlorinated dibenzo dioxin
PCDF	Polychlorinated dibenzo furan
PCE	Perchloroethylene/tetrachloroethene

PCP	Pentachlorophenol
PIC	Products of incomplete combustion
PNA	Polynuclear aromatics
POHC	Principal organic hazardous constituents
POTW	Publicly owned treatment work
PM	Particulate matter
PVC	Polyvinyl chloride
RBC	Rotating biological contactor
RCRA	Resource Conservation and Recovery Act
ROD	Record of Decision
ROI	Radius of influence
RPM	Superfund Remedial Project Manager
RREL	U.S. EPA Risk Reduction Engineering Laboratory (recently renamed National Risk Management Research Laboratory)
RSKERL-Ada	U.S. EPA Robert S. Kerr Environmental Research Laboratory, Ada, OK (recently renamed National Risk Management Research Laboratory)
SAIC	Science Applications International Corporation
SARA	Superfund Amendments and Reauthorization Act
SB	Soil bentonite (slurry mixture)
SBOD	Soluble biological oxygen demand
SCOD	Soluble chemical oxygen demand
SCWO	Supercritical water oxidation
SITE	Superfund innovative technology evaluation program
SIVE	Steam injection vapor extraction
SS	Suspended solids
S/S	Stabilization/solidification
Superfund	See CERCLA
SVE	Soil vapor extraction
SVOC	Semivolatile organic compound
TCA	Trichloroethane
TCE	Trichloroethylene
TCLP	Toxicity characteristic leaching procedure
TOC	Total organic carbon
TPH	Total petroleum hydrocarbons
TS	Treatability study
TSCA	Toxic Substances Control Act
TWA	Total waste analysis
UCS	Unconfined compressive strength
UDRI	University of Dayton Research Institute
U.S. EPA	U.S. Environmental Protection Agency
VC	Volatilized contaminants or vinyl chloride
VOC	Volatile organic compound

PART I

Contaminant, Pump-and-Treat, and In Situ Treatment

Chapter 1

Slurry Walls[1]

Cecil Cross, Science Applications International Corporation (SAIC), Cincinnati, OH

ABSTRACT

Slurry walls are used at Superfund sites to contain the waste or contamination and to reduce the potential of future migration of waste constituents. In many cases slurry walls are used in conjunction with other waste treatment technologies, such as covers and around water pump-and-treat systems.

The use of this well-established technology is a site-specific determination. Geophysical investigations and other engineering studies need to be performed to identify the appropriate measure or combination of measures (e.g., landfill cover and slurry wall) to be implemented and the necessary materials of construction based on the site conditions and constituents of concern at the site. Site-specific compatibility studies may be necessary to document the applicability and performance of the slurry wall technology. The EPA contact whose name is listed at the end of this bulletin can assist in the location of other contacts and sources of information necessary for such studies.

This bulletin discusses various aspects of slurry walls including their applicability, limitations on their use, a description of the technology including innovative techniques, and materials of construction including new alternative barrier materials, site requirements, performance data, the status of these methods, and sources of further information.

TECHNOLOGY APPLICABILITY

Slurry walls are applicable at Superfund sites where residual contamination or wastes must be isolated at the source in order to reduce possible harm to the public and environment by minimizing the migration of waste constituents present. These subsurface barriers are designed to serve a number of functions, including isolating wastes from the environment, thereby containing the leachate and contaminated ground water, and possibly returning the site to future land use.

Slurry walls are often used where a waste mass is too large for practical treatment, where residuals from the treatment are landfilled, and where soluble and mobile constituents pose an imminent threat to a source of drinking water. Slurry walls can generally be implemented quickly, and the construction requirements and practices associated with their installation are well understood.

The design of slurry walls is site specific and depends on the intended function(s) of the system. A variety of natural, synthetic, and composite materials and construction techniques are available for consideration when they are selected for use at a Superfund site.

Slurry walls can be used in a number of ways to contain wastes or contamination in the subsurface environment, thereby minimizing the potential for further contamination. Typical slurry wall construction involves soil-bentonite (SB) or cement-bentonite (CB) mixtures. These structures are often used in conjunction with covers and treatment technologies such as in situ treatment and ground-water collection and treatment systems. Source containment can be achieved through a number of mechanisms including diverting ground-water flow, capturing contaminated ground water, or creating an upward ground-water gradient within the area of confinement (e.g., in conjunction with a ground-water pump-and-treat system). Containment may also be achieved by lowering the ground-water level inside the containment area. This will help to reduce hydraulically driven transport (known as "advective transport") from the containment area. However, even if the hydraulic gradient is directed toward

---

[1] EPA/540/S-92/008.

the containment area, transport of the contaminants (although thought to be minimal) is still possible. In many cases slurry walls are expected to be in contact with contaminants; therefore, chemical compatibility of the barrier materials and the contaminants may be an issue [1, pp. 373–374].[2]

The effectiveness of slurry walls and high density polyethylene (HDPE) geomembranes on soils and ground water contaminated with general contaminant groups is shown in Table 1-1. Examples of constituents within contaminant groups are provided in the Technology Screening Guide for Treatment of CERCLA Soils and Sludges [2]. This table is based on current available information or on professional judgment where no information was available. The proven effectiveness of the technology for a particular site or waste does not ensure that it will be effective at all sites or that the containment efficiencies achieved will be acceptable at other sites. For ratings used in this table, demonstrated effectiveness means that, at some scale, compatibility tests showed that the technology was effective or compatible with that particular contaminant and matrix. The ratings of potential effectiveness and no expected effectiveness are both based on expert judgment. Where potential effectiveness is indicated, the technology is believed capable of successfully containing the contaminant groups in a particular matrix. When the technology is not applicable or will probably not work for a particular combination of contaminant group and matrix, a no-expected-effectiveness rating is given.

LIMITATIONS

In the construction of most slurry walls it is important that the barrier is extended and properly sealed into a confining layer (aquitard) so that seepage under the wall does not occur. For a light, nonaqueous phase liquid, a hanging slurry may be used. Similarly, irregularities in the wall itself (e.g., soil slumps) may also cause increased hydraulic conductivity.

Slurry walls also are susceptible to chemical attack if the proper backfill mixture is not used. Compatibility of slurry wall materials and contaminants should be assessed in the project design phase.

Slurry walls also may be affected greatly by wet/dry cycles which may occur. The cycles could cause excessive desiccation, which can significantly increase the porosity of the wall.

Once the slurry walls are completed, it is often difficult to assess their actual performance. Therefore, long-term ground-water monitoring programs are needed at these sites to ensure that migration of waste constituents does not occur.

TECHNOLOGY DESCRIPTION

Low-permeability slurry walls serve several purposes including redirecting ground-water flow, containing contaminated materials and contaminated ground water, and providing increased subsurface structural integrity. The use of vertical barriers in the construction business for dewatering excavations and building foundations is well established.

The construction of slurry walls involves the excavation of a vertical trench using a bentonite-water slurry to hydraulically shore up the trench during construction and seal the pores in the trench walls via formation of a "filter cake" [3, pp. 2–17]. Slurry walls are generally 20 to 80 feet deep, with widths 2 to 3 feet. These dimensions may vary from site to site. There are specially designed "long stick" backhoes that dig to 90 foot depths. Generally, there will be a substantial cost increase for walls deeper than 90 feet. Clamshell excavators can reach depths of more than 150 feet. Slurry walls constructed at water dam projects have extended to 400 feet using specialized milling cutters. Depending on the site conditions and contaminants, the trench can be either excavated to a level below the water table to capture chemical "floaters" (this is termed a "hanging wall") or extended ("keyed") into a lower confining layer (aquitard) [3, p. 3–1]. Similarly, on the horizontal plane the slurry wall can be constructed around the entire perimeter of the waste material/site or portions thereof (e.g., upgradient, downgradient). Figure 1-1 diagrams a waste area encircled by a slurry wall with extraction and monitoring wells inside and outside of the waste area, respectively, along with a cross-section view of a slurry wall being used with the landfill cover technology [4, p. 1].

---

[2] [Reference number, page number.]

Table 1-1. Effectiveness of HDPE Geomembranes and Slurry Walls on General Contaminant Groups for Soil and Ground Water

Contaminant Groups	Effectiveness		
	HDPE Geomembranes	Slurry Walls	
		SB	CB
Organic			
Halogenated volatiles	■	▼	▼
Halogenated semivolatiles	■	▼	▼
Nonhalogenated volatiles	■	▼	▼
Nonhalogenated semivolatiles	■	▼	▼
PCBs	■	▼	▼
Pesticides (halogenated)	■	▼	▼
Dioxins/furans	▼	▼	▼
Organic cyanides	■	▼	▼
Organic corrosives	▼	●	●
Inorganic			
Volatile metals	■	▼	▼
Nonvolatile metals	■	▼	▼
Asbestos	■	▼	▼
Radioactive materials	▼	▼	▼
Inorganic corrosives	●	●	●
Inorganic cyanides	■	▼	▼
Reactive			
Oxidizers	●	●	●
Reducers	▼	▼	▼

■   Demonstrated Effectiveness: Short-term effectiveness demonstrated at some scale.
▼   Potential Effectiveness: Expert opinion that technology will work.
●   No Expected Effectiveness: Expert opinion that technology will not work.

The principal distinctions among slurry walls are differences in the low-permeability materials used to fill the trenches. The ultimate permeability of the wall is controlled by water content and ratios of bentonite/soil or bentonite/cement. In the case of a SB wall, the excavated soil is mixed with bentonite outside of the trench and used to backfill the trench. During the construction of a CB slurry wall, the CB mixture serves as both the initial slurry and the trench backfill. When this backfill gels (SB) or sets (CB), the result is a continuous barrier with lower permeability than the surrounding soils. A landfill cover, if employed, must extend over the finished slurry wall to complete the containment and to avoid desiccation.

Soil-bentonite slurry walls are the most popular since they have a lower permeability than CB walls, and are less costly [3, pp. 1–6; 5, p. 2]. Attapulgite may also be used in situations where the bentonite is not compatible with the waste [5, p. 16]. A newer development is the use of fly ash as a high carbon additive, not only to lower the permeability of the SB but also to increase the adsorption capacity of the SB with respect to the transport of organic chemicals [6, p. 1; 7, p. 444]. Permeabilities of SB walls as low as 5.0×10^{-9} cm/sec have been reported, although permeabilities around 1×10^{-7} cm/sec are more typical [3, pp. 2–28]. The primary advantage of the CB wall is its greater shear strength and lower compressibility. CB walls are often used on unstable slopes and steep terrain or where soils of low permeability are not accessible [3, pp. 2–40]. The lowest permeabilities of CB walls are typically 1×10^{-6} cm/sec or greater [3, p. 242; 5, p. 14]. It should be noted that organic and inorganic contaminants in ground water/leachate can have a detrimental effect on bentonite and the trench backfill material in both SB and CB walls. Therefore, it is imperative that a compatibility testing program be conducted in order to determine the appropriate backfill mixture.

Composite slurry walls incorporate an additional barrier, such as a geomembrane, within the trench to improve impermeability and chemical resistance. The geomembranes often are plastic screens that are comprised of HDPE pile plank sections which lock together. The locking mechanism is designed to

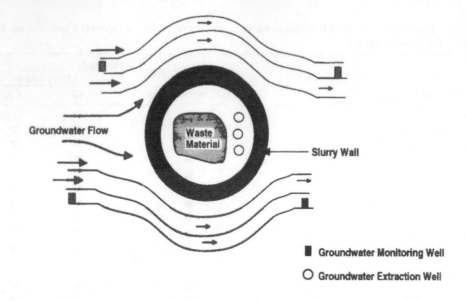

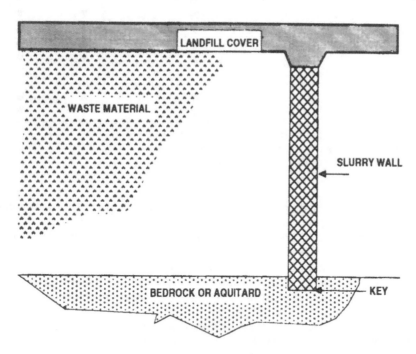

Figure 1-1. Aerial and cross-section view showing implementation of slurry walls [4].

minimize the leakage of the contaminated ground water. Table 1-2 shows one vendor's experience in using HDPE as a geomembrane [8]. The membrane: is easy to install; has a long life; and is resistant to animal and vegetation intrusion, microorganisms, and decay. Combining the membrane with a bentonite slurry wall may be the most effective combination. It is usually effective to construct the bentonite-cement slurry wall and then install the membrane in the middle of the wall. The toe of the membrane sheet is stabilized in the backfill material, cement, or in a special grout [5, p. 4]. The installation is reported to be effective in most every type of soil, is watertight, and may be constructed to greater depths.

A relatively new development in the construction of slurry walls is the use of mixed-in-place walls (also referred to as soil-mixed walls). The process was originally developed in Japan. A drill rig with multi-shaft augers and mixing paddles is used to drill into the soil. During the drilling operation a fluid

Table 1-2. Relative Chemical Resistivity of an HDPE Geomembrane [8][a]

Aromatic Compounds		Inorganic Contamination	
Benzene	+	NH_4	++
Ethylene Benzene	++	Fluorine	++
Toluene	+	CN	++
Xylene	++	Sulfides	++
Phenol	++	PO_4	++
Polycyclic Hydrocarbons		**Other Sources of Contamination**	
Naphthalene	++	Tetrahydrofuran	+
Anthracene	++	Pyrides	++
Phenanthrene	++	Tetrahydrothiophene	++
Pyrene	++	Cyclohexanone	++
Benzopyrene	++	Styrene	++
		Petrol	++
		Mineral Oil	++
Chlorinated Hydrocarbons		**Pesticides**	
Chlorobenzenes	+	Organic Chlorine	
Chlorophenols	++	Compounds	++
PCBs	++	Pesticides	++

Key: ++ = Good Resistance; + = Average

[a] Adapted from vendor's marketing brochure.

slurry or grout is injected and mixed with the soil to form a column. In constructing a mixed-in-place wall the columns are overlapped to form a continuous barrier. This method of vertical barrier construction is recommended for sites where contaminated soils will be encountered, soils are soft, traditional trenches might fail due to hydraulic forces, or space availability for construction equipment is limited. Both this method and a modified method termed "dry jet mixing" are usually more expensive than traditional slurry walls [5, p. 7; 9].

Another application of traditional slurry wall construction techniques is the construction of permeable trenches called bio-polymer slurry drainage trenches [10; 11]. Figure 1-2 diagrams a slurry wall and a bio-polymer slurry drainage trench constructed around a waste source; this will typically involve the use of a landfill cover in conjunction with the wall. Rather than restricting ground-water flow, these trenches are constructed as interceptor drains or extraction trenches for collecting or removing leachate, ground water, and ground-waterborne contaminants. These trenches also can be used as recharge systems. The construction sequence is the same as the traditional method described above. However, a biodegradable material (i.e., bio-polymer) with a high gel strength is used in the place of bentonite in the slurry, and the trench is backfilled with permeable materials such as sand or gravel. Once the trench is completed, the bio-polymer either degrades or is broken with a breaker solution that is applied to the trench. Once the bio-polymer filter cake is broken, the surrounding soil formation returns to its original hydraulic conductivity. Ground water collected in the trench can be removed by use of an extraction well or other collection system installed in the trench [10]. A bio-polymer trench can be used in conjunction with an SB or CB slurry wall to collect leachate or a contaminated plume within the wall (similar to the function of a well-point collection system). A geomembrane also can be installed with the bio-polymer wall to restrict ground-water flow beyond the bio-polymer wall.

Grouting, including jet grouting, employs high pressure injection of a low-permeability substance into fractured or unconsolidated geologic material. This technology can be used to seal fractures in otherwise impermeable layers or construct vertical barriers in soil through the injection of grout into holes drilled at closely spaced intervals (i.e., grout curtain) [5, p. 8; 12, pp. 5–97]. A number of substances can be used as grout including cement, alkali silicates, and organic polymers [12, pp. 5-97 to 5-101]. However, concerns surround the use of grouting for the construction of vertical barriers in soils because it is difficult to achieve and verify complete permeation of the soil by the grout. Therefore, the desired low permeabilities may not be achieved as expected [5, p. 8; 13, p. 7].

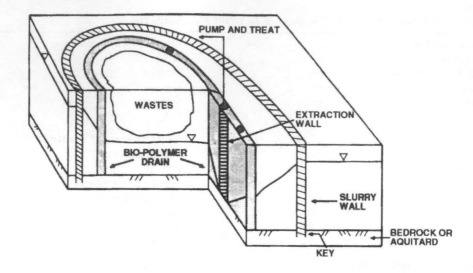

Figure 1-2. Schematic diagram of typical slurry wall and bio-polymer slurry trench [9]. (Drawing not to scale.)

SITE REQUIREMENTS

Treatment of contaminated soils or other waste materials requires that a site safety plan be developed to provide for personnel protection and special handling measures.

The construction of slurry walls requires a variety of construction equipment for excavation, earth moving, mixing, and pumping. Knowledge of the site, local soil, and hydrogeologic conditions is necessary. The identification of underground utilities is especially important during the construction phase [8].

In slurry wall construction, large backhoes, clamshell excavators, or multi-shaft drill rigs are used to excavate the trenches. Dozers or graders are used for mixing and placement of backfill. Preparation of the slurry requires batch mixers, hydration ponds, pumps, and hoses. An adequate supply of water and storage tanks is needed as well as electricity for the operation of mixers, pumps, and lighting. Areas adjacent to the trench need to be available for the storage of trench spoils (which could potentially be contaminated) and the mixing of backfill. If excavated soils will not be acceptable for use in the slurry wall backfill, suitable backfill material must be imported from off the site. In the case of CB walls, plans must be made for the disposal of the spoils since they are not backfilled. In marked contrast, deep soil mixing techniques require less surface storage area, use less heavy equipment, and may produce a smaller volume of trench spoils.

PERFORMANCE DATA

Performance data presented in this bulletin should not be considered directly applicable to all sites. A number of variables such as geographic region, topography, and material availability can affect the walls performance. A thorough characterization of the site and a compatibility study is highly recommended.

At the Hill Air Force Base in northern Utah the installation of a slurry wall, landfill covers, groundwater extraction and treatment, and monitoring was implemented to respond to ground water and soil contamination at the site. The slurry wall was installed along the upgradient boundary on three sides of Operable Unit No. 1 to intercept and divert ground water away from the disposal site. Operable Unit No. 1 consists of Landfill No. 3, Landfill No. 4, Chem Pits No. 1 and 2, and Fire Training Area No. 1. Shallow perched ground water and soils present were contaminated with halogenated organics and heavy metals. The performance of the slurry wall had been questioned because it was not successfully keyed into the underlying clay layer. This oversight was attributed to both the inadequate number and depth of soil borings. The combination of landfill caps, slurry wall, and ground-water extraction and treatment has resulted in a significant reduction in the concentrations of organics and inorganics detected seeping at the toe of Landfill No. 4. Organics were reduced to levels below 5 percent of their pre-remedial action levels and iron was reduced to 20 percent of its original observed concentration. Three separate QA/QC projects were implemented to assess the in situ effectiveness of the slurry wall. The determination of ground-water levels in monitoring wells on the inside and outside of the wall provided the most the useful data [14].

At the Lipari Landfill Superfund Site in New Jersey, a SB slurry wall was installed to encircle the landfill. A landfill cover, incorporating a 40 mil HDPE geomembrane, also was installed at the site. Heavy rains and snowmelt prior to the complete cap installation resulted in the need to perform an emergency removal (i.e., dewatering). Several years after completion of the slurry wall and landfill cover their effectiveness was evaluated during a subsequent feasibility study. The study concluded that the goal of an effective permeability of 1×10^{-7} cm/sec had been achieved in the slurry wall. Monitoring wells will be located at least 5 feet from the slurry wall on the upgradient side and 7 feet on the down gradient side [15]. The combination of technologies being used along with the slurry wall appears to be effectively containing the waste and its constituents.

A SB slurry wall, up to 70 feet deep, was installed at a municipal landfill Superfund site in Gratiot County, Michigan. The slurry wall was needed to prevent leachate from migrating into the local ground water. Approximately 250,000 ft^2 of SB slurry wall was installed at the site. The confirmation of achieving a goal of a laboratory permeability of less than 1×10^{-7} cm/sec for the soil-bentonite backfill was reported by an independent laboratory [16].

A SB slurry wall, extending through three aquifers, was installed at the Raytheon NPL site in Mountain View, California. Soil and ground water at the site were contaminated with industrial solvents. Permeability tests performed on the backfilled material achieved the goal of 1×10^{-7} cm/sec or less. Associated activities at the site included the rerouting of underground utilities, construction of 3-foot-high earthen berms around all work areas, construction of two bentonite slurry storage ponds, and construction of three lined ponds capable of storing 300,000 gallons of storm water. A ground-water extraction and stripping/filtration system is also in place at the site. The slurry wall, purposely, was not keyed into an aquitard so that the ground-water extraction program would create an upward gradient, thus serving to further contain the contaminants. The system appears to be functioning properly with the implementation of the combination of the technologies [17; 18]. However, this is the exception rather than the rule.

TECHNOLOGY STATUS

The construction and installation of slurry walls is considered a well-established technology. Several firms have experience in constructing this technology. Similarly, there are several vendors of geosynthetic materials, bentonitic materials, and proprietary additives for use in these barriers.

In EPA's FY 1989 ROD Annual Report [19], 26 RODs specified slurry walls as part of the remedial action. Of the RODs specifying slurry walls, 22 also indicated that covers would be used. Table 1-3 presents the status of selected Superfund sites employing slurry walls.

While site-specific geophysical and engineering studies (e.g., compatibility testing of ground water and backfill materials) are needed to determine the appropriate materials and construction specifications, this technology can effectively isolate wastes and contain migration of hazardous constituents. Slurry walls also may be implemented rather quickly in conjunction with other remedial actions. Long-term monitoring is needed to evaluate the effectiveness of the slurry wall.

EPA CONTACT

Technology-specific questions regarding slurry walls may be directed to Eugene Harris, RREL-Cincinnati, (513) 569-7862 (see Preface for mailing address).

ACKNOWLEDGMENTS

This bulletin was prepared for the U.S. Environmental Protection Agency, Office of Research and Development (ORD), Risk Reduction Engineering Laboratory (RREL), Cincinnati, Ohio, by Science

Table 1-3. Selected Superfund Sites Employing Slurry Walls [19]

Site	Location (Region)	Status
Ninth Avenue Dump	Gary, IN (5)	In design phase
Outboard Marine	Waukegan, IL (5)	In operation
Liquid Disposal	Utica, MI (5)	In design phase
Industrial Waste Control	Fort Smith, AR (6)	In operation since 3/91
E.H. Shilling Landfill	Ironton, OH (5)	In design phase
Allied/Ironton Coke	Ironton, OH (5)	In pre-design phase
Florence Landfill	Florence Township, NJ (2)	Design completed; remedial action beginning soon
South Brunswick	New Brunswick, NJ (2)	In operation since 1985
Sylvester	Nashua, NH (1)	In operation since 1983
Waste Disposal Engineering	Andover, MN (5)	In design phase
Diamond Alkali	Newark, NJ (2)	In pre-design phase
Hooker - 102nd St.	Niagara Falls, NY (2)	In remedial design phase
Scientific Chemical Processing	Carlstadt, NJ (2)	Completed 1992

Applications International Corporation (SAIC) under contract No. 68-C8-0062. Mr. Eugene Harris served as the EPA Technical Project Monitor. Mr. Gary Baker was SAIC's Work Assignment Manager. This bulletin was written by Mr. Cecil Cross of SAIC. The author is especially grateful to Mr. Eric Saylor of SAIC, who contributed significantly during the development of the document.

The following contractor personnel have contributed their time and comments by participating in the expert review meetings and/or peer-reviewing the document:

Dr. David Daniel, University of Texas
Dr. Charles Shackelford, Colorado State University
Ms. Mary Boyer, SAIC

REFERENCES

1. Gray, D.H. and W.J. Weber. Diffusional Transport of Hazardous Waste Leachate Across Clay Barriers. Seventh Annual Madison Waste Conference, Sept. 11–12, 1984.
2. Technology Screening Guide for Treatment of CERCLA Soils and Sludges. EPA/540/2-88/004 (NTIS PB89-132674). U.S. Environmental Protection Agency. 1988.
3. Slurry Trench Construction for Pollution Migration Control. EPA-540/2-84-001. U.S. Environmental Protection Agency. February 1984.
4. Waste Containment: Soil-Bentonite Slurry Walls. NEESA Document No. 20.2-051.1, November 1991.
5. Ryan, C.R. Vertical Barriers in Soil for Pollution Containment. Presented at the ASCE-GT Specialty Conference—Geotechnical Practice for Waste Disposal. Ann Arbor, MI. June 15–17, 1987.
6. Bergstrom, W.R. and D.H. Gray. Fly Ash Utilization in Soil-Bentonite Slurry Trench Cutoff Walls. Presented at the Twelfth Annual Madison Waste Conference, Sept. 20–21, 1989.
7. Gray, D.H., W.R. Bergstrom, H.V. Mott, and W.J. Weber. Fly Ash Utilization in Cutoff Wall Backfill Mixes. Proceedings from the Ninth Annual Symposium, Orlando, FL, January 1991.
8. Gundle Lining Systems, Inc. Geolock Vertical Watertight Plastic Screen for Isolating Ground Contamination. Marketing Brochure. 1991.
9. Geo-Con, Inc. Deep Soil Mixing, Case Study No. 1. Marketing Brochure. 1989.
10. Geo-Con, Inc. Deep Draining Trench by the Bio-Polymer Slurry Trench Method, Technical Brief. Marketing Brochure. 1991.
11. Hanford, R.W. and S.W. Day. Installation of a Deep Drainage Trench by the Bio-Polymer Slurry Drain Technique. Presented at the NWWA Outdoor Action Conference, Las Vegas, Nevada. May 1988.
12. Handbook—Remedial Action at Waste Disposal Sites (Revised). EPA-625/6-85/066 (NTIS PB87-201034). U.S. Environmental Protection Agency. 1985.

13. Technological Approaches to the Cleanup of Radiologically Contaminated Superfund Sites. EPA/540/ 2-88/002. U.S. Environmental Protection Agency. August 1988.

14. Dalpais, E.A., E. Heyse, and W.R. James. Overview of Contaminated Sites at Hill Air Force Base, Utah, and Case History of Actions Taken at Landfills No. 3 and 4, Chem. Pits 1 and 2. Utah Geol. Assoc. Publication 17. 1989.

15. U.S. Environmental Protection Agency. On-Site FS for Lipari Landfill, Final Draft Report. Prepared for U.S. EPA by CDM, Inc. et al. August 1985.

16. Geo-Con, Inc. Slurry Walls, Case Study No. 3, Marketing Brochure. 1990.

17. GKN Hayward Baker, Inc. Case Study Slurry Trench Cutoff Wall, Raytheon Company, Mountain View, CA. Marketing Brochure. 1988.

18. Burke, G.K. and F.N. Achhomer, Construction and Quality Assessment of the In Situ Containment of Contaminated Groundwater. In Proceedings of the 5th National Conference on Hazardous Wastes and Hazardous Materials. April 1988.

19. ROD Annual Report: FY 90. EPA/540/8-91/067. U.S. Environmental Protection Agency. July 1991.

Chapter 2

Landfill Covers[1]

Cecil Cross, Science Applications International Corporation (SAIC), Cincinnati, OH

ABSTRACT

Landfill covers are used at Superfund sites to minimize surface water infiltration and to prevent exposure to the waste. In many cases, covers are used in conjunction with other waste treatment technologies, such as slurry walls, ground-water pump-and-treat systems, and in situ treatment.

This bulletin discusses various aspects of landfill covers, their applicability, and limitations on their use, and describes innovative techniques, site requirements, performance data, current status, and sources of further information regarding the technology.

TECHNOLOGY APPLICABILITY

Covers may be applied at Superfund sites where contaminant source control is required. They can serve one or more of the following functions:

- isolate untreated wastes and treated hazardous wastes to prevent human or animal exposure
- prevent vertical infiltration of water into wastes that would create contaminated leachate
- contain waste while treatment is being applied
- control gas emissions from underlying waste
- create a land surface that can support vegetation and/or be used for other purposes

Covers may be interim (temporary) or final. Interim covers can be installed before final closure to minimize generation of leachate until a better remedy is selected. They are usually used to minimize infiltration when the underlying waste mass is undergoing most of its settlement. A more stable base will thus be provided for the final cover, reducing the cost of post-closure maintenance.

Covers also may be applied to waste masses that are so large that other treatment is impractical. At mining sites, for example, covers can be used to minimize the entrance of water to contaminated tailings piles and to provide a suitable base for the establishment of vegetation. In conjunction with water diversion and detention structures, covers may be designed to route surface water away from the waste area while minimizing erosion.

The effectiveness of covers on underlying soils and ground-water containing contaminants is shown in Table 2-1. Effectiveness is defined as the ability of the cover to perform its function over the long term without being damaged by the chemical characteristics of the underlying waste. Examples of constituents within contaminant groups are provided in the Technology Screening Guide for Treatment of CERCLA Soils and Sludges [1, p. 10].[2]

The degree of effectiveness shown in Table 2-1 is based on currently available information or on professional judgment when no information was available. The effectiveness of the technology for a particular site or waste does not ensure that it will be effective at all sites. Demonstrated effectiveness means that, at some scale, chemical resistance tests showed that landfill covers were resistant to that particular contaminant in a soil or ground-water matrix. The ratings of potential effectiveness and no expected effectiveness are based on expert judgment. Where potential effectiveness is indi-

---

[1] EPA/540/S-93/500.

[2] [Reference number, page number].

Table 2-1. Effectiveness of Covers on General Contaminant Groups for Soil and Ground Water

Contaminant Groups	Effectiveness of Covers
Organic	
Halogenated volatiles	▼
Halogenated semivolatiles	■
Nonhalogenated volatiles	▼
Nonhalogenated semivolatiles	■
PCBs	■
Pesticides (halogenated)	■
Dioxins/furans	■
Organic cyanides	■
Organic corrosives	■
Inorganic	
Volatile metals	■
Nonvolatile metals	■
Asbestos	■
Radioactive materials	■
Inorganic corrosives	■
Inorganic cyanides	■
Reactive	
Oxidizers	■
Reducers	■

■ Demonstrated Effectiveness: Short-term effectiveness demonstrated at some scale.

▼ Potential Effectiveness: Expert opinion that technology will work.

cated, the technology is believed capable of successfully containing the contaminant groups so indicated in a soil or ground-water matrix. If the technology were not applicable or probably would not work for a particular combination of contaminant group and matrix, a no expected effectiveness rating is given. Note that this rating does not occur in Table 2-1 for any of the contaminant groups.

LIMITATIONS

Landfill covers are part of landfilling technology, which is generally considered a technology of last resort in remediating hazardous waste sites. Landfilling of hazardous waste is not permitted without first applying the best available treatment. Landfilling technology does not lessen toxicity, mobility, or volume of hazardous wastes. However, when properly designed and maintained, landfills can isolate the wastes from human and environmental exposure for very long periods of time.

Covers are most effective where most of the underlying waste is above the water table. A cover, by itself, cannot prevent the horizontal flow of ground water through the waste, only the vertical entry of water into the waste. Other procedures (e.g., landfill liners, slurry walls, extraction wells) may be needed to exclude, contain, or treat contaminated ground water.

It is generally conceded that landfill components (liners and covers) will fail eventually, even though failure may occur after many tens or hundreds of years. Their effective life can be extended by long-term (30 years or more) inspection and maintenance [20]. Vegetation control and repairs associated with construction errors, cover erosion, settlement and subsidence are likely to be required. The need for cover repairs can be lessened considerably by adherence to a rigorous quality assurance program during construction.

TECHNOLOGY DESCRIPTION

The U.S. Environmental Protection Agency (EPA) has published several documents that provide guidance on the technology of cover construction at land disposal facilities [2–7]. Other documents specifically address remediation of radiologically-contaminated Superfund sites, including the use of covers [8; 9]. Design and construction of clay liners (not covers, specifically), properties of clay, testing methods, soil permeabilities, liner performance, and failure mechanisms are discussed at length in Reference 10.

The design of covers is site-specific and depends on the intended functions of the system. Many natural, synthetic, and composite materials and construction techniques are available. The effectiveness of covers (and other structural components of engineered landfills) has been shown to be primarily a function of the attention given to quality in choosing, installing, and inspecting those materials and techniques [24].

Covers can range from a one-layer system of vegetated soil to a complex multi-layer system of soils and geosynthetics. In general, less complex systems are required in dry climates and more complex systems are required in wet climates. The most complex systems are usually found on engineered landfills in the humid eastern United States, where the cover must meet the erosion and moisture requirements of the associated liner designed to contain the waste. Figure 2-1 depicts a vertical section of such a cover. Table 2-2 summarizes the function, materials of construction, and purpose of each of the components. Covers on Superfund sites usually contain some, but not necessarily all, of these components.

The materials used in the construction of covers include low-permeability and high-permeability soils and geosynthetic products. The low-permeability materials (geomembrane/soil layer) divert water and prevent its passage into the waste. The high-permeability materials (drainage layer) carry water away that percolates into the cover. Other materials may be used to increase slope stability.

The most critical components of a cover in respect to selection of materials are the barrier layer and the drainage layer. The barrier layer can be a geomembrane or low-permeability soil (clay), or both (composite).

Geomembranes are supplied in large rolls and are available in several thicknesses (20 to 140 mil), widths (15 to 100 ft), and lengths (180 to 840 ft). The polymers currently used include polyvinyl chloride (PVC) and polyethylenes of various densities. Geomembranes are much less permeable than clays; measurable leakage generally occurs because of imperfections created during their installation; however, the imperfections can be minimized [15].

Soils used as barrier materials generally are clays that are compacted to a hydraulic conductivity (usually referred to as permeability) no greater than 1×10^{-6} cm/sec or a combination of bentonite and other soil that will achieve a comparable or even lower permeability. Compacted soil barriers are generally installed in 6-inch minimum lifts to achieve a thickness of 2 feet or more.

A composite barrier uses both soil and a geomembrane, taking advantage of the properties of each. The geomembrane is essentially impermeable, but, if it develops a leak, the soil component prevents significant leakage into the underlying waste. A composite liner has proved to be the most effective in decreasing hydraulic conductivity [2, p. A-2].

Geosynthetic clay barriers are beginning to be used in place of both the geomembrane and clay components. The geosynthetic clay barriers are constructed of a thin layer of bentonite sandwiched between two geosynthetic materials. In use, the bentonite expands to create a low-permeability, resealable ("self-healing") barrier. It is supplied in rolls, but does not require seaming, as geomembranes do [21].

Other identified alternative barrier materials are flyash-bentonite-soil mixtures; super absorbent geotextiles; sprayed-on geomembranes and soil-particle binders; and custom-made bentonite composites with geomembranes or geotextiles [11, p. 63; 12, p. 6]. Potential advantages of alternative barriers include quick and easy installation, better quality control, cost savings potentially greater than use of compacted soil or standard soil/geomembrane composite, reduction in volume of material, lighter construction equipment required, and some self-healing capabilities [11, p. 65; 12, p. 6; 13, p. 225].

The following discussion briefly describes the construction of a multi-layer cover. It does not attempt to describe all of the possible configurations and materials.

Covers are usually constructed in a crowned or domed shape with side slopes as low as is consistent with good runoff characteristics. The bottom layer, which may be a granular gas collection layer, forms the base on top of the waste mass for the remainder of the cover. The clay component of the barrier layer is constructed on this base layer. The clay is spread and compacted in "lifts" a few inches thick until the desired barrier thickness is reached (usually 24 inches or more).

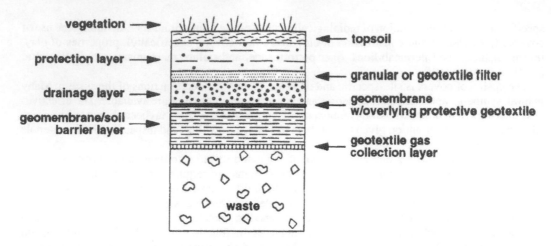

Figure 2-1. Cross-section of multi-layer landfill cover.

Each lift is scarified (roughed up) after compaction so there will be no discernible surface between it and the next higher lift when the latter is compacted. The top lift is compacted and rolled smooth so the geomembrane may be laid on it in direct and uniform contact. During the entire process the clay must be maintained at a near-optimum moisture content in order to attain the necessary low permeability upon compaction.

Low hydraulic conductivity is the most important property of the clay/soil barrier. Hydraulic conductivity is significantly influenced by the method of compaction, moisture content during compaction, compactive energy, clod size, and the degree of bonding between lifts [11, p. 6].

Geomembranes require a great deal of skill in their installation. They must be laid down without wrinkles or tension. Their seams must be fully and continuously welded or cemented, and they must be installed before the underlying clay surface can desiccate and crack. If vent pipes protrude through the cover, boots must be carefully attached to the membrane to prevent tearing if the cover subsides later. Care must be taken that the membrane is not accidentally punctured by workers or tools.

Extremes of temperature can adversely affect geomembrane installation, e.g., stiffness and brittleness are associated with low temperatures and expansion is associated with high temperatures. Thus, air temperature and seasonal variation are important design considerations [15].

A geotextile may be laid on the surface of the geomembrane for the geomembrane's protection, particularly if relatively coarse and sharp granular materials are applied as the drainage layer. Another geotexile can then be put on top of the drainage layer to prevent clogging of the drainage layer by soil from above. Fill soil and topsoil are then applied (compaction is not so critical) and the topsoil seeded with grass or other vegetation adapted to local conditions.

The drainage layer in a cover is designed to carry away water that percolates down to the barrier layer. It may be either a granular soil with high permeability or a geosynthetic drainage grid or geonet sandwiched between two porous geotextile layers. A geotextile may be used as a filter at the top of a granular soil drainage material to separate it from an overlying soil of different characteristics to prevent the drainage layer from becoming plugged with fine soil. A geotextile may also be used at the bottom of a granular drainage layer to protect the underlying geomembrane barrier from abrasion or puncture by sharp particles.

Other component layers may be used in landfill covers. Wider tolerances are generally acceptable in the material and construction requirements for these layers. Topsoil and subsoil from the vicinity are likely to be suitable for the surface and protection layers, respectively. The gas collection layer may be similar to the drainage layer in its characteristics, but it does not need to be. For example, gravel or coarse sand may be appropriate. Geosynthetic drainage materials may be used here too, but the chemical resistance to volatile wastes may be of greater concern due to the proximity of the waste and possibility for contact with it. However, EPA has no data that suggest damage to covers by volatiles.

Many laboratory tests are needed to ensure that the materials being considered for each of the cover components are suitable. Tests to determine the suitability of soil include grain size analysis

Table 2-2. Configuration of Cover Systems

Layer	Primary Function	Usual Materials	General Considerations
1. Surface Layer	Promotes vegetative growth (Most covers); Decreases erosion; Promotes evapotranspiration	Topsoil (humid site); cobbles (arid site); geosynthetic erosion control systems	Usually required for control of water and/or wind erosion
2. Protection Layer	Protects underlying layers from intrusion and barrier layer from desiccation and freeze/thaw damage; Maintains stability; storage of water	Mixed soils; cobbles	Usually required; May be combined with the protective layer into a single "cover soil" layer
3. Drainage Layer	Drains away infiltrating water to dissipate seepage forces	Sands; gravels; geotextiles; geonets; geocomposites	Optional; Necessary where excessive water passes through protection layer or seepage forces are excessive
4. Barrier Layer	Reduces further leaching of waste by minimizing infiltration of water into waste; Aids in directing gas to the emissions control system by reducing the amount leaving through the top of the cover	Compacted clay liners; geomembranes; geosynthetic clay liners; composites	Usually required; May not be needed at extremely arid sites
5. Gas Collection Layer	Transmits gas to collection points for removal and/or cogeneration	Sand; geotextiles; geonets	Usually required if waste produces excessive quantities of gas

(ASTM D422), Atterberg limits (ASTM D4318), and compaction characteristics (ASTM D698 or D1557). These tests generally are performed on the source material (called "borrow" material) before and during construction at predetermined intervals. Additional information on field and laboratory tests and procedures for installation quality verification are contained in an EPA Technical Guidance Document [23].

The major engineering soil properties that must be defined are shear strength and hydraulic conductivity. Shear strength may be determined with the unconfined compression test (ASTM D2166), direct shear test (ASTM D3080), or triaxial compression test (ASTM D2850). Hydraulic conductivity of soils may be measured in the laboratory with either ASTM D2434 or D5083. Field hydraulic conductivity tests are generally recommended and may be performed, prior to actual cover construction on test pads to ensure that the low-permeability requirements can actually be met under construction conditions. EPA strongly encourages the use of test pads [3; 4].

Laboratory tests are also needed to ensure that geosynthetic materials will meet the cover requirements. For example, geosynthetics in covers may be subjected to tensile stresses caused by subsidence and by the gravitational tendency of a geomembrane or material adjacent to it to slide or be pulled down slopes. Hydraulic conductivity of geomembranes is not defined, but leakage should not be significant in undamaged materials. Geosynthetic drainage materials (reinforcement-type products such as geonets and geotextiles) can become clogged or compressed under pressure and lose some or all of their drainage capacity.

The geosynthetics in a cover generally are not in direct contact with the underlying waste, so chemical resistance to the waste is not often a limitation [14, p. 79; 3, p. 109]. On the other hand, vapors from volatile contaminants have the potential to degrade cover materials. Note in Table 2-1 that although the organic volatiles are the only chemical groups with less than demonstrated effectiveness, the opinion of experts is that the use of geosynthetics in cover systems will work. The EPA has no evidence to suggest damage to covers by volatile organic compounds.

High-quality seams are essential to geomembrane integrity. Test-strip seaming, in which the actual seaming process is imitated on narrow pieces of excess membrane, can help to ensure high seam quality. The test strips should be prepared and subjected to strength (shear and peel) testing whenever equipment, personnel, or climatic changes are significant [15, p. 14]. Failure to meet specifications with the test strips indicates the necessity for destructive testing of actual field seams and correction of deficiencies in the seaming process.

Although construction quality assurance, including testing, will increase the installation cost about 10 to 15 percent and the time required to complete the project, it has been shown to improve the performance of the installation [22].

Steeply mounded landfills can have a negative effect on the construction and stability of the cover. A steep slope can make it difficult to compact soil properly due to the limited mobility and reduction of compacting effort of some compaction equipment. The rate of erosion is also a function of slope. Difficulty may arise in anchoring a geomembrane to prevent it from sliding along the interfaces of the geomembrane and soils. In some instances, geosynthetic reinforcement grids may be used to increase slope stability. Engineering design guidance addressing geomembrane stability can be found in Reference 16.

When constructing a new landfill or when covering an existing landfill where the surface of the waste mass can be graded, the EPA suggests that side slopes of a landfill cover not be less than 3 percent or exceed 5 percent [4, p. 24].

High air temperatures and dry conditions during construction may result in the loss of moisture from a clay barrier layer, causing desiccation cracking that can increase hydraulic conductivity. Desiccation cracking can be prevented by adding moisture to the clay surface and by installing the geomembrane in a composite barrier quickly after completion of the clay layer.

The hydraulic conductivity of compacted soil is also significantly influenced by the method of compaction, soil moisture content during compaction, compactive energy, clod size, and the degree of bonding between the individual lifts of soil in the barrier layer [11, p. 6].

Geomembranes are negatively influenced by different factors than soils during the construction process. Generally, more care must be taken to prevent accidental punctures. Sunlight can heat the material, causing it to expand. If installed while hot, the geomembrane can then shrink to the point of seam rupture if compensating actions are not taken. Seams must be carefully constructed to ensure continuity and strength. They should run up and down slopes rather than horizontally in order to reduce seam stress. Details of geomembrane installation can be found in Reference 15.

SITE REQUIREMENTS

The construction of covers requires a variety of construction equipment for excavating, moving, mixing, and compacting soils. The equipment includes bulldozers, graders, various rollers, and vibratory compactors Additional equipment is required in moving, placing, and seaming geosynthetic materials; e.g., forklifts and various types of seaming devices.

Storage areas are necessary for the materials to be used in the cover. If site soils are adequate for use in the cover, a borrow area needs to be identified and the soil tested and characterized. If site soils are not suitable, other low-permeability soils may have to be trucked in. An adequate supply of water may also be needed for application to the soil to achieve optimum soil density.

PERFORMANCE DATA

Once a cover is installed, it may be difficult to monitor or evaluate the performance of the system. Monitoring well systems or infiltration monitoring systems can provide some information, but it is often not possible to determine whether the water or leachate originated as surface water or ground water. Few reliable data are available on cover performance other than records of cover condition and repairs.

The difficulty in monitoring the performance of covers accentuates the need for strict quality assurance and control for these projects during construction. It is important to note that no landfill cover is completely impervious. It is also important to note that small perforations or poorly seamed or jointed materials can increase leakage potential significantly.

TECHNOLOGY STATUS

The construction of landfill covers is a well-established technology. Several firms have experience in constructing covers. Similarly, there are several vendors of geosynthetic materials, bentonitic materials, and proprietary additives for use in constructing these barriers.

In the EPA's FY 1989 ROD Annual Report [17], 154 RODs specified covers as part of the remedial action. Table 2-3 shows a selected number of Superfund sites employing landfill cover technology. While site-specific geophysical and engineering studies are needed to determine the appropriate materials and construction specifications, covers can effectively isolate wastes from rainfall and thus reduce leachate and control gas emissions. They can also be implemented rather quickly in conjunction with other anticipated remedial actions. Long-term monitoring is needed to ensure that the technology continues to function within its design criteria.

EPA CONTACT

Technology-specific questions regarding landfill covers may be directed to David A. Carson, RREL-Cincinnati, (513) 569-7527 (see Preface for mailing address).

ACKNOWLEDGMENTS

This bulletin was prepared for the U.S. Environmental Protection Agency, Office of Research and Development (ORD), Risk Reduction Engineering Laboratory (RREL), Cincinnati, Ohio, by Science Applications International Corporation (SAIC), under contract No. 68-C8-0062. Mr. Eugene Harris served as the EPA Technical Project Monitor. Mr. Gary Baker was SAIC's Work Assignment Manager. This bulletin was written by Mr. Cecil Cross of SAIC. The author is especially grateful to Mr. Eric Saylor of SAIC, who contributed significantly to the development of this bulletin.

Table 2-3. Selected Superfund Sites Employing Landfill Covers

Site	Location (Region)	Status
Chemtronics	Swannada, NC (4)	In design phase
Mid-State Disposal Landfill	Cleveland Township, WI (5)	In pre-design phase
Bailey Waste Disposal	Bridge City, TX (6)	In design phase
Cleve Reber	Sorrento, LA (6)	In design phase
Northern Engraving	Sparta, WI (5)	In operation since 1988
Ninth Avenue Dump	Gary, IN (5)	In design phase
Charles George Reclamation	Tyngsborough, MA (1)	In operation
E.H. Shilling Landfill	Ironton, OH (5)	In design phase
Henderson Road	PA (3)	In design phase
Ordinance Works Disposal	WV (3)	In design phase
Industri-Plex	Woburn, MA (1)	In design phase
Combe Fill North	Mount Olive Township, NJ (2)	Completed in 1991
Combe Fill South	Chester and Washington Township, NJ (2)	In design phase

The following contractor personnel have contributed their time and comments by participating in the expert review meetings or in peer-reviewing the document:

Dr. David Daniel, University of Texas
Mr. Robert Hartley, Private Consultant
Ms. Mary Boyer, SAIC

REFERENCES

1. U.S. Environmental Protection Agency. Technology Screening Guide for Treatment of CERCLA Soils and Sludges. EPA/540/288/004 (NTIS PB89-132674). 1988.
2. U.S. Environmental Protection Agency. Seminar Publication: Design and Construction of RCRA/CERCLA Final Covers. EPA/625/4-91/025. May 1991.
3. U.S. Environmental Protection Agency. Seminar Publication: Requirements for Hazardous Waste Design, Construction, and Closure. EPA/625/4-89/022. August 1989.
4. U.S. Environmental Protection Agency. Technical Guidance Document: Final Covers on Hazardous Waste Landfills and Surface Impoundments. EPA/530-SW-89-047 (NTIS PB89-233480). July 1989.
5. U.S. Environmental Protection Agency. Guide to Technical Resources for the Design of Land Disposal Facilities. EPA/625/6-88/018. December 1988
6. Lutton, R.J. Design, Construction, and Maintenance of Cover Systems for Hazardous Waste: An Engineering Guidance Document. EPA-600/2-87-039 (NTIS PB87-191656). 1987.
7. McAneny, C.C., P.G. Tucker, J.M. Morgan, C.R. Lee, M.F. Kelley, and R.C. Horz. Covers for Uncontrolled Hazardous Waste Sites. EPA-540/2-85-002 (NTIS PB87-152328). 1985.
8. U.S. Environmental Protection Agency. Technological Approaches to the Cleanup of Radiologically Contaminated Superfund Sites. EPA/540/2-88/002. August 1988.
9. U.S. Environmental Protection Agency. Assessment of Technologies for the Remediation of Radioactivity Contaminated Superfund Sites. EPA/540/2-90/001 (NTIS PB90-204140). January 1990.
10. U.S. Environmental Protection Agency. Design, Construction, and Evaluation of Clay Liners for Waste Management Facilities. EPA/530-SW-86/007F (NTIS PB89-181937). November 1988.
11. Daniel, D.E. and P.M. Estomell. Compilation of Information on Alternative Barriers for Liner and Cover Systems. EPA/600/2-91/002 (NTIS PB91-141846). October 1990.
12. Grube, W.E. and D.E. Daniel. Alternative Barrier Technology for Landfill Liner and Cover Systems. Presented at the AWWA 84th Annual Meeting and Exhibition, Vancouver, BC. June 1991.
13. Eith, A.W., J. Boschuk, and R.M. Koerner. Prefabricated Bentonite Clay Liners. Journal of Geotextiles and Membranes. 1991.

14. Koerner, R.M. and G.N. Richardson. Design of Geosynthetic System for Waste Disposal. ASCE-GT Specialty Conference Proceedings, Geotechnical Practice for Waste Disposal, Ann Arbor, MI. June 1987.

15. U.S. Environmental Protection Agency. Technical Guidance Document Inspection Techniques for the Fabrication of Geomembrane Field Seams. EPA/530/SW-91/051 (NITS PB92-109057). May 1991.

16. U.S. Environmental Protection Agency. Geosynthetic Design Guidance for Hazardous Waste Landfill Cells and Surface Impoundments. EPA/600/2-87/097. Cincinnati, OH. 1987.

17. U.S. Environmental Protection Agency. ROD Annual Report: FY 89. EPA/540/8-90/006. April 1990.

18. U.S. Environmental Protection Agency. Prediction and Mitigation of Subsidence Damage to Hazardous Waste Landfill Covers. EPA/600/2-87/025 (NTIS PB87-175378). 1987.

19. Daniel, D.E. and R.M. Koerner. Final Cover Systems. In: Geotechnical Practice for Waste Disposal, D.E. Daniel, Ed., Chapman & Hall, London, 1993, Chapter 18.

20. Bennett, R.D. and R.C. Horz. Recommendations to the NRC for Soil Cover Systems Over Uranium Mill Tailings and Low Level Radioactive Wastes. NU-REG/CR-5432. U.S. Army Engineer Waterways Experiment Station. 1991.

21. U.S. Environmental Protection Agency. Report of Workshop on Geosynthetic Clay Liners. EPA/600/R-93/171. August 1993.

22. Bonaparte, R. and B.A. Gross. Field Behavior of Double Liner Systems. Am. Soc. Civil Eng. Geotech. Publ. No. 26. November 1990.

23. U.S. Environmental Protection Agency. Technical Guidance Document: Construction Quality Management for Remedial Action and Remedial Design Waste Containment Systems. EPA/540/R-92/073. October 1992.

24. U.S. Environmental Protection Agency. Construction Quality Assurance for Hazardous Waste Land Disposal Facilities. EPA/530/SW-86-031 (NTIS PB87-132825). 1986.

Chapter 3

Control of Air Emissions from Materials Handling During Remediation[1]

Gary Baker, Science Applications International Corporation (SAIC), Cincinnati, OH

ABSTRACT

This bulletin presents an overview discussion on the importance of and methods for controlling emissions into the air from materials handling processes at Superfund or other hazardous waste sites. It also describes several techniques used for dust and vapor suppression that have been applied at Superfund sites.

Air emission control techniques have been utilized for Superfund cleanups at the McColl site (CA) and at the LaSalle Electric site (IL). Foam suppression has been used at Rocky Mountain Arsenal (CO), Texaco Fillmore (CA), and at a petroleum refinery (CA) site. A number of temporary vapor suppression techniques have also been applied at other sites. Additionally, the experience gained in the mining industry and at hazardous waste treatment, storage, and disposal sites will yield applicable methods for Superfund sites.

This bulletin provides information on the applicability of air emission controls for materials handling at Superfund sites, limitations of the current systems, a description of the control methods that have found application to date, site requirements, a summary of the performance experience, the status of the existing techniques and identification of future development expectations, and sources of additional information.

APPLICABILITY OF MATERIALS HANDLING CONTROLS

Estimation of the potential releases to the air and an analysis of the impacts to the air pathway are applicable to every activity in the Superfund process. Since nearly every Superfund site has a potential air emissions problem, the focus of this bulletin is to assist remedial project managers (RPMs) and on-site coordinators (OSCs) in considering the appropriate methods for material handling at Superfund sites. To do that, the first step is to estimate the potential releases using the air pathway analysis (APA) process.

The amended National Contingency Plan expands upon the requirement to conduct and fully document a regimented process called an air pathway analysis (APA). The process is defined as a "systematic approach involving a combination of modeling and monitoring methods to assess actual or potential receptor exposure to air contaminants" [1, p. 1-1].[2] When considering removal or remedial responses (i.e., technologies), an APA detailing emission estimate is useful for determining the potential compliance with applicable or relevant and appropriate requirements (ARARs) during remedial action, particularly at a state or local level. Compliance with National Ambient Air Quality Standards during a remediation or the excavation and processing of the contaminated media must be addressed. With the passage of the Clean Air Act Amendments in November 1990 and the advent of numerous state air toxics programs, remediation of Superfund sites must address the media transfer that excavation and materials handling (before and after treatment) will create, and the ARARs these regulations represent. Figure 3-1 [1, pp. 1-4] indicates the applicability of the guidance study series documents on the air pathway analysis to RPMs/OSCs and to contractors and other technical staff.

---

[1] EPA/540/2-91/023.
[2] [Reference number, page number].

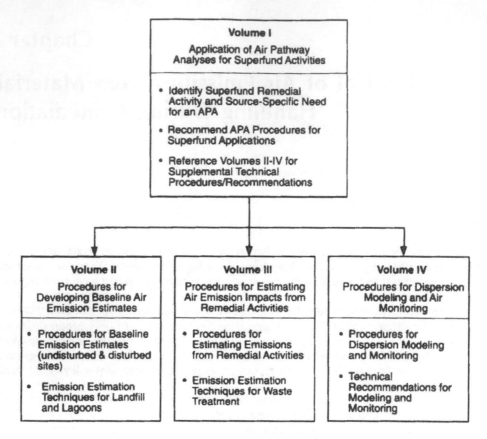

Figure 3-1. Procedures for Conducting APA for Superfund Application–Overview [1, p. 1–4].

The potential for short-term risk (i.e., during the remedial action) is a major criterion when selecting the best remedial alternative. The general classes of contaminants of concern are gaseous and particulate emissions. Particulate matter (PM) becomes airborne via wind erosion, mechanical disturbances (such as excavation and material processing), combustion, and desorption. Gaseous species are primarily volatilized contaminants (VCs), but natural processes such as biodegradation and photo-decomposition can result in releases once the site has been disturbed. Since volatilization is the primary mechanism for gaseous emissions, any volatile contaminant in the soil, a lagoon, a landfill, or even in open containers may be released to the air. The carcinogenic and noncarcinogenic hazards that gases and particulates present in the air pathway must be assessed.

When initially considering remediation technologies applicable to a site, the APA process can play an integral role in estimating the risk that excavation and materials processing pose to the receptors in the area. Any ex situ process that requires such excavation and material sizing, screening, or other pretreatment processing will result in losses of particulate and volatile contaminants.

Similarly, emissions generated during the operation of the technology (i.e., losses from air pollution control equipment or fugitive losses from the treatment process itself) must be estimated in order to complete the air emissions source assessment prior to final selection of the remedial technology. The ambient concentrations of air contaminants may have to be monitored during the remediation process to ensure compliance with local air toxics regulations. All of these considerations should be assessed, a cost estimate prepared, and the results should become an integral input to the selection of alternative technologies, according to the National Contingency Plan process. Of these criteria, overall protection of human health and the environment, ARAR compliance, implementability, cost, short-term effectiveness, and state and community acceptance become paramount concerns for the air pathway impact.

Results of a recently published study [16] indicate significant during typical soil excavation, transport, and feed/preparation operations. The contribution of each remedial step to the VC emis-

Table 3-1. Remedial Step Fractional Contribution to VCs [16, p. 39]

Remedial Activity	Overall Site
Excavation	0.0509
Bucket	0.0218
Truck Filling	0.0905
Transport	0.3051
Dumping	0.5016
Incinerator	0.0014
Exposed Soil	0.0287
Total	1.0000

sions was examined. Table 3-1 presents the results for each step. Although different chemical constituents and concentrations were present in two different site zones, the contribution of each remedial step to the VC emissions during the excavation process remained constant. This contribution was dependent on the parameters of the soil and the remedial activity pattern. At this site, dumping and temporary storage at the incinerator accounted for 50 percent of the VC emissions; transport from the excavation zone was the second highest contributor of emissions. All activities were assumed to be uncontrolled. The use of tarps and/or foam suppressants could substantially reduce these emissions from transport and storage.

LIMITATIONS

The control methods for dust and vapor suppression rarely remove 100 percent of the contaminants from the air. These releases have to be estimated, along with the cost estimate for application of the control method to properly assess the feasibility of implementing the remediation technology being considered. Site conditions determine the effectiveness of specific control methods.

Some methods have very limited periods of effectiveness, making multiple applications or specialized formulations necessary. The scheduling of media excavation and processing may be impacted, for example, in matching the length of effectiveness of a foam or spray suppression technique being used.

If gaseous emissions are expected to be high, or local fugitive limitations apply, costly areal containment methods may be required. If a very large site is to be excavated and the materials classified or preprocessed, portable versions will have to be designed for local air emission control. The use of such portable containment strategies will affect the overall schedule of the remediation and will mandate unique worker safety plans to ensure that the proper level of protective apparel and monitoring devices are used during the excavation process.

CONTROL METHODS

A list of the most commonly used control technologies applicable to VCs and PMs released during soils handling is presented in Table 3-2 [1, pp. 5-31].

Volatilization of contaminants from a hazardous waste site may be controlled by reducing soil vapor pore volume or using physical/chemical barriers [2, p. 116]. The rate of volatilization can be reduced by adding water to reduce the air-filled pore spaces or by reduction of the spaces themselves through compaction techniques. Compaction, however, would displace the volatiles occupying the free spaces (soil venting); water suppression might result in mobilizing the contaminant into a groundwater medium if not properly applied. Wastes amenable to this form of suppression include most volatile organic (e.g., benzene, gasoline, phenols) and inorganic (e.g., hydrogen sulfide, ammonia, radon, methyl mercury) compounds in soil. Contaminants with a high vapor phase mobility and low water phase partition potential are particularly amenable to this vapor control technique. However, the initial application of water will force VCs from the soil-free spaces.

Physical/chemical barriers have found broad utility in temporary vapor and particulate control from hazardous waste sites [3, pp. 4-1 to 4-10]. Evaporation retardants such as foams may be applied, while

Table 3-2. Common Control Technologies Available for Materials Handling[a]

Remedial Operation	Control Technology
Excavation	Water sprays of active areas Dust suppressants Surfactants Foam coverings Enclosures Aerodynamic considerations
Transportation	Water sprays of active areas Dust suppressants Surfactants Road carpets Road oiling Speed reduction Coverings for loads
Dumping	Water sprays of active areas Water spray curtains over bed during dumping Dust suppressants Surfactants
Storage (waste/residuals)	Windscreens Orientation of pile Slope of pile Foam covering and other coverings Dust suppressants Aerodynamic considerations Cover by structure with air displacement and control
Grading	Light water sprays Surfactants
Waste feed/preparation	Cover by structure with air displacement and control

[a] Adapted from [1].

simpler windscreens, synthetic covers, and water/surfactant sprays have been used during excavation and transportation operations. The most exotic system applied to a Superfund site included a special domed structure erected over the excavation area and equipped with carbon adsorption beds through which the internal vapors were drawn [4]. The domed structure was designed to limit emissions through the structure and was capable of being transported to the next excavation site when required. A similar structure may be necessary at the point of materials processing, prior to a proposed incinerator for the site. This facility might be fixed, provided a centralized location for the incinerator can be established.

Sound engineering practices include a multitude of methods for vapor and dust suppression; these techniques are shown in Table 3-3 [5, p. vi]. More than a dozen different techniques have been identified. Several of the methods in Table 3-3 can be used collectively to achieve fugitive emissions control. Application of foams during excavation operations and tarps for overnight storage can achieve a greater overall control efficiency at significantly lower cost than the use of an enclosure with carbon adsorption control. Good engineering practices employing the use of windscreens or other aerodynamic considerations may provide adequate control at some sites; other sites may require application of nearly every method in the list. Cost estimates of many control techniques for VCs are presented in Reference 6 [6, p. 68]. The cost estimates in Reference 6 are not specific to any particular Superfund site. Cost estimates vary significantly according to the site conditions, contaminant type, and ARARs to be met. Table 3-3 presents a relative cost index for illustrative purposes.

Table 3-3. Relative Effectiveness and Cost of PM/VC Suppression Technologies

Suppression Technique	Relative Effectiveness			Relative Cost
	Low	Medium	High	
Minimize waste surface area	★	★	★	1
Aerodynamic considerations	★			1
• windscreens	★			1
• wind blocks	★			1
• orientation of activities	★			1
Covers, mats, membranes, and fill materials	★	★		2–3
Water application	★	★		2–3
Water/additives	★	★		2–3
Inorganic control agents	★	★		2–3
Organic dust control		★		2–3
Foam suppressants		★	★	7–10
Enclosures			★	10

SITE REQUIREMENTS

General site conditions that dictate the estimated magnitude of air emissions are provided in Table 3-4 [7, p. 16]. The requirements for implementation of the dust/vapor control techniques are a function of the estimated emissions once these site conditions have been assessed. Baseline estimation techniques are available for both undisturbed and disturbed sites, as well as mathematical modeling and actual direct measurement methods to verify estimates. Consideration of the particular weather conditions relative to the proposed remediation schedule is critical to efficient control of air emissions. Tables 3-3 and 3-4 should be considered concurrently when structuring an air emissions control strategy for the site and the remediation activities.

PERFORMANCE EXPERIENCE

A study of fugitive dust control techniques conducted with test plots at an active cleanup area documented decreasing effectiveness of foam suppressants within 2 to 4 weeks of application. The effectiveness of water sprays on dump trucks and at the loading site was in the 40 to 60 percent range for the site and 60 to 70 percent range for the truck [8, p. 2]. Surfactants increased the effectiveness of the water sprays.

Foam suppressants have been thoroughly studied by at least two vendors: 3M and Rusmar Foam Technology [9; 10]. Laboratory data for highly volatile organics, such as benzene and trichloroethylene contaminated sand, indicated more than 99 percent suppression effectiveness for several days. Complementary data indicated better barrier performance of foams over 10-mil polyethylene film in controlling volatilization [11, pp. 7 and 8]. A burning landfill was doused and the vapors suppressed by more than 90 percent using foam at a site in Jersey City [12, p. 3]. Similarly, vapors from a petroleum waste site were compared using three different test agents: temporary foam, rigid urea-formaldehyde foam, and a stabilized foam. The temporary foam yielded an average 81 percent control for 20 minutes, rigid foam produced 73 percent control for about 2 hours, and the stabilized foam was 99 percent effective for 24 hours after application [13, pp. 4-7].

The performance data reported are specific to the sites and contaminants controlled. There is no direct applicability of the performance data to general Superfund sites or conditions.

Table 3-5 presents a summary of VC air emissions control technologies for landfills [14, p. 38]. Many of the techniques used can control fugitive particulate emissions as well.

TECHNOLOGY STATUS

The use of vapor and particulate control techniques has been directly applied to at least three Superfund sites: McColl (California), Purity Oil Site (California), and LaSalle Electric (Illi-

Table 3-4. Important Parameters Affecting Baseline Air Emission Levels [7]

| Parameter | Qualitative Effect[a] | |
	Volatiles	Particulate Matter
Site Conditions		
Size of landfill or lagoon	Affects overall magnitude of emissions, but not per area.	Affects overall magnitude of emissions, but not per area.
Amount of exposed waste	High	High
Depth of cover on landfills	Medium	High
Presence of oil layer	High	High
Compaction of cover on landfills	Medium	Low
Aeration of lagoons	High	High
Ground cover	Medium	High
Weather Conditions		
Wind speed	Medium	High
Temperature	Medium	Low
Relative humidity	Low	Low
Barometric pressure	Medium	Low
Precipitation	High	High
Solar radiation	Low	Low
Soil/Waste Characteristics		
Physical properties of waste	High	High
Adsorption/absorption properties of soil	Medium	Low
Soil moisture content	High	High
Volatile fraction of waste	High	Low
Semivolatile/nonvolatile fraction of waste	Low	High
Organic content of soil and microbial activity	High	Low

[a] High, medium, and low in this table refer to the qualitative effect that the listed parameter typically has on baseline emissions.

nois). The McColl work is available as a Superfund Innovative Technology Evaluation demonstration of excavation techniques. Although the domed structure used controlled sulfur dioxide and VOC releases to the atmosphere, working conditions within the dome were difficult. High concentrations of dust and contaminants mandated use of a high level of personal protective apparel. Consequently, personnel were able to work within the dome for only short periods of time [15].

A variety of dust and vapor control techniques may be applied at Superfund sites. A systematic approach to estimate the quantities of air emissions to be controlled, the ambient impact, and the selection of the most appropriate control technique requires a thorough understanding of the site, wastes, emissions potential, and the most relevant combinations of control methods.

EPA CONTACT

Technology-specific questions regarding air emissions may be directed to Paul dePercin, RREL-Cincinnati, (513) 569-7797 (see Preface for mailing address).

ACKNOWLEDGMENTS

This bulletin was prepared for the U.S. Environmental Protection Agency, Office of Research and Development (ORD), Risk Reduction Engineering Laboratory (RREL), Cincinnati, Ohio, by Science

Table 3-5. Summary of VOC Air Emissions Control Technologies for Landfills[a]

Control	Advantages	Disadvantages
Foams	Easy to apply Effective Allow for control of working faces Can reduce decontamination	Moderately expensive Requires trained operators
Complete enclosure/ treatment system	May provide the highest degree of control for some applications	High cost Air scrubbing required High potential risk Must work inside enclosure
Fill material	Inexpensive Equipment usually available	Hard to seal air-tight No control for working face Creates more contaminated soil
Synthetic membrane	Simple approach	Worker contact with waste on application Hard to seal air-tight
Aerodynamic modification	Simple Lower cost Low maintenance	Variable control Requires additional controls
Fugitive VC/PM collection systems	Can be used in active areas	Limited operational data exist Effective range limited Maintenance required
Minimum surface area, shape	Inexpensive Can be included in plan	Must maintain Cannot always dictate shape
Water	Easy to apply	A potential exists for leaching to ground water
Inorganic/organic control agents	Similar to foams	Not as effective as foams for working areas

[a] Adapted from [14].

Applications International Corporation (SAIC) under contract No. 68-C8-0062. Mr. Eugene Harris served as the EPA Technical Project Monitor. Mr. Gary Baker was SAIC's Work Assignment Manager and primary author. The author is especially grateful to Mr. Michael Borst of EPA-RREL, who contributed significantly by serving as a technical consultant during the development of this document.

The following other Agency and contractor personnel have contributed their time and comments by participating in the expert review meetings and/or peer-reviewing the document:

Mr. Edward Bates, EPA-RREL
Mr. Jim Rawe, SAIC
Dr. Chuck Schmidt, Environmental Consultant
Mr. Joe Tessitore, Cross, Tessitore & Associates

REFERENCES

1. Office of Air Quality Planning and Standards, Air Superfund National Technical Guidance Study Series, Volume 1: Application of Air Pathway Analysis for Superfund Activities. Interim Final EPA/450/1-89/001, 1989.
2. Review of In-Place Treatment Techniques for Contaminated Surface Soils, Volume 1: Technical Evaluation. EPA/540/2-84/003a (NTIS PB85-124881), 1984.
3. Handbook—Remedial Action at Waste Disposal Sites (Revised). EPA/625/6-85/006 (NTIS PB87-201034), 1985.

4.   Superfund Innovative Technology Evaluation (SITE) Program, Fourth Edition. EPA/540/8-91/005, 1991. [Seventh Edition published as EPA/540/R-94/526, November, 1994.]

5.   U.S. Environmental Protection Agency, Dust and Vapor Suppression Technologies for Excavating Contaminated Soils, Sludges, and Sediments—Draft Report, Contract No. 68-03-3450, 1987.

6.   Shen, T., et al. Assessment and Control of VOC Emissions from Waste Disposal Facilities Critical Reviews in Environmental Control, 20 (1), 1990.

7.   Office of Air Quality Planning and Standards, Air Superfund National Technical Guidance Study Series, Volume 2: Estimation off Baseline Air Emissions at Superfund Sites. Interim Final EPA/450/1-89/002, 1989.

8.   Fugitive Dust Control Techniques at Hazardous Waste Sites: Results of Three Sampling Studies to Determine Control Effectiveness (Project Summary). EPA/540/S2-85/003, 1988.

9.   Marketing Brochure, Rusmar Foam Technology, January 1991.

10.   Alm, R., et al. The Use of Stabilized Aqueous Foams to Suppress Hazardous Vapors—Study of Factors Influencing Performance. Presented at the HMCRI Symposium, November 16–18, 1987.

11.   Olson, K. Emission Control at Hazardous Waste Sites Using Stable, Non-Draining Aqueous Foams. Presented at the 80th Annual Meeting of the Air & Waste Management Association, June 20–24, 1988.

12.   Alm, R. Using Foam to Maintain Air Quality During Remediation of Hazardous Waste Sites. Presented at the Annual Meeting of the Air Pollution Control Association, June 1987.

13.   Radian Corporation 3M Foam Evaluation for Vapor Mitigation—Technical Memorandum. August 1986.

14.   Radian Corporation. Air Quality Engineering Manual for Hazardous Waste Site Mitigation Activities—Revision #2 November 1987.

15.   Schmidt, C.E. for USEPA-AEERL. The Effectiveness of Foam Products for Controlling the Contaminants Emissions from the Waste at McColl Site in Fullerton, California, Technical Paper Draft. November, 1989.

16.   Development of Example Procedures for Evaluating the Air Impacts of Soil Excavation Associated with Superfund Remedial Actions. Draft Report, July 1990. U.S. Environmental Protection Agency, Office of Air Quality Planning and Standards.

Chapter 4

Performance Evaluations of Pump-and-Treat Remediations[1]

Joseph F. Keely, Robert S. Kerr Environmental Research Laboratory, Ada, OK

SUMMARY

Pump-and-treat remediations are complicated by a variety of factors. Variations in ground-water flow velocities and directions are imposed on natural systems by remediation wellfields, and these variations complicate attempts to evaluate the progress of pump-and-treat remediations. This is in part because of the tortuosity of the flowlines that are generated and the concurrent redistribution of contaminant pathways that occurs. An important consequence of altering contaminant pathways by remediation wellfields is that historical trends of contaminant concentrations at local monitoring wells may not be useful for future predictions about the contaminant plumes.

An adequate understanding of the true extent of a contamination problem at a site may not be obtained unless the site's geologic, hydrologic, chemical, and biological complexities are appropriately defined. By extension, optimization of the effectiveness and efficiency of a pump-and-treat remediation may be enhanced by the utilization of sophisticated site characterization approaches to provide more complete, site-specific data for use in remediation design and management efforts.

INTRODUCTION

Pump-and-treat remediations of ground-water contamination are planned or have been initiated at many sites across the country. Regulatory responsibilities require that adequate oversight of these remediations be made possible by structuring appropriate monitoring criteria for monitoring and extraction wells. These efforts are nominally directed at answering the question: What can be done to show whether a remediation is generating the desired control of the contamination? Recently, other questions have come to the forefront, brought on by the realization that many pump-and-treat remediations may not function as well as has been expected: What can be done to determine whether the remediation will meet its timelines? and What can be done to determine whether the remediation will stay within budget?

Conventional wisdom has it that these questions can be answered by the use of sophisticated data analysis tools such as computerized mathematical models of ground-water flow and contaminant transport. Computer models can indeed be used to make predictions about future performance but such predictions are highly dependent on the quality and completeness of the field and laboratory data utilized. This is also true of models used for performance evaluations of pump-and-treat remediations. In most instances an accurate performance evaluation can be made simply by comparing data obtained from monitoring wells during remediation to the data generated prior to the onset of remediation. Historical trends of contaminant levels at local monitoring wells are often not useful for comparisons with data obtained during the operation of pump-and-treat remediations. This is a consequence of complex flow patterns produced locally by the extraction and injection wells where previously there was a comparatively simple flow pattern.

Complex ground-water flow patterns present great technical challenges in terms of characterization and manipulation (management) of the associated contaminant transport pathways. In Figure 4-1, for example, waters moving along the flowline that proceeds directly into a pumping well from upgradient

---

[1] EPA/540/4-89/005.

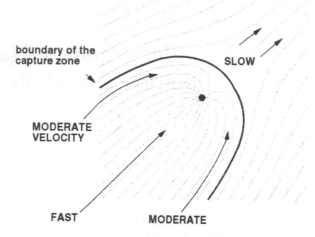

Figure 4-1. Flowline pattern generated by an extraction well. Ground water within the bold line will be captured by the well. Prior to pumping, the flowlines were straight diagonals.

are moving the most rapidly, whereas those waters at the limits of the capture zone move much more slowly. One result is that certain parts of the aquifer are flushed quite well and other parts poorly. Another result of the pumpage is that previously uncontaminated portions of the aquifer at the outer boundary of the contaminant plume may become contaminated by the operation of an extraction well that is located too close to the plume boundary because the flowline pattern extends downgradient of the well.

The latter is not a trivial situation that can be avoided without repercussions by simply locating the extraction well far enough inside the plume boundary so that its flowline pattern does not extend beyond the edge of the plume. Such actions would result in very poor cleansing of the aquifer between the extraction well and the plume boundary because of the stagnation of flow that occurs downgradient of the well. Detailed field investigations are required during remediation to determine the locations of the various flowlines generated by a pump-and-treat operation. Consequently, there may be a need for more data to be generated during the site remediation (especially inside the contamination plume boundaries) than were generated during the site investigation and for interpretations of those data to require highly sophisticated tools. For most settings it is likely that interpretations of the data that are collected during a pump-and-treat remediation will require the use of mathematical and statistical models to organize and analyze those data.

CONTAMINANT BEHAVIOR AND PLUME DYNAMICS

Ground water flows from recharge zones to discharge zones in response to the hydraulic gradient (the drop in hydraulic pressure) along that path. The hydraulic gradient may be obtained from water-level elevation contours for ground water that has constant fluid density but it must be obtained from water pressure contours when the fluid density varies. This is because hydraulic pressure is created by the combined effects of elevation, fluid density and gravity. Additions to the dissolved solids content of a fluid increase its density. For example, synthetic seawater can be prepared by adding mineral salts to fresh water. Landfill leachate is often so laden with dissolved contaminants that its density approaches that of seawater.

As ground water flows through the subsurface it may dissolve some of the materials it contacts and may also transport viruses and small bacteria. This gives rise to natural water quality—a combined chemical, biological, and physical state that may or may not be suitable for man's uses. Brines and brackish waters are examples of natural ground waters that are unsuitable for man's use. It is this same power of water to solubilize minerals and decayed plant and animal residues that causes contamination when ground water is brought into contact with man-made solids and liquids (Figure 4-2). Once contaminated, ground water also provides a medium for potentially destructive interactions

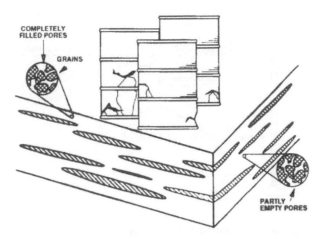

Figure 4-2. Aboveground spill of chemicals from storage drums. Spilled fluids initially fill the uppermost soil pores. As much as half of the fluids remain in each pore after drainage.

between contaminants and subsurface formations, such as the dissolution of limestone and dolomite strata by acidic wastewaters. Contaminated ground water is a major focus of many hazardous waste site cleanups. At these sites, a large number of EPA's Records of Decision (RODs) call for pump-and-treat remediations.

The mechanism by which a source introduces contaminants to ground water has a profound effect on the duration and areal extent of the resulting contamination. Aboveground spills (Figure 4-2) are commonly attenuated over short distances by the moisture retention capacity of surface soils. By contrast, there is much less opportunity for attenuation when the contaminant is introduced below the surface, such as occurs through leaking underground storage tanks, injection wells, and septic tanks.

The hydraulic impacts of some sources of ground-water contamination, especially injection wells and surface impoundments, may impart a strongly three-dimensional character to local flow directions. The water-table mounding that takes place beneath surface impoundments (Figure 4-3), for instance, is often sufficient to reverse ground-water flow directions locally and commonly results in much deeper penetration of contaminants into the aquifer than would otherwise occur. Interactions with streams and other surface water bodies may also impart three-dimensional flow characteristics to contaminated ground water (e.g., a losing stream creates local mounding that forces ground-water flow downward). In addition, contaminated ground water may move from one aquifer to another through a leaky aquitard, such as a tight silt layer that is sandwiched between two sand or gravel aquifers.

As ground water moves, contaminants are transported by advection and dispersion (Figure 4-4). Advection, or velocity, estimates can be obtained from Darcy's law, which states that the amount of water flowing through porous sediments in a given period of time is found by multiplying together values of the hydraulic conductivity of the sediments, the cross-sectional area through which flow occurs, and the hydraulic gradient along the flowpath through the sediments. The hydraulic conductivities of subsurface sediments vary considerably over small distances. It is primarily this spatial variability in hydraulic conductivity that results in a corresponding distribution of flow velocities and contaminant transport rates.

The plume spreading effects of spatially variable velocities can be confused with hydrodynamic dispersion (Figure 4-5), if the details of the velocity distribution are not adequately known. Hydrodynamic dispersion results from the combination of mechanical and chemical phenomena at the microscopic level.

The mechanical component of dispersion derives from velocity variations among water molecules traveling through the pores of subsurface sediments (e.g., the water molecules that wet the surfaces of the grains that bound each pore move little or not at all, whereas water molecules passing through the center of each pore move most rapidly) and from the branching of flow into the accessible pores around each grain. By contrast, the chemical component of dispersion is the result of molecular

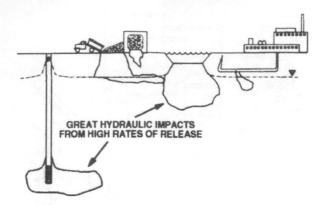

Figure 4-3. Hydraulic impacts of contaminant sources. Injection wells and surface impoundments may release fluids at a high rate, resulting in local mounding of the water table.

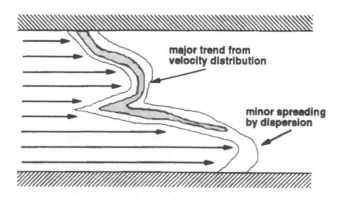

Figure 4-4. Birds-eye view of contaminant plume spreading. Advection causes the majority of plume spreading in most cases. Dispersion adds only marginally to the spreading.

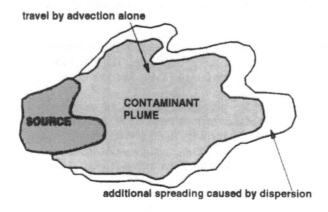

Figure 4-5. Cross-sectional view of contaminant spreading. Permeability differences between strata cause comparable differences in advection and, hence, plume spreading.

diffusion. At modest ground-water flow velocities, the chemical (or diffusive) component of dispersion is negligible and the mechanical component creates a small amount of spreading about the velocity distribution. At very slow ground-water flow velocities, such as occurring in clays and silts, the mechanical component of dispersion is negligible and contaminant spreading occurs primarily by molecular diffusion.

In some geologic settings, most of the ground-water flow occurs through fractures in low permeability rock formations. The flow in the fractures often responds quickly to rainfall events and other fluid inputs, whereas the flow through the bulk matrix of the rock is extremely slow—so slow that contaminant movement by molecular diffusion may be much quicker by comparison. On the other end of the ground-water flow velocity spectrum is the flow in karst aquifers, since it may occur mostly through large channels and caverns. In these situations, ground-water flow is often turbulent, and the advection and dispersion of dissolved contaminants are not adequately describable by Darcy's law and other porous media concepts. Dye tracers have been used to study contaminant transport in fractured rock and karst aquifers, but such studies have yet to yield relationships that can be transferred from the study site to other sites.

Regardless of the character of ground-water flow, contaminants may not be transported at the same rate as the water itself. Variations in the rate of contaminant movement occur as a result of sorption, ion-exchange, chemical precipitation, and biotransformation. The movement of a specific contaminant may be halted completely by precipitation or biotransformation, because these processes alter the chemical structure of the contaminant. Unfortunately, the resulting chemical structure may be more toxic and more mobile than the parent compound, such as in the anaerobic degradation of tetrachloroethene (PCE), which yields, successively, trichloroethene (TCE), dichloroethene (DCE), and monochloroethene (vinyl chloride).

Sorption and ion-exchange (Figure 4-6), conversely, are completely reversible processes that release the contaminant unchanged after temporarily holding it on or in the aquifer solids. This effect is commonly termed retardation and is quantified by projecting or measuring the mobility of the contaminant relative to the average flow velocity of the ground water. Projections of retardation effects on the mobility of contaminants are based on equations that incorporate physical (e.g., bulk density) and chemical (e.g., partition coefficients) attributes of the real system. Direct measurement of the effective mobility of contaminants can be made by observations of plume composition and spreading over time. Alternatively, samples of soils or sediments from the contamination site may be used in laboratory studies to determine the effective partitioning of contaminants between mobile (water) and immobile (solids) phases.

Retardation effects can be short circuited by facilitated transport, a term that refers to the combined effects of two or more discrete physical, chemical, or biological phenomena that act in concert to materially increase the transport of contaminants. Examples of facilitated transport include particle transport, cosolvation, and phase shifting.

Particle transport (Figure 4-7) involves the movement of colloidal particles to which contaminants have adhered by sorption, ion-exchange, or other means. Contaminants that otherwise exhibit moderate to extreme retardation may travel far greater distances than projected from their nominal retardation values. Pumping often removes many colloidal particles from the subsurface. This fact can complicate remediations, and is also relevant to public water supply concerns.

Cosolvation is the process by which the solubility and mobility of one contaminant is increased by the presence of another (Figure 4-8), usually a solvent present at levels of a few percent (note: 1 percent = 10,000 parts per million). Such phenomena are most likely to occur close to contamination sources, where pure solvents and high dissolved concentrations are often found.

Those who design treatment strategies should anticipate the need to remove from ground water certain contaminants that are normally immobile, if the ground water is to be extracted in areas that are close to a source of contamination. Those who make health risk estimates should attempt to factor in the increased mobility and exposure potential generated by cosolvation.

Shifts between chemical phases (Figure 4-9) involve a large change in the pH or redox (reaction) potential of water, and can increase contaminant solubilities and mobilities by ionizing neutral compounds, reversing precipitation reactions, forming complexes with other chemical species, and limiting bacterial activity. Phase shifts may occur as the result of biological depletion of the dissolved oxygen normally present in ground water, or as the result of biological mediation of oxidation reduction reactions (e.g., oxidation of iron II to iron III). Phase shifts may also result from raw chemical releases to the subsurface.

Some ground-water contaminants are components of immiscible solvents, which may be either floaters or sinkers (Figure 4-10). The floaters generally move along the upper surface of the saturated

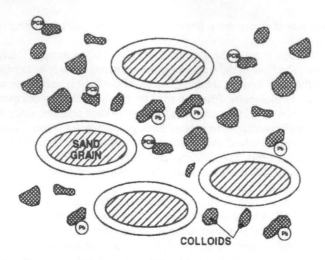

Figure 4-6. Retardation of metals by ion exchange. Metal ions carrying positive charges are attracted to negatively charged surfaces, where they may replace existing ions.

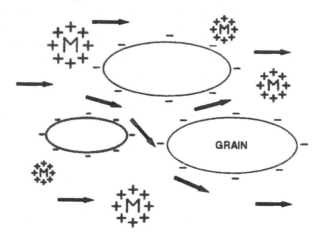

Figure 4-7. Contaminant transport facilitated by particles. Sorption of organics (e.g., PCBs) or metals (e.g., Pb) onto particles may be effective in increasing their transport.

zone, although they may depress this surface locally, and the sinkers tend to move downward under the influence of gravity. Both kinds of immiscible fluids leave residual portions trapped in pore spaces by capillary tension. This is particularly troublesome when an extraction well is utilized to control local gradients such that free product (drainable gasoline) flows into its cone of depression.

The point of concern is that the cone of depression will contain trapped residual gasoline below the water-table (Figure 4-11). That residual will become a continuous source of contamination, which will persist even when the extraction well is turned off. The extent of the contamination that is generated by the residual gasoline in the cone of depression may exceed that generated by the gasoline resting in place above the saturated zone prior to the onset of pumping.

Reliable prediction of the future movement of contaminant plumes under natural flow conditions is difficult because of the need to evaluate properly the many processes that affect contaminant transport in a particular situation. Remediation evaluations are even more difficult because of extensive redirection of preremediation transport pathways by pump-and-treat wellfields. Hence, to prepare for remediation, it is important to determine the potential transport pathways during the site investigation.

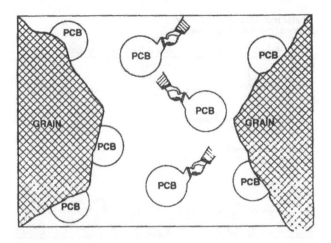

Figure 4-8. Conceptualization of transport by cosolvation. Insoluble contaminants may dissolve in ground water that contains solvents at high concentrations.

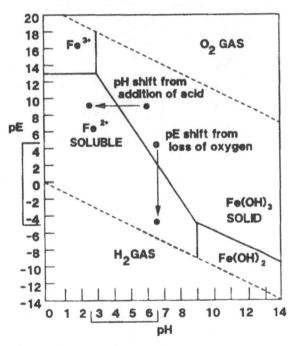

Figure 4-9. Facilitated transport by phase diagram shift. Releases of acidic contaminants, or depletion of oxygen by biota, may solubilize precipitated metals or ionize organics.

MONITORING FOR REMEDIATION PERFORMANCE EVALUATIONS

Ground-water data are collected during remediations to evaluate progress toward goals specified in a ROD. The key controls on the quality of these data are the monitoring criteria that are selected and the locations at which those criteria are to be applied. Ideally, the criteria and the locations would be selected on the basis of a detailed site characterization, from which transport pathways prior to remediation could be identified, and from which the probable pathways during remediation could be predicted.

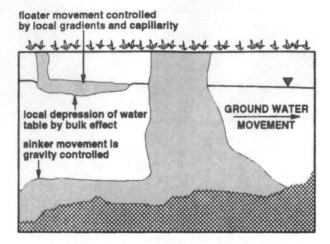

Figure 4-10. Dynamics of immiscible floater and sinker plumes. Buoyant plumes migrate laterally on top of the water-saturated zone. Dense plumes sink and follow bedrock slopes.

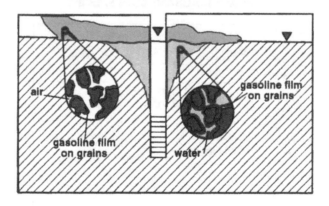

Figure 4-11. Zone of contaminant residuals caused by pumping. Pumping creates a cone of depression to trap gasoline for removal but also leaves residues below the water table.

The monitoring criteria and locations should also be chosen in such a way as to provide information on what is happening both downgradient of the plume boundary and inside the plume. Monitoring within the plume makes it possible to determine which parts of the plume are being effectively remediated and how quickly. This facilitates management of the remediation wellfield for greatest efficiency; for example, by reducing the flowrates of extraction wells that pump from relatively clean zones and increasing the flowrates of extraction wells that pump from zones that are highly contaminated. By contrast, the exclusive use of monitoring points downgradient of the plume boundary does not allow one to gain any understanding about the behavior of the plume during remediation, except to indicate out-of-control conditions when contaminants are detected.

There are many kinds of monitoring criteria and locations in use today. The former are divided into three categories: chemical, hydrodynamic, and administrative control. Chemical criteria are based on standards reflecting the beneficial uses of ground water (e.g., MCLs or other health-based standards for potential drinking water). Hydrodynamic monitoring criteria are such things as:

(1) prevention of infiltration through the unsaturated zone,
(2) maintenance of an inward hydraulic gradient at the boundary of a plume of ground-water contamination, and
(3) providing minimum flows in a stream.

Administrative controls may be codified governmental rules and regulations, but also include:

(1) effective implementation of drilling bans and other access-limiting administrative orders,
(2) proof of maintenance of site security, and
(3) reporting requirements, such as frequency and character of operational and post-operational monitoring.

Combinations of chemical, hydrodynamic, and administrative control criteria are generally necessary for specific monitoring points, depending on location relative to the source of contamination.

Natural Water Quality Monitoring Points

Natural water quality (or background) sampling locations are the most widely used monitoring points, and are usually positioned a short distance downgradient of the plume. The exact location is chosen so that:

(1) It is neither in the plume nor in adjacent areas that may be affected by the remediation,
(2) It is in an uncontaminated portion of the aquifer through which the plume would migrate if the remediation failed, and
(3) Its location minimizes the possibility of detecting other potential sources of contamination (e.g., relevant to the target site only).

Data gathered at a natural water quality monitoring point indicate out-of-control conditions when a portion of the plume escapes the remedial action. The criteria typically specified for this kind of monitoring point are known natural water quality concentrations, usually established with water quality data from wells located upgradient of the source.

Public-Supply Monitoring Points

Public water supply wells located downgradient of a plume are another kind of monitoring point. The locations of these points are not negotiable; they have been drilled in locations that are suitable for water supply purposes, and were never intended to serve as plume monitoring wells. The purpose of sampling these wells is to assure the quality of water delivered to consumers, as related to specific contaminants associated with the target site. The criteria typically specified for this kind of monitoring point are MCLs or other health-based standards.

Gradient Control Monitoring Points

A third kind of off-plume monitoring point frequently established is one for determinations of hydraulic gradients. This kind may be comprised of a cluster of small diameter wells that have very short screened intervals, and is usually located just outside the perimeter of the plume. Water level measurements are obtained from wells that have comparable screened intervals and are then used to prepare detailed contour maps from which the directions and magnitudes of local horizontal hydraulic gradients can be determined. It is equally important to evaluate vertical gradients, by comparison of water level measurements from shallow and deeper screened intervals, because a remediation wellfield may control only the uppermost portions of a contaminant plume if remediation wells are too shallow or have insufficient flow rates.

[Internal] Plume Monitoring Points

Less often utilized is the kind of monitoring point represented by monitoring wells located within the perimeter of the plume. Most of these are installed during the site investigation phase, prior to the remediation, but others may be added subsequent to implementation of the remediation; they are used to monitor the progress of the remediation within the plume. These can be subdivided into onsite plume monitoring points located within the property boundary of the facility that contains the source of the contaminant plume, and offsite plume monitoring points located beyond the facility boundary, but within the boundary of the contamination plume.

Interdependencies of Monitoring Point Criteria

Each kind of monitoring point has a specific and distinct role to play in evaluating the progress of a remediation. The information gathered is not limited to chemical identities and concentrations, but includes other observable or measurable items that relate to specific remedial activities and their attributes. In choosing specific locations of monitoring points, and criteria appropriate to those locations, it is essential to recognize the interdependency of the criteria for different locations.

In addition to the foregoing, one must decide the following: Should evaluations of monitoring data incorporate allowances for statistical variations in the reported values? If so, then what cutoff (e.g., the average value plus two standard deviations) should be used? Should evaluations consider each monitoring point independently or use an average? Finally, what method should be used to indicate that the maximum clean-up has been achieved? The zero-slope method, for example, holds that one must demonstrate that contaminant levels have stabilized at their lowest values prior to cessation of remediation—and that they will remain at that level subsequently, as shown by a flat (zero slope) plot of contaminant concentrations versus time.

LIMITATIONS OF PUMP-AND-TREAT REMEDIATIONS

Conventional remediations of ground-water contamination often involve continuous operation of an extraction-injection wellfield. In these remedial actions, the level of contamination measured at monitoring wells may be dramatically reduced in a moderate period of time, but low levels of contamination usually persist. In parallel, the contaminant load discharged by the extraction wellfield declines over time and gradually approaches a residual level in the latter stages (Figure 4-12). At that point, large volumes of water are treated to remove small amounts of contaminants.

Depending on the reserve of contaminants within the aquifer, this may cause a remediation to be continued indefinitely, or it may lead to premature cessation of the remediation and closure of the site. The latter is particularly troublesome because an increase in the level of ground-water contamination may follow (Figure 4-13) if the remediation is discontinued prior to removal of all residual contaminants.

There are several contaminant transport processes that are potentially responsible for the persistence of residual contamination and the kind of post-operational effect described in Figure 4-13. To cause such effects, releases of contaminant residuals must be slow relative to pumpage-induced water movement through the subsurface.

Transport processes that generate this kind of behavior during continuous operation of a remediation wellfield include:

(1) diffusion of contaminants in low permeability sediments,
(2) hydrodynamic isolation (dead spots) within wellfields,
(3) desorption of contaminants from sediment surfaces, and
(4) liquid-liquid partitioning of immiscible contaminants.

Advection vs. Diffusion

Localized variations in the rate of ground-water flow (advection) arise in heterogeneous settings because of interlayering of high- and low-permeability sediments. When operating a remediation wellfield, these advection variations result in rapid cleansing of the higher permeability sediments which conduct virtually all of the flow (Figure 4-14). By contrast, contaminants are removed from the lower permeability sediments very slowly, by diffusion. The specific rate at which this diffusive release occurs is dependent on the difference in contaminant concentrations within and external to the low permeability sediments.

When the higher permeability sediments are cleaned up, the chemical force drawing contaminants from the lower permeability sediments is at its greatest. This force is exhausted only when the chemical concentrations are nearly equal everywhere.

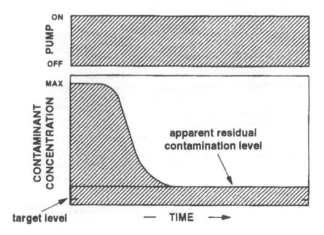

Figure 4-12. Apparent cleanup by pump-and-treat remediation. Contamination concentrations in pumped water decline over time, to an apparent irreducible level.

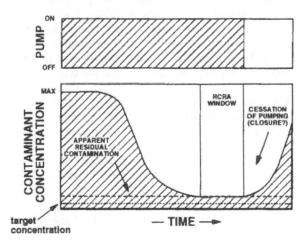

Figure 4-13. Contaminant increases after remediation stops. Contaminant levels may rebound when pump-and-treat operations cease because of contaminant residuals.

Low permeability sediments have orders-of-magnitude greater surface area per volume of material than high permeability sediments. Much greater amounts of contaminants may thus accumulate in a given volume of low permeability sediments as compared with contaminant accumulations in a like volume of high permeability sediments. The thicker the low permeability stratum, the more contaminant reserves it can hold and the more diffusion controls contaminant movement overall. Thus the majority of contaminant reserves in heterogeneous settings may be available only under just such diffusion-controlled conditions.

The situation is similar, though reversed, for in-situ remediations that require the injection and delivery of nutrients or reactants to the zone of intended action: access to contaminants in low permeability sediments is restricted to that provided by diffusion.

Hydrodynamic Isolation

The operation of any wellfield in an aquifer containing moving fluid results in the formation of stagnation zones downgradient of extraction wells and upgradient of injection wells. The stagnation

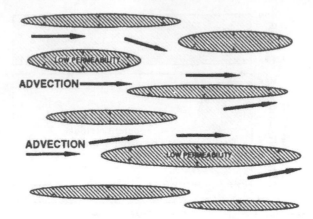

Figure 4-14. Permeably variations limit remediations. High permeability sediments conduct most of the flow; low permeability sediments act as leaky contaminant reservoirs.

zones are hydrodynamically isolated from the remainder of the aquifer, so mass transport into or out of the isolated water may occur only by diffusion. If remedial action wells are located within the bounds of a contaminant plume such as for the removal of contaminant hot-spots, the portion of the plume lying within their associated stagnation zone(s) will not be effectively remediated. The flowline pattern must be altered radically by major changes in the locations of pumping wells or by altering the balance of flowrates among the existing wells, or both, if the original stagnation zone(s) are to be remediated.

Another form of hydrodynamic isolation is the physical creation of enlarged zones of residual hydrocarbon (Figure 4-11). This occurs when deep wells are used to create cones of drawdown into which underground storage tank and pipeline leaks of gasoline can flow, so that skimmer pumps can remove the accumulated product. When the deep water pump is turned off, the water table will rise to its prepumping position. This will allow the aquifer waters that then fill the cone of depression and any subsequent ground-water flow through the former cone of depression to become highly contaminated with BTEX compounds (benzene, toluene, the ethyl benzenes, and the xylenes) as a result of contact with the gasoline remaining there on the aquifer solids. Gasoline in residual saturation may occupy as much as 40 percent of the pore space of the sediments.

Sorption Influences

The number of pore volumes of contaminated water to be removed during a remediation depends on the sorptive tendencies of the contaminant. The number of pore volumes to be removed also depends on whether ground-water flow velocities during remediation are too rapid to allow contaminant levels to build up to equilibrium concentrations locally (Figure 4-15). If insufficient contact time is allowed, the affected water is advected away from sorbed contaminant residuals prior to achieving a state of chemical equilibrium and is replaced by fresh water from upgradient.

Hence, continuous operation of pump-and-treat remediations may result in steady releases of contaminants at substantially less than their chemical equilibrium levels. With less contamination being removed per volume of water brought into contact with the affected sediments, it is clear that large volumes of mildly contaminated water are recovered where small volumes of highly contaminated water would otherwise be recovered.

Unfortunately, this is all too likely to occur with conventional pump-and-treat remediations and with those in-situ remediations that depend upon injection wells for delivery of nutrients and reactants. This is because ground-water flow velocities within wellfields may be many times greater than natural (nonpumping) flow velocities. Depending on the sorptive tendencies of the contaminant, the time to reach maximum equilibrium concentrations in the ground water may simply be too great compared with the average residence time in transit through the contaminated sediments.

Liquid-Liquid Partitioning

Subsequent to gravity drainage of free product that has been discharged to the subsurface, immiscible or nonaqueous phase liquids (NAPLs) remain trapped in the pores of subsurface sediments by surface tension to the grains that bound the pores. Liquid-liquid partitioning controls the dissolution of NAPL residuals into ground water.

As with sorbing compounds, flow rates during remediation may be too rapid to allow aqueous saturation levels of partitioned NAPL residuals to be reached locally (Figure 4-16). If insufficient contact time is allowed, the affected water is advected away from the NAPL residuals prior to reaching chemical equilibrium and is replaced by fresh water from upgradient.

Again, this process generates large volumes of mildly contaminated water where small volumes of highly contaminated water would otherwise result, and this means that it will be necessary to pump-and-treat far more water than would otherwise be the case. The efficiency loss is generally twofold because much of the pumped water will contain contaminant concentrations that are below the level at which optimal treatment is obtained.

DESIGN AND ANALYSIS COMPLICATIONS

Contaminant concentrations and ground-water flow velocities can be highly variable throughout the zone of action, which is that portion of an aquifer actively affected by the remediation wellfield. Consequently, monitoring strategies should be focused on detection of rapid sporadic changes in contaminant concentrations and flow velocities at any specific point in the zone of action. In practice, this means that tracking the effectiveness of pump-and-treat remediations by chemical sampling is quite complicated.

Decisions regarding the frequency and density of chemical sampling should take into account the detailed flowpaths generated by the remediation wellfield, including changes in contaminant concentrations that result from variations in the influences of transport processes along those flowpaths. The need to reposition extraction wells occasionally to remediate portions of the contaminated zone that were previously subject to slow flowlines means that the chemical sampling may generate results that are not easily understood. It also means that it may be necessary to move the chemical monitoring points during the course of a remediation.

Evaluations of the hydrodynamic performance of remediation wellfields are also data intensive. For example, it is usually required that an inward hydraulic gradient be maintained at the periphery of a contaminant plume undergoing pump-and-treat remediation. This requirement is imposed to ensure that no portion of the plume is free to migrate away from the zone of action. To assess this performance adequately, the hydraulic gradient should be measured accurately in three dimensions between each pair of adjacent pumping or injection wells. The design of an array of piezometers (small diameter wells with very short screened intervals that are used to measure the hydraulic head of selected positions in an aquifer) for this purpose is not as simple as one might first imagine. Many points are needed to define the convoluted water-table surface that develops between adjacent pumping or injection wells. Not only are there velocity divides in the horizontal dimension near active wells, but in the vertical dimension too, because the hydraulic influence of each well extends to only a limited depth in practical terms.

INNOVATIONS IN PUMP-AND-TREAT REMEDIATIONS

One of the promising innovations in pump-and-treat remediations is pulsed pumping. Pulsed operation of hydraulic systems is the cycling of extraction or injection wells on and off in active and resting phases (Figure 4-17). The resting phase of a pulsed-pumping operation can allow sufficient time for contaminants to diffuse out of low permeability zones and into adjacent high permeability zones until maximum concentrations are achieved in the higher permeability zones. For sorbed contaminants and NAPL residuals, sufficient time can be allowed for equilibrium concentrations to be reached in local ground water. Subsequent to each resting phase the active phase of the cycle removes the minimum volume of contaminated ground water at the maximum possible concentra-

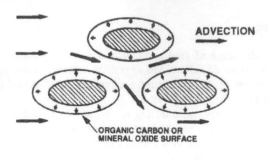

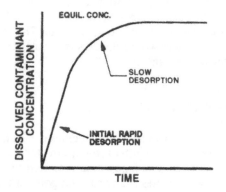

Figure 4-15. Sorption limitation of pump-and-treat remediations. Increased flow velocities caused by pumpage may not allow for sufficient time to reach chemical equilibrium locally.

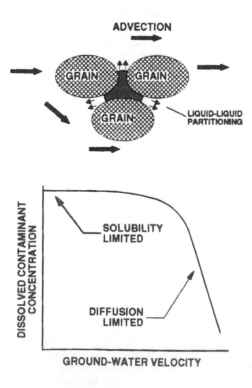

Figure 4-16. Partitioning limits pump-and-treat effectiveness. Less than solubility levels of contaminants may be released from trapped solvents if pumpage increases flow velocities.

tions for the most efficient treatment. By occasionally cycling only select wells, stagnation zones may be brought into active flowpaths and remediated.

Pulsed operation of remediation wellfields incurs certain additional costs and concerns that must be compared with its advantages for site-specific applications. During the resting phase of pulsed-pumping cycles, peripheral gradient control may be needed to ensure adequate hydrodynamic control of the plume. In an ideal situation, peripheral gradient control would be unnecessary. Such might be the case where there are no active wells, major streams, or other significant hydraulic stresses nearby to influence the contaminant plume while the remedial action wellfield is in the resting phase. The plume would migrate only a few feet during the tens to hundreds of hours that the system was at rest, and that movement would be rapidly recovered by the much higher flow velocities back toward the extraction wells during the active phase.

When significant hydraulic stresses are nearby, however, plume movement during the resting phase may be unacceptable. Irrigation or water-supply pumpage, for example, might cause plume movement on the order of several tens of feet per day. It might then be impossible to recover the lost portion of the plume when the active phase of the pulsed-pumping cycle commences. In such cases, peripheral gradient control during the resting phase would be essential. If adequate storage capacity is available, it may be possible to provide gradient control in the resting phase by injection of treated waters downgradient of the remediation wellfield. Regardless of the mechanics of the compensating actions, their capital and operating expenses must be added to those of the primary remediation wellfield to determine the complete cost.

Pump-and-treat remediations are underway today that incorporate some of the principles of pulsed pumping. For instance, pumpage from contaminated bedrock aquifers and other low permeability formations results in intermittent wellfield operations by default; the wells are pumped dry even at low flow rates. In such cases the wells are operated on-demand with the help of fluid-level sensors that trigger the onset and cessation of pumpage. This simultaneously accomplishes the goal of pumping ground water only after it has reached chemical equilibrium, since equilibrium occurs on the same time frame as the fluid recharge event in low permeability settings. In settings of moderate to high permeability, the onset and cessation of pumpage could be keyed to contaminant concentration levels in the pumped water independent of flow changes required to maintain proper hydrodynamic control. Peripheral hydrodynamic controls may or may not be necessary during the resting phase of the cycle.

Other strategies to improve the performance of pump-and-treat remediations include:

(1) scheduling of wellfield operations to satisfy simultaneously hydrodynamic control and contaminant concentration trends or other performance criteria,
(2) repositioning of extraction wells to change major transport pathways, and
(3) integration of wellfield operations with other subsurface technologies such as barrier walls that limit plume transport and minimize pumping of fresh water or infiltration ponds that maintain saturated flow conditions for flushing contaminants from previously unsaturated soils and sediments.

Flexible operation of a remediation wellfield, such as occasionally turning off individual pumps, allows for some flushing of stagnant zones. That approach may not be as hydrodynamically efficient as one that involves permanently repositioning or adding pumping wells to new locations at various times during remediation. Repositioned and new wells require access for drilling, however, and that necessarily precludes capping of the site until after completion of the pump-and-treat operations. The third approach cited above, combining pump-and-treat with subsurface barrier walls, trenching, or in-situ techniques, all of which may occur at any time during remediation, may also require postponement of capping until completion of the remediation.

The foregoing discussion may raise latent fears of lack of control of the contaminant source, something almost always mitigated by isolation of the contaminated soils and subsoils that remain long after man-made containers have been removed from the typical site. Fortunately, vacuum extraction of contaminated air/vapor from soils and subsoils has recently emerged as a potentially effective means of removing volatile organic compounds (VOCs). Steam flooding has shown promise for removal of the more retarded organics and in-situ chemical fixation techniques are being tested for the isolation of metals wastes.

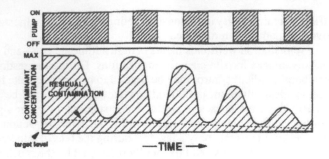

Figure 4-17. Pulsed pumping removal of residual contaminants. Repetitive removal of pulses of highly contaminated water ensures effective depletion of contaminant residuals.

Vacuum extraction techniques are capable of removing several pounds of VOCs per day, whereas air stripping of VOCs from comparable volumes of contaminated ground water typically results in the removal of only a few grams of VOCs per day because VOCs are so poorly soluble in water. Similarly, steam flooding is an economically attractive means of concentrating contaminant residuals as a front leading the injected body of steam. Steam flooding or chemical fixation have potential for control of fluid and contaminant movement in the unsaturated zone and should thus be considered a potentially significant addition to the list of source control options. In addition, soils engineering and landscape maintenance techniques can minimize infiltration of rainwater in the absence of a multilayer RCRA-style cap.

In terms of evaluation of the performance of a remediation, the presence of a multilayer RCRA-styled cap could pose major limitations. The periodic removal of core samples of subsurface solids from the body of the plume and the source zone with subsequent extraction of the chemical residues on the solids is the only direct means of evaluating the true magnitude of the residuals and their depletion rate. Since this must be done periodically, capping would conflict unless postponed until closure of the site. If capping can be postponed or foregone, one great flexibility for management of pump-and-treat remediations (e.g., concurrent operation of a soil vapor extraction wellfield, and sampling of subsoils) can be used to improve effectiveness.

MODELING AS A PERFORMANCE EVALUATION TECHNIQUE

Subsurface contaminant transport models incorporate a number of theoretical assumptions about the natural processes governing the transport and fate of contaminants. In order for solutions to be made tractable, simplifications are made in applications of theory to practical problems. A common simplification for wellfield simulations is to assume that all flow is horizontal so that a two-dimensional model can be applied rather than a three dimensional model, which is much more difficult to create and more expensive to use. Two-dimensional model representations are obviously not faithful to the true complexities of real world pump-and-treat remediations since most of these are in settings where three-dimensional flow is the rule. Moreover, most pump-and-treat remediations use partially penetrating wells which effect significant vertical flow components, whereas the two-dimensional models assume that the remediation wells are screened throughout the entire saturated thickness of the aquifer and therefore do not cause upconing of deeper waters.

Besides the errors that stem from simplifying assumptions, applications of mathematical models to the evaluation of pump-and-treat remediations are also subject to considerable error where the study site has not been adequately characterized. It is essential to have appropriate field determinations of natural process parameters and variables (Figure 4-18) because these determine the validity and usefulness of each modeling attempt. Errors arising from inadequate data are not addressed properly by mathematical tests such as sensitivity analyses or by the application of stochastic techniques for estimating uncertainty, contrary to popular beliefs, because such tests and stochastic simulations assume that the underlying conceptual basis of the model is correct. One cannot properly change the conceptual basis (e.g., from an isolated aquifer to one that has strong interaction with a stream or another underlying aquifer) without data to justify the change. The high degree of hydrogeological, chemical, and microbiological complexity typically present in field situations requires site-specific characterization of various natural processes by detailed field and laboratory investigations.

Hence, both the mathematics that describe models and the parameter inputs to those models should be subjected to rigorous quality control procedures. Otherwise, results from field applications of models are likely to be qualitatively as well as quantitatively incorrect. If done properly, however, mathematical modeling may be used to organize vast amounts of disparate data into a sensible framework that will provide realistic appraisals of which parts of a contaminant plume are being effectively cleansed, when the remediation will meet target contaminant reductions, and what to expect in terms of irreducible contaminant residuals. Models may also be used to evaluate changes in design or operation, so that the most effective and efficient pump-and-treat remediation can be attained.

OTHER DATA ANALYSIS METHODS FOR PERFORMANCE EVALUATIONS

Mathematical models are by no means the only methods available for use in evaluating the performance of pump-and-treat remediations. Two other major fields of analysis are statistical methods and graphical methods. The potential power of statistical methods has been tapped infrequently in ground-water contamination investigations, aside from their use in quality assurance protocols. The uses are many, however, as shown in Table 4-1. While data interpretation and presentation methods vary widely, many site documents lack statistical evaluations; many also present inappropriate simplifications of datasets, such as grouping or averaging broad categories of data, without regard to the statistical validity of those simplifications.

Graphical methods of data presentation and analysis have been used heavily in both ground-water flow problems (e.g., flowline plots and flownets) and water chemistry problems (e.g., Stiff kite diagrams). Figure 4-1, for example, is a flowline plot for a single well. From analysis of such plots, it is possible to estimate the number of pore volumes that will be removed over a set period of time of constant pumpage, at different locations in the contaminant plume. Figure 4-9 is a chemical phase diagram for iron, which may be used to relate pH and redox measurements to the most stable species of iron.

Figure 4-19 presents one means of producing readily recognizable patterns of the major ion composition of a water sample, so that it may be differentiated from other water types. At sites of subsurface contamination, the major ion composition often differs greatly from the natural quality of water in adjacent areas. Water quality specialists have used such plots for decades to differentiate zones of brackish water from zones of potable water, in studies of salt water intrusion into coastal aquifers, and in studies of the upconing of saline water from shales during pumping of overlying sandstone aquifers.

Figure 4-20 illustrates another means of producing readily recognizable patterns of the milli-equivalence values of major cations and anions in a ground-water sample. Geochemical prospectors have used this graphical technique as an aid in the identification of waters associated with mineral deposits. These graphical presentation techniques have been adapted recently to the display of organic chemical contaminants. For example, a compound of interest such as trichloroethene (TCE) may be evaluated in terms of its contribution to the total organic chemical contamination load, or against other specific contaminants so that some differentiation of source contributions to the overall plume can be obtained.

PERSPECTIVES FOR SITE CHARACTERIZATIONS

Concepts pertinent to investigating and predicting the transport and fate of contaminants in the subsurface are evolving. Additional effort devoted to site-specific characterizations of preferential pathways of contaminant transport and the natural processes that affect the transport behavior and ultimate fate of contaminants may significantly improve the timeliness and cost-effectiveness of remedial actions at hazardous waste sites.

Characterization Approaches

To underscore the latter point, it is useful to examine the principal activities, benefits, and shortcomings of increasingly sophisticated levels of site characterization approaches: conventional (Table

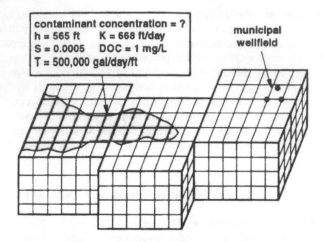

Figure 4-18. Grid of points for a contaminant transport model. Known values of water level and other inputs are used to predict concentration changes at each grid intersection (node).

Table 4-1. Statistical Methods Useful in Performance Evaluations

Analysis of Variance (ANOVA) Techniques

ANOVA techniques may be used to segregate errors due to chemical analyses from those errors that are due to sampling procedures and from the intrinsic variability of the contaminant concentrations at each sampling point.

Correlation Coefficients

Correlation coefficients can be used to provide justification for lumping various chemicals together (e.g., total VOCs), or for using a single chemical as a class representative, or to link sources by similar chemical behavior.

Regression Equations

Regression equations may be used to predict contaminant loads based on historical records and supplemental data and may be used to test cause-and-effect hypotheses about sources and contaminant release rates.

Surface Trend Analysis Techniques

Surface trend analysis techniques may be used to identify recurring and intermittent (e.g., seasonal) trends in contour maps of ground-water levels and contaminant distributions, which may be extrapolated to source locations or future plume trajectories.

4-2), state-of-the-art (Table 4-3), and state-of-the-science (Table 4-4). The conventional approach to site characterizations is typified by the description given in Table 4-2.

Each activity of the conventional approach can be accomplished with semi-skilled labor and off-the-shelf technology, with moderate to low costs. It may not be possible to characterize thoroughly the extent and probable behavior of a subsurface contaminant plume with the conventional approach.

Key management uncertainties regarding the degree of health threat posed by a site, the selection of appropriate remedial action technologies, and the duration and effectiveness of the remediations all should decrease significantly with the implementation of more sophisticated site characterization approaches.

It will probably cost substantially more to implement state-of-the-art and state-of-the-science approaches in site characterizations, but the increased value of the information obtained is likely to generate offsetting cost savings by way of improvements in the technical effectiveness and efficiency of the site cleanup.

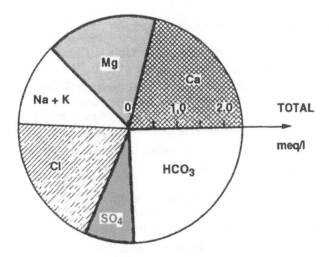

Figure 4-19. Pie chart of major ions in a ground-water sample. The milli-equivalence values of the major ions are computed and plotted to generate patterns specific to the source.

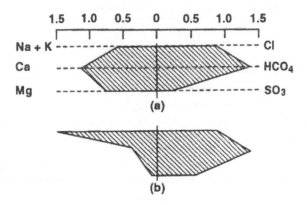

Figure 4-20. Stiff diagrams of major ions in two samples. The concentrations of the ions are plotted in the manner shown in (a); the uniqueness of another water type is shown by (b).

Obviously, it is not possible to test these conceptual relationships directly, because one cannot carry site characterization and remediation efforts to fruition along each approach simultaneously. One can infer many of the foregoing discussion points, however, by observing the changes in perceptions, decisions, and work plans that occur when more advanced techniques are brought to bear on a site that has already undergone a conventional level of characterization. The latter situation is a fairly common occurrence, because many first attempts at site characterization turn up additional sources of contamination or hydrogeologic complexities that were not suspected when the initial efforts were budgeted.

Hypothetical Example

It is helpful to examine possible scenarios that might result from the different site investigation approaches just outlined.

Figure 4-21 depicts a hypothetical ground-water and soil contamination situation located in a mixed residential and light industry section of a town in the Northeast. As illustrated, there are three major plumes: an acids plume (e.g., from electrolytic plating operations), a phenols plume (e.g., from a creosoting operation that used large amounts of pentachlorophenol), and a volatile organics plume

Table 4-2. Conventional Approach to Site Characterization

Actions Typically Taken

- Install a few dozen shallow monitoring wells
- Sample ground-water numerous times for 129+ priority pollutants
- Define geology primarily by driller's logs and drill cuttings
- Evaluate local hydrology with water level contour maps of shallow wells
- Possibly obtain soil and core samples for chemical analyses

Benefits

- Rapid screening of the site problems
- Costs of investigation are moderate to low
- Field and laboratory techniques used are standard
- Data analysis/interpretation is straightforward
- Tentative identification of remedial alternatives is possible

Shortcomings

- True extent of site problems may be misunderstood
- Selected remedial alternatives may not be appropriate
- Optimization of final remediation design may not be possible
- Cleanup costs remain unpredictable; tend to be excessive levels
- Verification of compliance is uncertain and difficult

Table 4-3. State-of-the-Art Approach to Site Characterization

Recommended Actions

- Install depth-specific clusters of monitoring wells
- Initially sample for 129+ priority pollutants, be selective subsequently
- Define geology by extensive coring/sediment sampling
- Evaluate local hydrology with well clusters and geohydraulic tests
- Perform limited tests on sediment samples (grain size, clay content, etc.)
- Conduct surface geophysical surveys (resistivity, EM, ground-penetrating radar)

Benefits

- Conceptual understandings of the site problems are more complete
- Prospects are improved for optimization of remedial actions
- Predictability of remediation effectiveness is increased
- Cleanup costs may be lowered, estimates are more reliable
- Verification of compliance is more soundly based

Shortcomings

- Characterization costs are higher
- Detailed understandings of site problems are still difficult
- Full optimization of remediation is still not likely
- Field tests may create secondary problems (disposal of pumped waters)
- Demand for specialists is increased, shortage is a key limiting factor

(e.g., from solvent storage leaks). In addition, soils onsite are heavily contaminated in one area with spilled pesticides, and in another area with spilled transformer oils that contained PCBs in high concentrations.

The hydrogeologic setting for the hypothetical site is a productive alluvial aquifer composed of an assortment of sands and gravels that are interfingered with clay and silt remnants of old stream-beds and floodplain deposits. The latter have been continually dissected by a central river as the valley matured over geologic time. The deeper portion of the sediments is highly permeable and is the zone most heavily used for municipal and industrial supply wells, whereas the shallow portion of the

Table 4-4. State-of-the-Science Approach to Site Characterization

Idealized Approach

- Assume state-of-the-art as starting point
- Conduct soil vapor surveys for volatiles, fuels
- Conduct tracer tests and borehole geophysical surveys (neutron and gamma)
- Conduct karst stream tracing and recharge studies, if appropriate to the setting
- Conduct bedrock fracture orientation and interconnectivity studies, if appropriate
- Determine the percent organic carbon and cation exchange capacity of solids
- Measure redox potential, pH, and dissolved oxygen levels of subsurface
- Evaluate sorption-desorption behavior by laboratory column and batch studies
- Assess the potential for biotransformation of specific compounds

Benefits

- Thorough conceptual understandings of site problems are obtained
- Full optimization of the remediation is possible
- Predictability of the effectiveness of remediation is maximized
- Clean-up costs may be lowered significantly, estimates are reliable
- Verification of compliance is assured

Shortcomings

- Characterization costs may be much higher
- Few previous applications of advanced theories and methods have been completed
- Field and laboratory techniques are specialized and are not easily mastered
- Availability of specialized equipment is low
- Need for specialists is greatly increased (it may be the key limitation overall)

---

sediments is only moderately permeable since it contains many clay and silt lenses. The predominant ground-water flow direction in the deeper zone parallels the river (which is also parallel to the axis of the valley), except in localized areas around municipal and industrial wellfields. The predominant direction of flow in the shallow zone is seasonally dependent, having the strongest component of flow toward the river during periods of low flow in the river, and being roughly parallel to the river during periods of high flow in the river.

Strong downward components of flow carry water from the shallow zone to the deeper zone throughout municipal and industrial wellfields, as well as along the river during periods of high flow. Slight downward components of flow exist elsewhere due to local recharge by infiltrating rainwaters.

Conventional Characterization Scenario

A conventional site characterization would define the horizontal extent of the most mobile, widespread plume. However, it would provide only a superficial understanding of variations in the composition of the sediments. An average hydraulic conductivity would be obtained from review of previously published geologic reports and would be assumed to represent the entire aquifer for the purpose of estimating flow rates. The kind of cleanup that would likely result from a conventional site investigation is illustrated in Figure 4-22. The volatile organics plume would be the most important to remediate, since it is the most mobile, and an extraction system would be installed. Extracted fluids would be air-stripped of volatiles and then passed through a treatment plant for removal of nonvolatile residues, probably by relatively expensive filtration through granular activated carbon.

Extraction wells would be placed along the downgradient boundary of the VOC plume to withdraw contaminated ground water. A couple of injection wells would be placed upgradient and would be used to return a portion of the extracted and treated waters to the aquifer. The remainder of the pumped and treated waters would be discharged to the tributary under an NPDES permit.

Information obtained from the drilling logs and samples of the monitoring wells would be inadequate to do more than position all of the screened sections of the remediation wells at the same

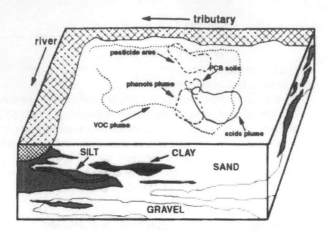

Figure 4-21. Hypothetical contamination site in an alluvial setting. Ground-water contamination at the site includes a variety of plumes; the setting is complex geologically and hydrologically.

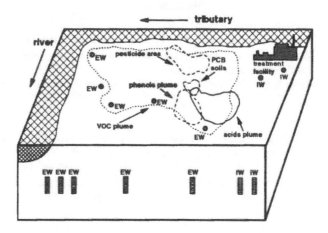

Figure 4-22. Conventional cleanup of the hypothetical site. EWs are extraction wells. IWs are injection wells; all are screened at the same elevation and have identical flowrates.

shallow depth. The remediation wellfield would operate for the amount of time needed to remove a volume of water somewhat greater than that estimated to reside within the bounds of the zone of contamination, allowing for average retardation values (from the scientific literature) for contaminants found at the site. The PCB-laden soils would be excavated and sent to an incinerator or approved waste treatment and disposal facility. The decision makers would have based their approval on the presumption that the plume had been adequately defined, and that if it had not, that the true magnitude of the problem does not differ substantially, except for the possibility of perpetual care.

State-of-the-Art Characterization Scenario

Incorporation of some of the more common state-of-the-art site investigation techniques, such as pump tests, installation of vertically-separated clusters of monitoring wells (shallow, intermediate, and deep) and river stage monitors, and chemical analysis of sediment and soil samples would likely result in the kind of remediation illustrated in Figure 4-23. Since a detailed understanding of the geology and hydrology would be obtained, optimal selection of well locations, wellscreen positions and flowrates (the values in the parentheses in Figure 4-23, in gallons per minute) for the remediation

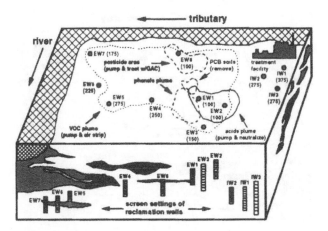

Figure 4-23. Moderate state-of-the-art remediation. Clusters of vertically-separated monitoring wells and an aquifer test are used to tailor the remedy to the site hydrogeology.

wells could be determined. A special program to recover the acid plume and neutralize it would be instituted. A special program could also be instituted for the pesticide plume. This approach would probably lower treatment costs overall, despite the need for separate treatment trains for the different plumes, because substantially lesser amounts of ground water would be treated with expensive carbon filtration for removal of nonvolatile contaminants.

The screened intervals of the extraction wells would be placed at deeper positions toward the river, if water quality data from monitoring well clusters show that the plume is migrating beneath shallow accumulations of clays and silts to the deeper, more permeable sediments. All or most of the extracted and treated ground water could be returned to the aquifer through injection wells. The wellscreens would need to be positioned (e.g., deeper) to avoid diminishing the effectiveness of nearby extraction wells. As in the conventional remedy, the remediation wellfield would operate for the amount of time needed to remove a volume of water that is determined from average contaminant retardation values and the rate of flushing of ground water residing in the zone of contamination. The detailed geologic and hydrologic information acquired would result in an expectation of more rapid cleansing of portions of the contaminated zone than others.

One could conclude that this remediation is optimized to the point of providing an effective cleanup, and decision makers would be reasonably justified in giving their approval. One should note, however, that the efficiency (esp., duration) of the remediation may be less than optimal.

Advanced State-of-the-Art Characterization Scenario

If all state-of-the-art investigation tools were used at the site there would be an opportunity to evaluate the desirability of using a subsurface barrier wall to enhance remediation efforts (Figure 4-24). The wall would not entomb the plumes, but would limit pumping to contaminated fluids rather than having the extracted waters diluted with fresh waters, as was true of the two previous approaches. The volume to be pumped could be lowered because the barrier wall will increase the drawdown at each well by hydraulic interference effects, thereby maintaining the same effective hydrodynamic control with less pumping (note the lower flowrates for each well in Figure 4-24). Treatment costs should be less also because the pumped waters should contain higher concentrations of contaminants, and treatment efficiencies are often greatest at moderate to high concentrations. Soil washing techniques could be used on the pesticide contaminated area to minimize future source releases to ground water.

Both the effectiveness and the efficiency of this remediation might therefore appear to be optimizable but that is a perception that is based on the presumption that the contaminants will be released readily. Given the potential limitations to pump-and-treat remediations discussed previously, it is doubtful that even this advanced state-of-the-art site investigation precludes further improvement. Much attention could be devoted to the chemical and biological peculiarities, just as has been given

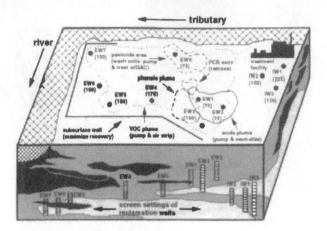

Figure 4-24. Advanced state-of-the-art remediation. Subsurface barrier walls or other technologies can be integrated with pump-and-treat operations.

to the geology and the hydrology. For example, detailed evaluation of sorption or other contaminant retardation processes at this site rather than the use of average retardation values from the literature should generate additional improvements in effectiveness and efficiency of the remediation. Likewise, detailed examination of the potential for biotransformation would be expected to lead to improved effectiveness and efficiency.

State-of-the-Science Characterization Scenario

At the state-of-the-science level of site characterization, tracer tests could be undertaken which would provide good information on the potential for diffusive restrictions in low permeability sediments and on anisotropic biases in the flow regime. Sorption behavior of the VOCs could be evaluated in part by determining the total organic carbon content of selected subsurface sediments. Similarly, the cation exchange capacities of subsurface sediment samples could be determined to obtain estimates of release rates and mobilities of toxic metals. The stabilities of various possible forms of elements and compounds could be evaluated with measurements of pH redox potential and dissolved oxygen. Contemporary research indicates that the acids plume and the phenols plume might be better addressed with such measurements (e.g., chemical speciation modeling). Finally, if state-of-the-science findings regarding potential biotransformations could be taken advantage of it might be possible to effect in-situ degradation of the phenols plume and remove volatile residues too (Figure 4-25).

ADDITIONAL CONSIDERATIONS

The foregoing discussion highlights generic gains in effectiveness and efficiency of remediation that should be expected by better defining ground-water contamination problems and using that information to develop site-specific solutions.

Because the complexities of the subsurface cannot fully be delineated even with "state-of-the-science" data collection techniques and many of these techniques are not fully developed nor widely available at this time, it is important to proceed with remediation in a phased process so that information gained from initial operation of the system can be incorporated into successive stages of the remedy. Some considerations that may help to guide this process include the following:

1. In many cases, it may be appropriate to initiate a response action to contain the contaminant plume before the remedial investigation is completed. Containment systems (e.g., gradient control) can often be designed and implemented with less information than required for full

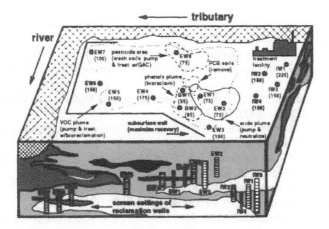

Figure 4-25. State-of-the-science remediation. Bioreclamation and other emerging technologies could be tested and implemented with reasonable certainty of effects.

remediation. In addition to preventing the contamination from migrating beyond existing boundaries, this action can provide valuable information on aquifer response to pumping.

2. Early actions might also be considered as a way of obtaining information pertinent to design of the final remedy. This might consist of installing an extraction system in a highly contaminated area and observing the response of the aquifer and contaminant plume as the system is operated.

3. The remedy itself might be implemented in a staged process to optimize system design. Extraction wells might be installed incrementally and observed for a period of time to determine their range of influence. This will help to identify appropriate locations for additional wells and can assure proper sizing of the treatment systems as the range of contaminant concentrations in extracted ground water is confirmed.

4. In many cases, ground-water response actions should be initiated even though it is not possible to assess the restoration time frames or ultimate concentrations achievable. After the systems have been operated and monitored over time, it should be possible to better define the final goals of the action.

EPA CONTACTS

For further information contact Randall R. Ross, RSKERL-Ada, (405) 436-8611 (see Preface for mailing address).

REFERENCES

Abriola, L.M. and G.F. Pinder. 1985. A Multiphase Approach to the Modeling of Porous Media Contamination by Organic Compounds. Water Resources Research 21(1):11–18.

Baehr, A.L., G.E. Hoag, and M.C. Marley. 1989. Removing Volatile Contaminants from the Unsaturated Zone by Inducing Advective Air-Phase Transport. Journal of Contaminant Hydrology 4(1):1–26.

Barker, J.F., G.C. Patrick, and D. Major. 1987. Natural Attenuation of Aromatic Hydrocarbons in a Shallow Sand Aquifer. Ground Water Monitoring Review 7(1):64–71.

Borden, R., M. Lee, J.M. Thomas, P. Bedient, and C.H. Ward. 1989. In Situ Measurement and Numerical Simulation of Oxygen Limited Biotransformation. Ground Water Monitoring Review 9(1):83–91.

Bouchard, D.C., A.L. Wood, M.L. Campbell, P. Nkedi-Kizza, and P.S.C. Rao. 1988. Sorption Nonequilibrium During Solute Transport. Journal of Contaminant Hydrology 2(3):209–223.

Chau, T.S. 1988. Analysis of Sustained Ground-Water Withdrawals by the Combined Simulation-Optimization Approach. Ground Water 26(4):454–463.

Cheng, Songlin. 1988. Computer Notes - Trilinear Diagram Revisited: Application, Limitation, and an Electronic Spreadsheet Program. Ground Water 26(4):505–510.

Curtis, G.P., P.V. Roberts, and Martin Reinhard. 1986. A Natural Gradient Experiment on Solute Transport in a Sand Aquifer: 4. Sorption of Organic Solutes and Its Influence on Mobility. Water Resources Research 22(13):2059–2067.

Enfield, C.G. and G. Bengtsson. 1988. Macromolecular Transport of Hydrophobic Contaminants in Aqueous Environments. Ground Water 26(1):64–70.

Faust, C.R. 1985. Transport of Immiscible Fluids Within and Below the Unsaturated Zone: A Numerical Model. Water Resources Research 21(4):587–596.

Feenstra, S., J.A. Cherry, E.A. Sudicky, and Z. Haq. 1984. Matrix Diffusion Effects on Contaminant Migration from an Injection Well in Fractured Sandstone. Ground Water 22(3):307–316.

Flathman, P., D. Jerger, and L. Bottomley. 1989. Remediation of Contaminated Ground Water Using Biological Techniques. Ground Water Monitoring Review 9(1):105–119.

Gelhar, L.W. 1986. Stochastic Subsurface Hydrology from Theory to Applications Water Resources Research 22(9):135S–145S.

Goltz, M.N. and P.V. Roberts. 1986. Interpreting Organic Solute Transport Data from a Field Experiment Using Physical Nonequilibrium Models. Journal of Contaminant Hydrology 1(1/2):77–94.

Goltz, M.N. and P.V. Roberts. 1988. Simulations of Physical Nonequilibrium Solute Transport Models: Application to a Large-Scale Field Experiment. Journal of Contaminant Hydrology 3(1):37–64.

Guven, O. and F.J. Molz. 1986. Deterministic and Stochastic Analyses of Dispersion in an Unbounded Stratified Porous Medium. Water Resources Research 22(11):1565–1574.

Hinchee, R. and H.J. Reisinger. 1987. A Practical Application of Multiphase Transport Theory to Ground Water Contamination Problems. Ground Water Monitoring Review 7(1):84–92.

Hossain, M.A. and M.Y. Corapcioglu. 1988. Modifying the USGS Solute Transport Computer Model to Predict High Density Hydrocarbon Migration. Ground Water 26(6):717–723.

Hunt, J.R., N. Sitar, and K.S. Udell. 1988. Nonaqueous Phase Liquid Transport and Cleanup: 1. Analysis of Mechanisms. Water Resources Research 24(8):1247–1258.

Hunt, J.R., N. Sitar, and K.S. Well. 1988. Nonaqueous Phase Liquid Transport and Cleanup: 2. Experimental Studies. Water Resources Research 24(8):1259–1269.

Jensen, B.K., E. Arvin, and A.T. Gundersen. 1988. Biodegradation of Nitrogen- and Oxygen-Containing Aromatic Compounds in Groundwater from an Oil-Contaminated Aquifer. Journal of Contaminant Hydrology 3(1):65–76.

Jorgensen D.G., T. Gogel, and D.C. Signor. 1982. Determination of Flow in Aquifers Containing Variable-Density Water. Ground Water Monitoring Review 2(2):40–45.

Keely, J.F. 1984. Optimizing Pumping Strategies for Contaminant Studies and Remedial Actions. Ground Water Monitoring Review 4(3):63–74.

Keely, J.F., M.D. Piwoni, and J.T. Wilson. 1986. Evolving Concepts of Subsurface Contaminant Transport. Journal Water Pollution Control Federation 58(5):349–357.

Keely, J.F. and C.F. Tsang. 1983. Velocity Plots and Capture Zones of Pumping Centers for Ground Water Investigations. Ground Water 22(6):701–714.

Kipp, K.L., K.G. Stollenwerk, and D.B. Grove. 1986. Groundwater Transport of Strontium 90 in a Glacial Outwash Environment. Water Resources Research 22(4):519–530.

Konikow, L.F. 1986. Predictive Accuracy of a Ground-Water Model - Lessons from a Postaudit. Ground Water 24(2):173–184.

Kueper, B.H. and E.O. Frind. 1988. An Overview of Immiscible Fingering in Porous Media. Journal of Contaminant Hydrology 2(2):95–110.

Macalady, D.L., P.G. Tratnyek, and T.J. Grundl. 1986. Abiotic Reduction Reactions of Anthropogenic Organic Chemicals in Anaerobic Systems: A Critical Review. Journal of Contaminant Hydrology 1(1/2):1–28.

Mackay, D.M., W.P. Ball, and M.G. Durant. 1986. Variability of Aquifer Sorption Properties in a Field Experiment on Groundwater Transport of Organic Solutes: Methods and Preliminary Results. Journal of Contaminant Hydrology 1(1/2):119–132.

Mackay, D.M., D.L. Freyberg, P.V. Roberts, and J.A. Cherry. 1986. A Natural Gradient Experiment on Solute Transport in a Sand Aquifer: 1. Approach and Overview of Plume Movement. Water Resources Research 22(13):2017–2029.

Major, D.W., C.I. Mayfield, and J.F. Barker. 1988. Biotransformation of Benzene by Denitrification in Aquifer Sand. Ground Water 26(1):8–14.

Mercado, Abraham. 1985. The Use of Hydrogeochemical Patterns in Carbonate Sand and Sandstone Aquifers to Identify Intrusion and Flushing of Saline Water. Ground Water 23(5):635–645.

Molz, F.J., O. Guven, J.G. Melville, and J.F. Keely. 1986. Performance and Analysis of Aquifer Tracer Tests with Implications for Contaminant Transport Modeling. EPA/600/2-86-062.

Molz, F.J., O. Guven, J.G. Melville, J.S. Nohrstedt, and J.K. Overholtzer. 1988. Forced Gradient Tracer Tests and Inferred Hydraulic Conductivity Distributions at the Mobile Field Site. Ground Water 26(5):570–579.

Molz, F.J., M.A. Widdowson, and L.D. Benefield. 1986. Simulation of Microbial Growth Dynamics Coupled to Nutrient and Oxygen Transport in Porous Media. Water Resources Research 22(8):1207–1216.

Novak, S.A. and Y. Eckstein. 1988. Hydrogeochemical Characterization of Brines and Identification of Brine Contamination in Aquifers. Ground Water 26(3):317–324.

Osiensky, J.L., K.A. Peterson, and R.E. Williams. 1988. Solute Transport Simulation of Aquifer Restoration After In Situ Uranium Mining. Ground Water Monitoring Review 8(2):137–144.

Osiensky, J.L, G.V. Winter, and R.E. Williams. 1984. Monitoring and Mathematical Modeling of Contaminated Ground-Water Plumes in Fluvial Environments. Ground Water 22(3):298–306.

Ophorl, D.U., and J. Toth. 1989. Patterns of Ground-Water Chemistry, Ross Creek Basin, Alberta, Canada. Ground Water 27(1):20–26.

Pinder, G.F. and L.M. Abriola. 1986. On the Simulation of Nonaqueous Phase Organic Compounds In the Subsurface. Water Resources Research 22(9):109S–119S.

Pollock, D.W. 1988. Semianalytical Computation of Path Lines for Finite Difference Models. Ground Water 26(6):743–750.

Roberts, P.V., M.N. Goltz, and D.M. Mackay. 1986. A Natural Gradient Experiment on Solute Transport in a Sand Aquifer: 3. Retardation Estimates and Mass Balances for Organic Solutes. Water Resources Research 22(13):2047–2058.

Satlin, R.L. and P.B. Bedient. 1988. Effectiveness of Various Aquifer Restoration Schemes Under Variable Hydrogeologic Conditions. Ground Water 26(4):488–499.

Siegrist, H. and P.L. McCarty. 1987. Column Methodologies for Determining Sorption and Biotransformation Potential for Chlorinated Aliphatic Compounds in Aquifers. Journal of Contaminant Hydrology 2(1):31–50.

Spain, J.C., J.D. Milligan, D.C. Downey, and J.K. Slaughter. 1989. Excessive Bacterial Decomposition of H_2O_2 During Enhanced Biodegradation. Ground Water 27(2):163–167.

Spayed, S.E. 1985. Movement of Volatile Organics Through a Fractured Rock Aquifer. Ground Water 23(4):496–502.

Srinivasan, P. and J.W. Mercer. 1988. Simulation of Biodegradation and Sorption Processes in Ground Water. Ground Water 26(4):475–487.

Staples, C.A and S.J. Geiselmann. 1988. Cosolvent Influences on Organic Solute Retardation Factors. Ground Water 26(2):192–198.

Starr, R C., R.W. Gillham, and E.A. Sudicky. 1985. Experimental Investigation of Solute Transport in Stratified Porous Media: 2. The Reactive Case. Water Resources Research 21(7):1043–1050.

Steinhorst, R.K. and R.E. Williams. 1985. Discrimination of Groundwater Sources Using Cluster Analysis, MANOVA, Canonical Analysis, and Discriminant Analysis. Water Resources Research 21(8):1149–1156.

Stover, E. 1989. Co-Produced Ground Water Treatment and Disposal Options During Hydrocarbon Recovery Operations. Ground Water Monitoring Review 9(1):75–82.

Sudicky, E.A., R.W. Gillham, and E.O. Frind. 1985. Experimental Investigation of Solute Transport in Stratified Porous Media: 1. The Nonreactive Case. Water Resources Research 21(7):1035–1041.

Testa, S. and M. Paczkowski. 1989. Volume Determination and Recoverability of Free Hydrocarbon. Ground Water Monitoring Review 9(1):120–128.

Thomsen, K., M. Chaudhry, K. Dovantzis, and R. Riesing. 1989. Ground Water Remediation Using an Extraction, Treatment, and Recharge System. Ground Water Monitoring Review 9(1):92–99.

Thorstenson, D.C. and D.W. Pollock. 1989. Gas Transport in Unsaturated Porous Media: The Adequacy of Ficke's Law. Reviews of Geophysics 27(1):61–78.

Usunoff, E.J., and A. Guzman-Guzman. 1989. Multivariate Analysis in Hydrochemistry: An Example of the Use of Factor and Correspondence Analyses. Ground Water 27(1):27–34.

Valocchi, A.J. 1988. Theoretical Analysis of Deviations from Local Equilibrium During Sorbing Solute Transport through Idealized Stratified Aquifers. Journal of Contamination Hydrology 2(3):191–208.

Watson, I. 1984. Contamination Analysis—Flow Nets and the Mass Transport Equation. Ground Water 22(1):31–37.

Zheng C., K.R. Bradbury, and M.P. Anderson. 1988. Role of Interceptor Ditches in Limiting the Spread of Contaminants in Ground Water. Ground Water 26(6):734–742.

Chapter 5

Chemical Enhancements to Pump-and-Treat Remediation[1]

Carl D. Palmer and William Fish, Oregon Graduate Institute of Science and Technology, Beaverton, OR

BACKGROUND

Conventional pump-and-treat technologies are among the most widely used systems for the remediation of contaminated ground water. Within recent years it has become recognized that these systems can require protracted periods of time to make significant reductions in the quantity of contaminants associated with both the liquid and solid phases which constitute the subsurface matrix. Recent research has led to a better understanding of the processes involved in the transport and transformation of contaminants in the subsurface. While some of these processes are not readily amenable to enhanced removal by ground-water extraction, others suggest that there are available techniques to increase the efficiency of these types of remediation systems. The intent of this chapter is to explore the use of chemical enhancement to improve ground-water remediation efficiencies using pump-and-treat technologies, and point out arenas of contamination where such techniques are not practical.

SUMMARY AND CONCLUSIONS

Recognition that conventional pump-and-treat remediation often requires lengthy periods of time to achieve cleanup objectives will encourage professionals involved in site remediation to contemplate alternative methods of aquifer restoration. Some form of chemical enhancement for pump-and-treat will likely be an alternative considered for many waste-site cleanups. Although chemical enhancement of pump-and-treat may be a means of accelerating aquifer remediation, there are many aspects of chemical enhancement that need to be known before these techniques can be successfully implemented.

Not all waste sites are amenable to chemical enhancement methods. In particular, if tailing in the concentration-versus-time curves for the extraction wells is dominated by physical processes, then chemical enhancement methods will have no advantage over conventional pump-and-treat. Knowledge of the relative contributions of chemical and physical processes limiting pump-and-treat are needed during the early stages of site Remedial Investigations to ascertain the general usefulness of chemical enhancement.

Even when it is known that physical processes contribute little to the tailing, specific knowledge is needed about the chemical processes that contribute to tailing at a particular site. Only then can potential chemical agents that are likely to influence these processes be identified. The reactive agents may be chosen to compete with the contaminants for adsorption sites, complex the contaminant, change the redox state of the contaminant, change the solvation properties of the ground water, act as a surfactant, ionize the contaminant, or substitute for the contaminant in a precipitate. If the reactive agents are chosen on the basis of incorrectly identified limiting processes, there is a risk that the reactive agents will provide no net benefit and may even prolong remediation.

Even when a reactive agent is found that specifically addresses the limiting chemical process, other considerations must be investigated to assure successful implementation. The key areas of concern in any chemical enhancement method are 1) the delivery of the reactive agent to those areas

---

[1] EPA/540/S-92/001.

of the aquifer where it is needed, 2) the enhanced removal of the target contaminants by the reactive agent, 3) the removal of the reactive agent from subsurface, 4) the impact of the reactive agent on the treatment of the target contaminant and the volume of sludges to be disposed. Additional site characterization, treatability tests, and design studies must be conducted to address each of these important aspects of a chemical-enhancement program. At some sites, implementation of chemical enhancement will require additional capital expenditures for wells and treatment facilities. The advantages and disadvantages, including the additional costs, of chemical enhancements need to be compared with other methods of remediation, such as conventional pump-and-treat.

While many individual components are associated with the implementation of chemical enhancement to pump-and-treat, they all should be investigated. If one aspect of this process fails, the entire system can fail. While such failure is not necessarily a disaster (conventional pump-and-treat can continue), it is a waste of resources that could be utilized for more beneficial uses. It is believed that these issues must be addressed and a reasonable probability of success demonstrated in all aspects of a chemical enhancement system before it is implemented.

INTRODUCTION

The recognition that ground water in many areas of the U.S. is contaminated has brought about demands that the quality of these aquifers be restored. At Superfund sites, the initial cleanup is accomplished in a relatively short time by removing sources of contamination from the surface, removing highly contaminated shallow soil, and in some cases installing a low-permeability cap. In contrast, remediation of the ground water beneath a site is often an inexact process requiring years to complete.

A common method for aquifer remediation is to withdraw the contaminated water from the aquifer and treat it onsite. The treated water may then be returned to the aquifer, discharged to surface water, or transferred to a public water treatment plant. Such "pump-and-treat" technology is widely used for remediating aquifers (Palmer et al., 1988) with about 68% of the Records of Decision identifying it as the method of remediation (Travis and Doty, 1989). However, at many sites pump-and-treat technology will require decades of costly operation to achieve the desired levels of cleanup. Extended periods for remediation are highly undesirable because the operation and maintenance costs associated with the remediation can be large, and, in many cases, otherwise valuable land cannot be used for any economic purpose.

The great costs of cleanup make it essential to investigate technologies that may speed up remediation. One such technology is the injection of chemical constituents, "reactive agents," that improve the rate of removal of contaminants from the subsurface. The applicability of such "chemical enhancement" technology and the specific chemicals that can be used depend on the processes that control the slow "tailing" of contaminant concentrations in the extraction wells. Not all processes leading to lengthy remediations can be corrected by chemical enhancement. However, certain problematic types of contamination may be amenable to well-conceived applications of reactive agents.

The limitations of aquifer remediation by conventional pump-and-treat will encourage engineers, scientists, and regulators to propose various chemical enhancement methods for the remediation of particular sites. While these proposed methods must be evaluated with regard to specific site conditions, there are general concepts applicable to all chemical enhancement methods. This chapter is intended to 1) outline these general concepts, 2) pose key questions that should be answered before any chemical-enhancement scheme is initiated, 3) stimulate discussion on the merits and limitations of chemical enhancement methods, and 4) focus research on particularly problematic areas of chemical enhancement.

PROCESSES AFFECTING PUMP-AND-TREAT REMEDIATION

A major concern in pump-and-treat operations is that contaminant concentrations within the extraction wells will decline at a progressively slower rate as pumping continues. When the rate of decline becomes small and the contaminant concentrations are still above the target cleanup levels, an extraction well is said to exhibit "tailing" (Figure 5-1). Contaminant concentrations may have

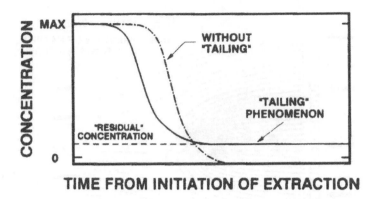

TIME FROM INITIATION OF EXTRACTION

Figure 5-1. Concentration-versus-time curve for an extraction well with continuous pumping (after Keely et al., 1987).

dropped several orders of magnitude, but they remain above the target clean-up level despite a considerable period of pumping. A great uncertainty in pump-and-treat operations is the time required for these tailing concentrations to decrease below the target clean-up levels. Reasonable estimates of clean-up times under these conditions require an understanding of the physical and chemical processes that can cause tailing: 1) the differing amounts of time required by contaminated waters to flow along different streamlines from the irregular boundary of the plume to the extraction wells, 2) multiple rates of mass transport within spatially variable sediments, 3) limited mass transfer from reserves of nonaqueous phase liquids and solid phase mineral precipitates, and 4) slow desorption reactions (Keely et al., 1987; see also Chapter 4).

Physical Causes of Tailing

Ground water entering an extraction well is a mixture of waters that have traveled along multiple subsurface pathways between the edge of the contaminant plume and the well. The time required for contaminated ground water to travel along these different flow paths is controlled by 1) placement of the extraction wells relative to the contaminant boundaries, 2) the extraction rate, 3) the aquifer porosity, 4) the magnitude and direction of the natural hydraulic gradient, and 5) the location and types of hydraulic boundaries. As an example, consider a single extraction well in an aquifer with a natural hydraulic gradient of 0.007 toward the east (Figure 5-2). If the edge of the contaminant plume is to the west, then only a portion of the plume's edge is captured by the extraction well. The flow paths along the outside of the capture zone for the well have a greater distance to travel and are influenced by lower hydraulic gradients than the flow paths near the center of the capture zone. As a consequence of these variable residence times within each stream tube, the concentration-versus-time curve for the extraction well exhibits substantial tailing even in the absence of chemical reaction (Figure 5-3).

In heterogeneous porous media, ground water in higher permeability layers has greater velocities than water within the lower permeability zones. The higher permeability pathways are not necessarily sand or gravel nor are the lower permeability zones necessarily silts or clays; it is sufficient if one region possesses greater hydraulic conductivity *relative* to the adjacent materials. When the contrast in hydraulic conductivity between these zones is large, the advective component of transport through the lower permeability lenses becomes small. As contaminants are transported through such a heterogeneous aquifer, they are advected along the high permeability layers and diffuse into the lower permeability layers. Such an advection diffusion process can affect the concentration of contaminants within higher permeability layers (Gillham et al., 1984; Sudicky et al., 1985). If aqueous contaminants have been present over many years, their concentration in the lower permeability layers can equal the concentrations in the higher permeability zones. During pump-and-treat remediation, contaminants in the high permeability layers are removed more quickly than from the lower permeability layers. These variable rates of advective transport create concentration gradients be-

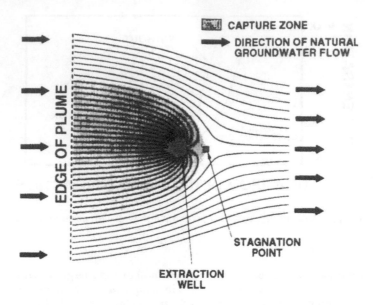

Figure 5-2. Flow lines from the edge of a contaminant plume toward an extraction well.

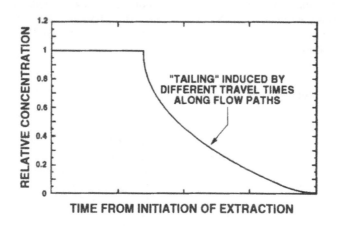

TIME FROM INITIATION OF EXTRACTION

Figure 5-3. Concentration-versus-time for the extraction well illustrated in Figure 5-2. The tailing is caused by the differential travel times along the stream lines.

tween zones of contrasting permeability and cause the slow diffusion of contaminants from the low permeability zones to the high permeability zones where they can be pumped to the surface (Figure 5-4). Thus, the contaminant concentrations in the extracted water are initially high as the more permeable layers are flushed. At later times, the concentration in the extracted water is limited by the rate of diffusion of the contaminants into the high permeability zones (Figure 5-5). If pumping is discontinued, the velocities within the high permeability layers decrease and the concentration of contaminants within these zones increase (Figure 5-6) because of the greater residence time of a parcel of water within the contaminated portion of the aquifer.

The main point is that, in most cases, lengthy tailing-off of contaminant concentrations in extraction wells is at least partly due to physical attributes of the system that cannot be ameliorated by injections of chemical agents. Thus, chemical enhancement cannot be expected to eliminate all unexpected delays in pump-and-treat removal. However, when the rates of chemical mass transfer from contaminant reserves in an aquifer are the primary limitations on removal, then the use of reactive agents may substantially enhance remediation.

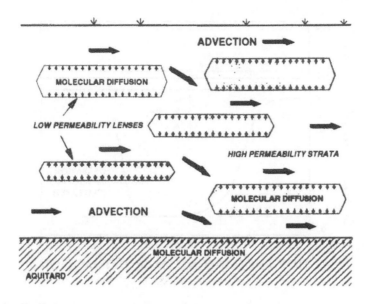

Figure 5-4. Heterogeneous porous medium with advection through the high permeability zones and mass transfer by molecular diffusion from the lower permeability lenses.

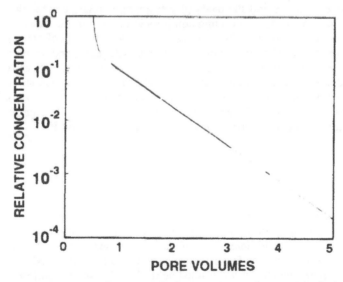

Figure 5-5. Concentration versus time for removal of contaminants from an ideally layered aquifer. The layers are assumed to be 10 cm thick, the length is 10 m, the retardation factor is unity, and the diffusion coefficient is 10^{-6} cm^2/s. Calculated using the equations given by Sudicky and Frind (1982).

Chemical Causes of Tailing

At many sites, some or even most of the contaminant mass will not be dissolved in the ground water but will be present as 1) adsorbed species, 2) precipitates, or 3) nonaqueous phase liquids (NAPLs). These reserves of matrix-associated contaminants and contaminants in the immobile fraction of the NAPLs cannot be directly extracted by pump-and-treat: they must transfer from the solid or NAPL to the ground water before they can be removed. If the equilibrium concentration in the ground water is small relative to the total mass of contaminant in the soil or if the rate of mass

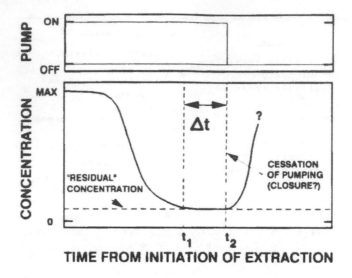

Figure 5-6. Concentration-versus-time for an extraction well that is turned off at time t_2.

transfer is small relative to the ground-water velocity, then large quantities of water must pass through contaminated sections of aquifer before it is remediated.

If reactions between solutes and stationary phases are rapid relative to the flow rate, equilibrium partitioning can be assumed. However, rapid equilibration does not translate to rapid removal rates if the equilibrium concentration in solution is very low. The retention of contaminants by mineral surfaces and microbial cell walls, ion exchange reactions with clays, and the partitioning of organic contaminants between soil organic matter and the ground water can significantly increase the time required for remediation of contaminated aquifers. For hydrophobic, nonpolar organic compounds, desorption can often be represented by linear isotherms (e.g., Chiou et al., 1979). In the absence of free product (NAPL), the number of pore volumes required to remove the organic contaminant from a homogeneous aquifer is approximately equal to the retardation factor, R,

$$R = 1 + \frac{\rho b}{n}(f_{oc}k_{oc}) \tag{1}$$

where ρb is the dry bulk density of the soil, n is porosity, f_{oc} is the fraction of organic carbon in the soil (mass of carbon/mass of soil), and K_{oc} is the partition coefficient for the contaminant into soil organic carbon (mass per unit mass of carbon/equilibrium concentration in water). A compound with a large K_{oc} value can have a large retardation factor even in a soil with a small to a moderate amount of organic carbon. Thus, many pore volumes of water must be flushed through the soil to remove such hydrophobic organic contaminants.

In some cases, equilibrium partitioning may not be applicable. Laboratory tests had shown that weeks are required to achieve equilibrium concentrations in laboratory experiments with sediments (Hamaker and Thompson, 1972; Coates and Elzerman, 1986; Karickhoff, 1980). Desorption of pyrene, hexachlorobenzene, and pentachlorobenzene from river sediments requires days to weeks (Karickhoff and Morris, 1985). If such rates of desorption are slow relative to the rate of ground-water flow, then equilibrium concentrations may not be attained during pump-and-treat, and tailing in the concentration-versus-time curve can result.

Although the linear adsorption model is adequate for describing the adsorption equilibria of many nonpolar, hydrophobic organic contaminants (Chiou et al., 1979), it does not adequately describe the behavior of organic or inorganic ions over a wide range of pH and adsorbate concentrations. The adsorption of ionic solutes is often represented by an adsorption isotherm. An adsorption isotherm is a plot of the contaminant concentration on the soil versus the equilibrium solution concentration of the contaminant. Adsorption isotherms are defined according to their general shape and mathematical

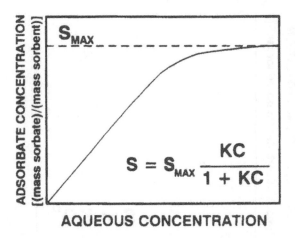

Figure 5-7. A Langmuir isotherm.

representation. For a Langmuir isotherm, the concentration on the soil increases with increasing concentration in the ground water until a maximum concentration on the soil is reached (Figure 5-7). The isotherm can be represented by the equation:

$$S = S_{max} \left(\frac{KC}{1 + KC} \right) \qquad (2)$$

where S (mass/mass) is the concentration on the soil, S_{max} (mass/mass) is the maximum concentration on the soil, K ((length)^3/mass) is the Langmuir adsorption constant, and C (mass/(length)^3) is the concentration in the ground water. A Freundlich (or Küster) isotherm is given by the equation:

$$S = KC^b \qquad (3)$$

where K is the Freundlich adsorption constant and b is a positive parameter. The shape of a Freundlich isotherm depends on the value of b. If b is greater than 1.0, the isotherm becomes steeper with increasing concentrations in the ground water. If b is less than 1.0, the isotherm becomes steeper at lower concentrations (Figure 5-8). A linear isotherm is a special case of the Freundlich isotherm where the parameter b is equal to unity. At constant pH, cations tend to follow Freundlich isotherms, while anions tend to follow Langmuir isotherms (Dzombak, 1986; Dzombak and Morel, 1990).

Adsorption isotherms are useful for illustrating the dependence of the solid phase concentration of the contaminant on the aqueous phase concentration of the contaminant at a given pH. However, adsorption of inorganic ions is pH-dependent and the form of the isotherm should be known over the entire pH range likely to be found at a site. Sometimes this pH-dependence is presented as the fraction of the contaminant adsorbed versus pH or a "pH-edge" (Figure 5-9). For cations, the pH-edge for most minerals show little or no adsorption at low pH. As pH increases, the portion of the contaminant that is adsorbed increases until the fraction is unity (provided that the mass of contaminant does not exceed the available adsorption sites). The pH-edges for anions are the opposite to those for cations. There is little or no adsorption at higher pH but as pH decreases, the fraction of the anion that is adsorbed increases to unity or to the ratio of the mass of sites to the mass of contaminant, if the amount of contaminant exceeds the number of available sites. For either cations or anions, the shape and position of the pH-edge depends on the specific mineral surface and ions under consideration.

Adsorption processes can also be modeled using surface complexation models (e.g., Stumm et al., 1976; Schindler, 1981; Schindler and Stumm, 1987; Dzombak and Morel, 1990). The key advantage of this type of approach is that it has a foundation in chemical theory allowing the results to be extended beyond the exact test conditions. The dependence of the amount of adsorption on the pH of the

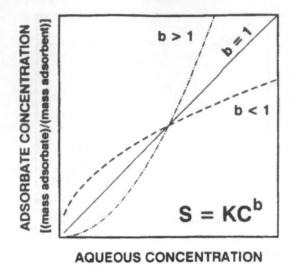

Figure 5-8. Freundlich isotherms. A linear isotherm is the special case for which the exponent, b, is equal to unity.

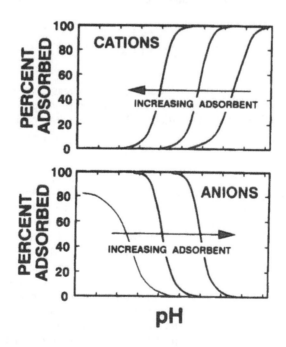

Figure 5-9. pH adsorption edges for cations and anions. The fraction of contaminant adsorbed will not reach unity if the sites are saturated.

solution and the competition between several adsorbates for the adsorption sites are, in principle, accounted for in such a model. The disadvantages of the surface complexation model are the lack of a consistent set of equilibrium constants and the potential lack of linear additivity when multiple adsorbents are present. The first limitation is being overcome through compilations of consistent data sets for oxide surfaces. At this time, there does exist a consistent set of adsorption constants for adsorption onto hydrous ferric oxide (Dzombak, 1986; Dzombak and Morel, 1990) based on a two-layer surface complexation model. Such data sets need to be derived for other oxide surfaces as well.

The general concepts of ion adsorption can be applied to anticipate some of the behavior of contaminants during pump-and-treat remediation. The rate of removal of ionic contaminants under acidic conditions can be substantially different than under neutral or alkaline conditions. Adsorption of anions, for example, is more likely to be a problem at lower pH than at more neutral or alkaline conditions. Furthermore, there may be changes in the amount of adsorption during remediation as acid or alkaline waters are returned to more neutral pH conditions. In all of these adsorption models, ionic contaminants follow nonlinear isotherms. The partition coefficient equals the slope of the adsorption isotherm. As aqueous contaminant concentrations decrease during remediation, the slope of the isotherm, hence the retardation, changes. In most cases, the retardation will increase with decreasing concentrations, making it more difficult to decrease intermediate concentrations below the maximum contaminant level (MCL) than to decrease the initially high concentrations to intermediate levels.

Large reserves of inorganic contaminants may be formed as the result of the precipitation of crystalline and amorphous materials within the soils. For example, one concern is the potential effect of a reserve of solid $BaCrO_4$ within aquifer systems contaminated with hexavalent chromium (Cr(VI)) (Palmer and Wittbrodt, 1990). As Cr(VI)-laden waters enter the subsurface, natural Ba^{2+} may react with aqueous chromate (CrO^-) to precipitate a reserve of $BaCrO_4$. In many cases, the size of the reserve will be limited by the availability of Ba^{2+} in the soil rather than the mass of chromate spilled. During the initial phases of a pump-and-treat remediation ground water containing high concentrations of Cr(VI) in excess of available Ba^{2+} are removed and the concentrations in the extraction wells quickly decrease (Figure 5-10). At some point, $BaCrO_4$ becomes the principal source of Cr(VI) in the pore water. The Cr(VI) concentration will remain relatively constant for as long as there is $BaCrO_4$ remaining in the soil. Using equilibrium concepts, Palmer and Wittbrodt (1990) estimated that 25 to 50 pore volumes are required to remove Cr(VI) from soils at a hard chrome plating facility. If equilibrium is not obtained and kinetic processes control the amount of $BaCrO_4$ dissolved as the ground water passes through the soil, then more pore volumes would be required.

Similar solubility limitations may occur for other inorganic contaminants. The effect of precipitates on efficacy of pump-and-treat remediation depends on the solubility of the mineral phase. The most troublesome mineral precipitates are those with solubilities low enough to create a relatively large contaminant reserve, yet with solubilities large enough to exceed the target cleanup levels. A complicating factor is substitution of contaminants in the crystalline structure of other minerals. The degree of substitution affects the equilibrium concentration of the contaminant. Regardless of whether the contaminant has been precipitated in pure or substituted mineral phases, if the rate of dissolution is slow relative to the velocity of the ground water, then the time required for the removal of the contaminant from the subsurface will be greater than when equilibrium conditions have been achieved.

Nonaqueous-phase liquids can also provide large reserves of contaminants in the subsurface. For example, if a cubic meter of soil with a 35% porosity contains trichloroethylene (TCE) at 20% residual saturation, then approximately 270 pore volumes must pass through the soil and reach equilibrium with the TCE (1100 mg/L) before the solvent is removed from the soil by dissolution. Sandbox experiments with perchloroethylene (Anderson, 1988, Anderson et al., 1992) suggest that this equilibrium is achieved very quickly as the water passes through fingers of residual solvent. Longer periods of time are required to remove solvents when they are present in pools (e.g., Johnson and Pankow, 1992; Anderson et al., 1992). Using the equations given by Hunt et al. (1988) and reasonable choices of transport parameters, it can be shown that only the water that passes very close to the edge of the solvent pool is likely to reach equilibrium concentrations with the solvent, while the concentrations further above the pool are limited by the rate of mass transfer from the pool to the bulk aquifer (Figure 5-11). The average concentration of the solvent measured in a monitoring well with a two-meter length of screen placed just above the pool will increase across the length of the pool over which the ground water has flowed (Figure 5-12). If the 103 kg of TCE used in the previous example is distributed in a 20 cm thick pool below the cubic meter of soil, the average concentration of TCE in the ground water exiting from the block of soil is 28.6 mg/L and 10,200 pore volumes must pass through the aquifer before the TCE is removed. Thus, it takes approximately 39 times longer to remove solvent from a pool than to remove the same mass of solvent from residual saturation.

When pump-and-treat remediation is predominantly limited by chemical processes that restrict the transfer of mass from these contaminant reserves to the ground water, chemical enhancement to pump-and-treat should be considered. Although the choice of a reactive agent that will greatly enhance contaminant removal is a primary concern, there are several other factors that must be considered before implementation of a chemical-enhancement program.

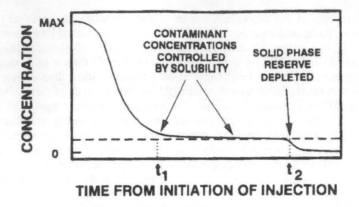

Figure 5-10. Concentration versus time for an extraction well in a formation that contains a solid phase precipitate.

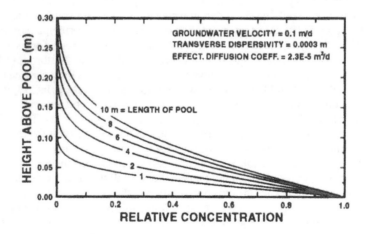

Figure 5-11. Concentration of a contaminant at different elevations above a DNAPL pool for different pool lengths (after Johnson and Pankow, 1992).

CHEMICAL ENHANCEMENTS FOR PUMP-AND-TREAT REMEDIATIONS

If chemical enhancement of pump-and-treat is to be successful, four key areas must be satisfactorily addressed in the design: 1) delivery of the reactive agent to where it is needed within the aquifer, 2) the interaction between the reactive agent and the contaminant, 3) the removal of the contaminant and the reactive agent from the subsurface, 4) the treatment of the extracted water and disposal of the resulting sludges (Figure 5-13).

Delivery

Delivery of the reactive agents to the areas within the aquifer where they are needed to enhance the removal of contaminants can be a complex process. Reactive agent solutions must be injected without clogging the aquifer near the injection well with particles and chemical precipitates. The ground water containing the reactive agents must then move in some reasonable period of time to the contaminated portion of the aquifer. The rate, mode, and scheduling of the injection and pumping must be designed such that the reactive agent reaches those areas in a relatively short period of time. Many of these processes are influenced by the heterogeneities within the aquifer.

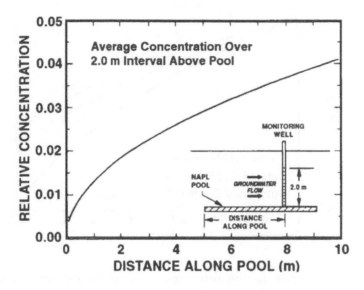

Figure 5-12. Average concentration of a contaminant over a 2-m interval above a DNAPL pool (after Johnson and Pankow, 1992).

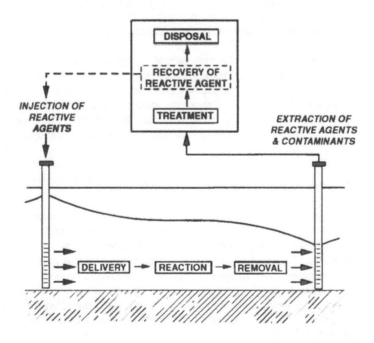

Figure 5-13. Schematic representation of chemical enhancement of a pump-and-treat operation. Key areas of concern are shown in boxes. In some cases, the reactive agent will be recovered and reused.

Clogging of Injection Wells

The clogging of injection wells is a common problem. Clogging can be the result of the physical filtration of suspended particles at the well interface, the formation of inorganic precipitates, or the growth of microorganisms. As more materials are entrapped, precipitated, or grown in the pore space adjacent to the well face, they occupy increasing amounts of the pore space and severely reduce the hydraulic conductivity (Palmer and Cherry, 1984). A plot of the ratio of the new permeability, k, to the original permeability, k_0, versus the new porosity, n, to original porosity, n_0, using the Carmen-Kozeny

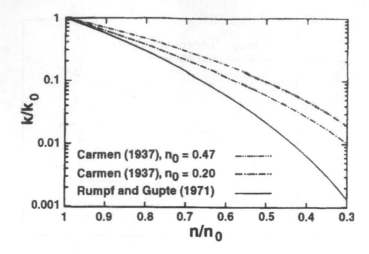

Figure 5-14. Ratio of permeability to initial permeability versus porosity to initial porosity using several empirical models (after Palmer and Cherry, 1984).

model (Carmen, 1937) and the Rumpf and Gupte (1971) model (Figure 5-14) illustrates that small changes in porosity can result in order of magnitude reductions in permeability. Because this reduction in permeability is immediately adjacent to the wellscreen, it is generally manifested as a reduction in well efficiency.

The mechanisms responsible for well clogging dictate the actions to avoid the problem. Additives may be required to prevent precipitation. Problems with particles in the injection water can be overcome by flocculation and removal with filter presses. Microbiological activity can be inhibited by the removal of nutrients or dissolved oxygen. Otherwise, periodic treatment such as surging, jetting, development, and acid treatment of the injection well may be required. These options need to be factored into estimates of capital costs of the remedial design, as well as the operation and maintenance costs.

Transport of the Reactive Agent to the Contaminated Areas

A fundamental problem with chemical enhancements to pump-and-treat remediation is getting the reactive agent to the contaminated portions of the aquifer so that it can interact with the contaminants to facilitate their removal. Two major problems should be considered: 1) differential flow paths governed by well hydraulics, and 2) mass transfer between heterogeneities that is governed by molecular diffusion. The paths the injected water follows depend on the hydraulic properties of the medium, the aquifer thickness, the natural hydraulic gradient, the rate of injection, the placement and pumping (injection) rate of other nearby wells, and the location and type of hydraulic boundaries in the vicinity of the site. Some of these effects are illustrated for homogeneous, isotropic aquifers with stream function calculations (Figure 5-15).

The two wells in Figure 5-15a are 50 feet apart. The upgradient well injects fluid at a rate of 0.5 gpm while the downgradient extraction well removes water at the same rate. Not all of the stream lines from the injection well are captured by the extraction well. In this particular case, 20% of the injection fluid continues to be transported through the aquifer. If the wells are placed 25 feet from one another (Figure 5-15b), then less of the injected fluid is lost (15%) but a much smaller volume of the aquifer is remediated with the chemical extractant. If the rate of injection is reduced to 0.25 gpm, all of the injected fluid is captured by the extraction well (Figure 5-15c); however, the volume of affected aquifer is diminished even further.

While, in principle, the distribution of wells and the rate of injection and extraction could be optimized, in practice, it may prove to be difficult. Firstly, the factors to be optimized are not universally agreed upon. Possibilities may include minimizing the number of wells, the total cost, or the time for cleanup, or maximizing the mass of contaminant removed per unit time. Secondly, none of

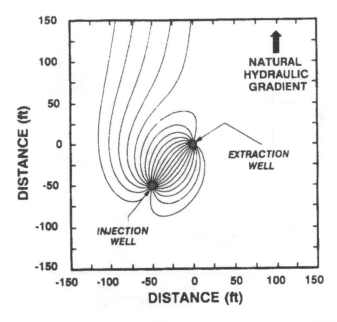

Figure 5-15a. Flow lines between an injection well and an extraction well located 50 feet apart and with a natural gradient of 0.007 to the north. Extraction rate and injection rate are both 0.5 gpm.

the necessary parameters are known with certainty. For example, concentration isopleths are known only within some range of distance, the variation in the hydraulic parameters have not been measured, and economic factors such as interest rates and costs over the next few years are based on extrapolation of current conditions.

Aquifer heterogeneity is an important factor in the rate of transport of dissolved constituents. As a reactive agent is injected, it is advected along the higher permeability pathways through the aquifer. When the hydraulic conductivity of the low permeability lenses is two or more orders of magnitude less than that of the high permeability zones, transport of the reactive agent into the lower permeability zones is controlled by molecular diffusion. The concentration of the reactive agent in the higher permeability zones should be maintained long enough to permit the diffusion of the reactive agent into the lower permeability lenses. The time it takes the reactive agent to diffuse into the lower permeability lenses depends on the diffusion coefficient, the concentration, and the soil-water interactions of the reactive agent.

The process of diffusion can be complex in a multicomponent system. If a complexing agent is used to enhance the pump-and-treat remediation, the complex initially forms in the higher permeability zone, creating a concentration gradient between the higher permeability layers and the lower permeability layers. Thus, not only does the reactive agent diffuse into the lower permeability zones, but so may the complex. The extent to which this affects the time for remediation depends upon the relative rates of diffusion of the solutes into and out of the lower permeability lenses and the relative rates of reaction for the important chemical processes.

Modes of Injection

The above discussion implicitly assumes that the reactive agent is continuously injected into the aquifer, but there are several options for injecting the fluid as well as for adding the reactive agent. The water may be injected continuously, as a series of pulses, or as a slug. With continuous fluid injection the reactive agent may be added to the injection fluid continuously or it may be added in a pulsed or slug mode. For pulsed injection of the fluid, the reactive agent may be injected with each pulse or as a slug into a single pulse of fluid. For a slug injection of fluid, the reactive agent can be added as a single slug. Thus, there are six general combinations of fluid and reactive agent injection modes (Figure 5-16). The advantages and disadvantages of these modes of injection (Table 5-1) were used to determine the suitability of each method for chemical enhancements to pump-and-treat (Figure 5-16).

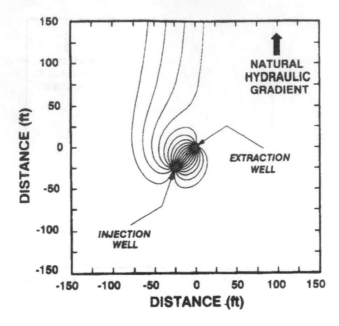

Figure 5-15b. Flow lines between an injection well and an extraction well located 25 feet apart and with a natural gradient of 0.007 to the north. Extraction rate and injection rate are both 0.5 gpm.

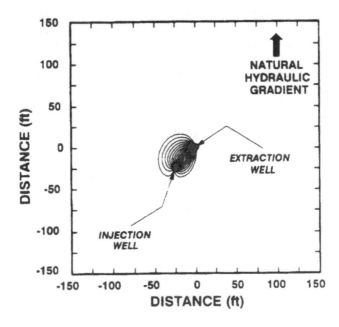

Figure 5-15c. Flow lines between an injection well and an extraction well located 25 feet apart and with a natural gradient of 0.007 to the north. Extraction rate is 0.5 gpm, while the injection rate is 0.25 gpm.

The modes most likely to be useful are continuous fluid injection with either continuous or pulsed addition of the reactive agent. The use of the slug mode for the injection of water is not practical; a relatively small slug of water injected into the aquifer does not affect a large volume of the aquifer. The major concern is the volume of aquifer over which the injected water is transported and how long it takes to accomplish this distribution. While continuous injection may provide the shortest time period over which it takes to distribute the fluid over a given volume of aquifer, there may be inherent economic benefits of using a pulsed mode. If staff must be present during injection, then coordinating the injection pulses with personnel shifts may be advantageous.

A key advantage of continuous addition of the reactive agent during continuous fluid injection is that relatively high concentrations of the agent can be maintained within the higher permeability zones of the aquifer. This creates a large concentration gradient between the higher permeability lenses and the lower permeability lenses, potentially reducing the time to remove the contaminants from those zones. However, this means that a larger mass of the reactive agent is required. A slug addition of the reactive agent would require less mass; however, the concentrations may decrease with time because of the nonlinear relationship between the adsorbed and aqueous concentrations. Pulsed addition would repeatedly increase the concentration gradients between the higher and lower permeability zones; however, the gradients may be locally reversed between the passing pulses. The system would have to be more carefully designed to ensure that the net direction of diffusion of the reactive agent is into the low permeability zones.

Timing of the Injection of the Reactive Agent

Chemical enhancement can be initiated at any time during the pump-and-treat remediation. Injection of the reactive agent may begin 1) as the extraction program begins, 2) after the concentration-versus-time curve significantly flattens, or 3) at some time intermediate to the first two.

An advantage of initiating injection of the reactive agent at the same time that extraction of the contaminated water begins is that it provides the earliest start to the chemical enhancements. The flow paths over which the reactive agent must travel remain approximately the same and therefore any delay in initiating the injection also delays aquifer cleanup. However, several problems may arise, suggesting that this method may not always be the most advantageous. If high concentrations of contaminant are still within the aquifer, then a greater concentration of the reactive agent may be required to remove this material. The cost of the additional reactive chemicals should be compared with the cost of a longer extraction time. Geochemical interactions within the contaminant plume can result in elevated concentrations of solutes other than the contaminant (e.g., Fe and Mn in anoxic plumes). High concentrations of these solutes may severely impede the effectiveness of the reactive agent, thereby requiring a greater mass of reactive agent to be injected into the aquifer. In some cases, these elevated levels of solutes may result in precipitation of the reactive agent and clogging of the aquifer.

Some of these problems can be avoided by using conventional pump-and-treat until the rate of decline in the concentration of the contaminant is low. As the concentrations of the contaminant decline, the high levels of other solutes present within the contaminated area are also likely to decline toward background levels. At lower concentrations of interfering solutes, the reactive agent interacts more efficiently with that portion of the contaminant that is most difficult to remove by conventional pump-and-treat. However, it may require the removal of several pore volumes over several years to reach these lower concentration levels, so added costs for operations and maintenance may be incurred.

The third possibility is to initiate chemical enhancement at some time between the initial startup of the pump-and-treat system and the time it takes for the concentrations to level off. Deciding the optimal time requires a more sophisticated analysis than either of the previous two choices; however, some simple criteria may serve as guides. For example, if there is concern over the precipitation of a solid phase, the criterion may be the time at which the concentration of the interfering ion decreases below the critical concentration computed from the solubility limit and the concentration of the reactive agent. While this is not necessarily the "optimal" time, it may serve as a practical estimate.

Retardation and the Rate of Transport of the Reactive Agent

The reactive agent must travel through the aquifer almost as quickly as the water. If the reactive agent is significantly retarded, then it may take longer for the reactive agent to be transported to the target areas of the aquifer than it takes to remove the contaminant from the subsurface without chemical enhancement. Thus, there is a possible paradox here: the reactive agent must react within the subsurface to enhance the removal of the contaminant, yet it must not be retarded. However, careful

REACTIVE AGENT ADDITION MODE	FLUID INJECTION MODE		
	CONTINUOUS	PULSED	SLUG
CONTINUOUS	good	N.A.	N.A.
PULSED	good	good	N.A.
SLUG	fair	fair	poor

N.A. = not applicable

Figure 5-16. Relative rating for different combinations of the modes of injection of the fluid and reactive agent.

Table 5-1. Modes of Injection for Chemical Enhancement to Pump-and-Treat

Fluid Injection Mode	Advantages	Disadvantages
Continuous	Fluid distributed over wide area. Less maintenance of pumping schedules.	Greater potential for clogging of screens. Greater pumping costs.
Slug	Less volume of water. Less potential clogging of wells.	Fluid distributed over very small volume of aquifer.
Pulsed	Can be developed around working schedules.	Requires more design to ensure injection and off periods are balanced relative to natural groundwater flow.
Active Agent Injection Mode		
Continuous	Maintain concentration in high permeability zones allowing for diffusion into low permeability zones.	Requires more mass of active agent.
Slug	Requires less mass of active agent.	May not allow sufficient time for diffusion into low permeability lenses. Concentration decreases with time/distance which can reduce effectiveness of the active agent.
Pulsed	Less total mass of active agent. Can be planned around work schedules. Allows for sufficient time for diffusion	Requires greater maintenance/ control. Requires more analysis to ensure that injection and off periods are of sufficient length.

consideration of the chemistry of the system may allow this paradox to be resolved so that both objectives are achieved.

If the reactive agent is chosen to compete with the contaminant for adsorption sites, both of these objectives can be realized by controlling the concentration of the reactive agent. For example, if the adsorption of the reactive agent follows a Langmuir-type adsorption isotherm, the amount of retardation is insignificant if the concentrations are high enough to saturate all of the available adsorption sites in the soil. If a complexing agent is utilized to enhance the removal of the contaminant, then it should be chosen so that neither the agent nor the complex are significantly retarded. Reactive agents that are reducing or oxidizing agents may be retarded as they react with materials other than the contaminant. Injection may have to simply continue until all of the material between the injection point and the extraction well that reacts faster than the contaminant is titrated from the aquifer. The amount of oxidant or reductant required to titrate the soil can be estimated from the oxidation and reduction capacities of the soil (Barcelona and Holm, 1991). In some cases, if something is known about the reaction rates for these redox reactions, some control can be obtained through control of the ground-water velocities (i.e., by controlling the rates of injection and extraction).

Reactive Agent—Contaminant Interactions

Different reactive agents can be chosen (Table 5-2), depending upon the processes that control the tailing in the concentration-versus-time curve for the extraction wells. The reactive agent may compete with the contaminant for adsorption sites, complex the contaminant, change the redox state of the contaminant, change the solvation properties of the ground water, act as a surfactant, ionize the contaminant, or substitute for the contaminant in a precipitate. These possibilities are not necessarily exclusive of one another. For example, a reactive agent may change the redox state of the contaminant and then form a complex with the altered form.

Competition for Adsorption Sites

If the tailing of the concentration-versus-time curves for the extraction wells are controlled by adsorption processes, the reactive agent can be chosen to compete for the adsorption sites. Such competition is most likely to be effective for ionic solutes, and least effective in displacing neutral organic molecules partitioned into soil organic matter.

The general concepts of ion adsorption can be applied to anticipate some of the constraints on the use of reactive agents to enhance pump-and-treat remediation. The adsorption of ions onto hydrous ferric oxide may be used as a model for appreciating the qualitative effects. Ultimately, laboratory tests utilizing the site contaminants and geologic materials should be performed. In general, we expect competition to be significant only when the adsorption sites are near saturation. An ionic contaminant can be easily displaced by a reactive agent with similar adsorption properties if the concentration of the reactive agent is sufficient to saturate the adsorption sites. Ionic contaminants can also be displaced by a reactive agent with a lower adsorption affinity, but only if the agent is present in great excess of the total number of adsorption sites in the soil.

Complexation

The reactive agent may be effective in forming aqueous complexes with an ionic contaminant. The aqueous complexes are not expected to be adsorbed as readily as the noncomplexed contaminant, therefore they are more mobile and relatively easy to remove by pump-and-treat technology. For example, James and Bartlett (1983) found that citric and diethylenetriaminepentaacetic (DTPA) acids complexed Cr(III) sufficiently to maintain elevated solution concentrations at pH 7.5 and 6.5, respectively. These organic acid anions can also contribute to the removal of chromium by competing with chromate for adsorption sites at oxide surfaces. Citrate may also contribute to the reduction of the Cr(VI) to Cr(III), which can then be complexed by another citrate molecule.

While there are many such potential chelators, those that are environmentally safe enough to be used in aquifers are relatively weak and nonspecific in their binding action. Two key consequences of these properties are that 1) the chelator must be present in great excess of the contaminant concentrations and 2) high concentrations of common nonhazardous soil constituents such as Fe, Mn, and Al may also be removed (e.g., Grove and Ellis, 1980; Norvell, 1984). The presence of these constituents in the waste stream may substantially increase the costs of treatment and disposal over conventional pump-and-treat. Some chelating agents, such as citrate, can be utilized as substrate by bacteria in the subsurface. As the concentrations of contaminants decrease below toxic levels around injection

Table 5-2. Reactive Agents - Contaminant Interactions

- Competition for adsorption sites
- Complexation of the contaminant
- Cosolvent effects
- Enhanced mobilization and solubilization by surfactants
- Oxidation
- Reduction
- Precipitation/Dissolution
- Ionization

wells, bacterial growth may increase to the level where the increased biomass clogs the aquifer and wells.

In some cases, the adsorption properties of the soil matrix may be altered by the use of chemical extractants. The removal of iron and aluminum oxide surfaces should decrease the adsorption density of the geologic materials. Zachara et al. (1988b), however, found the adsorption of chromate onto kaolinite increased with treatment with dithionate-citrate-bicarbonate (DCB) or hydroxylamine-hydrochloride ($NH_2OH \cdot HCl$) solutions. The reasons for this increase in adsorption are not clear.

Treatability studies should be conducted to determine not only the efficacy with which contaminants are removed by such chelators but also to estimate the total load of metals that must be treated and disposed, and the potential increases in biomass. The cost of these additional loads must then be compared with the costs of conventional pump-and-treat remediation.

Cosolvents

The rate of removal of hydrophobic organic contaminants is often limited by their relatively low solubility in water. However, the solubilities of many of these contaminants are much greater in other solvents. Theoretical models suggesting an exponential decrease in the amount of adsorbed organic contaminant with increasing fractions of water miscible solvents (Rao et al., 1985; Woodburn et al., 1986) have been substantiated in laboratory experiments for several organic compounds (Rao et al., 1985; Nkedi-Kizza et al., 1985, 1987; Mahmood and Sims, 1985 Woodburn et al., 1986; Fu and Luthy, 1986a, 1986b; Zachara et al., 1988a). For example, the adsorption coefficient for anthracene in methanol-water mixtures decreased by four orders of magnitude as the fraction of methanol was increased from 0 to 1 (Nkedi-Kizza et al., 1985). The injection of cosolvents may therefore be expected to drastically increase the solubility and decrease the retardation factors for these organic compounds, thereby facilitating their removal from the subsurface. Cosolvents that are used as substrate by microbes may have the added advantage of promoting cometabolism of primary contaminants. Small amounts of biodegradable cosolvent that are difficult to remove from the subsurface will be of less concern because of their eventual transformation. Thus, cosolvents, such as alcohols, are potentially effective reactive agents for chemical enhancement to pump-and-treat of hydrophobic organics. However, some consequences of cosolvent injection may be less desirable.

The order-of-magnitude decreases in adsorbed contaminants are generally achieved with cosolvent concentrations greater than 20%. Fluids containing this amount of cosolvent will have densities and viscosities that differ substantially from the ground water. Thus, the transport behavior of these fluids is more complex and more difficult to predict than for fluids with homogeneous properties. Cosolvent interaction with clays in the aquifer matrix may either increase or decrease the permeability of the soil. Cracks have appeared in soils treated with methanol (Brown and Anderson, 1982). The formation of such high permeability pathways may be particularly troublesome at sites where dense nonaqueous phase liquids (DNAPLs) are present. Cosolvents such as methanol can serve as substrate for subsurface microbes, resulting in biofouling of the aquifer. Biotransformation may substantially alter the geochemistry of the aquifer and promote the reductive dissolution of Fe and Mn oxides. These metals can create problems with well clogging, and interfere with surface treatment. Also, additional treatment facilities must be constructed for the separation of the cosolvent from the water. These facilities incur capital expenditures as well as operation and maintenance costs.

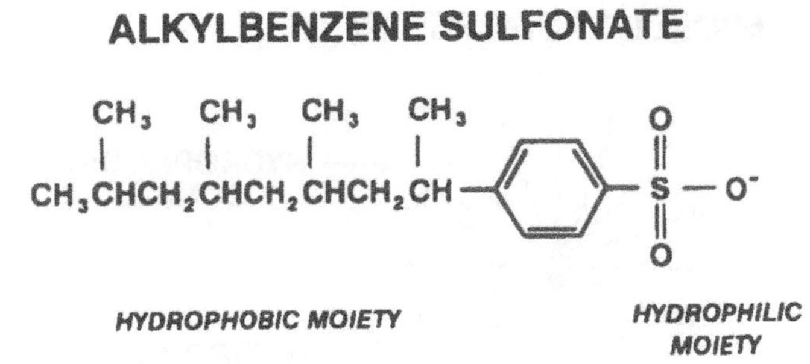

ALKYLBENZENE SULFONATE

$$CH_3CHCH_2CHCH_2CHCH_2CH$$

HYDROPHOBIC MOIETY

HYDROPHILIC
MOIETY

Figure 5-17. The surfactant alkylbenzene sulfonate.

Surfactants

A surfactant adsorbs to interfaces and significantly decreases the interfacial tension (Rosen, 1978). This property of surfactants has made these chemicals useful in enhanced oil recovery, and several researchers have proposed their use in the remediation of NAPL-contaminated sites (e.g., Ellis et al., 1985). In general, surfactants are composed of a hydrophobic moiety, often a long chain aliphatic (C_{10} to C_{20}) group, and a hydrophilic moiety (Figure 5-17) that can be anionic, cationic, nonionic, or zwitterionic (i.e., possess both positive and negative charges). The orientation of the surfactant molecules at an interface can reduce the interfacial tension and alter the wetting properties of the soil matrix. When the interface is a nonaqueous phase liquid, the lowering of the interfacial tensions decreases the capillary forces keeping the NAPL in place and results in greater mobility of the NAPL.

For enhanced oil recovery, increased mobility of the NAPL allows a greater fraction of the available oil to be pumped to the surface. In the case of NAPLs that have a greater density than water (DNAPLs), increased mobility is not necessarily desirable. Once mobilized, the DNAPL may migrate deeper into the aquifer. If the DNAPL migrates into areas that were previously uncontaminated, additional wells and pumps will be required and the costs of remediation will increase accordingly.

Surfactants can also promote the solubilization of hydrophobic organic contaminants. Above a critical concentration known as the "critical micelle concentration," colloidal-size micelles can form by the aggregation of the monomeric surfactant molecules. In water, the micelles form by the hydrophobic moieties grouping together in the core of the micelle, and the hydrophilic moieties orienting toward the surface of the micelle (Figure 5-18). Hydrophobic organic contaminants partition into the hydrophobic core of the micelle, thereby increasing the solubility of the organic contaminant.

Although the successful application of surfactants to enhanced oil recovery has been demonstrated, transfer of this knowledge to aquifer remediation is not direct. Surfactants used for enhanced oil recovery are chosen on the basis of temperatures and salinities that are much higher than those at most hazardous waste sites. To achieve the desired behavior, the surfactant must be chosen for the solvent under the conditions of use (Rosen, 1978). Incorrect surfactant formulations may result in high-viscosity macroemulsions that are difficult to remove. The surfactant can alter the wetting properties of the soil matrix and cause the NAPL to become the wetting phase. The NAPL would then occupy the smaller pores of the soil matrix, thereby exacerbating cleanup efforts. The toxicity and potential biodegradation of surfactants that will remain in the aquifer following NAPL removal is of great concern in shallow aquifers.

The use of surfactants for aquifer remediation looks promising; however, there is little experience in their application. Laboratory experiments have demonstrated enhanced removal of anthracene and biphenyls (Vignon and Rubin, 1989), petroleum hydrocarbons (Ellis et al., 1985), DDT and trichlorobenzene (Kile and Chiou, 1989), automatic transmission fluid (Abdul et al., 1990), and PCBs (Ellis et al., 1985; Abdul and Gibson, 1991). Surfactant mixtures that specifically address the needs for aquifer remediation need to be developed and tested in the field as well as in the laboratory. When DNAPLs are present, mixtures that increase solubilization more than mobilization may be desired.

MICELLE FORMATION

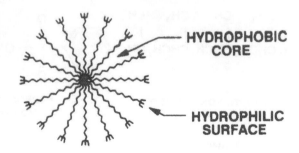

Figure 5-18. Aggregation of surfactant molecules into a micelle.

Oxidants-Reductants

The addition of a reactive agent that changes the oxidation state of a contaminant is potentially useful for 1) decreasing the toxicity of the contaminant, 2) increasing its mobility, or 3) increasing its susceptibility to complexing agents. For example, chromium can be reduced from the more toxic Cr(VI) to the less toxic Cr(III). The oxidation of selenite (Se(IV)) to selenate (Se(VI)) results in a solute that is less toxic and more mobile. However, oxidants and reductants are not specific and must, therefore, be in excess of the amount of contaminant. This will locally alter redox conditions within the aquifer and may result in the precipitation of solid phases that may clog the aquifer and injection/extraction wells, or mobilization of other metals that must be handled in the treatment train.

The rate of reaction is an important factor in considering an oxidation or reduction reaction to facilitate the removal of a contaminant from the subsurface. Often the rates are strongly dependent on pH. For example, rate of reduction of Cr(VI) by ferrous iron varies with $\{H^+\}^3$ (Wiberg, 1965).

Precipitation-Dissolution-Ionization

At metal-contaminated sites, it may be possible to add a chemical constituent that will cause the precipitation of the contaminant in a solid phase with very low solubility. For example, the neutralization of acid mine waters by carbonate-buffered solutions will cause the precipitation of metal-oxides, hydroxides, and carbonates. Pb^{2+} can precipitate as a relatively insoluble $PbCO_3$ phase. While this may reduce the risk of contaminant concentrations exceeding the MCL, it does not remove the metals from the site. The precipitates can continue to act as long-term, low level sources and the contaminants may still enter the biosphere through plant root systems or erosion. In addition, the precipitation of metal oxyhydroxides and carbonates can cause clogging of the aquifer and severe reduction in well efficiency.

Remediation of contaminated sites by conventional pump-and-treat may often be limited by dissolution of sparingly soluble mineral phases and nonaqueous phase liquids. Reactive agents that increase the solubility of these phases will release the contaminant into solution where it can be removed via an extraction well. For example, many phenolic compounds can be ionized at higher pH (e.g., Palmer and Johnson, 1991). The use of a base as a reactive agent will enhance the solubility of the phenolic phase and decrease the retardation factor of the dissolved compounds. If cadmium is being released into solution from $CdCO_3$, the addition of acid can dissolve the carbonate mineral phase and bring the Cd^{2+} into solution. However, such treatments are not selective and other ions including Fe, Al, and SiO_2 will be added to solution. These ions may interfere with treatment processes and increase the volume of sludge to be disposed. The natural buffering capacity of the aquifer will require that the concentration of injected acid or base be in excess of the amount of contaminant in the subsurface.

At metals-contaminated sites where remediation is limited by the presence of a sparingly soluble mineral phase, it may be possible to release the contaminant more rapidly by the addition of ion that will substitute for the contaminant within the mineral phase. This is most likely to be applicable where the availability of one of the counter ions in the solid phase is limited. By scavenging the counter ion into another solid phase, the contaminant will be released into solution where it can be easily removed. For example, if $BaCrO_4$ limits the remediation of chromium-contaminated sites, the injection of high levels of sulfate would precipitate $BaSO_4$ and increase the solubility of the $BaCrO_4$ component, thereby allowing the removal of the Cr(VI) in fewer pore volumes than by conventional pump-and-treat.

Such chemical enhancement methodologies require specific knowledge about the mineral phases limiting aquifer remediation and the geochemistry of the ground waters onsite. Detailed geochemical studies will increase the cost of site investigation and feasibility studies but may more than pay for themselves if efficient removal methods can be identified and problems associated with the implementation and operation of the clean-up effort are avoided.

Removal of the Contaminants and Reactive Agents from the Subsurface

The basic concept of chemical enhancement to pump-and-treat is to increase the mobility of the contaminants in the subsurface so they may be more easily removed via extraction wells. Removal of contaminants from the subsurface in a chemical enhancement scheme, therefore, requires decisions about the density, placement, and pumping rates of these extraction wells. Several aspects of such an enhanced extraction system design will be similar to those utilized in conventional pump-and-treat remediations. For example, the contaminants must be contained within the capture zones of the extraction wells. However, key differences between conventional pump-and-treat and chemical-enhancement extraction system designs need to be explored. For example, in conventional pump-and-treat, contaminants are expected to be retarded. Therefore, the density of the well system is typically chosen to shorten travel times and thereby decrease the time for remediation to the extent possible. In chemical-enhancement methods, the reactive agent is used to make the contaminants behave more like a nonreactive tracer. Therefore, it may be possible to utilize a lower density of extraction wells to achieve an effective removal of the mobilized contaminant. Another important consideration in the design of an extraction system for chemical enhancement is the coordination with the injection of the reactive agent. As described above, there can be advantages to initially removing high levels of contaminants by conventional pump-and-treat before initiating the injection of the reactive agent. Also, pumping rates for the extraction wells may be adjusted during the injection to aid in the distribution of the reactive agent within the aquifer.

Use of a reactive agent in a pump-and-treat scheme introduces one or more new chemical constituents into the subsurface. To be effective, reactive agents generally must be added to an aquifer at nontrace concentrations. Even if the reactive agent is harmless to human health, state and federal regulations will often require that concentrations of the reactive agent be lowered to some permissible level. Removal of the agent then involves all of the problems encountered in the removal of the original contaminant and in some cases the agent may even be more difficult to remove. For example, if the reactive agent is a solute that is used to compete with the contaminant for the adsorption sites, then the reactive agent must have a greater affinity for the adsorbent; but this greater affinity also makes it more difficult to remove from the subsurface. There still may be a net benefit if the target cleanup level for the reactive agent is greater than for the contaminant. It can also be argued that the net risk is reduced because the reactive agent must, by any reasonable choice, be less toxic than the original contaminant.

One complication that may arise during the removal of the reactive agent from the subsurface is clogging of the screen and filterpack as waters are mixed at the extraction wells. This problem is likely to be most acute when the reactive agent changes the redox conditions in the subsurface. As oxidized waters mix with reduced waters that contain iron, precipitates may clog the screen, pipes, and treatment tanks.

Treatment and Disposal

The previous sections have outlined many of the technical considerations that must be addressed to implement an effective chemical enhancement strategy. However, even a chemical enhancement plan that is completely satisfactory in terms of subsurface deployment and removal of solutions may still present technical difficulties in the treatment and handling of the extracted wastes. Three broad categories of post-extraction problems are discussed in this section: the effects of the reactive agent on the treatment of the target contaminants, the removal of the reactive agent from the waste stream before discharge, and the recovery and reuse of the reactive agent.

Removal of the Reactive Agent Before Discharge

As described above, the use of a reactive agent in a pump-and-treat scheme introduces one or more new chemical constituents in nontrace levels into the water brought to the surface. Extracted water will, therefore, contain substantial quantities of the reactive agent, probably in excess of the target contaminants. State and federal regulations will often require that concentrations of the reactive agent be lowered to some permitted discharge level. For example, if phosphate is used as an extractant, standards may restrict the permissible concentration in discharges from the treatment facility, even if the treated wastes are routed into a municipal sewage treatment system. Limits on phosphate discharges can be anticipated in localities in which eutrophication is a problem in waters receiving regional waste waters.

If the levels of reactive agent in discharges from the site are regulated, then the treatment plan must explicitly include a means of removing the agent. In many cases, the most efficient system will effect the removal of the reactive agent simultaneously with the treatment of the targeted contaminants. For example, if phosphate were used to enhance chromate removal, then the neutralization step in a treatment process could be modified to induce the precipitation of much of the phosphate. Phosphate and reduced Cr would precipitate in the same step and could be removed collectively in the sludge.

For the specific system mentioned, note that the treatment procedure would need to be modified. Removal of Cr^{3+} alone can be effected by addition of NaOH to achieve a basic pH. The resulting sludge then contains a mixture of Cr hydroxides, probably coprecipitated with by-products of the reductant step. For instance, if bisulfite is the reducing agent, sulfate and bisulfite will constitute part of the sludge.

If phosphate is present in the extracted water, some phosphate is likely to coprecipitate with Cr^{3+}. However, the concentration of phosphate remaining in solution in such a complex system would be difficult to accurately predict, and tests would be necessary to find the optimal pH for phosphate precipitation.

It is possible that pH adjustment alone would not precipitate sufficient phosphate. In that case, an additional treatment reagent would be needed; for example, substituting $Ca(OH)_2$ for some or all of the NaOH in the neutralization step. Ca-phosphates are relatively insoluble and would strip out much of the phosphate. However, pilot studies would be needed to ensure that the sludges produced behaved in the desired fashion. The presence of phosphate might decrease the density of the sludge, so that longer settling times are required. Furthermore, the presence of Ca^{2+} could lead to the buildup of scale in unexpected parts of the system. In general, it is wise to bear in mind that the treatment process will be efficient only if it is regarded as a coordinated chemical system in which the alteration of one part can cause dramatic changes in the behavior of another part.

If the reactive agent is successfully coprecipitated with the target contaminants, the total volume of sludge sent to disposal will increase correspondingly. Although the reactive agent may be harmless, once it is commingled with a toxic waste, the entire volume could be classified as hazardous and the cost of disposal assessed accordingly. Thus, while the removal of contaminant and reactive agent in a single step may save operation or capital costs, the increased volume of sludge to be landfilled over the case where no reactive agent is used, will generate costs that will offset some of the savings. The cost of testing and design of the removal system for the reactive agent must also be factored into the economic analysis.

In some cases, it will be desirable or essential to remove the reactive agent in a separate stage of the treatment system. One reason for a separate treatment step is to achieve a better removal of the reactive agent than could be conveniently achieved in a single step. In the example above, it might be desirable to optimize Cr^{3+} precipitation without regard for phosphate, and then strip phosphate out of the supernatant with $Ca(OH)_2$ or alum treatment in a subsequent step. If the Cr^{3+} precipitation step could be designed to minimize phosphate coprecipitation, and if the phosphate sludge were sufficiently free of Cr to be classified as nonhazardous, the two-stage removal would have the added advantage of minimizing the volume of hazardous solids for disposal.

Interference of Reactive Agent with Treatment Processes

Even if regulations do not require the removal of reactive agents, it may be necessary to remove them from the process stream. Some reactive agents may be harmless to humans or to the environment, but they nonetheless may have chemical properties that alter the behavior of the contaminants in the waste stream. For example, many conceivable reactive agents would function by complexing the target contaminants and enhancing their solubility. Specifically, citrate or oxalate salts might be used to bind up and mobilize metal ions. The same solubilization of metal ions that is desirable in the extraction step may become a major headache in the treatment step.

As discussed above, most chelators that are environmentally safe enough to be used in aquifers, such as malonate, succinate, and citrate, will be relatively weak and nonspecific in their binding action. A weak affinity for the target metal means that the chelator must be present in great excess and will be found in corresponding excess in the extracted water. The excess chelator may interfere with one or more segments of the treatment process by binding to the target contaminant.

The most obvious interference would be the inhibition of precipitation. The soluble metal-chelator complex will not readily precipitate out of solution. Lowering the pH will dissociate most metal complexes (such as citrate or oxalate), but metal ions will be soluble at the lower pH. Raising the pH will favor precipitation, but the higher pH also favors the binding of most chelators. Very caustic pH levels may be required to induce precipitation of metals in the presence of excess chelator. An alternative scheme is to remove the chelator from solution before the metal precipitation step. An organic chelator such as citrate could be degraded by biological treatment. An inorganic chelator such as polyphosphate would not biodegrade and would be difficult to remove economically by chemical means. Any process that requires a separate treatment step for the chelator will have greater operation and capital costs than the corresponding process in the absence of that step.

The nonspecificity of chelators used as reactive agents creates another substantial problem for treatment processes. Aquifer materials may contain large amounts of naturally occurring metals that may be solubilized by the chelator. Iron and manganese will be the most important in many aquifers, but copper, zinc, aluminum, and other metals may be extracted to varying degrees. The solution brought to the surface will contain not only free chelator, but also substantial quantities of chelator bound to nontarget metals. In some situations, the amount of chelator actually bound to the target metal(s) may be only a small fraction of the total chelator in the extracted water. The additional metals would not be solubilized in the absence of chelator, so these metals are a specific feature of chemical enhancement.

Large quantities of iron and manganese in the extracted water will require special attention in the treatment process. Precipitation of the target metals will be accompanied by the precipitation of substantial amounts of iron and manganese hydroxides. The sludge volume will be correspondingly increased, with the concomitant elevation of disposal costs. Furthermore, the behavior of iron in particular is apt to be different from that of many target metals. Iron is an especially insoluble metal (under aerobic conditions) and may therefore precipitate out of the waste stream before the target metal precipitates. If this precipitation can be anticipated and controlled, it may simplify the treatment process. However, a more likely scenario is that iron precipitation will be somewhat unpredictable and will occur in inopportune sections of the treatment facility, causing plugging or fouling of the equipment and interfere with the treatment process by coprecipitating the contaminants.

Recovery and Reuse of the Reactive Agent

In some cases, it may be beneficial to remove the reactive agent from the waste stream so that it can be reinjected into the aquifer and reused for additional extraction of the subsurface contaminants. Such reuse may be particularly advantageous when the reactive agent is expensive or when it must be removed because of concerns about interference with treatment or because of regulatory requirements on discharges of the treated water. However, the methods that may be used to extract the reactive agent from the waste stream for reuse may not be the optimal methods for removal for other purposes.

EFFECTS OF RAPID AND CONCENTRATED EXTRACTION

One of the assumptions underlying the use of chemical enhancement in a pump-and-treat operation is that the rapid extraction of concentrated waste solutions is beneficial to the cleanup operation. Of course, this will be true at many sites because annual operation and maintenance costs are directly reduced by a more rapid removal of contaminants.

Furthermore, more concentrated wastes may be easier to treat than dilute wastes. However, an accurate economic analysis of the various options available in restoring an aquifer should consider all costs associated with chemical enhancement. Besides the above-mentioned increases in research and development and operation and maintenance associated with chemical enhancement, there may be added costs due to the rapid extraction and concentration of wastes.

Careful consideration should be given to the capital costs of designing facilities that can efficiently handle and treat large volumes of a target contaminant in a short time. Injection/extraction wells, settling basins, sludge pumps, metering pumps, and other facilities may need to be greatly expanded to handle the concentrated waste load. Although the system will be operated for a shorter period of time there will be a tradeoff between increased capital costs and lower operation and maintenance costs. It may be cheaper to run a pump-and-treat operation for ten years with a small facility, rather than build a much larger facility that will only need to operate for one or two years.

Furthermore, planning in the pilot stages should give careful attention to the performance of the treatment process at different contaminant concentrations. If the process is especially efficient and reliable at the high concentrations attainable only with chemical enhancement, then the additional costs will be mitigated. If, however, the treatment process becomes more difficult or unreliable when contaminants are very concentrated, then the use of chemical enhancement may be contraindicated. Special care is needed if a biological treatment step is anticipated. Microbes generally thrive at higher substrate concentrations, but if higher contaminant levels lead to toxic levels of the target contaminant or some secondary constituent, then the biological treatment may fail altogether. Many of the problems of high concentration can be circumvented by appropriate dilution, but this is a feature that should be anticipated and incorporated into the design.

SITE CHARACTERIZATION FOR CHEMICAL ENHANCEMENTS

Rational implementation of a chemically enhanced pump-and-treat remediation will require many of the same characterization and testing methods required for conventional pump-and-treat operations. Physical hydrogeological parameters such as hydraulic conductivity, the potentiometric surface, and porosity (Table 5-3) can be obtained using the methods outlined by Mercer and Spalding (1991a,b), Palmer and Johnson (1989), Rehm et al. (1985) and Ford et al. (1984). These physical parameters can be used in modeling studies to ascertain the feasibility of getting the reactive agent to the contaminated areas within a reasonable period of time while maintaining a capture zone for the contaminant and the reactive agent. The results of such studies should help identify the optimum injection concentrations, the number of wells, and their location.

If chemical enhancements are to be considered, greater effort must be placed on the *chemical* characterization of the site. In particular, the key chemical processes that limit pump-and-treat remediation must be identified if the proper type of reactive agent is to be chosen. Important chemical processes and their characterization have been recently addressed by Boulding and Barcelona (1991a,b,c), Palmer and Johnson (1991) and Palmer and Fish (1991).

Several approaches must be used to determine the chemical processes limiting pump-and-treat remediation. For ionic solutes, adsorption tests are important for quantifying the fraction of the solute adsorbed onto the surfaces of the soil as a function of pH and the aqueous concentration of the contaminant. The potential for mineral controls can be identified by calculating mineral saturation indices using geochemical models such as MINTEQ (Felmy et al., 1984) and may be verified through X-ray diffraction or electron microscopy. Oxidation and reduction tests (e.g., Barcelona and Holm, 1991) are useful for determining the amounts of oxidant and reductant necessary to alter the redox

Table 5-3. Physical-Hydrogeological and Chemical Parameters That Should Be Identified During Site Characterization for Chemical Enhancements to Pump-and-Treat Remediation.

Physical-Hydrogeological Parameters

- bulk density
- porosity
- hydraulic conductivity
- storativity
- potentiometric surface
- site boundary conditions
- ground water/surface water interactions
- infiltration rates
- leakage from adjacent aquitards

Chemical Parameters

- pH
- redox conditions
- contaminant concentrations and spatial distribution
- noncontaminant concentrations
- oxidation capacity of the aquifer
- reduction capacity of the aquifer
- organic contaminant partition coefficients
- ionic adsorption parameters

state of a contaminant in the subsurface. Bench-scale tests to measure the increase in solute concentrations following the addition of proposed reactive agents can provide information about potential compositions of interfering solutes entering the treatment train. Treatment studies using water compositions based on these tests can be used to determine potential problems and test proposed solutions to the treatment process.

For neutral organic contaminants, batch-sorption tests can be conducted to determine the fraction of the contaminant partitioned into the soil organic matter. However, at most sites, the partitioning can be determined from published values of the K_{oc} of the contaminant (e.g., Mabey et al., 1982, Montgomery and Welkom, 1989) and the fraction of organic carbon in the soil. This approach is simpler than batch experiments; however, it does require that the f_{oc} of the soil be measured. When nonaqueous phase liquids are present, they are the limiting factor in site remediation; pools of NAPLs are more problematic than NAPLs retained at residual saturation in the soil. Again, proposed reactive agents should be tested at the bench scale and the treatability of the extracted water tested before pilot testing and implementation of the chemical enhancement operation.

The tests discussed above are generalizations of a few that can be conducted. The specific tests required at a site depend on the target contaminants and the nature of the soil materials from which they must be extracted. Utilizing the knowledge from laboratory studies and the experience from other hazardous waste sites will be important in directing the type of tests that need to be conducted. Unfortunately, at this time there have been few field demonstrations of chemical enhancement methods from which to obtain such experience.

EPA CONTACTS

For further information contact Scott Huling, RSKERL-Ada, (405) 436-8610 (see Preface for mailing address).

REFERENCES

Abdul, A.S. and T.L. Gibson. 1991. Laboratory Studies of Surfactant Enhanced Washing of Polychlorinated Biphenyl from Sandy Material. Eviron. Sci. Technol. 25(4):665–671.

Abdul, A.S., T.L. Gibson, and D.N. Rai. 1990. Selection of Surfactants for the Removal of Petroleum Products from Shallow Sandy Aquifers. Ground Water 28(6):920–926.

Anderson, M.R. 1988. The Dissolution of Residual Dense Non-Aqueous Phase Liquids in Saturated Porous Media. Ph.D dissertation, Oregon Graduate Institute of Science and Technology, Beaverton, OR, 260 pp. (Available from University Microfilms.)

Anderson, M.R., R.L. Johnson, and J.F. Pankow. 1992. Dissolution of Dense Chlorinated Solvents into Ground Water. 1. Dissolution from a Well-Defined Residual Source. Ground Water 30(2):250-256.

Barcelona, M.J. and T.R. Holm. 1991. Oxidation-Reduction Capacities of Aquifer Solids. Eviron. Sci. Technol. 25(9):1565–1572.

Boulding, J.R. and M.J. Barcelona. 1991a. Geochemical Characterization of the Subsurface: Basic Analytical and Statistical Concepts. In: Site Characterization for Subsurface Remediation, EPA/625/4-91/026, Chapter 7.

Boulding, J.R. and M.J. Barcelona. 1991b. Geochemical Variability of the Natural and Contaminated Subsurface Environment. In: Site Characterization for Subsurface Remediation, EPA/625/4-91/026, Chapter 8.

Boulding, J.R. and M.J. Barcelona. 1991c. Geochemical Sampling of Subsurface Solids and Groundwater. In: Site Characterization for Subsurface Remediation, EPA/625/4-91/026, Chapter 9.

Brown, K.W. and D.C. Anderson. 1982. Effects of Organic Solvents on the Permeability of Clay Soils. Draft Report. Texas A&M University, Texas Agric. Exp. Stn., College Station. (Cited by Sheets, P.J and W.H. Fuller, 1986. Transport of Cadmium by Organic Solvents through Soils. Soil Sci. Soc. Am. J. 50:24–28.)

Carmen, P.C. 1937. Fluid Flow through a Granular Bed. Transactions of the Institute of Chemical Engineers 15:150–156.

Chiou, C.T., L.J. Peters, and V.H. Freed. 1979. A Physical Concept of Soil-Water Equilibria for Nonionic Organic Compounds. Science 206:831–832.

Coates, J.T. and A.W. Elzerman. 1986. Desorption Kinetics for Selected PCB Congeners from River Sediments. J. Contaminant Hydrology 1:191–210.

Dzombak, D.M. 1986. Towards a Uniform Model for Sorption of Inorganic Ions on Hydrous Oxides. Ph.D. Dissertation, Department of Civil Engineering, Massachusetts Institute of Technology.

Dzombak, D.A. and F.M.M. Morel. 1990. Surface Complexation Modeling. Hydrous Ferric Oxide. Wiley Interscience Publication, New York.

Ellis, W.D, J.R. Payne, and G.D. McNabb. 1985. Treatment of Contaminated Soils with Aqueous Surfactants. EPA/600/2-85/129, 84 pp.

Felmy, A.R., D.C. Girvin, and E.A. Jenne. 1984. MINTEQ: A Computer Program for Calculating Aqueous Geochemical Equilibria. EPA/600/3-84-032 (NTIS PB84-157148).

Ford, P.J, P.J. Turina, and D.E. Seely. 1984. Characterization of Hazardous Waste Sites—A Methods Manual, II, Available Sampling Methods, 2nd ed. EPA600/4-84-076 (NTIS PB85-521596).

Fu, J.K. and R.G. Luthy. 1986a. Aromatic Compound Solubility in Solvent/Water Mixtures. J. Environ. Eng. 112:328–345.

Fu, J.K. and R.G. Luthy. 1986b. Effect of Organic Solvent on Sorption of Aromatic Solutes onto Soils. J. Environ. Eng. 112:346–366.

Gillham, R.W., E.A. Sudicky, J.A. Cherry, and E.O. Frind. 1984. An Advection-Diffusion Concept for Solute Transport in Heterogeneous Unconsolidated Geologic Deposits. Water Resources Research 20(3):369–378.

Grove, J.H. and B.G. Ellis. 1980. Extractable Iron and Manganese as Related to Soil pH and Applied Chromium. Soil Sci. Soc. Am. J. 44:243–246.

Hamaker, J.W. and J.M. Thompson. 1972. Adsorption. In: Organic Chemicals in the Soil Environment, C.A Going and J.W. Hamaker, (eds), pp. 49–143.

Hunt, J.R., M. Sitar, and K.S. Udell. 1988. Nonaqueous Phase Liquid Transport and Cleanup 1. Analysis of Mechanisms. Water Resources Research 24(8):1247–1259. (See correction, Water Resources Research 25(6):1450.)

James, B.R. and R.J. Bartlett. 1983. Behavior of Chromium in Soils: VII. Adsorption and Reduction of Hexavalent Forms. J. Environ. Quality 12(2):177–181.

Johnson, R.L. and J.F. Pankow. 1992. Dissolution of Dense Immiscible Solvents into Groundwater. 2. Dissolution from Pools of Solvent and Implications for the Remediation of Solvent-Contaminated Sites. Eviron. Sci. Technol. 26(5):896-901.

Karickhoff, S.W. 1980. Sorption Kinetics of Hydrophobic Pollutants in Natural Sediments. In: Contaminants and Sediments, Vol. 2, Robert Baker (ed.), Ann Arbor Science, pp. 193–205.

Karickhoff, S.W. and K.R. Morris. 1985. Sorption Dynamics of Hydrophobic Pollutants in Sediment Suspensions. Environmental Toxicology and Chemistry 1:469–479.

Keely, J.F., C.D. Palmer, R.J. Johnson, M.D. Piwoni, and C.G. Enfield. 1987. Optimizing Recovery of Contaminant Residuals by Pulsed Operation of Hydraulically Dependent Remediations. In: Proceedings of the Petroleum Hydrocarbons and Organic Chemicals in Ground Water Prevention, Detection and Restoration Conference and Exposition, National Ground Water Association, Dublin, OH.

Kile, D.E. and C.T. Chiou. 1989. Water Solubility Enhancements of DDT and Trichlorobenzene by Some Surfactants Below and Above the Critical Micelle Concentration. Eviron. Sci. Technol. 23(7):832–838.

Mabey, W.R., et al. 1982. Aquatic Fate Process Data for Organic Priority Pollutants. EPA 440/4-81-014 (NTIS PB87-169090).

Mahmood, R.J. and R.C. Sims. 1985. Enhanced Mobility of Polynuclear Aromatic Compounds in Soil Systems. In: Environmental Engineering Specialty Annual Conference (Boston MA), American Society of Civil Engineers, pp. 128–135.

Mercer, J.W. and C.P. Spalding. 1991a. Geologic Aspects of Site Remediation. In: Site Characterization for Subsurface Remediation, EPA/625/4-91/026, Chapter 3.

Mercer, J.W. and C.P. Spalding, 1991b. Characterization of Water Movement in the Saturated Zone. In: Site Characterization for Subsurface Remediation, EPA/625/4-91/026, Chapter 4.

Montgomery, J.H. and L.M. Welkom. 1989. Groundwater Chemicals Desk Reference. Lewis Publishers, Chelsea, MI, 640 pp.

Nkedi-Kizza, P., P.S.C. Rao, and A.G. Hornsby. 1985. Influence of Organic Cosolvents on Sorption of Hydrophobic Organic Chemicals by Soils. Eviron. Sci. Technol. 19:975–979.

Nkedi-Kizza, P., P.S.C. Rao, and A.G. Hornsby. 1987. Influence of Organic Cosolvents on Leaching of Hydrophobic Organic Chemicals through Soils. Eviron. Sci. Technol. 21(11):1107–1111.

Norvell, W.A. 1984. Comparison of Chelating Agents as Extractants for Metals in Diverse Soil Materials. Soil Sci. Soc. Am. J. 48:1285–1292.

Palmer, C.D. and J.A. Cherry. 1984. Geochemical Reactions Associated with Low Temperature Thermal Storage in Aquifers. Canadian Geotechnical Journal 21: 475–488.

Palmer, C.D., W. Fish, and J.F. Keely. 1988. Inorganic Contaminants: Recognizing the Problem. In: Proceedings of the Second National Outdoor Action Conference on Aquifer Restoration, Ground-Water Monitoring, and Geophysical Methods, National Ground Water Association, Dublin, OH, pp. 555–579.

Palmer, C.D. and W. Fish. 1991. Physicochemical Processes - Inorganic Contaminants. In: Site Characterization for Subsurface Remediation, EPA/625/4-91/026, Chapter 12.

Palmer, C.D. and R.J. Johnson. 1991. Physicochemical Processes - Organic Contaminants. In: Site Characterization for Subsurface Remediation, EPA/625/4-91/026, Chapter 10.

Palmer, C.D. and R.J. Johnson. 1989. Determination of Physical Transport Parameters. In: Transport and Fate of Contaminants in the Subsurface. EPA/625/4-89/019, Chapter 4.

Palmer, C.D. and P.R. Wittbrodt. 1990. Geochemical Characterization of the United Chrome Products Site. In: Stage 2 Deep Aquifer Drilling Technical Report, CH2M-Hill to U.S. Environmental Protection Agency.

Rao, P.S.C., A.G. Hornsby, D.P. Kilcrease, and P. Nkedi-Kizza. 1985. Sorption and Transport of Toxic Organic Substances in Aqueous and Mixed Solvent Systems. J. Environmental Quality 14:376–383.

Rehm, B.W., T.R. Stolzenburg, D.G. Nichols, R. Taylor, W. Kean, and B. Lowery. 1985. Field Measurements Methods for Hydrogeologic Investigations: A Critical Review of the Literature. EPRI Report EA-4301, Electric Power Research Institute, Palo Alto, CA.

Rosen, M.J. 1978. Surfactants and Interfacial Phenomena. John Wiley & Sons, New York, 304 pp.

Rumpf, H. and A.R. Gupte. 1971. Einflusse der Porosität und Korngroenverteilung im Widerstandsgesetz der Proenstromung. Chemie Ingenieur Technik 43:367–375.

Schindler, P.W. 1981. Surface Complexes at Oxide-Water-Interfaces. In: Adsorption of Inorganics at Solid-Liquid Interfaces, M.A. Anderson and A.J. Rubin (eds.), Ann Arbor Science, Ann Arbor, MI, pp. 1–49.

Schindler, P.W. and W. Stumm. 1987. The Surface Chemistry of Oxides, Hydroxides, and Oxide Minerals. In: Aquatic Surface Chemistry, W. Stumm (ed.), pp. 83–110.

Stumm, W., H. Hohl, and F. Dalang. 1976. Interaction of Metal Ions with Hydrous Oxide Surfaces. Croat. Chem. Acta 48(4):491–504.

Sudicky, E.A. and E.O. Frind. 1982. Contaminant Transport in Fractured Porous Media: Analytical Solutions for a System of Parallel Fractures. Water Resources Research 18(6):1634–1642.

Sudicky, E.A., R.W. Gillham, and E.O. Frind. 1985. Experimental Investigation of Solute Transport in Stratified Porous Media: 1. The Nonreactive Case. Water Resources Research 21(7):1035–1041.

Travis, C.C. and C.B. Doty. 1989. Superfund: A Program Without Priorities. Eviron. Sci. Technol. 23(11):1333–1334.

Vignon, B.W. and A.J. Rubin. 1989. Practical Considerations in the Surfactant-Aided Mobilization of Contaminants in Aquifers. J. Water Pollution Control Federation 61(7):1233–1240.

Wiberg, K.B. 1965. Oxidation by Chromic Acid and Chromyl Compounds. In: Oxidation in Organic Chemistry, K.B. Wiberg (ed.), Academic Press, New York, pp. 69–184.

Woodburn, K.B., P.S.C. Rao, M. Fukui, and P. Nkedi-Kizza. 1986. Solvaphobic Approach for Predicting Sorption of Hydrophobic Organic Chemicals on Synthetic Sorbents and Soils. J. Contaminant Hydrology 1:227–241.

Zachara, J.M, C.E. Cowan, R.L. Schmidt, and C.C. Ainsworth. 1988a. Chromate Adsorption by Kaolinite. Clays and Clay Minerals 36(4):317–326.

Zachara, J.M., C.C. Ainsworth, R.L. Schmidt, and C.T. Resch. 1988b. Influence of Cosolvents on Quinoline Sorption by Subsurface Materials and Clays. J. Contaminant Hydrology 2:343–364.

Chapter 6

TCE Removal from Contaminated Soil and Ground Water[1]

Hugh H. Russell, U.S. Army Corps of Engineers, Tulsa, OK
John E. Matthews (retired), and **Guy W. Sewell**, U.S. EPA Robert S. Kerr Environmental Research Laboratory, Ada, OK

INTRODUCTION

Trichloroethylene (TCE) is a halogenated aliphatic organic compound which, due to its unique properties and solvent effects, has been widely used as an ingredient in industrial cleaning solutions and as a "universal" degreasing agent. TCE, perchloroethylene (PCE), and trichloroethane (TCA) are the most frequently detected volatile organic chemicals (VOCs) in ground water in the United States (Fischer et al., 1987). Approximately 20% of 315 wells sampled in a New Jersey study contained TCE and/or other VOCs above the 1 ppb detection limit (Fusillo et al., 1985). The presence of TCE has led to the closure of water supply wells on Long Island, NY and in Massachusetts (Josephson, 1983). Detectable levels of at least one of 18 VOCs, including TCE, were reported in 15.9% of 63 water wells sampled in Nebraska, a state having a low population density and industrial base (Goodenkauf and Atkinson, 1986).

Trichloroethylene per se is not carcinogenic; it is thought to become a human health hazard only after processing in the human liver (Bartseh et al., 1979). Epoxidation by liver oxidase enzymes confers a suspected carcinogenic nature (Apfeldorf and Infante, 1981; Tu et al., 1985). However, processing in the human liver is not the only way in which TCE may become a health hazard. Reductive dehalogenation of TCE through natural or induced mechanisms may result in production of vinyl chloride (VC) which, in contrast to TCE, is a known carcinogen [*Fed. Register*, 1984, 49:114, 24334(11)].

Wastewater or municipal water supply treatment systems which utilize coagulation, sedimentation, precipitative softening, filtration and chlorination have been found ineffective for reducing concentrations of TCE to nonhazardous levels (Robeck and Love, 1983). Other methods are required for remediation of water contaminated with TCE if such water is to be used for human consumption. The purpose of this chapter is to: 1) present a synopsis of physicochemical properties and reactive mechanisms of TCE, and 2) delineate and discuss promising remediation technologies that have been proposed and/or demonstrated for restoring TCE-contaminated subsurface environmental media.

PHYSICAL AND CHEMICAL PROPERTIES OF TRICHLOROETHYLENE

Preliminary assessment of remediation technologies feasible for reclamation of subsurface environmental media contaminated with TCE must involve consideration of the compound's physical and chemical properties (i.e., distribution coefficients, reactivity, solubility, etc.). These properties are directly responsible for behavior, transport and fate of the chemical in the subsurface environment. Knowledge of a compound's physicochemical tendencies can be used to alter behavior and fate of that compound in the environment. Important considerations derived from physical and chemical properties of TCE presented in Table 6-1 are:

---

[1] EPA/540/S-92/002.

Table 6-1. Physicochemical Properties of Trichloroethylene

Density	1.46 g/mL
Water solubility	1000 mg/L
Henry's law constant (atm-m³/mol @ 20°C)	0.00892
Molecular weight	131.4
Boiling point	86.7°C
Organic carbon partition coefficient (K_{oc})	2.42

Density (1.46 g/mL)—Density can be defined as the concentration of matter, and is measured by the mass per unit volume. In relation to liquids, these units are grams per milliliter. The density of a substance is usually referenced to pure water, which is taken to be 1 gram per milliliter. TCE is heavier than water; therefore, a spill of sufficient magnitude is likely to move downward through the subsurface until lower permeability features impedes its progress. This often results in formation of a plume or pool(s) of dense nonaqueous phase liquid (DNAPL) in the aquifer, plus a trail of residual saturation within the downward path. Implicit in this statement is the fact that residual saturation also will serve as source areas for contamination and migration of TCE within an aquifer system. There is an inherent difficulty associated with location of DNAPL in the subsurface and subsequent removal of DNAPL pools or plumes using a standard pump-and-treat regime. A major reason for this difficulty of removal is that water coning using conventional extraction wells results in poor DNAPL water ratios.

A density difference of about 1% above or below that of water (1.0 mg/mL) can significantly influence movement of contaminants in saturated and unsaturated zones (Josephson, 1983). This does not mean that density differences of less than 1% do not influence movement of contaminants. These density differences may even be apparent at low solute concentrations (Short, 1991, personal communication). At a site in New Jersey, it was determined that although soil and shallow ground water at the source area were contaminated with benzene, toluene and other volatile organic compounds with relatively low specific gravities, the compounds did not migrate downward into deeper municipal supply wells (Spayed, 1985). This is in contrast with what could happen if the source area were contaminated with chlorinated aliphatics, including TCE. Areas containing insoluble TCE, i.e., DNAPL pools, can serve as source areas for spreading of contamination. As ground water moves through and/or around these source areas, equilibrium concentrations partition into the aqueous phase. Aqueous phase TCE is then spread through the aquifer by advection and dispersion. As a result, small source areas can serve to contaminate large portions of an aquifer to levels exceeding drinking water standards. At a site in Texas, for example, Freeberg et al. (1987) determined that 8 kg of nonaqueous phase TCE was responsible for contaminating 12.3×10^6 gallons of water at an average concentration of 176 ppb.

Water solubility—Water solubility can be defined as the maximum concentration of a solute which can be carried in water under equilibrium conditions, and is generally given as ppm (parts per million) or mg/L (milligrams per liter). The water solubility limit of TCE is 1000 mg/L, the maximum concentration of TCE that can be in aqueous solution at 20°C. Water solubility of a compound has a direct relation on distribution coefficients and biodegradability of a particular compound. A compound that is relatively insoluble in water will prefer to partition into another phase; i.e., volatilize into soil gas or sorb to organic material. Compounds that are relatively insoluble also are not as readily available for transport across the bacterial membrane, and thus less subject to biological action.

K_{oc} *value*—K_{oc} is defined as the amount of sorption on a unit carbon basis. The K_{oc} can be predicted from other chemical properties of compounds such as water solubility and octanol water coefficient. The low K_{oc} value of 2.42 for TCE translates into little retardation by soil or aquifer organic materials. Since the retardation is so low, pump-and-treat technologies appear attractive as remediation alternatives. Mehran et al. (1987) through field and laboratory investigations determined a retardation factor of approximately 2.0 for TCE, which agrees with the data of Wilson et al. (1981) and falls within the range of 1–10 given by Mackay et al. (1985) for sand and gravel aquifers with low organic content. Measured partition coefficients, however, may be considerably higher than calculated values, especially at lower aqueous concentrations. Johnson et al. (1989) found that at equilibrium concentrations of approximately 1 ppm, the measured partition coefficient of TCE was significantly higher than calculated values. These authors also report that other workers who have observed the same effect theorize that this higher value may be related to clay interactions.

If the retardation factor is considered to be 2.0, TCE should migrate at half the speed of water through soil and aquifer materials low in organic carbon content, and theoretically somewhat slower in material with a high organic carbon content.

Henry's Law Constant—Henry's law states that the amount of gas that dissolves in a given quantity of liquid at constant temperature and pressure, is directly proportional to partial pressure of the gas above the solution. Henry's coefficients, as a result, describe the relative tendency of a compound to volatilize from liquid to air. The Henry's law constant for TCE is 0.00892 which is high enough, when combined with its low solubility in water and high vapor pressure, for efficient transfer of TCE to the atmosphere. The evaporation half-life of TCE in water is on the order of 20 minutes at room temperature in both static and stirred vessels (Dilling, 1975; Dilling et al., 1975).

The chemical structure of TCE bestows chemical reactivity. Three chlorine atoms attached to the carbon-carbon double bond make TCE a highly oxidized compound. Oxidized molecules readily accept electrons (reduction) under appropriate conditions, but resist further oxidation. As a result, chemical reactivity of TCE is greatest under a reducing atmosphere, conditions that favor transfer of electrons to TCE.

Due to their size, the three carbon atoms surrounding the double bond are responsible for stearic hindrance. This lowers the rate at which large nucleophilic groups can approach or react with the carbon-carbon double bond.

Due to its chemical nature, TCE can (under the appropriate conditions) undergo a number of abiotic transformations, or interphase transfers. The ease with which any chemical can undergo such reactions is an important indicator of that chemical's susceptibility to abiotic remediation processes.

ENVIRONMENTAL DISTRIBUTION OF TRICHLOROETHYLENE

While it has been estimated that about 60% of the total TCE produced in the United States is lost to the atmosphere, with negligible discharge into water bodies (Cohen and Ryan, 1985), widespread contamination of terrestrial subsurface environmental media also has occurred. Given the wide use of TCE as a degreasing solvent plus its recalcitrance and chemical properties, this type of distribution pattern is not surprising.

Trichloroethylene contamination exists in both vadose and saturated zones of the subsurface environment. This contamination results from spills, leaking transfer lines, storage tanks and poor environmental awareness. Because of its density and low K_{oc}, TCE will ultimately move downward in the vadose zone until an impermeable barrier is reached. Such a scenario occurs with a TCE spill of sufficient magnitude or deep enough in the vadose zone for volatilization to be restricted.

Once in the vadose zone, TCE can become associated with soil pore water, enter the gas phase because of its Henry's constant, or exist as nonaqueous phase liquid (NAPL). It is therefore conceivable that upward or downward movement of TCE can occur in each of these three phases, thereby increasing areal extent of the original spill in both the vadose and saturated zones. While movement of large concentrations of TCE through the vadose zone may be rapid, surface tension exerted at the capillary fringe may retard further downward movement of smaller spills.

Nonaqueous phase concentrations of TCE which are large enough to overcome capillary forces will move downward into the aquifer. Once the water table is penetrated, lateral flow may be mediated by the regional ground-water flow. Due to its high density, the movement of free-phase TCE is still directed vertically until lower permeability features are encountered. Once an impermeable layer is encountered, horizontal movement will occur. Such movement may even be directed against the natural ground-water flow by the effects of gravity.

Since permeability is a function of the liquid as well as the medium, the vertical movement of TCE through an aquifer is determined by geological properties of the aquifer material; i.e., granular size of sand or clay lenses. Trichloroethylene will tend to pool near these impermeable features. Water passing over and around these pools solubilize TCE that can be spread throughout the aquifer.

REACTION MECHANISMS FOR TRICHLOROETHYLENE

Oxidation

Highly oxidized chemicals such as TCE have a high reduction potential and are thereby resistant to further oxidation. This is the reason that aerobic biological mediated degradation of TCE was once thought to be illogical.

It is, however, possible to oxidize TCE using chemicals such as potassium ferrate. Delucca et al. (1983) determined in laboratory experiments that 30 ppm potassium ferrate would completely oxidize 100 ppb TCE in less than fifty minutes (after an initial 5 to 20 minute delay) at 23 degrees Celsius and a pH of 8.3. These experiments, however, were conducted using water that had been double distilled, deionized and passed through resin columns. These purification steps effectively removed all organic and inorganic species that might have inhibited the oxidation of TCE through a competitive process. The presence of other solutes with lower oxidation states or more accessible electrophilic binding sites would effectively lower the concentration of the radicals by reacting first, thus leaving fewer radicals to combine with TCE.

In wastewater treatment, ozone has been shown effective at the removal of organic material. Ozonation, however, is generally used to remove less oxidized chemicals than chlorinated aliphatics. The actual mechanism in this case is either the direct reaction of ozone with the carbon-carbon double bond or nucleophilic substitution by hydroxyl radials which are generated upon the decomposition of ozone. Destruction of TCE by ozone has been reported by Glaze and Kang (1988) and Francis (1987), and with some success by Cheng and Olvey (1972); however, the low concentration and reactivity with other solutes of hydroxyl radicals is still a problem. To overcome this problem, investigators have tried to couple chemical reactions designed to increase production of hydroxyl radicals. Glaze and Kang (1988) have described four ways to enhance the oxidation potential of ozone: 1) variation in pH, 2) addition of hydrogen peroxide, 3) addition of ultraviolet light and 4) addition of a combination of peroxide and ultraviolet radiation. Their results indicate that direct ozonation of TCE in high alkalinity ground water is a slow process. The addition of hydrogen peroxide at a ratio of 0.5–0.7 (w/w hydrogen peroxide:ozone) accelerated the oxidation rate by a factor of two to three. They concluded that oxidation of TCE under certain conditions is a promising destruction process.

Reduction

Compounds such as TCE also are susceptible to reduction. While transition metals may play a role in abiotic reductions, the major reductive components may be electrons or reducing equivalents produced from biological reactions or molecular hydrogen (Barbash and Roberts, 1986). Reduction of TCE may be possible by any organic compound that has a sufficiently low oxidation potential for efficient hydrogen transfer under ambient conditions.

Dehalohydrolysis

Natural dehalohydrolysis of TCE (or other halogenated VOCs), with the subsequent production of an alcohol, is possible. Half-lives of such reactions are on the order of days to centuries (Barbash and Roberts, 1986). Natural dehalohydrolysis proceeds by either hydrolysis in an aqueous phase or nucleophilic substitution and elimination reaction at unsaturated carbons (ethene bond). Removal of a chlorine atom from one carbon coincides with removal of a hydrogen atom from the adjacent carbon. Given the number of soil and aquifer systems contaminated with TCE, however, natural dehalohydrolysis is not considered to be a significant mechanism of degradation.

The rate determining step in nucleophilic attack is hydroxyl ion (or other nucleophile) reaction with the double bond. The rates of unimolecular and bimolecular nucleophilic substitution depend on: 1) structure of the carbon bearing the leaving group (general order of reactivity in bimolecular reactions methyl > primary > secondary > tertiary), 2) concentration and reactivity of the leaving group, 3) nature of the leaving group (I > Br > Cl > F), 4) nature and polarity of the solvent (concentration of ions such as bicarbonate) and, 5) in the case of S_n2 type reactions, the concentration of the radical. It is quite evident then, should nucleophilic substitution occur, the daughter product or products (dichloroethylene or vinyl chloride) would be increasingly resistant to nucleophilic displacement, yet still be of environmental consequence. This fact, coupled with the low rates of natural dehydrohalogenation of TCE, point to the need for some type of catalyst or enhancement of reactivity.

In regards to catalysts, a number of chemical compounds or processes have been developed, or borrowed from the wastewater treatment industry. Ultraviolet radiation has been suggested as one method to enhance reaction rates through the formation of reactive hydroxyl radicals. Ultraviolet light may be an important mechanism for the degradation of recalcitrant chemicals in surface waters and the photolytic zone (top 1–2 mm) of soil (Miller et al., 1987). In a laboratory setting with a six-fold molar excess of dissolved oxygen, under an oxygen atmosphere, the photooxidizable half-life of TCE has been determined to be 10.7 months (Dilling et al., 1975). Given the interphase transfer potential of TCE, photodegradation should not occur to any extent before transfer to the atmosphere.

SURFACE TREATMENT TECHNOLOGIES

Air Stripping

Air stripping is an applicable technology for removal of TCE from contaminated water. A constant stream of air is used to drive TCE from solution, taking advantage of the low Henry's law constant and water solubility. Since TCE is fairly recalcitrant to other remediation efforts, air stripping seems to be the current method of choice to return water to potability. A number of pilot-scale demonstration and full-scale case studies have shown this alternative to be effective for removal of TCE from contaminated water. This process, however, only shifts the compound to another medium. New restrictions on venting of volatile organic compounds to the atmosphere may preclude the use of such technologies or require treatment of the air-stripped off-gases by carbon adsorption.

The typical air stripper is designed to allow for percolation of large volumes of air through contaminated water. This has the effect of changing conditions to favor volatilization of TCE. Application limitations usually occur as the concentration of TCE falls below a threshold level where volumes of air larger than logistically possible are required to continue the stripping process.

In 1982, 10–12 United States utilities were using some form of aeration technique to strip volatile organic chemicals such as TCE from water supplies. The air strippers used were of three principal types: redwood slat aerators, packed towers, and spray towers, all of which are designed to allow maximum contact of water, air, and TCE (Robeck and Love, 1983). Since that time many sites and municipal supply systems have utilized air strippers (generally packed towers) as a method for removing VOCs from contaminated ground water.

Wurtsmith AFB—At Wurtsmith AFB, Oscoda, Michigan, TCE was detected in drinking water at a concentration of 6,000 ppb, with a concentration of 10,000 ppb in the centerline of the plume (Gross and Termath, 1985). The remedial design selected was for two packed towers that could be run parallel or in series. The towers were run constantly for one year, consistently obtaining effluent concentrations below 1.5 ppb. The one problem encountered was bacterial growth which plugged the towers. This had the effect of lowering the surface area available for transfer of TCE into the gas phase. Constant chlorination was found to retard bacterial growth, while not affecting operational efficiency of the tower.

Savannah River, Georgia—TCE was found in ground water at the Savannah River Plant in 1981 near an abandoned settling basin which had been used for storage of process waters (Boone et al., 1986). The horizontal area of the plume contour at 100 ppb total concentration chlorocarbons was estimated to be 360 acres. The total mass of chlorocarbons was thought to be 360,000 pounds. In 1983, air stripping was thought to be the best available technology for remediation; therefore, two pilot air stripping units of 20 and 50 gpm capacity were designed. The two pilot strippers, when operated properly, reduced the concentration of total chlorinated hydrocarbons in ground water from 120,000 ppb to less than detection limits of 1 ppb. Data from these pilot units were used to design a production tower to process contaminated water at a rate of 400 gpm.

In 1985, the production tower (400 gpm column) was placed on line and fed by a network of 11 recovery wells. As of 1986, some 65,000 pounds of chlorinated solvents had been recovered from the aquifer through the combined operation of pilot and production towers. In 13 months, the production tower alone, which was reported to operate at 90% efficiency had removed some 33,500 pounds of chlorocarbons from 115 million gallons of water.

Refrigerator Manufacturing Facility—In the late 1960s, a refrigerator manufacturing facility was granted permission for a waste disposal area in the upper of two peninsulas formed by the "S" meander of a river. This practice resulted in the contamination of the underlying aquifer with chlorinated aliphatics. TCE concentrations were approximately 35 ppm.

The treatment system consisted of installation of 14 wells for extraction of ground water which was processed by air-stripping. Effluent from the "stripper" was then discharged through a recharge basin that was capped with 1.5 feet of pea gravel. Recharge through the pea gravel was thought to serve two purposes: 1) additional removal of TCE by percolation through the gravel, and 2) flushing residual TCE contamination from the vadose zone.

It was reported that the system operated at an influent flow rate of 210 gpm (4000 ppb TCE) and a TCE removal efficiency of 78% (Thomsen et al., 1989). Additional removal attributed to spraying of process water over and percolation through the gravel raised the TCE removal efficiency to greater than 95%. Full scale operation began in the summer of 1987. From July 1987 to October 1987, approximately 24.5 million gallons of ground water were treated, and 775 pounds of TCE were removed.

COMBINED AIR STRIPPING AND CARBON ADSORPTION

A currently popular method for remediating water contaminated with TCE is to combine the technologies of air stripping and granular activated carbon (GAC) adsorption. This treatment train is attractive because it ameliorates shortcomings of both technologies.

In the case of air stripping, the residual concentration of TCE in treated effluent may be above local or regional drinking water standards, thereby necessitating a "polishing" step prior to use or discharge. Limitations associated with GAC adsorption are:

1) a given sorbent has a finite capacity for sorbtion of a given contaminant; once this limit is reached, contaminant breakthrough occurs. This breakthrough results from competition between contaminants and normal solutes for unbound sites; less tightly bound solutes are displaced by those solutes having a greater affinity. When loss of binding efficiency becomes great enough, breakthrough of contaminants occurs. Once breakthrough occurs, the sorbent must be changed or cleaned. Adsorption of TCE to GAC, based on equilibrium concentrations of 1 ppm at neutral pH and 20°C, is 28 mg/g (Amy et al., 1987).

2) high dissolved organic carbon (DOC) and other contaminants can compete with TCE for binding sites available on the sorbent, thus increasing the likelihood of breakthrough. A concentration of 10 ppm natural organic matter in river water has been shown to reduce TCE adsorption by up to 70% (Amy et al., 1987).

Recognizing the limitations of sorption (the first may only be economically limiting), use of this technology as a polishing step following air stripping can be a viable link in a surface treatment process for TCE. Air stripping can be used to remove the majority of TCE, followed by adsorption which is used to polish the stripper effluent. This approach will lower construction costs of the air stripper and increase life expectancy of the adsorbent.

Rockaway Township—In 1979, ground-water contamination by TCE and lesser amounts of diisopropyl ether and methyl tertiary butyl ether was discovered in three of the water supply wells serving Rockaway, New Jersey. The first technology used to treat the ground water was sorption utilizing GAC. Quick breakthrough of ether contaminants resulted in rapid loss of GAC sorptive effectiveness. This led to consideration and ultimate use of air stripping as a primary treatment technology (McKinnon and Dykson, 1984). A countercurrent packed tower with a water flow rate of 1400 gpm and an air flow rate of 37,500 cfm was installed and placed on line February 4, 1982. By July 1983, the GAC system was taken off line because of the excellent performance of the packed tower, in addition to reduced influent levels of the ether compounds.

SUBSURFACE REMEDIATION TECHNOLOGIES

Soil Venting

Soil venting is an in situ air stripping technique used to remove volatile contaminants from the vadose zone. Air is forced into the soil subsurface through a series of air inlets, then vented or extracted under vacuum through extraction pipes. Air laden with organic vapors moves along an induced flow path toward withdrawal wells, where it is removed from the unsaturated zone and treated and/or released to the atmosphere (Baeher et al., 1988). Success of the method depends on rate of contaminant mass transfer from immiscible and water phases to the air phase, and on ability to establish an air-flow field that intersects the distributed contaminant. In many cases, treatment of off-gases may be required, due to air quality standards.

Trichloroethylene, due to its high potential for interphase transfer to the gaseous phase, should be an excellent candidate for soil venting technologies. Cary et al. (1989) suggested that by forcing air into the water table below the contamination and maintaining sufficient air entry pressure throughout a significant volume of soil, TCE should be trapped at the soil water interface and released to the gas phase. This release to the gaseous phase would favor transfer to the soil surface.

Field results reported by Mehran et al. (1987) link the TCE concentration in soil gas with current levels in ground water; therefore, the theory suggested by Cary et al. would appear to be correct for contamination at or near the soil-water interface. This also may point to the possibility that movement of gaseous TCE from the saturated zone into the soil gas phase may act to further spread contamination. Soil venting technology for removal of TCE has not been fully tested in the field or, at least, has not been reported in refereed literature; however, results from two pilot-scale tests demonstrate the potential applicability of the technology for removal of TCE from the vadose zone (Coia et al., 1985 and Danko et al., 1989).

The first test site was an area where open burning of solvents such as TCE had occurred for over thirty years (Coia et al., 1985). The site consisted of sandy glacial soils with a TCE concentration of 5,000–7,000 mg/kg. A closed, forced ventilation system was installed. The system consisted of perforated PVC pipes extended vertically into the vadose zone, and separate extraction pipes. For 14 weeks, two pilot systems were run, one designated the high-contaminated zone (50–5,000 ppm TCE) and the second, the low (5–50 ppm TCE) zone. In the low-contaminated zone, extraction pipes were installed on 20 foot spacing, and air was forced into the soil at 50 cfm. In contrast, in the high-contaminated zone, extraction pipes were installed on 50 foot spacing, and air was forced into the soil at 50–225 cfm. At the end of the study, 1 kg of TCE was removed from the low-contaminated zone and 730 kg (10–20%) from the high-contaminated zone.

The second test site was the Verona Superfund site in Battle Creek, Michigan (Danko et al., 1989). A number of private and city wells in the Verona Field were discovered to be contaminated with VOCs, in August 1981. The predominant contaminants based on total mass were perchloroethylene (PCE), cis/trans-dichloroethylene (DCE), TCE, 1,1,1-trichloroethane (TCA) and toluene. A Record of Decision (ROD) was issued in 1985 specifying corrective action which would include a network of groundwater extraction wells, followed by air stripping plus a soil vapor extraction system for vadose zone contamination.

A soil vapor extraction system was installed to specifically remove VOCs from the most contaminated vadose zone source area. The system design consisted of a network of 4" PVC wells screened from approximately 5 ft. below grade to 3 ft. below the water table. A surface collection manifold connected to a centrifugal air/water separator was attached to a carbon adsorption system. The outlet end of the carbon adsorption system was connected to a vacuum system which pulls air from the subsurface into the extraction wells. The pilot system began operation in November 1987. By August 1989, total mass of VOCs removed was reported to be 40,000 lbs.

In-Well Aeration

Another remediation method based on interphase transfer potential of TCE is in-well aeration. In-well aeration can be accomplished using either an air lift or electric submersible pump and sparger. The air lift pump may or may not be used with a sparger.

In-well aeration has been tested at sites near Collegeville, PA and at Glen Cove, NY (Coyle et al., 1988). At the Pennsylvania site, the air lift pump used with a sparger removed 78% of the TCE; the electric pump run concurrently with a sparger removed 82%. At the New York site, the air lift pump without a sparger removed 65% of the TCE; with sparging, the same pump removed approximately 73%.

Technologies based strictly on volatilization of TCE, however effective at subsurface remediation, do not satisfy the intent of Section 121 of CERCLA. This Section states that "remedial actions in which treatment permanently and significantly reduces the volume, toxicity or mobility of the hazardous substances, pollutants, and contaminants, in a principal element, are to be preferred over remedial actions not involving such treatment." Air stripping simply transfers TCE from one medium (soil or water) to another (atmosphere) without any significant reduction in volume or toxicity.

BIOREMEDIATION

Biological remediation is one methodology which has the potential to satisfy the intent of Section 121 of CERCLA. Although bioremediation of environments contaminated with TCE can best be described as in its infancy, this technology remains attractive since the possibility exists for complete mineralization of TCE to CO_2, water, and chlorine instead of simply a transfer from one medium to another. In the case of biological degradation, bacteria may produce necessary enzymes and cofactors which act as catalysts for some of the chemical processes already described.

Anaerobic degradation—Because of the oxidized state of TCE, the ecological condition under which degradation is most likely to occur is a reducing environment. Biological degradation (and possibly mineralization) of TCE under anaerobic conditions has been studied for a number of years (Bouwer and McCarty, 1983; Bouwer et al., 1981). The first reported biological attenuation of TCE was that of the obligate anaerobic methanogenic bacteria through a process known as reductive dehalogenation. Under anaerobic conditions, oxidized TCE can function as an electron sink and is readily reduced by electrons (or reducing equivalents) formed as a result of metabolism (oxidation) of organic electron donors by members of methanogenic consortia. Volatile fatty acids and toluene may serve as oxidizable substrates (electron donors) which can be coupled to reduction of haloorganic molecules such as TCE with the resultant removal of a chlorine atom (Sewell et al., 1990). Freedman and Gossett (1989) reported conversion of PCE and TCE to ethylene without significant conversion to carbon dioxide. The conversion occurred in an anaerobic system stimulated with electron donors such as hydrogen, methanol, formate and acetate, with added yeast extract. Presumably, the electron donors provided the electrons or reducing equivalents necessary for complete chlorine removal by reductive dehalogenation. PCE mineralization has been reported by other authors (Vogel and McCarty, 1985), although it seems that in the absence of sufficient oxidizable organic compounds a buildup of DCE(s) or vinyl chloride also will occur (Bouwer and McCarty, 1983; Bouwer et al., 1981). Theoretically, under anaerobic conditions, with sufficient quantities of other readily oxidizable substrates and the necessary auxiliary nutrients, methanogenic consortia may be capable of converting TCE to harmless end-products. More research is needed, however, to determine how effective this remediation may be, and what actual requirements are needed to drive the process. Advantages of anaerobic processes are that there appears to be no apparent lower concentration limit to activity nor is there a need to perfuse the subsurface with copious amounts of oxygen.

Aerobic Degradation—It has long been thought that TCE is resistant to degradation under aerobic conditions due to its already oxidized state. Recently, a number of monooxygenases produced under aerobic conditions have been shown to degrade TCE (Nelson et al., 1987; Harker and Kim, 1990). Under these conditions, there is no buildup of vinyl chloride, and complete mineralization is possible. An aromatic compound such as toluene or phenol is required, however, for induction of the enzymes responsible. As a result, any application of this system would require the presence of a suitable aromatic compound or other inducer. The inducer requirement might be alleviated by manipulation of the proper genetic sequence for monooxygenase production, but other carbon and/or energy sources would probably be required for growth.

Methylotrophic degradation—The enzyme methane monooxygenase (MMO) produced by methylotrophic bacteria growing in the presence of oxygen at the expense of methane has a wide range of growth substrates and pseudosubstrates, one of which is TCE. This enzyme epoxidates TCE. The resulting chemical complex is unstable and quickly hydrolyzes to various products dependent on the pH of the menstruum. TCE epoxide in phosphate buffer at pH 7.7 has a half-life of 12 seconds (Miller and Guengerich, 1982). If TCE epoxidation by MMO follows enzyme kinetics similar to that of a true growth substrate and inhibitor rather than some enzyme modification due to its reaction with MMO, the question arises as to whether or not methane necessary for production of MMO will competitively inhibit TCE epoxidation (or more correctly, overcome the TCE inhibition). If such is the case, only a certain percentage of TCE may be epoxidated before it is transported away from the bacteria in a flow situation. Since this process is cometabolic, methane is a strict requirement. Removal of the true substrate will result in rapid loss of production of MMO, thereby reducing the ability to epoxidate TCE. Methods to enhance this ability are genetic in nature. Genetic recombination resulting in constitutive expression of the MMO genes is required in order that methane will no longer be required for induction. Even if constructive expression is achieved, small amounts of methane still may be needed as a carbon and energy source.

Mixed Consortia Degradation

Recently, it has been shown that a mixed consortia of bacteria can effectively mineralize TCE (Henson et al., 1988; Wilson and Wilson, 1985). This involves co-metabolism of TCE (epoxidation) by bacteria that oxidize gaseous hydrocarbons such as methane, propane and butane, followed by hydrolysis of the TCE epoxide. The hydrolysis products are then utilized by other naturally occurring bacteria. Wackett et al. (1989) surveyed a number of propane oxidizing bacteria for their ability to degrade TCE. While TCE oxidation was not common among the bacteria surveyed, unique members could oxidize TCE. High concentrations (>15% v/v) of the gaseous hydrocarbons was found to inhibit cometabolism of TCE. Oxygen concentrations also could be limiting in aqueous treatment systems, since oxygen is required by the gaseous, alkane-utilizing and heterotrophic population.

BIOREMEDIATION OF EXTRACTED GROUND WATERS/SUBSURFACE AIR STREAMS

Surface bioreactors could be used to remediate TCE-laden ground water or subsurface air streams extracted using any type of pump-and-treat system. Water treatment technology would consist of passing contaminated water, along with suitable concentrations of methane or other gases such as propane through the reactor. The methylotrophic or gaseous hydrocarbon-utilizing portion of the mixed population would epoxidate the TCE. This would spontaneously hydrolyze the TCE to glyoxylate, formate and carbon monoxide, which would then be utilized by the heterotrophic population. Laboratory work at the Robert S. Kerr Environmental Research Laboratory in Ada, Oklahoma (RSKERL-Ada) has shown that TCE can be removed from water utilizing surface bioreactors (Wilson and Pogue, 1987; Wilson and White, 1986). A trailer-mounted pilot-scale system has been designed and constructed for use in field trials (Miller and Callaway, 1991).

Treatment of extracted air streams would consist of passing off-gases stripped from pumped ground waters or collected from soil venting systems through surface bioreactors. A suitable concentration of hydrocarbon gases would be added to the feed stream to initiate the cometabolic process.

Systems developed for treatment of contaminated air streams would not be limited by the oxygen concentration. Treatment of contaminated air streams such as those resulting from air stripping operations or soil venting processes would have the additional advantage of eliminating any role water chemistry may play on the treatment process. Variations in water chemistry from site to site could be virtually ignored if contaminants are first transferred into an air stream prior to treatment.

In Situ Bioremediation

In situ bioremediation for removal or reduction of TCE contamination in the saturated zone is still in the research mode at this time. The RSKERL-Ada has an active research program directed toward this promising technology.

Anaerobic processes without support of secondary organic compounds often lead to the production of vinyl chloride as the final end product. Aerobic processes, on the other hand, require presence of an inducer compound which may not be available. Since stimulation of populations of mixed consortia and methylotrophs will usually require injection of methane, oxygen and other nutrients, aerobic processes also are subject to mass transport limitations. For example, given a retardation factor of 2.0 for TCE as the standard, it is possible that injection of required stimulants may result in transport of TCE partitioned into the aqueous phase away from appropriate bacteria before such bacteria are stimulated and begin actively producing the enzyme MMO. If injection does not result in complete transport of aqueous-phase TCE, its concentration may be diluted such that the TCE:methane or TCE:substrate ratio is too low for effective treatment. In situ processes utilizing methanotrophs or mixed consortia require modification of the currently conceived design of in situ bioremediation in the saturated zone to account for the possibility of aqueous-phase TCE transport away from the biologically activated portion of the aquifer.

Coupling of anaerobic in situ processes to surface based aerobic reactors is a promising variation of this technology also under consideration. In such a system, in situ anaerobic bacteria would be stimulated to reductively remove chlorine atoms from TCE. Resultant daughter products more susceptible to aerobic degradation could then be pumped to the surface and effectively treated in a bioreactor.

Moffet Field Study—A form of in situ aquifer bioreclamation has been studied at a test site on the Moffet Field Air Station in California (Roberts et al., 1989; Semprini et al., 1987). The test aquifer was shallow, confined, and composed of coarse-grained alluvial sands. To create the test zone, an extraction well and injection well 6 meters apart with 3 intermediate monitoring wells were installed. The 3 monitoring wells were used to gather data for tracer and degradation studies. Parameters including methane, oxygen and halogenated solvent concentrations were continuously monitored.

Test zone microbiota were stimulated by injecting ground water containing methane and oxygen in alternating pulses. Within a few weeks, complete methane utilization was observed. This confirmed the presence of indigenous methanotrophic bacteria. Although methanotrophic bacteria utilize methane as a carbon source, atmospheric oxygen also is required for growth.

During the initial phase of the field test, TCE was injected at an average concentration of 100 g/L; concentrations were then lowered to 60 g/L to determine the effect of sorption on observed concentration losses. Early breakthrough results at the first monitoring well, after normalizing concentrations to account for adsorption, indicated that microbial degradation could have been as high as 30%.

A second phase of this study involved injection of approximately 1 ppm normalized concentrations of vinyl chloride, trans-DCE, cis-DCE and TCE. At 2–4 meters from the injection well, transformation rates were found to be 95%, 90%, 45% and 29%, respectively. The recalcitrance of TCE was attributed to a number of factors, including degree of chlorination and low solubility of methane and oxygen.

Results from the Moffet Field study do not provide a wholesale license for initiation of in situ bioremediation of aquifers contaminated with TCE. Injection of methane, oxygen and other nutrients into an aquifer to stimulate a methylotrophic population would, in aquifers having little TCE sorption potential, push aqueous-phase TCE from the microbially stimulated portion of the aquifer before the appropriate bacteria could adapt or be stimulated. This technological limitation may be overcome in the future by redesigning the injection protocols currently utilized for in situ biological remediation. Such scenarios have been developed by Roberts et al. (1989). However, given the low rates of TCE degradation in initial field tests, this scenario awaits documentation.

SUMMARY

Due to its chemical structure and unreactive nature, TCE is not easily transformed to environmentally safe compounds. The most efficient and cost-effective method for removal of TCE from polluted ground water at this time appears to be air stripping, especially for aquifers which are utilized as potable water sources. Aquifers subjected to bioremediation would require further treatment of the ground water before human consumption, thus increasing the overall cost.

The primary alternative now being studied for removal of TCE from contaminated soil in the vadose zone is soil venting. Soil venting appears to be both efficient and cost-effective for VOC removal from the vadose zone in many instances. Limitations will occur in soil contaminated with a mixture of waste (TCE mixed with low volatility waste), as well as in highly impermeable soils. These limitations highlight the need for continued research into soil bioventing, including anaerobic and aerobic techniques based on cometabolism of TCE driven by the less volatile constituents.

With new restricted air emissions standards, the use of air stripping and soil venting technologies may require that off-gases from them be treated in some manner before discharge to the atmosphere. Use of biological reactors to treat air or water streams from air strippers or soil venting operations may be a cost-effective alternative to GAC. The movement toward clean air will require that research and demonstration studies be conducted pertaining to development of efficient cost-effective technologies.

Biological remediation, though unproven at this point, could destroy TCE completely, thus meeting the intent of Section 121 of CERCLA. More research is required, however, before full-scale implementation of such technology. Anaerobic treatment of aquifers contaminated with TCE may be a viable option. Extensive monitoring would be necessary to assure that the final product of this remediation is not vinyl chloride. Extensive protocols also may be developed for utilization of native methylotrophs in aquifers through stimulation with methane and oxygen before contact with TCE contaminated ground water. Additionally, biological methods may be designed to effectively treat off-gases from air strippers or soil vacuum extraction systems.

EPA CONTACTS

For further information contact Guy W. Sewell, RSKERL-Ada, (405) 436-8566 (see Preface for mailing address).

REFERENCES

Amy, G.L., R.M. Narbaitz, and W.J. Cooper. 1987. Removing VOCs from Groundwater Containing Humic Substances by Means of Coupled Air Stripping and Adsorption. J. Am. Water Works Ass. 49:54.

Apfeldorf, R. and R.F. Infante. 1981. Review of Epidemiology Study Results of Vinyl Chloride-Related Compounds. Environ. Health Perspective 41:221–226.

Baeher, A.L., G.E. Hoag, and M.C. Marley. 1988. Removing Volatile Contaminants from the Unsaturated Zone by Inducing Advective Air-phase Transport. J. Cont. Hydrol. 4:1–26.

Barbash, J. and P.V. Roberts. 1986. Volatile Organic Chemical Contamination of Groundwater Resources in the U.S. J. Water Pollut. Control Fed. 58:343–348.

Bartseh, H., C. Malaveille, A. Barbin, and G. Planche. 1979. Mutagenic and Alkylating Metabolism of Haloethylenes, Chlorobutadienes and Dichlorobutenes Produced by Rodent or Human Liver Tissues. Evidence for Oxirane Formation by Cytochrome P-450 Linked Microsomal Monooxygenases. Arch. Toxicol. 41:249–278.

Boone, L.F., R. Lorenz, C.F. Muska, J.L. Steele, and L.P. Fernandez. 1986. A Large Scale High Efficiency Air Stripper and Recovery Well for Removing Volatile Organic Chlorocarbons from Groundwater. In: Proceedings of the 6th National Symposium on Aquifer Restoration and Groundwater Monitoring. National Water Well Association, Dublin, OH, pp. 608-622.

Bouwer, E.J. and P.L. McCarty. 1983. Transformations of 1- and 2-Carbon Halogenated Aliphatic Organic Compounds Under Methanogenic Conditions. Appl. Environ. Microbiol. 45:1286–1294.

Bouwer, E.J., B.E. Rittman, and P.L. McCarty. 1981. Anaerobic Degradation of Halogenated 1- and 2-Carbon Organic Compounds. Environ. Sci. Technol. 15:596–599.

Cary, J.W., J.F. McBride, and C.S. Simmons. 1989. Trichloroethylene Residuals in the Capillary Fringe as Affected by Air-Entry Pressures. J. Environ. Qual. 18:72–77.

Cheng, P.W. and C.E. Olvey. 1972. Removal of Halomethanes from Water with Ozonation. Natural Resources Library. U.S. Dept. of the Interior. Final Report Project #A-072-RI.

Cohen, Y. and P.A. Ryan. 1985. Multimedia Modeling of Environmental Transport: Trichloroethylene Test Case. Environ. Sci. Technol. 19:412–417.

Coia, M.F., M.H. Corbin, and G. Arastas. 1985. Soil Decontamination Through In Situ Air Stripping of Volatile Organics - A Pilot Demonstration. In: Proc. the NWWA/API Conf. on Petroleum Hydrocarbons and Organic Chemicals in Groundwater - Prevention, Detection and Restoration, National Water Well Association, Dublin, OH, pp. 555-564.

Coyle, J.A., H.J. Borchers, and R.J. Miltmer. 1988. Control of Volatile Organic Contaminants in Groundwater by In-Well Aeration. EPA 600/2-88-020.

Danko, J.P., W.D. Byers, and J.E. Thorn. 1989. Remediation at the Verona Well Field Superfund Site. In: Superfund 89, Hazardous Materials Control Research Institute, Silver Spring, MD, pp. 479-484.

Delucca, S.J., A.C. Chao, M. Asce, and C. Smallwood, Jr. 1983. Removal of Organic Priority Pollutants by Oxidation Coagulation. J. Environ. Eng. 109:36–46.

Dilling, W.L. 1975. Interphase Transfer Processes. II. Evaporation Rates of Chloromethanes, Ethanes, Ethylenes, Propanes and Propylenes from Dilute Aqueous Solutions, Comparisons with Theoretical Predictions. Environ. Sci. Technol. 11:405–409.

Dilling, W.L., N.B. Tefertiller, and G.J. Kollos. 1975. Evaporation Rates and Reactivities of Methylene Chloride, Chloroform, 1,1,1-Trichloroethane, Trichloroethylene, Tetrachloroethylene and Other Chlorinated Compounds in Dilute Aqueous Solutions. Environ. Sci. Technol. 9:833–838.

Fischer, A.J., E.A. Rowan, and R.F. Spalding. 1987. VOCs in Groundwater Influenced by Large Scale Withdrawals. Ground Water 25:407–413.

Francis, P.D. 1987. Oxidation by UV and Ozone of Organic Contaminants Dissolved in Deionized and Raw Mains Water. Ozone Sci. and Eng. 9:369–390.

Freeberg, K.M., P.B. Bedient, and J.A. Connor. 1987. Modeling of TCE Contamination and Recovery in a Shallow Sand Aquifer. Ground Water 25:70–80.

Freedman, D.L. and J.M. Gossett. 1989. Biological Reductive Dechlorination of Tetrachloroethylene and Trichloroethylene to Ethylene Under Methanogenic Conditions. Appl. and Environ. Microbiol. 55:2144–2151.

Fusillo, T.V., J.J. Hochreiter, and D.G. Lord. 1985. Distribution of Volatile Organic Compounds in a New Jersey Coastal Plain Aquifer System. Ground Water 23:354–360.

Glaze, W.H. and J. Kang. 1988. Advanced Oxidation Processes for Treating Ground Water Contaminated with TCE and PCE: Laboratory Studies. J. Am. Water Works Ass. 57:63.

Goodenkauf, O. and J.C. Atkinson. 1986. Occurrence of Volatile Organic Chemicals in Nebraska Ground Water. Ground Water 24:231–233.

Gross, R.L. and S.G. Termath. 1985. Packed Tower Aeration Strips TCE from Groundwater. Environ. Prog. 4:119–124.

Harker, A.R. and Y. Kim. 1990. Trichloroethylene Degradation by Two Independent Aromatic Degrading Pathways in *Alcaligenes eutrophus* JMP 134. Appl. Environ. Microbiol. 56:1179–1181.

Henson, J.M., M.V. Yates, J.W. Cochran, and D.L. Shackleford. 1988. Microbial Removal of Halogenated Methanes, Ethanes and Ethylenes in an Aerobic Soil Exposed to Methane. FEMS Micro. Ecol. 53:193–201.

Johnson, R.L., J.A. Cherry and J.F. Pankow. 1989. Diffusive Contaminant Transport in Natural Clay: A Field Example and Implications for Clay-lined Waste Disposal Sites. Environ. Sci. Technol. 23:340–349.

Josephson, J. 1983. Subsurface Contaminants. Environ. Sci. Technol. 17:518A–521A.

Mackay, D.M., P.V. Roberts and J.A. Cherry. 1985. Transport of Organic Contaminants in Groundwater. Environ. Sci. Technol. 19:384–392.

McKinnon and J.E. Dyksen. 1984. Removing Organics from Groundwater Through Aeration plus GAC. J. Am. Water Works Ass. 42:47.

Mehran, M., R.L. Olsen, and B.M. Rector. 1987. Distribution Coefficient of Trichloroethylene in Soil Water Systems. Ground Water 25:275–282.

Miller, D.E. and R.W. Callaway. 1991. Biological Treatment of Trichloroethylene Vapor Streams. Submitted to Environmental Science & Technology. In Review. [Editor's Note: A search of several science periodical databases did not yield a citation for a final published paper.]

Miller, R.E. and F.P. Guengerich. 1982. Oxidation of Trichloroethylene by Liver Microsomal Cytochrome P-450: Evidence for Chlorine Migration in a Transition State Not Involving Trichloroethylene Oxide. Biochem. 21:1090–1097.

Miller, G.C., V.R. Hebert, and R.G. Zepp. 1987. Chemistry and Photochemistry of Low-Volatility Organic Chemicals on Environmental Surfaces. Environ. Sci. Technol. 21:1164–1167.

Nelson, M.J.K., S. Montgomery, W.R. Mahaffey, and P.H. Pritchard. 1987. Biodegradation of Trichloroethylene and Involvement of an Aromatic Biodegradation Pathway. Appl. Environ. Microbiol. 53:949–954.

Robeck, G.G. and O.T. Love. 1983. Removal of Volatile Organic Contaminants from Groundwater. Environ. Micro. 53:949–954.

Roberts, P., L. Semprini, G. Hopkins, P. McCarty, and D. Grbic-Galic. 1989. Biostimulation of Methanotrophic Bacteria to Transform Halogenated Alkenes for Aquifer Restoration. EPRI/EPA Environmental Conference on Groundwater Quality.

Semprini, L., P.V. Roberts, G.D. Hopkins, and D.M. Mackay. 1987. A Field Evaluation of In Situ Biodegradation for Aquifer Restoration. EPA/600/2-87/096.

Sewell, G.W., S.A. Gibson, and H.H. Russell. 1990. Anaerobic In Situ Treatment of Chlorinated Ethenes. In: In-Situ Bioremediation of Ground Water and Contaminated Soils, Water Pollution Control Federation, pp. 67–79.

Spayd, S.E. 1985. Movement of Volatile Organics Through a Fractured Rock Aquifer. Ground Water 23:496–502.

Thomsen, K.O., M.A. Chaudhry, K. Dovantzis, and R.R. Riesing. 1989. Ground Water Remediation Using an Extraction Treatment, and Recharge System. Ground Water Monitoring Review 9(1):92–99.

Tu, A.S., T.A. Murray, K.A. Hutch, A. Sivak, and H.A. Milman. 1985. In Vitro Transformation of Balb C-3T3 Cells by Chlorinated Ethenes and Ethylenes. Cancer Lett. 28:85–92.

Vogel, T.M. and P.L. McCarty. 1985. Biotransformation of Tetrachloroethylene to Trichloroethylene, Dichloroethylene, Vinyl Chloride, and Carbon Dioxide Under Methanogenic Conditions. Appl. Environ. Microbiol. 49:1080–1083.

Wackett, L.P., G.A. Brusseau, S.R. Householder, and R.S. Hanson. 1989. Survey of Microbial Oxygenases: Trichlorethylene Degradation by Propane-Oxidizing Bacteria. Appl. Environ. Microbiol. 55:2960–2964.

Wilson, J.T., C.G. Enfield, W.J. Dunlap, R.L. Cosby, D.A. Foster, and L.B. Boskin. 1981. Transport and Fate of Selected Organic Pollutants in a Sandy Soil. J. Environ. Qual. 10:501–506.

Wilson, B.H. and D.W. Pogue. 1987. Biological Removal of Trichloroethylene from Contaminated Groundwater presented at Chemical and Biochemical Detoxification of Hazardous Waste. ACS New Orleans. Aug. 30–Sept. 4.

Wilson, B.H. and M.V. White. 1986. A Fixed-Film Bioreactor to Treat Trichloroethylene-Laden Waters from Interdiction Wells. In: Proceedings, Sixth National Symposium and Exposition on Aquifer Restoration and Groundwater Monitoring, National Water Well Association, Dublin, OH, pp. 425–435.

Wilson, J.T. and B.H. Wilson. 1985. Biotransformation of Trichloroethylene in Soil. Appl. Environ. Microbiol. 49:242–243.

ADDITIONAL REFERENCES

Barat, R.B. and J.W. Bozzelli. 1989. Reaction of Chlorocarbons to HCL and Hydrocarbons in a Hydrogen Rich Microwave-Induced Plasma Reactor. Environ. Sci. Technol. 23:666–671.

Barrio-Lage, G., F.Z. Parsons, R.S. Nassar, and P.A. Lorenzo. 1986. Sequential Dehalogenation of Chlorinated Ethenes. Environ. Sci. Technol. 20:96–99.

Mergia, G., B. Larsen, and W.E. Kelly. 1989. Three-dimensional Flow and Transport Model for Groundwater Remediation in Western Nebraska. Hazardous Waste Research Conference. Kansas State Univ. May 23–24.

Morrison, A. 1981. If Your City Well Water Has Chemical Pollutants, Then What? Civil Eng. 51:65–67.

Petura, J.C. 1981. Trichloroethylene and Methyl Chloroform in Groundwater: A Problem Assessment. J. Am. Water Works Ass. 25:25.

Chapter 7

In Situ Soil Flushing[1]

Jim Rawe, Science Applications International Corporation (SAIC), Cincinnati, OH

ABSTRACT

In situ soil flushing is the extraction of contaminants from the soil with water or other suitable aqueous solutions. Soil flushing is accomplished by passing the extraction fluid through in-place soils using an injection or infiltration process. Extraction fluids must be recovered and, when possible, are recycled. The method is potentially applicable to all types of soil contaminants. Soil flushing enables removal of contaminants from the soil and is most effective in permeable soils. An effective collection system is required to prevent migration of contaminants and potentially toxic extraction fluids to uncontaminated areas of the aquifer. Soil flushing, in conjunction with in situ bioremediation, may be a cost-effective means of soil remediation at certain sites [1, p. vi; 2, p. 11].[2] Typically, soil flushing is used in conjunction with other treatments that destroy contaminants or remove them from the extraction fluid and ground water.

Soil flushing is a developing technology that has had limited use in the United States. Typically, laboratory and field treatability studies must be performed under site-specific conditions before soil flushing is selected as the remedy of choice. To date, the technology has been selected as part of the source control remedy at 12 Superfund sites. This technology is currently operational at only one Superfund site; a second is scheduled to begin operation in 1991 [3; 4]. The EPA completed construction of a mobile soil-flushing system, the In Situ Contaminant/Treatment Unit, in 1988. This mobile soil-flushing system is designed for use at spills and uncontrolled hazardous waste sites [5].

This chapter provides information on the technology applicability, the technology limitations, a description of the technology, the types of residuals resulting from the use of the technology, site requirements, the latest performance data, the status of the technology, and sources of further information.

TECHNOLOGY APPLICABILITY

In situ soil flushing is generally used in conjunction with other treatment technologies such as activated carbon, biodegradation, or chemical precipitation to treat contaminated ground water resulting from soil flushing. In some cases, the process can reduce contaminant concentrations in the soil to acceptable levels, and thus serve as the only soil treatment technology. In other cases, in situ biodegradation or other in situ technologies can be used in conjunction with soil flushing to achieve acceptable contaminant removal efficiencies. In general, soil flushing is effective on coarse sand and gravel contaminated with a wide range of organic, inorganic, and reactive contaminants. Soils containing a large amount of clay and silt may not respond well to soil flushing, especially if it is applied as a stand-alone technology.

A number of chemical contaminants can be removed from soils using soil flushing. Removal efficiencies depend on the type of contaminant as well as the type of soil. Soluble (hydrophilic) organic contaminants often are easily removed from soil by flushing with water alone. Typically, organics with octanol/water partition coefficients (K_{ow}) of less than 10 (log K_{ow}<1) are highly soluble. Examples of such compounds include lower molecular weight alcohols, phenols, and carboxylic acids [6].

---

[1] EPA/540/2-91/021.

[2] [Reference number, page number.]

Low solubility (hydrophobic) organics may be removed by selection of a compatible surfactant [7]. Examples of such compounds include chlorinated pesticides, polychlorinated biphenyls (PCBs), semivolatiles (chlorinated benzenes and polynuclear aromatic hydrocarbons), petroleum products (gasoline, jet fuel, kerosene, oils and greases), chlorinated solvents (trichloroethene), and aromatic solvents (benzene, toluene, xylenes and ethylbenzene) [8]. However, removal of some of these chemical classes has not yet been demonstrated.

Metals may require acids, chelating agents, or reducing agents for successful soil flushing. In some cases, all three types of chemicals may be used in sequence to improve the removal efficiency of metals [9]. Many inorganic metal salts, such as carbonates of nickel, zinc, and copper, can be flushed from the soil with dilute acid solutions [6]. Some inorganic salts such as sulfates and chlorides can be flushed with water alone.

In situ soil flushing has been considered for treating soils contaminated with hazardous wastes, including pentachlorophenol and creosote from wood-preserving operations, organic solvents, cyanides and heavy metals from electroplating residues, heavy metals from some paint sludges, organic chemical production residues, pesticides and pesticide production residues, and petroleum/oil residues [10, p. 13; 11, p. 8; 7; 12].

The effectiveness of soil flushing for general contaminant groups [10, p. 13] is shown in Table 7-1. Examples of constituents within contaminant groups are provided in Reference 10, Technology Screening Guide for Treatment of CERCLA Soils and Sludges. Table 7-1 is based on currently available information or professional judgment where definitive information is currently inadequate or unavailable. The demonstrated effectiveness of the technology for a particular site or waste does not ensure that it will be effective at all sites or that the treatment efficiency achieved will be acceptable at other sites. For the ratings used in this table, demonstrated effectiveness means that, at some scale, treatability was tested to show that, for that particular contaminant and matrix, the technology was effective. The ratings of potential effectiveness and no expected effectiveness are based upon expert judgment. Where potential effectiveness is indicated, the technology is believed capable of successfully treating the contaminant group in a particular matrix. When the technology is not applicable or will probably not work for a particular combination of contaminant group and matrix, a no-expected-effectiveness rating is given. Other sources of general observations and average removal efficiencies for different treatability groups are the Superfund LDR Guide #6A, Obtaining a Soil and Debris Treatability Variance for Remedial Actions (OSWER Directive 9347.3-06FS) [13], and Superfund LDR Guide #6B, Obtaining a Soil and Debris Treatability Variance for Removal Actions (OSWER Directive 9347.3-07FS) [14].

Information on cleanup objectives, as well as the physical and chemical characteristics of the site soil and its contaminants, is necessary to determine the potential performance of this technology. Treatability tests are also required to determine the feasibility of the specific soil-flushing process being considered. If bench-test results are promising, pilot-scale demonstrations should be conducted before making a final commitment to full-scale implementation. Table 7-2 contains physical and chemical soil characterization parameters that should be established before a treatability test is conducted at a specific site. The table contains comments relating to the purpose of the specific parameter to be characterized and its impact on the process [15, p. 715; 16, p. 90; 17].

Soil permeability is a key physical parameter for determining the feasibility of using a soil-flushing process. Hydraulic conductivity (K) is measured to assess the permeability of soils. Soils with low permeability (K $<1.0 \times 10^{-5}$ cm/sec) will limit the ability of flushing fluids to percolate through the soil in a reasonable time frame. Soil flushing is most likely to be effective in permeable soils (K $>1.0 \times 10^{-3}$ cm/sec), but may have limited application to less permeable soils (1.0×10^{-5} cm/sec $<$ K $<1.0 \times 10^{-3}$ cm/sec). Since there can be significant lateral and vertical variability in soil permeability, it is important that field measurements be made using the appropriate methods.

Prior to field implementation of soil flushing, a thorough ground-water hydrologic study should be carried out. This should include information on seasonal fluctuations in water level, direction of ground-water flow, porosity, vertical and horizontal hydraulic conductivities, transmissivity and infiltration (data on rainfall, evaporation, and percolation).

Moisture content can affect the amount of flushing fluids required. Dry soils will require more flushing fluid initially to mobilize contaminants. Moisture content is also used to calculate pore volume to determine the rate of treatment [15].

Table 7-1. Effectiveness of Soil Flushing on General Contaminant Groups

Contaminant Groups	Effectiveness
Organic	
Halogenated volatiles	■
Halogenated semivolatiles	▼
Nonhalogenated volatiles	▼
Nonhalogenated semivolatiles	■
PCBs	▼
Pesticides (halogenated)	▼
Dioxins/furans	▼
Organic cyanides	▼
Organic corrosives	▼
Inorganic	
Volatile metals	▼
Nonvolatile metals	■
Asbestos	●
Radioactive materials	▼
Inorganic corrosives	▼
Inorganic cyanides	▼
Reactive	
Oxidizers	▼
Reducers	▼

■ Demonstrated Effectiveness: Successful treatability test at some scale completed.

▼ Potential Effectiveness: Expert opinion that technology will work.

● No Expected Effectiveness: Expert opinion that technology will not work.

The concentration and distribution of organic contaminants and metals are key chemical parameters. These parameters determine the type and quantity of flushing fluid required, as well as any post-treatment requirements. The solubility and partition coefficients of organics in water or other solutions are also important in the selection of the proper flushing fluids. The species of metal compounds present will affect the solubility and leachability of heavy metals.

High humic content and high cation exchange capacity tend to reduce the removal efficiency of soil flushing. Some organic contaminants may adsorb to humic materials or clays in soils and, therefore, are difficult to remove during soil flushing. Similarly, the binding of certain metals with clays due to cationic exchange makes them difficult to remove with soil flushing. The buffering capacity of the soil will affect the amount required of some additives, especially acids. Precipitation reactions (resulting in clogging of soil pores) can occur, due to pH changes in the flushing fluid caused by the neutralizing effect of soils with high buffering capacity. Soil pH can affect the speciation of metal compounds, resulting in changes in the solubility of metal compounds in the flushing fluid.

LIMITATIONS

Generally, remediation times with this technology will be lengthy (one to many years) due to the slowness of diffusion processes in the liquid phase. This technology requires hydraulic control to avoid movement of contaminants offsite. The hydrogeology of some sites may make this difficult or impossible to achieve.

Contaminants in soils containing a high percentage of silt- and clay-sized particles typically are strongly adsorbed and difficult to remove. Also, soils with silt and clay tend to be less permeable. In such cases, soil flushing generally should not be considered as a stand-alone technology.

Hydrophobic contaminants generally require surfactants or organic solvents for their removal from soil. Complex mixtures of contaminants in the soil (such as a mixture of metals, nonvolatile organics, and semivolatile organics) make it difficult to formulate a single suitable flushing fluid

Table 7-2. Characterization Parameters

Parameter	Purpose and Comment

Soil permeability
 $\geq 1.0 \times 10^{-3}$ cm/sec
 $<1.0 \times 10^{-5}$ cm/sec
	Affects treatment time and efficiency of contaminant removal
	    Effective soil flushing
	    Limited soil flushing

Soil structure	Influences flow patterns (channeling, blockage)
Soil porosity	Determines moisture capacity of soil at saturation (pore volume)
Moisture content	Affects flushing fluid transfer requirements
Groundwater hydrology	Critical in controlling the recovery of injected fluids and contaminants

Organics
  Concentration
  Solubility
  Partition coefficient
	Determine contaminants and assess flushing fluids required, flushing fluid compatibility, changes in flushing fluid with changes in contaminants.

Metals
  Concentration
  Solubility products
  Reduction potential
  Complex stability constants
	Concentration and species of constituents will determine flushing fluid compatibility, mobility of metals, post treatment.

Total Organic Carbon (TOC)	Adsorption of contaminants on soil increases with increasing TOC. Important in marine wetland sites, which typically have high TOC.
Clay content	Adsorption of contaminants on soil increases with increasing clay content.
Cation Exchange Capacity (CEC)	May affect treatment of metallic compounds.
pH, buffering capacity	May affect treatment additives required, compatibility with equipment materials of construction, wash fluid compatibility.

that will consistently and reliably remove all the different types of contaminants from the soil. Frequent changes in contaminant concentration and composition in the vertical and horizontal soil profiles will complicate the formulation of the flushing fluid. Sequential steps with frequent changes in the flushing formula may be required at such complex sites [10, p. 77].

Bacterial fouling of infiltration and recovery systems and treatment units may be a problem, particularly if high iron concentrations are present in the groundwater, or if biodegradable reagents are being used.

While flushing additives such as surfactants and chelants may enhance some contaminant removal efficiencies in the soil flushing process, they also tend to interfere with the downstream wastewater treatment processes. The presence of these additives in the washed soil and in the wastewater treatment sludge may cause some difficulty in their disposal. Costs associated with additives, and the management of these additives as part of the residuals/wastewater streams, must be carefully weighed against the incremental improvements in soil-flushing performance that they may provide.

TECHNOLOGY DESCRIPTION

Figure 7-1 is a general schematic of the soil flushing process [18, p. 7]. The flushing fluid is applied (1) to the contaminated soil by subsurface injection wells, shallow infiltration galleries, surface flooding, or aboveground sprayers. The flushing fluid is typically water and may contain additives to improve contaminant removal.

The flushing fluid percolates through the contaminated soil, removing contaminants as it proceeds. Contaminants are mobilized by solubilization into the flushing fluid, formation of emulsions, or through chemical reactions with the flushing fluid [19].

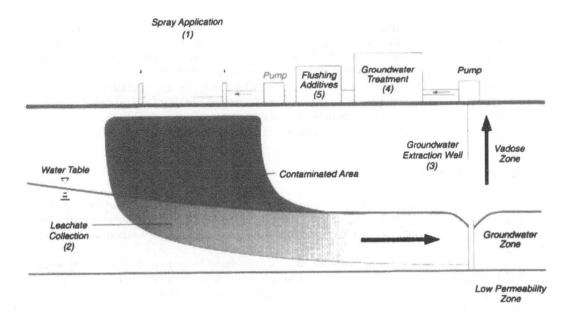

Figure 7-1. Schematic of soil flushing system.

Contaminated flushing fluid or leachate mixes with ground water and is collected (2) for treatment. The flushing fluid delivery and the ground-water extraction systems are designed to ensure complete contaminant recovery [7]. Ditches open to the surface, subsurface collection drains, or ground-water recovery wells may be used to collect flushing fluids and mobilized contaminants. Proper design of a fluid recovery system is very important to the effective application of soil flushing.

Contaminated ground water and flushing fluids are captured and pumped to the surface in a standard ground-water extraction well (3). The rate of ground-water withdrawal is determined by the flushing fluid delivery rate, the natural infiltration rate, and the ground-water hydrology. These will determine the extent to which the ground-water removal rate must exceed the flushing fluid delivery rate to ensure recovery of all reagents and mobilized contaminants. The system must be designed so that hydraulic control is maintained.

The ground water and flushing fluid are treated (4) using the appropriate wastewater treatment methods. Extracted ground water is treated to reduce the heavy metal content, organics, total suspended solids, and other parameters until they meet regulatory requirements. Metals may be removed by lime precipitation or by other technologies compatible with the flushing reagents used. Organics are removed with activated carbon, air stripping, or other appropriate technologies. Whenever possible, treated water should be recycled as makeup water at the front end of the soil-flushing process.

Flushing additives (5) are added, as required, to the treated ground water, which is recycled for use as flushing fluid. Water alone is used to remove hydrophilic organics and soluble heavy-metal salts [9]. Surfactants may be added to remove hydrophobic and slightly hydrophilic organic contaminants [12]. Chelating agents, such as ethylene-diaminetetra-acetic acid (EDTA), can effectively remove certain metal compounds. Alkaline buffers such as tetrasodium pyrophosphate can remove metals bound to the soil organic fraction. Reducing agents such as hydroxylamine hydrochloride can reduce iron and manganese oxides that can bind metals in soil. Insoluble heavy-metal compounds also can be reduced or oxidized to more soluble compounds. Weak acid solutions can improve the solubility of certain heavy metals [9]. Treatability studies should be conducted to determine compatibility of the flushing reagents with the contaminants and with the site soils.

PROCESS RESIDUALS

The primary waste stream generated is contaminated flushing fluid, which is recovered along with ground water. Recovered flushing fluids may need treatment to meet appropriate discharge standards prior to release to a local, publicly-owned wastewater treatment works or receiving streams. To the maximum extent practical, this water should be recovered and reused in the flushing process. The separation of surfactants from recovered flushing fluid, for reuse in the process, is a major factor in the cost of soil flushing. Treatment of the flushing fluid results in process sludges and residual solids, such as spent carbon and spent ion exchange resin, which must be appropriately treated before disposal. Air emissions of volatile contaminants from recovered flushing fluids should be collected and treated, as appropriate, to meet applicable regulatory standards. Residual flushing additives in the soil may be a concern and should be evaluated on a site-specific basis.

SITE REQUIREMENTS

Access roads are required for transport of vehicles to and from the site. Stationary or mobile soil-flushing process systems are located onsite. The exact area required will depend on the vendor system selected and the number of tanks or ponds needed for washwater preparation and wastewater treatment.

Because contaminated flushing fluids are usually considered hazardous, their handling requires that a site safety plan be developed to provide for personnel protection and special handling measures during wastewater treatment operations. Fire hazard and explosion considerations should be minimal, since the soil-flushing fluid is predominantly water.

An Underground Injection Control (UIC) Permit may be necessary if subsurface infiltration galleries or injection wells are used. When ground water is not recycled, a National Pollution Discharge Elimination System (NPDES) or State Pollution Discharge Elimination System (SPDES) permit may be required. Federal, state, and local regulatory agencies should be contacted to determine permitting requirements before implementing this technology.

Slurry walls or other containment structures (see Chapter 1) may be needed, along with hydraulic controls to ensure capture of contaminants and flushing additives. Climatic conditions such as precipitation cause surface runoff and water infiltration. Berms, dikes, or other runoff control methods may be required. Impermeable membranes may be necessary to limit infiltration of precipitation, which could cause dilution of flushing solution and loss of hydraulic control. Cold weather freezing must also be considered for shallow infiltration galleries and aboveground sprayers.

PERFORMANCE DATA

Some of the data presented for specific contaminant removal effectiveness were obtained from publications developed by the respective soil-flushing-system vendors. The quality of this information has not been determined; however, it does give an indication of the effectiveness of in situ soil flushing.

Tetrachloroethylene was discharged into the aquifer at the site of a spill in Sindelfingen, Germany. The contaminated aquifer is a high-permeability ($k = 5.10 \times 10^{-4}$ m/sec) layer overlaying a clay barrier. Soil flushing was accomplished by infiltrating water into the ground through ditches. The leaching liquid and polluted ground water were pumped out of eight wells and treated with activated carbon. The treated water was recycled through the infiltration ditches. Within 18 months, 17 metric tons of chlorinated hydrocarbons were recovered [19, p. 565].

Two percolation basins were installed to flush contaminated soil at the United Chrome Products site near Corvallis, Oregon. Approximately 1,100 tons of soil containing the highest chromium concentrations were excavated and disposed of offsite. The resulting pits from the excavations were used as infiltration basins to flush the remaining contaminated soil. The soil-flushing operation for the removal of hexavalent chromium from an estimated 2.4 million gallons of contaminated ground water began in August 1988. No information on the site soils was provided, but preliminary estimates were that a ground-water equilibrium concentration of 100 mg/L chromium would be reached in 1 to 2 years, but that final cleanup to 10 mg/L would take up to 25 years [20, p. H-1]. Since that time, over 8-million gallons of ground water, containing over 25,000 pounds of chromium, have been removed from the 23 extraction wells in the shallow aquifer. Average monthly chromium concentrations in the ground water decreased from 1,923 mg/L in August 1988 to 96 mg/L in March 1991 [4].

Waste-Tech Services, Inc. performed two tests of soil-flushing techniques to remove creosote contamination at the Laramie Tie Plant site in Wyoming. The first test involved slowly flooding the soil surface with water to perform primary oil recovery (POR). Soil flushing reduced the average concentration of total extractable organics (TEO) from an estimated initial concentration of 93,000 mg/kg to 24,500 mg/kg, a 74 percent reduction. The second test involved sequential treatment with alkaline agents, polymers, and surfactants. During the 8-month treatment period, average TEO concentrations were reduced to 4,000 mg/kg. This represents an 84 percent reduction from the post-POR concentration (24,500 mg/kg) and a 96 percent reduction from the estimated initial concentration (93,000 mg/kg). The tests were performed in alluvial sands and gravels. The low permeability of adjacent silts and clays precluded soil flushing [22].

Laboratory tests were conducted on contaminated soils from a fire-training area at Volk Air Force Base. Initial concentrations of oil and grease in the soils were reported to be 10,000 and 6,000 mg/kg. A 1.5-percent surfactant solution in water was used to flush soil columns. The tests indicated that 75 to 94 percent of the initial hydrocarbon contamination could be removed by flushing with 12-pore volumes of liquid. However, field tests were unsuccessful in removing the same contaminants. Seven soil-flushing solutions, including the solution tested in the laboratory studies, were tested in field studies. The flushing solutions were delivered to field test cells measuring 1 foot deep and 1 to 2 feet square. Only three of the seven tests achieved the target delivery of 14-pore volumes. Two of the test cells plugged completely, permitting no further infiltration of flushing solutions. There was no statistically significant removal of soil contaminants due to soil flushing. The plugging of test cells may be related to the use of a surfactant solution. By hydrolyzing in water, surfactants may block soil pores by forming either flocs or surfactant aggregates called micelles. In addition, if the surfactant causes fine soil particles to become suspended in the flushing fluid, narrow passages between soil particles could be blocked. If enough of these narrow passages are blocked along a continuous front, a "mat" is said to have formed, and fluid flow is halted in that area [23; 7].

Resource Conservation Recovery Act (RCRA) Land Disposal Restrictions (LDRs) that require treatment of wastes to best demonstrated available technology (BDAT) levels prior to land disposal may sometimes be determined to be applicable or relevant and appropriate requirements (ARARs) for CERCLA response actions. The soil-flushing technology can produce a treated waste that meets treatment levels set by BDAT, but may not reach these treatment levels in all cases. The ability of the technology to meet required treatment levels is dependent upon the specific waste constituents and the waste matrix. In cases where soil flushing does not meet these levels, it still may, in certain situations, be selected for use at the site if a treatability variance establishing alternative treatment levels is obtained. The EPA has made the treatability variance process available in order to ensure that LDRs do not unnecessarily restrict the use of alternative and innovative treatment technologies. Treatability variances may be justified for handling complex soil and debris matrices. The following guides describe when and how to seek a treatability variance for soil and debris: Superfund LDR Guide #6A, Obtaining a Soil and Debris Treatability Variance for Remedial Actions (OSWER Directive 9347.3-06FS) [13], and Superfund LDR Guide #6B, Obtaining a Soil and Debris Treatability Variance for Removal Actions (OSWER Directive 9347.3-07FS) [14]. Another approach could be to use other treatment techniques in conjunction with soil flushing to obtain desired treatment levels.

TECHNOLOGY STATUS

In situ soil flushing is a developing technology that has had limited application in the United States. In situ soil flushing technology has been selected as one of the source control remedies at the 12 Superfund sites listed in Table 7-3 [3].

EPA CONTACT

Technology-specific questions regarding soil flushing may be directed to Don Draper, (405) 436-8603.

Table 7-3. Superfund Sites Using In Situ Soil Flushing

Site	Location (Region)	Primary Contaminants	Status
Byron Barrel & Drum	Genesee County, NY (2)	VOCs (BTX, PCE, and TCE)	Pre-design: finalizing workplan
Goose Farm	Plumsted Township, NJ (2)	VOCs (Toluene, Ethylbenzene, Dichloromethane, and TCE), SVOCs, and PAHs	In design: 30% design phase
Lipari Landfill	Gloucester, NJ (2)	VOCs (Benzene, Ethylbenzene, Dichloromethane, and TCE), SVOCs, PAHs and Chlorinated Ethers (bis-2-chloroethylether)	Operational, summer '91
Vineland Chemical	Vineland, NJ (2)	Arsenic and VOCs (Dichloromethane)	Pre-design
Harvey-Knott Drum	_____, DE (3)	Lead	In design: re-evaluating alternative
L.A. Clarke & Son	Spotsylvania, VA (3)	Creosote, PAHs, and Benzene	In design
Ninth Avenue Dump	Gary, IN (5)	VOCs (BTEX, TCE), PAHs, Phenols, Lead, PCBs, and Total Metals	In design: pilot failed
U.S. Aviex	Niles, MI (5)	VOCs (Carbon Tetrachloride, DCA, Ethylbenzene, PCE, TCE, Toluene, TCA, Freon, Xylene, and Chloroform)	Pre-design: re-evaluating alternatives
South Calvacale Street	Houston, TX (6)	PAHs	In design
United Chrome Products	Corvallis, OR (10)	Chromium	Operational since 8/88
Cross Brothers Pail	Pembroke, IL (5)	VOCs (Benzene, PCE, TCE, Toluene, and Xylenes) and PCBs	In design: developing workplan
Bog Creek Farm	Howell Township, NJ (2)	VOCs, Organics	In design: treatment plant completed, pump-and-treat not installed

ACKNOWLEDGMENTS

This chapter was prepared for the U.S. Environmental Protection Agency, Office of Research and Development (ORD), Risk Reduction Engineering Laboratory (RREL), Cincinnati, Ohio, by Science Applications International Corporation (SAIC) under contract No. 68-C8-0062. Mr. Eugene Harris served as the EPA Technical Project Monitor. Mr. Gary Baker was SAIC's Work Assignment Manager. This chapter was authored by Mr. Jim Rawe of SAIC. The author is especially grateful to Ms. Joyce Perdek of EPA, RREL, who has contributed significantly by serving as a technical reviewer during the development of this chapter.

The following other Agency and contractor personnel have contributed their time and comments by participating in the expert review meeting and/or peer reviewing the chapter.

Mr. Benjamin Blaney, EPA-RREL
Ms. Sally Clement, Bruck, Hartman and Esposito
Mr. Clyde Dial, SAIC
Ms. Linda Fiedler, EPA-TIO
Dr. David Wilson, Vanderbilt University
Ms. Tish Zimmerman, EPA-OSWER

REFERENCES

1. Handbook: In Situ Treatment of Hazardous Waste Contaminated Soils. EPA/540/2-90/002 (NTIS PB90-155607), 1990.
2. A Compendium of Technologies Used in the Treatment of Hazardous Wastes. EPA/625/8-87/014 (NTIS PB90-274093), 1987.
3. Innovative Treatment Technologies: Semi-Annual Status Report. EPA/540/2-91/001, U.S. Environmental Protection Agency, 1991. [Annual Status Report (Sixth Edition) was published as EPA 542-R-94-005, September, 1994.]
4. Personal communications of SAIC staff with RPMs, 1991.
5. In Situ Containment/Treatment System, Fact Sheet. U.S Environmental Protection Agency, 1988.
6. Sanning, D.E., et al. Technologies for In Situ Treatment of Hazardous Wastes. EPA/600/D-87/014, U.S. Environmental Protection Agency, 1987.
7. Nash, J. and R.P. Traver. Field Evaluation of In Situ Washing of Contaminated Soils with Water/Surfactants. In: Overview-Soils Washing Technologies for: Comprehensive Environmental Response, Compensation, and Liability Act, Resource Conservation and Recovery Act, Leaking Underground Storage Tanks, Site Remediation, U.S. Environmental Protection Agency, 1989. pp. 383–392.
8. Wilson, D.J., et al., Soil Washing and Flushing with Surfactants. Tennessee Water Resources Research Center September, 1990.
9. Ellis, W.D., T.R. Fogg, and A.N. Tafuri. Treatment of Soils Contaminated with Heavy Metals. In: Overview-Soils Washing Technologies for: Comprehensive Environmental Response, Compensation, and Liability Act, Resource Conservation and Recovery Act, Leaking Underground Storage Tanks, Site Remediation, U.S. Environmental Protection Agency, 1989. pp. 127–134.
10. Technology Screening Guide for Treatment of CERCLA Soils and Sludges. EPA/540/2-88/004 (NTIS PB89-132674), 1988.
11. Nunno, T.J., J.A. Hyman, and T. Pheiffer. Development of Site Remediation Technologies in European Countries. Presented at Workshop on the Extractive Treatment of Excavated Soil. U.S. Environmental Protection Agency, Edison, New Jersey, 1988.
12. Ellis, W.D., J.R. Payne, and G.D. McNabb, Project Summary: Treatment of Contaminated Soils with Aqueous Surfactants. EPA/600/S2-85/129, 1985.
13. Superfund LDR Guide #6A: Obtaining a Soil and Debris Treatability Variance for Remedial Actions. OSWER Directive 9347.3-06FS (NTIS PB91-921327), 1990.
14. Superfund LDR Guide #6B: Obtaining a Soil and Debris Treatability Variance for Removal Actions. OSWER Directive 9347.3-07FS (NTIS PB91-921310), 1990.
15. Sims, R.C. Soil Remediation Techniques at Uncontrolled Hazardous Waste Sites, A Critical Review. J. Air & Waste Management Association 40:704–732.

16. Guide for Conducting Treatability Studies Under CERCLA, Interim Final. EPA/540/2-89/058, 1989.
17. Connick, C.C. Mitigation of Heavy Metal Migration in Soil. In: Overview-Soils Washing Technologies for: Comprehensive Environmental Response, Compensation, and Liability Act, Resource Conservation and Recovery Act, Leaking Underground Storage Tanks, Site Remediation, U.S. Environmental Protection Agency, 1989. pp. 155–165.
18. Handbook: Remedial Action at Waste Disposal Sites (Revised). EPA/625/6-85/006 (NTIS PB87-201034), 1985.
19. Stief, K. Remedial Action for Groundwater Protection Case Studies Within the Federal Republic of Germany. Presented at the 5th National Conference on Management of Uncontrolled Hazardous Waste Sites. Washington, DC, 1984.
20. Young, C., et al. Innovative Operational Treatment Technologies for Application to Superfund Site—Nine Case Studies, Final Report. EPA 540/2-90/006 (NTIS PB90-202656), 1990.
21. United Chrome Groundwater Extraction and Treatment Facility. Monthly Report—March 1991. U.S. Environmental Protection Agency, Region 10, 1991.
22. Marketing Brochure, Waste-Tech Services, Inc., Waste Minimization Division, 1990.
23. Sale, T., K. Piontek, and M. Pitts. Chemically Enhanced In Situ Soil Washing. In: Proceedings of the Conference on Petroleum Hydrocarbons and Organic Chemicals in Ground Water: Prevention, Detection and Restoration. National Water Well Association, 1989, pp. 487–503.

Chapter 8

In Situ Soil Vapor Extraction Treatment[1]

Peter Michaels, Foster Wheeler, Enviresponse, Inc.

ABSTRACT

Soil vapor extraction (SVE) is designed to physically remove volatile compounds, generally from the vadose or unsaturated zone. It is an in situ process employing vapor extraction wells alone or in combination with air injection wells. Vacuum blowers supply the motive force, inducing air flow through the soil matrix. The air strips the volatile compounds from the soil and carries them to the screened extraction well.

Air emissions from the systems are typically controlled by adsorption of the volatiles onto activated carbon, thermal destruction (incineration or catalytic oxidation), or condensation by refrigeration [1, p. 26].[2]

SVE is a developed technology that has been used in commercial operations for several years. It was the selected remedy for the first Record of Decision (ROD) to be signed under the Superfund Amendments and Reauthorization Act of 1986 (the Verona Well Field Superfund Site in Battle Creek, Michigan). SVE has been chosen as a component of the ROD at over 30 Superfund sites [2–6].

Site-specific treatability studies are the only means of documenting the applicability and performance of an SVE system. The EPA Contact indicated at the end of this chapter can assist in the location of other contacts and sources of information necessary for such treatability studies.

The final determination of the lowest cost alternative will be more site-specific than process equipment dominated. This bulletin provides information on the technology applicability, the limitations of the technology, the technology description, the types of residuals produced, site requirements, the latest performance data, the status of the technology, and sources for further information.

TECHNOLOGY APPLICABILITY

In situ SVE has been demonstrated effective for removing volatile organic compounds (VOCs) from the vadose zone. The effective removal of a chemical at a particular site does not, however, guarantee an acceptable removal level at all sites. The technology is very site-specific. It must be applied only after the site has been characterized. In general, the process works best in well drained soils with low organic carbon content. However, the technology has been shown to work in finer, wetter soils (e.g., clays), but at much slower removal rates [7, p. 5].

The extent to which VOCs are dispersed in the soil—vertically and horizontally—is an important consideration in deciding whether SVE is preferable to other methods. Soil excavation and treatment may be more cost-effective when only a few hundred cubic yards of near-surface soils have been contaminated. If volume is in excess of 500 cubic yards, if the spill has penetrated more than 20 or 30 feet, or the contamination has spread through an area of several hundred square feet at a particular depth, then excavation costs begin to exceed those associated with an SVE system [8; 9; 10, p. 6].

The depth to ground water is also important. Ground-water level in some cases may be lowered to increase the volume of the unsaturated zone. The water infiltration rate can be controlled by placing

---

[1] EPA/540/2-91/006.
[2] [Reference number, page number.]

an impermeable cap over the site. Soil heterogeneities influence air movement as well as the location of chemicals. The presence of heterogeneities may make it more difficult to position extraction and inlet wells. There generally will be significant differences in the air permeability of the various soil strata which will affect the optimum design of the SVE facility. The location of the contaminant on a property and the type and extent of development in the vicinity of the contamination may favor the installation of an SVE system. For example, if the contamination exists beneath a building or beneath an extensive utility trench network, SVE should be considered.

SVE can be used alone or in combination with other technologies to treat a site. SVE, in combination with ground-water pumping and air stripping, is necessary when contamination has reached an aquifer. When the contamination has not penetrated into the zone of saturation (i.e., below the water table), it is not necessary to install a ground-water pumping system. A vacuum extraction well will cause the water table to rise and will saturate the soil in the area of the contamination. Pumping is then required to draw the water table down and allow efficient vapor venting [11, p. 169].

SVE may be used at sites not requiring complete remediation. For example, a site may contain VOCs and nonvolatile contaminants. A treatment requiring excavation might be selected for the nonvolatile contaminants. If the site required excavation in an enclosure to protect a nearby populace from VOC emissions, it would be cost-effective to extract the volatiles from the soil before excavation. This would obviate the need for the enclosure. In this case it would be necessary to vent the soil for only a fraction of the time required for complete remediation.

Performance data presented in this chapter should not be considered directly applicable to other Superfund sites. A number of variables such as the specific mix and distribution of contaminants affect system performance. A thorough characterization of the site and a well-designed and conducted treatability study are highly recommended.

The effectiveness of SVE on general contaminant groups for soils is shown in Table 8-1. Examples of constituents within contaminant groups are provided in the Technology Screening Guide for Treatment of CERCLA Soils and Sludges [12]. This table is based on the current available information or professional judgment where no information was available. The proven effectiveness of the technology for a particular site or waste does not ensure that it will be effective at all sites or that the treatment efficiencies achieved will be acceptable at other sites. For the ratings used in this table, demonstrated effectiveness means that, at some scale, treatability tests showed that the technology was effective for that particular contaminant and matrix. The ratings of potential effectiveness, or no expected effectiveness are both based upon expert judgment. Where potential effectiveness is indicated, the technology is believed capable of successfully treating the contaminant group in a particular matrix. When the technology is not applicable or will probably not work for a particular combination of contaminant group and matrix, a no-expected-effectiveness rating is given. Another source of general observations and average removal efficiencies for different treatability groups is contained in the Superfund Land Disposal Restrictions (LDR) Guide #6A, Obtaining a Soil and Debris Treatability Variance for Remedial Actions, (OSWER Directive 9347.3-06FS, July 1989) [13] and Superfund LDR Guide #6B, Obtaining a Soil and Debris Treatability Variance for Removal Actions, (OSWER Directive 9347.3-07FS, December 1989) [14].

LIMITATIONS

Soils exhibiting low air permeability are more difficult to treat with in situ SVE. Soils with a high organic carbon content have a high sorption capacity for VOCs and are more difficult to remediate successfully with SVE. Low soil temperature lowers a contaminant's vapor pressure, making volatilization more difficult [11].

Sites that contain a high degree of soil heterogeneity will likely offer variable flow and desorption performance, which will make remediation difficult. However, proper design of the vacuum extraction system may overcome the problems of heterogeneity [7, p. 19; 15].

It would be difficult to remove soil contaminants with low vapor pressures and/or high water solubilities from a site. The lower limit of vapor pressure for effective removal of a compound is 1 mm Hg abs. Compounds with high water solubilities, such as acetone, may be removed with relative ease from arid soils. However, with normal soils (i.e., moisture content ranging from 10 percent to 20 percent), the likelihood of successful remediation drops significantly because the moisture in the soil acts as a sink for the soluble acetone.

Table 8-1. Effectiveness of SVE on General Contaminant Groups for Soil

Contaminant Groups	Effectiveness
Organic	
Halogenated volatiles	■
Halogenated semivolatiles	▼
Nonhalogenated volatiles	■
Nonhalogenated semivolatiles	■
PCBs	●
Pesticides (halogenated)	●
Dioxins/furans	●
Organic cyanides	●
Organic corrosives	●
Inorganic	
Volatile metals	●
Nonvolatile metals	●
Asbestos	●
Radioactive materials	●
Inorganic corrosives	●
Inorganic cyanides	●
Reactive	
Oxidizers	●
Reducers	▼

■ Demonstrated Effectiveness: Successful treatability test at some scale completed.

▼ Potential Effectiveness: Expert opinion that technology will work.

● No Expected Effectiveness: Expert opinion that technology will not work.

TECHNOLOGY DESCRIPTION

Figure 8-1 is a general schematic of the in situ SVE process. After the contaminated area is defined, extraction wells (1) are installed. Extraction well placement is critical. Locations must be chosen to ensure adequate vapor flow through the contaminated zone while minimizing vapor flow through other zones [11, p. 170]. Wells are typically constructed of PVC pipe that is screened through the zone of contamination [11]. The screened pipe is placed in a permeable packing; the unscreened portion is sealed in a cement/bentonite grout to prevent a short-circuited air flow direct to the surface. Some SVE systems are installed with air injection wells. These wells may either passively take in atmospheric air or actively use forced air injection [9]. The system must be designed so that any air injected into the system does not result in the escape of VOCs to the atmosphere. Proper design of the system can also prevent offsite contamination from entering the area being extracted.

The physical dimensions of a particular site may modify SVE design. If the vadose zone depth is less than 10 feet and the area of the site is quite large, a horizontal piping system or trenches may be more economical than conventional wells.

An induced air flow draws contaminated vapors and entrained water from the extraction wells through headers—usually plastic piping—to a vapor-liquid separator (2). There, entrained water is separated and contained for subsequent treatment (4). The contaminant vapors are moved by a vacuum blower (3) to vapor treatment (5).

Vapors produced by the process are typically treated by carbon adsorption or thermal destruction. Other methods—such as condensation, biological degradation, and ultraviolet oxidation—have been applied, but only to a limited extent.

PROCESS RESIDUALS

The waste streams generated by in situ SVE are vapor and liquid treatment residuals (e.g., spent granular activated carbon [GAC]), contaminated ground water, and soil tailings from drilling the

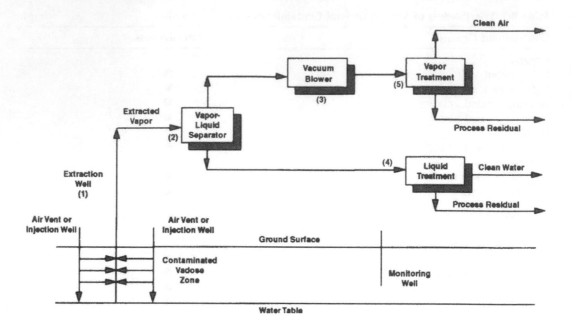

Figure 8-1. Process schematic of the in situ soil vapor extraction system.

wells. Contaminated ground water may be treated and discharged onsite [12, p. 86] or collected and treated offsite. Highly contaminated soil tailings from drilling must be collected and may be either cleaned onsite or sent to an offsite, permitted facility for treatment by another technology such as incineration.

SITE REQUIREMENTS

SVE systems vary in size and complexity depending on the capacity of the system and the requirements for vapor and liquid treatment. They are typically transported by vehicles ranging from trucks to specifically adapted flatbed semitrailers; therefore, a proper staging area for these vehicles must be incorporated in the plans.

Adequate access roads must be provided to bring mobile drilling rigs onsite for construction of wells and to deliver equipment required for the process (e.g., vacuum blowers, vapor-liquid separator, emission control devices, GAC canisters).

A small commercial-size SVE system would require about 1,000 square feet of ground area for the equipment. This area does not include space for the monitoring wells which might cover 500 square feet. Space may be needed for a forklift truck to exchange skid-mounted GAC canisters when regeneration is required. Large systems with integrated vapor and liquid treatment systems will need additional area based on vendor-specific requirements.

Standard 440V, three-phase electrical service is needed. For many SVE applications, water may be required at the site. The quantity of water needed is vendor- and site-specific.

Contaminated soils or other waste materials are hazardous, and their handling requires that a site safety plan be developed to provide for personnel protection and special handling measures. Storage should be provided to hold the process product streams until they have been tested to determine their acceptability for disposal or release. Depending upon the site, a method to store soil tailings from drilling operations may be necessary. Storage capacity will depend on waste volume.

Onsite analytical equipment, including gas chromatographs and organic vapor analyzers capable of determining site-specific organic compounds for performance assessment, make the operation more efficient and provide better information for process control.

PERFORMANCE DATA

SVE, as an in situ process (no excavation is involved), may require treatment of the soil to various cleanup levels mandated by federal and state site-specific criteria. The time required to meet a target cleanup level (or performance objective) may be estimated by using data obtained from bench-scale and pilot-scale tests in a time-predicting mathematical model. Mathematical models can estimate cleanup time to reach a target level, residual contaminant levels after a given period of operation and can predict location of hot spots through diagrams of contaminant distribution [16].

Table 8-2 shows the performance of typical SVE applications. It lists the site location and size, the contaminants and quantity of contaminants removed, the duration of operation, and the maximum soil contaminant concentrations before treatment and after treatment. The data presented for specific contaminant removal effectiveness were obtained, for the most part, from publications developed by the respective SVE system vendors. The quality of this information has not been determined.

Midwest Water Resources, Inc. (MWRI) installed its VAPORTECH™ pumping unit at the Dayton, Ohio site of a spill of uncombusted paint solvents caused by a fire in a paint warehouse [19]. The major VOC compounds identified were acetone, methyl isobutyl ketone (MIBK), methyl ethyl ketone (MEK), benzene, ethylbenzene, toluene, naphtha, xylene, and other volatile aliphatic and alkyl benzene compounds. The site is underlain predominantly by valley-fill glacial outwash within the Great Miami River Valley, reaching a thickness of over 200 feet. The outwash is composed chiefly of coarse, clean sand and gravel, with numerous cobbles and small boulders. There are two outwash units at the site separated by a discontinuous till at depths of 65 to 75 feet. The upper outwash forms an unconfined aquifer with saturation at a depth of 45 to 50 feet below grade. The till below serves as an aquitard between the upper unconfined aquifer and the lower confined to semiconfined aquifer. Vacuum withdrawal extended to the depth of ground water at about 40 to 45 feet. During the first 73 days of operation, the system yielded 3,720 pounds of volatiles and after 56 weeks of operation, had recovered over 8,000 pounds of VOCs from the site. Closure levels for the site were developed for ground water VOC levels of ketones only. These soil action levels (acetone, 810 µg/L; MIBK, 260 µg/ L, and MEK, 450 µg/L) were set so that waters recharging through contaminated soils would result in ground water VOC concentrations at or below regulatory standards. The site met all the closure criteria by June 1988.

A limited amount of performance data is available from Superfund sites. The EPA Superfund Innovative Technology Evaluation (SITE) Program's Groveland, Massachusetts, demonstration of the Terra Vac Corporation SVE process produced data that were subjected to quality assurance/ quality control tests. These data appear in Table 8-2 [7, p. 29] and Table 8-3 [7, p. 31]. The site is contaminated by trichloroethylene (TCE), a degreasing compound which was used by a machine shop that is still in operation. The subsurface profile in the test area consists of medium sand and gravel just below the surface, underlain by finer and silty sands, a clay layer 3 to 7 feet in depth, and—below the clay layer—coarser sands with gravel. The clay layer or lens acts as a barrier against gross infiltration of VOCs into subsequent subsoil strata. Most of the subsurface contamination lay above the clay lens, with the highest concentrations adjacent to it. The SITE data represent the highest percentage of contaminant reduction from one of the four extraction wells installed for this demonstration test. The TCE concentration levels are weighted average soil concentrations obtained by averaging split spoon sample concentrations every 2 feet over the entire 24-foot extraction well depth. Table 8-3 shows the reduction of TCE in the soil strata near the same extraction well. The Groveland Superfund Site is in the process of being remediated using this technology [2].

The Upjohn facility in Barceloneta, Puerto Rico, is the first and, thus far, the only Superfund site to be remediated with SVE. The contaminant removed from this site was a mixture containing 65 percent carbon tetrachloride (CCl_4) and 35 percent acetonitrile [20]. Nearly 18,000 gallons of CCl_4 were extracted during the remediation, including 8,000 gallons that were extracted during a pilot operation conducted from January 1983 to April 1984. The volume of soil treated at the Upjohn site amounted to 7,000,000 cubic yards. The responsible party originally argued that the site should be considered clean when soil samples taken from four boreholes drilled in the area of high pretest contamination show nondetectable levels of CCl_4. The EPA did not accept this criterion, but instead required a cleanup criteria of nondetectable levels of CCl_4 in all the exhaust stacks for 3 consecutive months [21]. This requirement was met by the technology and the site was considered remediated by the EPA.

Table 8-2. Summary of Performance Data for In Situ Soil Vapor Extraction

Site	Size	Contaminants	Quantity Removed	Duration of Operation	Soil Concentrations (mg/kg)	
					Max. Before Treatment	After Treatment
Industrial, CA [17]	—	TCE	30 kg	440 days	0.53	0.06
Sheet Metal Plant, MI [18]	5,000 cu yds	PCE	59 kg	35 days	5600	0.70
Prison Const. Site, MI [19]	165,000 cu yds	TCA	—	90 days	3.7	0.01
Sherwin-Williams Site, OH [19]	425,000 cu yds	Paint solvents	4,100 kg	6 mo	38	0.04
Upjohn, PR [20; 21]	7,000,000 cu yds	CCl_4	107,000 kg	3 yr	2200	<0.005
UST Bellview, FL [7]	—	BTEX	9,700 kg	7 mo	97	<0.006
Verona Wellfield, MI [7; 22]	35,000 cu yds	TCE, PCE, TCA	12,700 kg	Over 1 yr	1380	Ongoing
Petroleum Terminal Owensboro, KY [19]	12,000 cu yds	Gasoline, diesel	—	6 mo	>5000	1.0 (target)
SITE Program, Groveland, MA [7]	6,000 cu yds	TCE	590 kg	56 days	96.1	4.19

Table 8-3. Extraction Well 4: TCE Reduction in Soil Strata—EPA Site Demonstration (Groveland, MA)
[7, p. 31]

Depth	Description of Strata	Hydraulic Conductivity (cm/s)	Soil TCE Concentrations (mg/kg)	
			Pre-Treatment	Post-Treatment
0–2	Med. sand w/gravel	10^{-4}	2.94	ND
2–4	Lt. brown fine sand	10^{-4}	29.90	ND
4–6	Med. stiff lt. brown fine sand	10^{-5}	260.0	39.0
6–8	Med. stiff brown sand	10^{-5}	303.0	9.0
8–10	Med. stiff brown sand	10^{-4}	351.0	ND
10–12	V. stiff lt. brown med. sand	10^{-4}	195.0	ND
12–14	V. stiff brown fine sand w/silt	10^{-4}	3.14	2.3
14–16	M. stiff grn-brn clay w/silt	10^{-8}	ND	ND
16–18	Soft wet clay	10^{-8}	ND	ND
18–20	Soft wet clay	10^{-8}	ND	ND
20–22	V. stiff brn med-coarse sand	10^{-4}	ND	ND
22–24	V. stiff brn med-coarse w/gravel	10^{-3}	6.17	ND

Approximately 92,000 pounds of contaminants have been recovered from the Tyson's Dump site (Region 3) between November 1988 and July 1990. The site consists of two unlined lagoons and surrounding areas formerly used to store chemical wastes. The initial Remedial Investigation identified no soil heterogeneities and indicated that the water table was 20 feet below the surface. The maximum concentration in the soil (total VOCs) was approximately 4 percent. The occurrence of dense nonaqueous-phase liquids (DNAPLs) was limited in areal extent. After over 18 months of operation, a number of difficulties have been encountered. Heterogeneities in soil grain size, water content, permeability, physical structure and compaction, and in contaminant concentrations have been identified. Soil contaminant concentrations of up to 20 percent and widespread distribution of DNAPLs have been found. A tar-like substance, which has caused plugging, has been found in most of the extraction wells. After 18 months of operation, wellhead concentrations of total VOCs have decreased by greater than 90 percent [23, p. 28].

As of December 31, 1990, approximately 45,000 pounds of VOCs had been removed from the Thomas Solvent Raymond Road Operable Unit at the Verona Well Field site (Region 5). A pilot-scale system was tested in the fall of 1987 and a full-scale operation began in March, 1988. The soil at the site consists of poorly-graded, fine-to-medium-grained loamy soils underlain by approximately 100 feet of sandstone. Groundwater is located 16 to 25 feet below the surface. Total VOC concentrations in the combined extraction well header have decreased from a high of 19,000 µg/L in 1987 to approximately 1,500 µg/L in 1990 [22].

An SVE pilot study has been completed at the Colorado Avenue Subsite of the Hastings (Nebraska) Groundwater Contamination site (Region 7). Trichloroethylene (TCE), 1,1,1-trichloroethane (TCA), and tetrachloroethylene (PCE) occur in two distinct unsaturated soil zones. The shallow zone, from the surface to a depth of 50 to 60 feet, consists of sandy and clayey silt. TCE concentrations as high as 3,600 µg/L were reported by the EPA in this soil zone. The deeper zone consists of interbedded sands, silty sands, and gravelly sands extending from about 50 feet to 120 feet. During the first 630 hours of the pilot study (completed October 11, 1989), removal of approximately 1,488 pounds of VOCs from a deep zone extraction well and approximately 127 pounds of VOCs from a shallow zone extraction well were reported. The data suggest that SVE is a viable remedial technology for both soil zones [24].

As of November 1989, the SVE system at the Fairchild Semiconductor Corporation's former San Jose site (Region 9) has reportedly removed over 14,000 pounds of volatile contaminants. Total contaminant mass removal rates for the SVE system fell below 10 pounds per day on October 5, 1989 and fell below 6 pounds per day in December, 1989. At that time, a proposal to terminate operation of the SVE system was submitted to the Regional Water Quality Control Board for the San Francisco Bay Region [25, p. 3].

Resource Conservation and Recovery Act (RCRA) LDRs that require treatment of wastes to best demonstrated available technology (BDAT) levels prior to land disposal may sometimes be determined to be applicable or relevant and appropriate requirements for CERCLA response actions. SVE can produce a treated waste that meets treatment levels set by BDAT but may not reach these treatment levels in all cases. The ability to meet required treatment levels is dependent upon the specific waste constituents and the waste matrix. In cases where SVE does not meet these levels, it still may, in certain situations, be selected for use at the site if a treatability variance establishing alternative treatment levels is obtained. The EPA has made the treatability variance process available in order to ensure that LDRs do not unnecessarily restrict use of alternative and innovative treatment technologies. Treatability variances are justified for handling complex soil and debris matrices. The following guides describe when and how to seek a treatability variance for soil and debris: Superfund LDR Guide #6A, Obtaining a Soil and Debris Treatability Variance for Remedial Actions (OSWER Directive 9347.3-06FS, July 1989) [13], and Superfund LDR Guide #6B, Obtaining a Soil and Debris Treatability Variance for Removal Actions (OSWER Directive 9347.3-07FS, December 1989) [14]. Another approach could be to use other treatment techniques in series with SVE to obtain desired treatment levels.

TECHNOLOGY STATUS

During 1989, at least 17 RODs specified SVE as part of the remedial action [5]. Since 1982, SVE has been selected as the remedial action, either alone or in conjunction with other treatment technologies, in more than 30 RODs for Superfund sites [2–6]. Table 8-4 presents the location, primary contaminants, and status for these sites [3–5]. The technology also has been used to clean up numerous underground gasoline storage tank spills.

A number of variations of the SVE system have been investigated at Superfund sites. At the Tinkhams Garage Site in New Hampshire (Region 1), a pilot study indicated that SVE, when used in conjunction with ground-water pumping (dual extraction), was capable of treating soils to the 1 ppm cleanup goal [26, 3–7; 27]. Soil dewatering studies have been conducted to determine the feasibility of lowering the water table to permit the use of SVE at the Bendix, PA site (Region 3) [28]. Plans are underway to remediate a stockpile of 700 cubic yards of excavated soil at the Sodeyco site in Mt. Holly, NC using SVE [29].

With the exception of the Barceloneta site, no Superfund site has yet been cleaned up to the performance objective of the technology. The performance objective is a site-specific contaminant concentration, usually in soil. This objective may be calculated with mathematical models by which EPA evaluates delisting petitions for wastes contaminated with VOCs [30]. It also may be possible to use a TCLP test on the treated soil with a corresponding drinking water standard contaminant level on the leachate.

Most of the hardware components of SVE are available off the shelf and represent no significant problems of availability. The configuration, layout, operation, and design of the extraction and monitoring wells and process components are site specific. Modifications may also be required as dictated by actual operating conditions.

On-line availability of the full-scale systems described in this bulletin is not documented. System components are highly reliable and are capable of continuous operation for the duration of the cleanup. The system can be shut down, if necessary, so that component failure can be identified and replacements made quickly for minimal downtime.

Based on available data, SVE treatment estimates are typically $50/ton for treatment of soil. Costs range from as low as $10/ton to as much as $150/ton [7]. Capital costs for SVE consist of extraction and monitoring well construction; vacuum blowers (positive displacement or centrifugal); vapor and liquid treatment systems piping, valves, and fittings (usually plastic); and instrumentation [31]. Operations and maintenance costs include labor, power, maintenance, and monitoring activities. Off-gas and collected ground-water treatment are the largest cost items in this list; the cost of a cleanup can double if both are treated with activated carbon. Electric power costs vary by location (i.e., local utility rates and site conditions). They may be as low as 1 percent or as high as 2 percent of the total project cost.

Caution is recommended in using these costs out of context, because the base year of the estimates vary. Costs also are highly variable due to site variations as well as soil and contaminant

Site	Location (Region)	Primary Contaminants	Status
Groveland Wells 1 & 2	Groveland, MA (1)	TCE	SITE demonstration complete [2; 7] Full-scale remediation in design
Kellogg-Deering Well Field	Norwalk, CT (1)	PCE, TCE, and BTX	Pre-design [3; 5; 6]
South Municipal Water Supply Well	Peterborough, NH (1)	PCE, TCE, Toluene	Pre-design completion expected in the fall of 1991 [3; 5; 6]
Tinkhams Garage	Londonderry, NH (1)	PCE, TCE	Pre-design pilot study completed [26; 27]
Wells G & H	Woburn, MA (1)	PCE, TCE	In design [3; 5]
FAA Technical Center	Atlantic County, NJ (2)	BTX, PAHs, Phenols	In design [3; 5]
Upjohn Manufacturing Co.	Barceloneta, PR (2)	CCl_4	Project completed in 1988 [20; 21]
Allied Signal Aerospace-Bendix Flight Systems Div.	South Montrose, PA (3)	TCE	Pre-design tests and dewatering study completed [28]
Henderson Road	Upper Merion Township, PA (3)	PCE, TCE, Toluene, Benzene	Pre-design [3; 4]
Tyson's Dump	Upper Merion Township, PA (3)	PCE, TCE, Toluene, Benzene, Trichloropropane	In operation (since 11/88) [23]
Stauffer Chemical	Cold Creek, AL (4)	CCl_4, pesticides	Pre-design [5; 6]
Stauffer Chemical	Lemoyne, AL (4)	CCl_4, pesticides	Pre-design [5; 6]
Sodyeco	Mt. Holly, NC (4)	PCE, PAHs	Design approved [29]
Kysor Industrial	Cadillac, MI (5)	PCE, TCE, Toluene, Xylene	In design; pilot studies in progress [3; 5; 6]
Long Prairie	Long Prairie, MN (5)	PCE, TCE, DCE, Vinyl chloride	SVE construction expected in the fall of 1991 [3; 6]
MIDCO 1	Gary, IN (5)	BTX, TCE, Phenol, Dichloromethane, 2-Butanone, Chlorobenzene	In design [3; 5; 6]
Miami County Incinerator	Troy, OH (5)	PCE; TCE; Toluene	Pre-design [3; 5; 6]
Pristine	Cincinnati, OH (5)	Benzene; Chloroform; TCE; 1,2-DCA; 1,2-DCE	Pre-design [3; 6]
Seymour Recycling	Seymour, IN (5)	TCE; Toluene; Chloromethane; cis-1,2-DCE; 1,1,1-DCA; Chloroform	Pre-design investigation completed [32]
Verona Well Field	Battle Creek, MI (5)	PCE, TCA	Operational since 3/81 [22]

Table 8-4. Continued

Site	Location (Region)	Primary Contaminants	Status
Wausau Groundwater Contamination	Wausau, WI (5)	PCE, TCE	Pre-design [3; 5; 6]
South Valley/General Electric	Albuquerque, NM (6)	Chlorinated solvents	Pilot studies scheduled for summer of 1991 [4; 6]
Hastings Groundwater Contamination	Hastings, NE (7)	CCl₄ Chloroform	Pilot studies completed for Colorado Ave. & Far-Marco subsites [24]
Sand Creek Industrial	Commerce City, CO (8)	PCE, TCE, pesticides	Pilot study completed [33]
Fairchild Semiconductor	San Jose, CA (9)	PCE, TCA, DCE, DCA, Vinyl chlorides, Phenols, and Freon	Operational since 1988, currently conducting resaturation studies [25]
Fairchild Semiconductor/MTV-1	Mountain View, CA (9)	PCE, TCA, DCE, DCA, Vinyl chlorides, Phenols, and Freon	Pre-design [3; 5]
Fairchild Semiconductor/MTV-2	Mountain View, CA (9)	PCE, TCA, DCE, DCA, Vinyl chlorides, Phenols, and Freon	Pre-design [3; 5]
Intel Corporation	Mountain View, CA (9)	PCE, TCA, DCE, DCA, Vinyl chlorides, Phenols, and Freon	Pre-design [3; 5]
Raytheon Corporation	Mountain View, CA (9)	PCE, TCA, DCE, DCA, Vinyl chlorides, Phenols, and Freon	Pre-design [3; 5]
Motorola 52nd Street	Phoenix, AZ (9)	TCA, TCE, CCl₄ Ethylbenzene	Pre-design [3; 4; 6]
Phoenix-Goodyear Airport Area (also Litchfield Airport Area)	Goodyear, AZ (9)	TCE, DCE, MEK	North Unit - In design [34] South Unit - pilot study completed

characteristics that impact the SVE process. As contaminant concentrations are reduced, the cost-effectiveness of an SVE system may decrease with time.

EPA CONTACT

Technology-specific questions regarding SVE may be directed to Michelle Simon, RSKERL-Cincinnati, (513) 569-7469 (see Preface for mailing address).

ACKNOWLEDGMENTS

This chapter was prepared for the U.S. Environmental Protection Agency, Office of Research and Development (ORD), Risk Reduction Engineering Laboratory (RREL), Cincinnati, Ohio, by Science Applications International Corporation (SAIC), and Foster Wheeler Enviresponse Inc. (FWEI) under contract No. 68-C8-0062. Mr. Eugene Harris served as the EPA Technical Project Monitor. Gary Baker was SAIC's Work Assignment Manager. This chapter was authored by Mr. Pete Michaels of FWEI. The author is especially grateful to Mr. Bob Hillger and Mr. Chi-Yuan Fan of EPA, RREL, who have contributed significantly by serving as technical consultants during the development of this chapter.

The following other Agency and contractor personnel have contributed their time and comments by participating in the expert review meetings and/or peer-reviewing the document:

Dr. David Wilson, Vanderbilt University
Dr. Neil Hutzler, Michigan Technological University
Mr. Seymour Rosenthal, FWEI
Mr. Jim Rawe, SAIC
Mr. Clyde Dial, SAIC
Mr. Joe Tillman, SAIC

REFERENCES

1. Cheremesinoff, P.N. Solvent Vapor Recovery and VOC Emission Control. Pollution Engineering, 1986.
2. Records of Decision System Database, Office of Emergency and Remedial Response, U.S. Environmental Protection Agency, 1989.
3. Innovative Treatment Technologies: Semi-Annual Status Report. EPA/540/2-91/001, January 1991. [Annual Status Report (Sixth Edition) was published as EPA 542-R-94-005, September, 1994.]
4. ROD Annual Report, FY 1988. EPA/540/8-89/006, July 1989.
5. ROD Annual Report, FY 1989. EPA/540/8-90/006, April 1990.
6. Personal Communications with Regional Project Managers, April 1991.
7. Applications Analysis Report—Terra Vac In Situ Vacuum Extraction System. EPA/540/A5-89/003, 1989. (SITE Report).
8. CH2M Hill, Inc. Remedial Planning/Field Investigation Team. Verona Well Field-Thomas Solvent Co. Operable Unit Feasibility Study. U.S. Environmental Protection Agency, Chicago, Illinois, 1985.
9. Payne, F.C., et al. In Situ Removal of Purgeable Organic Compounds from Vadose Zone Soils. Presented at Purdue Industrial Waste Conference, May 14, 1986.
10. Hutzler, N.J., B.E. Murphy, and J.S. Gierke. State of Technology Review: Soil Vapor Extraction Systems. EPA/600/2-89/024 (NTIS PB89-195184), 1989.
11. Johnson, P.C., et al. A Practical Approach to the Design, Operation, and Monitoring of In Situ Soil Venting Systems. Groundwater Monitoring Review, Spring, 1990.
12. Technology Screening Guide for Treatment of CERCLA Soils and Sludges. EPA/540/2-88/004 (NTIS PB89-132674), 1988, pp. 86–89.
13. Superfund LDR Guide #6A: Obtaining a Soil and Debris Treatability Variance for Remedial Actions. OSWER Directive 9347.3-06FS, (NTIS PB91-921327), 1990.
14. Superfund LDR Guide #6B: Obtaining a Soil and Debris Treatability Variance for Removal Actions. OSWER Directive 9347.3-07FS, (NTIS PB91-921310), 1990.

15. Michaels, P.A. and M.K. Stinson. Terra Vac In Situ Vacuum Extraction Process SITE Demonstration. In: Proceedings of the Fourteenth Annual Research Symposium. EPA/600/9-88/021, 1988.

16. Mutch, R.D., Jr., A.N. Clarke, and D.J. Wilson. In Situ Vapor Stripping Research Project: A Progress Report—Soil Vapor Extraction Workshop. USEPA Risk Reduction Engineering Laboratory, Releases Control Branch, Edison, New Jersey, 1989.

17. Ellgas, R.A. and N.D. Marachi. Vacuum Extraction of Trichloroethylene and Fate Assessment in Soils and Groundwater: Case Study in California. Joint Proceedings of Canadian Society of Civil Engineers ASCE National Conferences on Environmental Engineering, 1988.

18. Groundwater Technology Inc., Correspondence from Dr. Richard Brown.

19. Midwest Water Resource, Inc., Correspondence from Dr. Frederick C. Payne.

20. Geotec Remedial Investigation Report and Feasibility Study for Upjohn Manufacturing Co. Barceloneta, Puerto Rico, 1984.

21. Geotec Evaluation of Closure Criteria for Vacuum Extraction at Tank Farm. Upjohn Manufacturing Company, Barceloneta, Puerto Rico, 1984.

22. CH2M Hill, Inc. Performance Evaluation Report Thomas Solvent Raymond Road Operable Unit. Verona Well Field Site, Battle Creek, MI, April 1991.

23. Terra Vac Corporation. An Evaluation of the Tyson's Site On-Site Vacuum Extraction Remedy Montgomery County, Pennsylvania, August 1990.

24. IT Corporation. Final Report-Soil Vapor Extraction Pilot Study, Colorado Avenue Subsite, Hastings Ground Water Contamination Site, Hastings, Nebraska, August, 1990.

25. Canonie Environmental. Supplement to Proposal to Terminate In-Situ Soil Aeration System Operation at Fairchild Semiconductor Corporation's Former San Jose Site, December 1989.

26. Malcom Pirnie, Tinkhams Garage Site, Pre-Design Study, Londonderry, New Hampshire—Final Report, July 1988.

27. Terra Vac Corp., Tinkhams Garage Site Vacuum Extraction Pilot Test, Londonderry, New Hampshire, July 20, 1988.

28. Environmental Resources Management, Inc. Dewatering Study for the TCE Tank Area—Allied Signal Aerospace, South Montrose, PA, December 1990.

29. Letter Correspondence from Sandoz Chemicals Corporation to the State of North Carolina Department of Environmental Health and Natural Resources, RE: Remediation Activities in CERCLA C Area (Sodeyco) Superfund Site, March 28, 1991.

30. Federal Register, Volume 50, No. 229, Wednesday, November 27, 1985, pp. 48886-48910.

31. Assessing UST Corrective Action Technologies: Site Assessment and Selection of Unsaturated Zone Treatment Technologies. EPA/600/2-90/011, U.S. Environmental Protection Agency, 1990.

32. Hydro Geo Chem, Inc. Completion Report, Pre-Design Investigation for a Vapor Extraction at the Seymour Site, Seymour, Indiana, February 1990.

33. Groundwater Technology, Inc. Report of Findings, Vacuum Extraction Pilot Treatability at the Sand Creek Superfund Site (OU-1), Commerce City, Colorado, March 1990.

34. Hydro Geo Chem, Inc. Results and Interpretation of the Phoenix Goodyear Airport SVE Pilot Study, Goodyear, Arizona, May 1989.

ADDITIONAL REFERENCES

Loden, M.E. 1992. A Technology Assessment of Soil Vapor Extraction and Air Sparging. EPA/600/R-92/173 (NTIS PB93-100154).

Pedersen, T.A. and J.T. Curtis. 1991. Soil Vapor Extraction Technology Reference Handbook. EPA/540/2-91/003 (NTIS PB91-168476).

U.S. Environmental Protection Agency. 1991. AWD Technologies Integrated AquaDetox/SVE Technology Applications Analysis Report. EPA/540/A5-91/002. (SITE Report)

U.S. Environmental Protection Agency. 1994. Soil Vapor Extraction (SVE) Treatment Technology Resource Guide. EPA/542-B-94-007.

U.S. Environmental Protection Agency. 1995. Subsurface Volatilization and Ventilation System (SVVS) Innovative Technology Evaluation Report. EPA/540/R-94/529. (SITE Report).

Chapter 9

Evaluation of Soil Venting Application[1]

Dominic C. DiGiulio, U.S. EPA Robert S. Kerr Environmental Research Laboratory, Ada, OK

INTRODUCTION

The ability of soil venting to inexpensively remove large amounts of volatile organic compounds (VOCs) from contaminated soils is well established. However, the time required using venting to remediate soils to low contaminant levels often required by state and federal regulators has not been adequately investigated. Most field studies verify the ability of a venting system to circulate air in the subsurface and remove, at least initially, a large mass of VOCs. They do not generally provide insight into mass transport limitations which eventually limit performance, nor do field studies generally evaluate methods such as enhanced biodegradation which may optimize overall contaminant removal. Discussion is presented to aid in evaluating the feasibility of venting application. Methods to optimize venting application arc also discussed.

DETERMINING CONTAMINANT VOLATILITY

The first step in evaluating the feasibility of venting application at a hazardous waste site is to assess contaminant volatility. If concentrations of VOCs in soil are relatively low and the magnitude of liquid hydrocarbons present in the soil is negligible, VOCs can be assumed to exist in a three-phase system (i.e., air, water, and soil), as illustrated in Figure 9-1. If soils are sufficiently moist, relative volatility in a three-phase system can be estimated using Equation 1 which incorporates the effects of air-water partitioning (Henry's constant) and sorption (soil-water partition coefficient).

$$\frac{C_g}{C_t} = \frac{1}{(\rho K_{oc} f_{oc} / K_h) + \theta / K_h + \phi} \tag{1}$$

where C_g = vapor concentration of VOCs in gas phase (mg/cm3 air)
C_t = total volatile organic concentration (mg/cm^3 soil)
ρ_g = bulk density (g/cm^3)
K_{oc} = organic carbon-water partition coefficient (cm^3/g)
f_{oc} = fraction of organic carbon content (g/g)
K_h = Henry's constant (mg/cm^{3}air/mg/cm^{3}water)
θ = volumetric moisture content (cm^3/cm^3)
ϕ = volumetric air content (cm^3/cm^3)

Caution must be exercised when using this approach since this relationship is based on the assumption that solid phase sorption is dominated by natural organic carbon content. This assumption is frequently invalid in soils below the root zone where soil organic carbon is less than 0.1%.

Equation 1 can be used to evaluate individual VOC contaminant reduction trends and attainment of soil-based remediation standards. Vapors should be collected from dedicated vapor probes under static (venting system not operating) conditions. This estimate is valid only for soils in the immediate vicinity of the probe intake. This approach minimizes sample dilution and collection of vapor

---

[1] EPA/540/S-92/004.

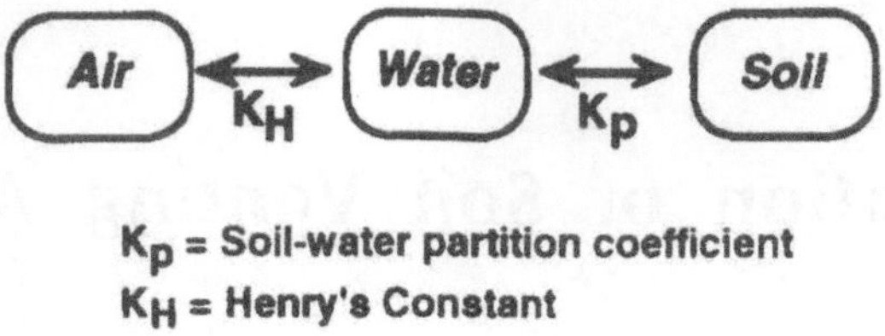

K_p = Soil-water partition coefficient
K_H = Henry's Constant

Figure 9-1. Three phase system.

samples under nonequilibrium conditions. However, it necessitates periodic cessation of venting. When the vapor concentration for a VOC approaches a corresponding total soil concentration, actual soil samples can be collected to confirm remediation. This approach has several benefits over conventional soil samples collection and analysis. At lower VOC concentration levels, collection of static vapor samples is likely to be more sensitive than soil collection and analysis, due to VOC loss in the latter procedure. Siegrist and Jenssen (1990) demonstrated substantial VOC loss during normal soil sample collection, storage, and analysis. Also, comparing contaminant reduction trends strictly with soil samples is difficult due to spatial variability in soils. No two soil samples can be collected at the exact same location. In addition, soil gas analyses can be accomplished more quickly and inexpensively than soil sample collection, thus enabling more frequent evaluation of trends. A potential disadvantage of using this approach is inability to distinguish VOC vapors emanating from soils as opposed to ground water. Hypothetically, soils could be remediated to desired levels with probes still indicating contamination above remediation standards. This concern could be alleviated to some degree by determining the presence of a diffusion vapor gradient from the water table using vertically placed vapor probes.

If soils are visibly contaminated or the presence of nonaqueous phase liquids (NAPLs) is suspected in soils based on high contaminant, total organic carbon, or total petroleum hydrocarbon analysis, contaminants are likely present in a four-phase system as illustrated in Figure 9-2. Under these circumstances, most of the VOC mass will be associated with the immiscible fluid, and assuming that the fluid acts as an ideal solution, volatilization will be governed by Raoult's law.

$$P_a = X_a P_a^o \qquad (2)$$

where P_a = vapor pressure of component over solution (mm Hg)
  X_a = mole fraction of component in solution
  P^o_a = saturated vapor pressure of pure component (mm Hg)

In a four-phase system, contaminant volatility will be governed by the VOC's vapor pressure and mole fraction within the immiscible fluid. The vapor pressure of all compounds increases substantially with an increase in temperature, while solubility in a solvent phase is much less affected by temperature. This suggests that soil temperature should be taken into account when evaluating VOC recovery for contaminants located near the soil surface (seasonal variations in soil temperature quickly dampen with depth). For instance, if conducting a field test to evaluate potential remediation of shallow soil contamination in the winter, one should realize that VOC recovery could be substantially higher during summer months, and low recovery should not necessarily be viewed as venting system failure.

As venting proceeds, lower molecular weight organic compounds will preferentially volatilize and degrade. This process is commonly described as weathering and has been examined by Johnson (1989) in laboratory experiments. Samples of gasoline were sparged with air and the concentration and composition of vapors were monitored. The efficiency of vapor extraction decreased to less than 1% of its initial value even though approximately 40% of the gasoline remained. Theoretical and experimental work on product weathering indicate the need to monitor temporal variation in specific VOCs of concern in extraction and observation wells.

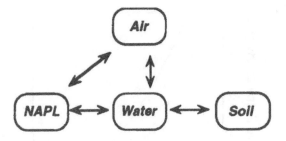

Figure 9-2. Four phase system.

EVALUATING AIR FLOW

Air permeability (k_a) in soil is a function of a soil's intrinsic permeability (k_i) and liquid content. At hazardous waste sites, liquid present in soil pores is often a combination of soil water and immiscible fluids. Air permeability (k_a) can be estimated by multiplying a soil's intrinsic permeability (k_i) by the relative permeability (k_r).

$$k_a = k_i\, k_r \tag{3}$$

The dimensionless ratio k_r varies from one to zero and describes the variation in air permeability as a function of air saturation. Equations developed by Brooks and Corey (1964) and Van Genuchten (1980) are useful in estimating air permeability as a function of air saturation or liquid content. The Brooks-Corey equation to estimate relative permeability of a nonwetting fluid (i.e., air) is given by:

$$k_r = (1 - S_e)^2\,(1 - S_e^{(2+\lambda)/\lambda}) \tag{4}$$

where S_e = effective saturation
       λ = a pore distribution parameter

The effective saturation is given by:

$$S_e = \frac{\left(\dfrac{\theta}{\varepsilon} + \dfrac{\theta_r}{\varepsilon}\right)}{\left(1 - \dfrac{\theta_r}{\varepsilon}\right)} \tag{5}$$

where θ = volumetric moisture content
       ε = total porosity
       θ_r = residual saturation

The pore size distribution parameter and residual water content can be estimated using soil-water characteristic curves which relate matric potential to volumetric water content. When initially developing an estimate of relative permeability for a given soil texture and liquid content, values for ε, θ_r, S_e, and λ can be obtained from the literature. Rawls et al. (1982) summarized geometric and arithmetic means for Brooks-Corey parameters for various USDA soil textural classes. Figure 9-3 illustrates relative permeability as a function of volumetric moisture content for clayey soils assuming $\varepsilon = 0.475$, $\theta_r = 0.090$, and $\lambda = 0.131$.

The most effective method of measuring air permeability is by conducting a field pneumatic pump test. Using permeameters or other laboratory measurements provides information on a relatively small scale. Information gained from pneumatic pump tests is vital in determining site-specific design considerations (e.g., spacing of extraction wells). Selection of the placement and screened

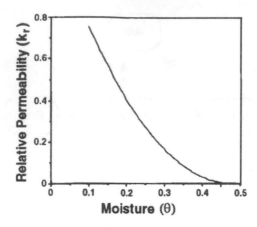

Figure 9-3. Relative permeability vs. moisture content of clay.

intervals of extraction and observation wells and applied vacuum rates during a pump test is often based on preliminary mathematical modeling.

EVALUATING MASS TRANSFER LIMITATIONS AND REMEDIATION TIME

The effects of mass transport limitations are usually manifested by a substantial drop in soil vapor contaminant concentrations as illustrated in Figure 9-4 or by an asymptotic increase in total mass removal with operation time. Typically, when venting is terminated, an increase in soil gas concentration is observed over time. Slow mass transfer with respect to advective air flow is most likely caused by diffusive release from porous aggregate structures or lenses of lesser permeability, as illustrated in Figure 9-5. The time required for the remediation of heterogeneous and fractured soils depends on the proportion of contaminated material exposed to direct bulk airflow. It would be expected that long-term performance of venting would be limited to a large degree by gaseous and liquid diffusion from soil regions not exposed to direct airflow.

Regardless of possible causes, the significance of mass transport limitations should be evaluated during venting field tests. This can be achieved by pneumatically isolating a small area of a site and aggressively applying vacuum extraction until mass transport limitations are realized. Isolation can be achieved by surrounding extraction wells with passive inlet or air injection wells as shown in Figure 9-6. Quantifying the effects of mass transport limitations on remediation time might then be attempted by utilizing models incorporating mass transfer rate coefficients.

The discrepancy frequently observed between mass removal predicted from equilibrium conditions using Henry's law constants and that observed from laboratory column and field studies is sometimes reconciled by the use of "effective or lumped" soil-air partition coefficients. These parameters are determined from laboratory column tests and are then used for model input to determine required remediation times. While this method does indirectly account for mass transport limitations, problems may arise when one attempts to quantitatively describe several processes with lumped parameters. The primary concern is whether the lumped parameter is suitable for use only under the laboratory conditions from which it was determined, or whether it can be transferred for modeling use in the field. Perhaps the most direct method of accounting for mass transport limitations would be to incorporate diffusive transfer directly into convective-dispersive vapor transport models.

ENHANCED AEROBIC BIODEGRADATION

With the exception of a few field research projects, soil vacuum extraction has been applied primarily for removal of volatile organic compounds from the vadose zone. However, circulation of air in soils can be expected to enhance the aerobic biodegradation of both volatile and semivolatile

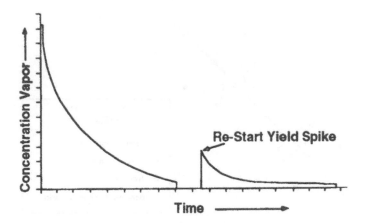

Figure 9-4. Concentration vs. time.

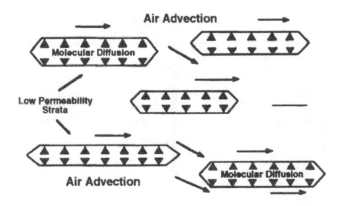

Figure 9-5. Effect of low permeability lenses. (Palmer et al., 1988.)

organic compounds. One of the most promising uses of this technology is in manipulating subsurface oxygen levels to maximize in-situ biodegradation. Bioventing can reduce vapor treatment costs and can result in the remediation of semivolatile organic compounds which cannot be removed by physical stripping alone.

Venting circulates air in soils at depths much greater than are possible by tilling, and oxygen transport via the gas phase is much more effective than injecting or flooding soils with oxygen saturated liquid solutions.

Hinchee (1989) described the use of soil vacuum extraction at Hill AFB, Utah for oxygenation of the subsurface and the enhancement of biodegradation of petroleum hydrocarbons in soils contaminated with JP-4 jet fuel. Figures 9-7 and 9-8 illustrate subsurface oxygen profiles at the Hill site prior to and during venting. It is evident that soil oxygen levels dramatically increased following one week of venting. Soil vapor samples collected from observation wells during periodic vent system shutdown revealed rapid decreases in oxygen concentration and corresponding CO_2 production suggesting that aerobic biodegradation was occurring at the site. Laboratory treatability studies using soils from the site demonstrated increased carbon-dioxide evolution with increasing moisture content when enriched with nutrients. It is worthwhile to note that soils at Hill AFB were relatively dry at commencement of field vacuum extraction, indicating that the addition of moisture could perhaps stimulate aerobic biodegradation even further under field operating conditions.

When conducting site characterization and field studies, it is recommended that CO_2 and O_2 levels be monitored in soil vapor probes and extraction well off-gas to allow the assessment of basal soil respiration and the effects of site management on subsurface biological activity. These measurements are simple and inexpensive to conduct and can yield a wealth of information regarding:

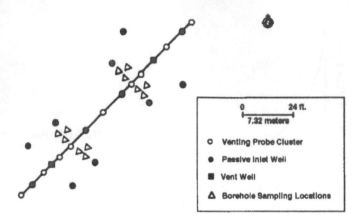

Figure 9-6. Proposed pilot test design.

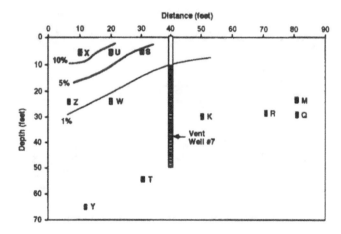

Figure 9-7. Oxygen concentration in vadose zone before venting.

1. The mass of VOCs and semivolatiles which have undergone biodegradation versus volatilization. This information is crucial if subsurface conditions (e.g., moisture content) are to be manipulated to enhance biodegradation to reduce VOC off-gas treatment costs and maximize semivolatile removal.
2. Factors limiting biodegradation. If O_2 and CO_2 monitoring reveals low O_2 consumption and CO_2 generation while readily biodegradable compounds persist in soils, further characterization studies could be conducted to determine if biodegradation is being limited by insufficient moisture content, toxicity (e.g., metals), or nutrients.
3. Subsurface air flow characteristics. Observation wells which indicate persistent low O_2 levels may indicate an insufficient supply of oxygenated air at that location, suggesting the need for air injection, higher extraction well vacuum, additional extraction wells, or additional soils characterization, which may indicate high moisture content or the presence of immiscible fluids impeding the flow of air.

LOCATION AND NUMBER OF VAPOR EXTRACTION WELLS

One of the primary objectives in conducting a venting field test is to evaluate the initial placement of extraction wells to optimize VOC removal from soil. Placement of extraction wells and selected applied vacuum is largely an iterative process requiring continual reevaluation as additional

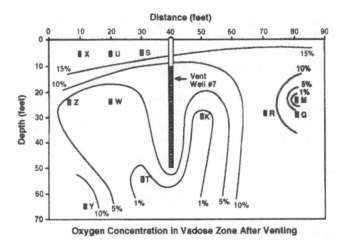

Figure 9-8. Oxygen concentration in vadose zone after venting.

data are collected during remediation. Vacuum extraction wells produce complex three-dimensional reduced pressure zones in affected soils. The size and configuration of this affected volume depends on the applied vacuum, venting geometry (e.g., depth to water table), soil heterogeneity, and intrinsic (e.g., permeability) and dynamic (e.g., moisture content) properties of the soil. The lateral extent of this reduced pressure zone (beyond which static vacuum is no longer detected) is often termed the radius or zone of influence (ROI). Highly permeable sandy soils typically exhibit large zones of influence and high air flow rates, whereas less permeable soils, such as silts and clays, exhibit smaller zones of influence and low air flows.

Measured or anticipated Radii of Influence are often used to space extraction wells. For instance, if a ROI is measured at 10 feet, extraction wells are placed 20 feet apart. However, this strategy is questionable since vacuum propagation and air velocity decrease substantially with distance from an extraction well. Thus, only a limited volume of soil near an extraction well will be effectively ventilated, regardless of the ROI. Johnson (1988) describes how the addition of 13 extraction wells within the ROI of other extraction wells increased blower VOC concentration by 4000 ppmv and mass removal by 40 kg/day. He concluded that the radius of influence was not an effective parameter for locating extraction wells and that operation costs could be reduced by increasing the number of extraction wells, as opposed to pumping at higher rates with fewer wells.

Determining the propagation of induced vacuum requires conducting pneumatic pump tests in which variation in static vacuum is measured in vapor observation wells at depth and distance from extraction wells. Locating extraction and observation wells along transects as illustrated in Figure 9-6 minimizes the number of observation wells necessary to evaluate vacuum propagation at linear distances from extraction wells. Pressure differential can be observed at greater distances than would otherwise be possible in other configurations.

Propagation of vacuum in soils as a function of applied vacuum can be determined by conducting pneumatic pump tests with incrementally increasing flow or applied vacuum. Vacuum is increased after steady state conditions (relatively constant static vacuum measurements in observation wells) exist in soils from the previously applied vacuum. A step pump test will indicate a significant increase in static vacuum or air velocity with increasing applied vacuum near an extraction well. However, at distance from an extraction well, a significant increase in static vacuum will not be observed with an increase in applied vacuum. Pneumatic pump tests allow determination of radial distances from extraction wells in which air velocity is sufficient to ensure remediation.

After the initial placement of extraction wells has been established based on the physics of air flow, an initial applied vacuum must be selected to ensure optimal VOC removal. In regard to mass transfer considerations, the vent rate should be increased if a significant corresponding mass flux is observed. Even though an increased venting rate may not substantially increase the propagation of vacuum with distance, air velocity will increase near the extraction well. If most contaminants are in more permeable deposits, an increase in applied vacuum will increase mass removal eventually to a point of diminishing returns or until the system is limited by diffusion. Note that this strategy is for optimization of volatilization, not biodegradation. Optimizing in-situ biodegradation often necessitates reducing air velocity in soil. As a result, vapor treatment costs are minimized but overall mass flux decreases Thus, in-situ biodegradation of VOCs minimizes overall costs but may extend venting operation time.

During a field test, it is desirable to operate until mass transport limitations are realized in order to evaluate the long-term performance of the technology. This can be achieved by isolating small selected areas of a site by the use of passive air inlet wells. When attempting to evaluate diffusion limited mass removal in isolated areas, applied vacuum should remain high and the distance between passive inlet and extraction wells should be minimized. Too often venting field tests are conducted for relatively short periods of time (e.g., 2–21 days), which only results in assessment of air permeability and initial mass removal. Longer field studies (e.g., 6 months–12 months) enable better insight into mass transfer limitations which eventually govern venting effectiveness.

SCREENED INTERVAL

The screened interval of extraction wells will play a significant role in directing air flow through contaminated soils. Minimum depths are recommended by some practitioners for venting operation to avoid short-circuiting of air flow. However, the application of venting need not be limited by depth to water table since horizontal vents can be used in lieu of vertically screened extraction wells to remediate soils with shallow contamination. Often, it is desirable to dewater contaminated shallow aquifer sediments for venting application. For remediation of more permeable soils with deep contamination, an extraction well should be screened at the maximum depth of contamination or to the seasonal low water table, whichever is shallowest, to direct air flow and reduce short-circuiting. For less permeable soils, or for more continuous vertical contamination, a higher and longer screened interval may be useful. In stratified systems, such as in the presence of clay layers between more permeable deposits, more than one well will be required, each venting a distinct strata. Screening an extraction well over two strata of significantly different permeability will result in most air flow being directed only in the strata of greater permeability. It is important to screen extraction wells over the interval of highest soil contamination to avoid extracting higher volumes of air at lower vapor concentration.

During venting, the reduced pressure in the soil will cause an upwelling of the water table. The change in water table elevation can be determined from the predicted radial pressure distribution. Johnson et al. (1988) indicated that upwelling can be significant under typical venting conditions. Water table rise will cause contaminated soil lying above the water table to become saturated, resulting in decreased mass removal rates. Ground-water upwelling due to venting system operation can be minimized with concurrent water table dewatering.

PLACEMENT OF OBSERVATION WELLS

Observation wells are essential in determining whether contaminated soils are being effectively ventilated and in the evaluation of interactions among extraction wells. The more homogeneous and isotropic the unsaturated medium, the fewer the number of vapor monitoring probes required. To adequately describe vacuum propagation during a field test, usually at least three observation well clusters are needed within the ROI of an extraction well. At least one of these clusters should be placed near an extraction well because of the logarithmic decrease in vacuum with distance. The depth and number of vapor probes within a cluster depends on the screened intervals of extraction wells and soil stratigraphy. However, vertical placement of vapor probes might logically be near the soil-water table interface, soil horizon interfaces, and near the soil surface. As previously mentioned, the use of air flow modeling can assist in optimizing the depth and placement of vapor observation wells and in the interpretation of data collected from these monitoring points.

When constructing observation wells it is desirable to minimize vapor storage volume in the screened interval and sample transfer line. This will minimize purging volumes and ensure a representative vapor sample in the vicinity of each observation well. Analysis of soil gas in an onsite field laboratory is preferred to provide real time data for implementation of engineering controls and process modifications. It is recommended that steel canisters, sorbent tubes, or direct

GC injection be used in lieu of Tedlar bags when possible because of potential VOC loss through bag leakage or diffusion within the Teflon material itself. This problem may lead to erroneous analytical results and the potential of a false negative indication of soil remediation at low soil gas concentrations.

SUMMARY/CONCLUSIONS

While the application of soil vacuum extraction is conceptually simple, its success depends on understanding complex subsurface physical, chemical, and biological processes which provide insight into factors limiting venting performance. Optimizing venting performance is critical when attempting to meet stipulated soil-based cleanup levels required by regulators. The first step in evaluating a venting application is to assess contaminant volatility. Volatility is a function of a contaminant's soil-water partition coefficient and Henry's constant if present in a three-phase system, and a contaminant's vapor pressure and mole fraction in an immiscible fluid, if present in a four-phase system. Volatility is greatly decreased when soils are extremely dry. As vacuum extraction proceeds, lower molecular weight organic compounds preferentially volatilize and biodegrade. Decreasing mole fractions of lighter compounds and increasing mole fractions of heavier compounds affects observed off-gas concentrations. Understanding contaminant volatility is necessary when attempting to utilize off-gas vapor concentrations as an indication of venting progress.

The significance of mass transport limitations should be evaluated during venting field tests. Long-term performance of venting will most likely be limited by diffusion from soil regions of lesser permeability which are not exposed to direct airflow. Mass transport limitations can be assessed by isolating a small area of a site and aggressively applying vacuum extraction. Simplistic methods to evaluate remediation time should be avoided. One of the most promising uses of vacuum extraction is in manipulating subsurface oxygen levels to enhance biodegradation. When conducting field studies, it is recommended that CO_2 and O_2 levels be monitored in vapor probes to evaluate the feasibility of VOC and semivolatile contaminant biodegradation.

Air permeability in soil is a function of a soil's intrinsic permeability and liquid content. Relative permeability of air can be estimated using relationships developed by Brooks and Corey (1964) and Van Genuchten (1980). The most effective method of measuring air permeability is by conducting pneumatic pump tests. Information gained from pneumatic pump tests can be used to determine site-specific design considerations such as the spacing of extraction wells. Measured or anticipated zones of influence are not particularly useful in spacing extraction wells. Extraction wells should be located to maximize air velocity in contaminated soils. Pneumatic pump tests with increasing applied vacuum may be useful in determining radial distances from extraction wells in which air velocity is sufficient to ensure remediation.

Screened intervals should be located at or below the depth of contamination. In stratified soils, more than one well is necessary to ventilate each strata. At least three observation well clusters are usually necessary to observe vacuum propagation within the radius of influence of an extraction well. Logical vertical placement of vapor probes might be near the soil-water table interface, soil horizon interfaces, and near the soil surface.

EPA CONTACT

For further information contact Dominic DiGiulio, RSKERL-Ada, (405) 436-8605 (see Preface for mailing address).

REFERENCES

Brooks, R.H. and A.T. Corey. 1964. Hydraulic Properties of Porous Media. Colorado State University, Fort Collins, CO, Hydrol. Pap. No. 3, 27 pp.

Hinchee, R.E. 1989. Enhanced Biodegradation through Soil Venting. In: Proceedings of the Workshop on Soil Vacuum Extraction, Robert S. Kerr Environmental Research Laboratory, Ada, Oklahoma, April 27–28, 1989.

Johnson, J.J. 1988. In Situ Air Stripping: Analysis of Data from a Project Near Benson, Arizona, Master of Science Thesis, Colorado School of Mines, Colorado.

Johnson, P.C., M.W. Kemblowski, and J.D. Colthart. 1988. Practical Screening Models for Soil Venting Applications. In: NWWA/API Conference on Petroleum Hydrocarbons and Organic Chemicals in Groundwater, National Water Well Association, Dublin, OH, pp. 521-546.

Johnson, R.L. 1989. Soil Vacuum Extraction: Laboratory and Physical Model Studies. In: Proceedings of the Workshop on Soil Vacuum Extraction, Robert S. Kerr Environmental Research Laboratory, Ada, Oklahoma, April 27–28, 1989.

Rawls, W.J., D.L. Brakensiek, and K.E. Saxton. 1982. Estimation of Soil Water Properties. Trans. Am. Soc. Agric. Eng. 25:1316-1320, 1328.

Siegrist, R.L., and P.C. Jenssen. 1990. Evaluation of Sampling Method Effects on Volatile Organic Compound Measurements in Contaminated Soils. Environ. Sci. Technol. 24(9):1387–1392.

Van Genuchten, M.T. 1980. A Closed-Form Equation for Predicting the Hydraulic Conductivity of Unsaturated Soils. Soil Sci. Soc. Am. J. 44:982–898.

Chapter 10

In Situ Steam Extraction Treatment[1]

Kyle Cook, Science Applications International Corporation (SAIC), Cincinnati, OH

ABSTRACT

In situ steam extraction removes volatile and semivolatile hazardous contaminants from soil and ground water without excavation of the hazardous waste. Waste constituents are removed in situ by the technology and are not actually treated. The use of steam enhances the stripping of volatile contaminants from soil and can be used to displace contaminated ground water under some conditions. The resultant condensed liquid contaminants can be recycled or treated prior to disposal. The steam extraction process is applicable to organic wastes but has not been used for removing insoluble inorganics and metals. Steam is injected into the ground to raise the soil temperature and drive off volatile contaminants. Alternatively, steam can be injected to form a displacement front by steam condensation to displace ground water. The contaminated liquid and steam condensate are then collected for further treatment.

In situ steam extraction is a developing technology that has had limited use in the United States. In situ steam extraction is currently being considered as a component of the remedy for only one Superfund site, the San Fernando Valley (Area 1), California site [1][2] [2].[3] However, a limited number of commercial-scale in situ steam extraction systems are in operation. Two types of systems are discussed in this chapter: the mobile system and the stationary system. The mobile system consists of a unit that volatilizes contaminants in small areas in a sequential manner by injecting steam and hot air through rotating cutter blades that pass through the contaminated medium. The stationary system uses steam injection as a means to volatilize and displace contaminants from the undisturbed subsurface. Each system has specific applications; however, the lowest cost alternative will be determined by site-specific considerations. This chapter provides information on the technology applicability, limitations, a description of the technology, types of residuals produced, site requirements, the latest performance data, the status of the technology, and sources for further information.

TECHNOLOGY APPLICABILITY

In situ steam extraction has been shown to be effective in treating soil and ground water containing such contaminants as volatile organic compounds (VOCs) including halogenated solvents and petroleum wastes. The technology has been shown to be effective for extracting soluble inorganics (i.e., acids, bases, salts, heavy metals) on a laboratory scale [3]. The presence of semivolatile organic compounds (SVOCs) does not interfere with extraction of the VOCs [4, p. 12]. This process has been shown to be applicable for the removal of VOCs including chlorinated organic solvents [4, p. 9][5, p. i], gasoline [6, p. 1265], and diesel [7, p. 506]. It has been shown to be particularly effective on alkanes and alkane-based alcohols such as octanol and butanol [8].

---

[1] EPA/540/2-91/005. EPA SITE program reports covering other innovative in situ thermal extraction techniques (radio frequency heating and heated compressed air) are also listed at the end of this chapter.

[2] [Reference number, page number.]

[3] Editor's Note: EPA's 7th Superfund Innovative Technology Evaluation Program Report (EPA/540/R-94/526) reports that mobile steam extraction system described in this chapter developed by Toxic Treatments USA (now NOVATERRA, Inc.), has been successfully used at five other contaminated sites.

Table 10-1. RCRA Codes for Wastes Applicable to Treatment by In Situ Steam Extraction

Spent Halogenated Solvents used in Degreasing	F001
Spent Halogenated Solvents	F002
Spent Nonhalogenated Solvents	F003
Spent Nonhalogenated Solvents	F004
Spent Nonhalogenated Solvents	F005

Steam extraction applies to less volatile compounds than ambient vacuum extraction systems. By increasing the temperature from initial conditions to the steam temperature, the vapor pressures of most contaminants will increase, causing them to become more volatile. Semivolatile components can volatilize at significant rates only if the temperature is increased [3, p. 3]. Steam extraction also may be used to remove low boiling point VOCs more efficiently.

Table 10-1 lists specific Resource Conservation and Recovery Act (RCRA) wastes that are applicable to treatment by this technology. The effectiveness of the two steam extraction systems (mobile and stationary) on general contaminant groups for soil and ground water is shown in Table 10-2. Examples of constituents within contaminant groups are provided in Reference 9, Technology Screening Guide for Treatment of CERCLA Soils and Sludges. Table 10-2 is based on the current available information or professional judgment where no information was available. The proven effectiveness of the technology for a particular site or waste does not ensure that it will be effective at all sites or that the treatment efficiencies achieved will be acceptable at all sites. For the ratings used for this table, demonstrated effectiveness means that, based on treatability studies at some scale, the technology was effective for that particular contaminant and matrix. The ratings of potential effectiveness or no expected effectiveness are based upon expert judgment. Where potential effectiveness is indicated, the technology is believed capable of successfully treating the contaminant group in a particular matrix. When the technology is not applicable or will probably not work for a particular combination of contaminant group and matrix, a no-expected-effectiveness rating is given. The table shows that the stationary system shows potential effectiveness for inorganic and reactive contaminants. This is only true if the compounds are soluble.

LIMITATIONS

Soil with high silt and clay content may become malleable and unstable when wet, potentially causing problems with support and mobility of the mobile steam extraction system. Remediation of low permeability soil (high clay content) requires longer treatment times [4, p. 8]. The soil must be penetrable by the augers and free of underground piping, wiring, tanks, and drums. Materials of this type must be relocated before treatment can commence. Surface and subsurface obstacles greater than 12 inches in diameter (e.g., rocks, concrete, wooden piles, trash, and metal) must be removed to avoid damage to the equipment. Substantial amounts of subsurface obstacles may preclude the use of a mobile system. A climate temperature range of 20–100°F is desirable for best operation of the mobile system [4, p. 18].

Mobile steam extraction systems can treat large contaminated areas but are limited by the depth of treatment. One system that has been evaluated can treat to a depth of 30 feet.

To be effective, the stationary steam extraction system requires a site with predominately medium- to high-permeability soil. Sites with homogeneous physical soil conditions are more amenable to the system. If impermeable lenses of contaminated soil exist, the stationary system may not remediate these areas to desired cleanup levels [5, p. 19]. However, a combination of steam injection followed by vacuum extraction (drying) may be effective on sites with heterogeneous soil conditions [10]. Steam extraction may be effective for remediation of contaminated ground water near the source of contamination [5, p. 14; 10].

There may be residual soil contamination after application of in situ steam extraction. Study of a mobile system showed the average removal efficiency for volatile contaminants was 85%; 15% of the volatile compound contamination remained in the soil [4, p. 4]. If other organic or inorganic contamination exists, the cleaned soil may need subsequent treatment by some other technique (i.e., stabilization).

Table 10-2. Effectiveness of In Situ Steam Extraction on General Contaminant Groups for Soil and Ground Water

| Contaminant Groups | Effectiveness Mobile System | | Stationary System |
	Soil	Ground Water	Soil/Ground Water
Organic	■	▼	■
Halogenated volatiles	▼	▼	▼
Halogenated semivolatiles	■	▼	■
Nonhalogenated volatiles	▼	▼	▼
Nonhalogenated semivolatiles	●	●	▼
PCBs	●	●	▼
Pesticides (halogenated)	●	●	▼
Dioxins/furans	●	●	▼
Organic cyanides	●	●	▼
Organic corrosives	●	●	▼
Inorganic			
Volatile metals	●	●	▼
Nonvolatile metals	●	●	▼
Asbestos	●	●	●
Radioactive materials	●	●	▼
Inorganic corrosives	●	●	▼
Inorganic cyanides	●	●	▼
Reactive			
Oxidizers	●	●	▼
Reducers	●	●	▼

■ Demonstrated Effectiveness: Successful treatability test at some scale completed.
▼ Potential Effectiveness: Expert opinion that technology will work.
● No Expected Effectiveness: Expert opinion that technology will not work.

In situ steam extraction may not remove SVOCs and inorganics effectively. The operational costs of steam extraction are greater than ambient vacuum extraction, but may be offset by higher recovery and/or reduction in time required to remediate the site due to more efficient removal of contaminants.

In situ steam extraction requires boilers to generate steam and a sophisticated process to capture and treat extracted steam and contaminants. Because the mobile system is mechanically complex, its equipment may fail and shut down frequently; however, mechanical problems may be corrected fairly quickly. Equipment failure and shutdown are less frequent for the stationary system.

The increase in soil temperature may adversely affect other soil properties such as microbial populations, although some microbial populations can withstand soil temperatures up to 140°F.

TECHNOLOGY DESCRIPTION

Figure 10-1 is a general schematic of a mobile steam extraction system [4, p. 48]. A process tower supports and controls a pair of cutter blades which bore vertically through the soil. The cutter blades are rotated synchronously in opposite directions during the treatment process to break up the soil and ensure flow-through of gases. Steam (at 400°F) and compressed air (at 275°F) are piped to nozzles located on the cutter blades. Heat from the injected steam and hot air volatilizes the organics. A steel shroud covers the area of soil undergoing treatment. Suction produced by the blower keeps the area underneath the shroud at a vacuum to pull gases from the soil and to protect against leakage to the outside environment. The off-gases are pulled by the blower from the shroud to the treatment train, where water and organics are removed by condensation in coolers. The airstream is then treated by carbon adsorption, compressed, and returned to the soil being treated. Water is

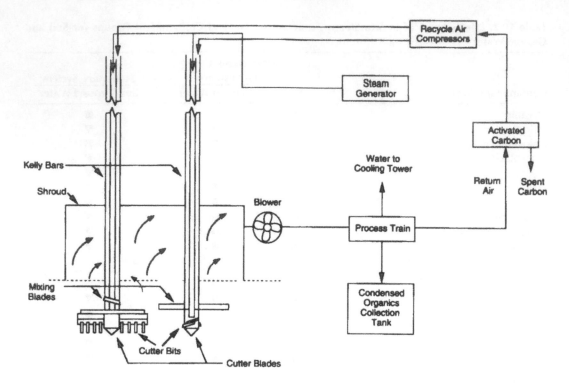

Figure 10-1. Schematic of the mobile steam extraction system.

removed from the liquid stream with a gravity separator, followed by batch distillation and carbon adsorption, and is then recycled to a cooling tower. The condensed organics are collected and held for removal and transportation.

Mobile systems treat small areas of contamination until an entire site is remediated. The action of the cutter blades enables the process to treat low-permeability zones (high clay content) by breaking up the soil. Current systems treat blocks of soil measuring 7'4" × 4' by up to 30' deep.

Figure 10-2 is a schematic of a stationary steam extraction system [5, p. 9]. High-quality steam is delivered through individual valves and flow meters to the injection wells from the manifold. Gases and liquids are removed from the soil through the recovery wells. Gases flow through a condenser and into a separation tank where water and condensed gases are separated from the contaminant phase. Liquid organics are pumped from the separation tank through a meter and into a holding tank. The water may require treatment by carbon adsorption or another process to remove remaining contaminants. Noncondensible gases are passed through activated carbon tanks where contaminants are adsorbed before the cleaned air is vented to the atmosphere. A vacuum pump maintains the subatmospheric pressure on the recovery well and drives the flow of recovered gases. Contaminated liquids are pumped out of the recovery well to a wastewater tank.

PROCESS RESIDUALS

At the conclusion of both processes, the contaminants are recovered as condensed organics in the produced water and on the spent carbon. Residual contamination will also remain in the soil. The recovered contaminants are temporarily stored onsite and may require analysis to determine the need for further treatment before recycling, reuse, or disposal.

Separated, cleaned water is used as cooling tower makeup water in the mobile system. Also in this system, cleaned gas is heated and returned as hot air to the soil. Separated water from the stationary system must be treated to remove residual contaminants before disposal or reuse. The cleaned gas from this system is vented to the atmosphere. Both systems produce contaminated granular activated carbon from the gas cleaning. The carbon must be regenerated or disposed. There may be

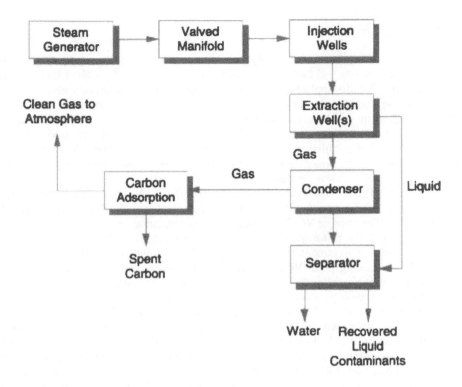

Figure 10-2. Process schematic of the stationary steam extraction system.

minor fugitive emissions of VOCs from the soil during treatment by the steam stripping systems and from the gas-phase carbon beds [4, p. 2].

SITE REQUIREMENTS

Power and telephone lines or other overhead obstacles must be removed or rerouted to avoid conflict with the 30-foot treatment tower on the mobile steam extraction system. Access roads must be available for transporting the mobile system. Sufficient land area must be available around the identified treatment zone to maneuver the unit and to place support equipment and trailers. The area to be treated by the mobile steam extraction system must be capable of supporting the treatment rig so that it does not sink or tip. The ground must be flat and gradable to less than 1% slope. A minimum treatment area of approximately 0.5 acre (20,000 ft²) is necessary for economical use of the mobile system. Rectangular shaped treatment areas are most efficient. The mobile system requires a water supply of at least 8 to 10 gpm at 30 psig. Power for the process can be provided by onboard diesel generators [4, p. 18].

Boilers that generate steam for the stationary steam extraction system use No. 2 fuel oil or other hydrocarbon fuels. Water and electricity must be available at the site. The site must have sufficient room for a drilling rig to install the injection and extraction wells and for steam generation and waste treatment equipment to be set up, as well as room for support equipment and trailers.

Contaminated soils or waste materials are hazardous and their handling requires that a site safety plan be developed to provide for personnel protection and special handling measures. Storage should be provided to hold the process product streams until they have been tested to determine their acceptability for disposal, reuse, or release. Depending on the site, a method to store waste that has been prepared for treatment may be necessary. Storage capacity will depend on waste volume.

Onsite analytical equipment capable of determining site-specific organic compounds for performance assessment make the operation more efficient and provide better information for process control.

PERFORMANCE DATA

Toxic Treatments (USA) Inc. used a prototype of its mobile system to remediate a site in Los Angeles, California. The site soil had been contaminated by diesel and gasoline fuel from underground storage tanks. For this application, the steam stripping was augmented with potassium permanganate to promote oxidation of hydrocarbons in the highly contaminated zones [7, p. 506]. Table 10-3 summarizes the results of the treatment by steam stripping. The level of petroleum hydrocarbons was reduced overall by an average of 91%. The mobile system was reported to have effectively reduced the level of petroleum hydrocarbon compounds found in the soil at a wide range of concentrations. However, the system's ability to remove the higher molecular weight, less volatile components of the diesel fuel was limited.

Under the Superfund Innovative Technology Evaluation (SITE) program, Toxic Treatments demonstrated an average VOC removal rate of 85 percent for a test area of 12 soil blocks [4, p. 10] as shown in Table 10-4. The average VOC post-treatment concentration was 71 ppm; the cleanup level for the site was 100 ppm. The primary VOCs were trichloroethene, tetrachloroethene, and chlorobenzene. The test achieved a treatment rate of 3 cu. yds./hr. in soils having high clay content and containing some high-boiling-point VOCs. Toxic Treatments obtained similar results in tests conducted throughout the site; baseline testing demonstrated an average post-treatment concentration of 61 ppm. The mobile technology also demonstrated the ability to diminish the level of SVOCs by approximately 50%, as shown in Table 10-5, although the fate of these SVOCs could not be determined [4, p. 45]. These tests were conducted on contamination in the unsaturated zone. A follow-up test was conducted on six soil blocks where treatment extended into the saturated zone. Pretreatment data from the vendor indicated significant VOC contamination in this area. Post-treatment results showed that the average level of VOC contamination in the unsaturated zone was reduced to 53 ppm. Ketones (specifically, acetone, 2-methyl-4-pentanone, and 2-butanone) were found to be the primary contaminants in the post-treatment soil. Data from the vendor indicated that similar reduction of VOCs occurred in the saturated zone.

The stationary steam extraction system using steam injection alone decreased soil contaminant concentrations by 90 percent in a recent pilot study [5]. High concentrations of individual contaminants were found in a low permeability zone by use of temperature logs. The residual high contaminant concentrations are thought to have been caused by: 1) retention of highly contaminated steam condensate found ahead of the condensation front in the dry, low-permeability zones and 2) the decreased evaporation rate of the high-boiling-point compounds due to the high water content in the low permeability zones [5, p. 19]. This issue is currently under study at the University of California, Berkeley [10]. Experimental testing has shown that a combination of steam injection and vacuum extraction can effectively remove volatile contaminants from a heterogeneous soil type [10]. Steam injection followed by vacuum extraction produces an effective drying mechanism. The process achieves greater contaminant removals by enhancing the vapor flow from low- to high-permeability regions.

Performance data may be forthcoming from full-scale stationary system steam extraction projects being conducted by Solvent Service, Inc. and Hydro-Fluent, Inc. Data from laboratory-scale studies are also available [6; 3].

RCRA Land Disposal Restrictions (LDRs) that require treatment of wastes to best demonstrated available technology (BDAT) levels prior to land disposal may sometimes be determined to be applicable or relevant and appropriate requirements for CERCLA response actions. The in situ steam extraction technology produces liquid contaminants which may be recyclable or may require treatment to meet treatment levels set by BDAT. A common approach to treating liquid waste may be to use other treatment techniques in series with in situ steam extraction.

TECHNOLOGY STATUS

In situ extraction is being considered as a component of the selected remedy for the San Fernando Valley (Area 1) site in Burbank, California. The Area 1 site consists of an aquifer contaminated with VOCs, including TCE and PCE [1, p. 145]. Toxic Treatments' mobile steam extraction technology (Detoxifier™) was used in 1986 to remediate 4,700 cu. yds. of soil contaminated with diesel fuel at the Pacific Commerce Center site in Los Angeles, California [7, p. 506].

Table 10-3. Total Petroleum Hydrocarbons Removed by Toxic Treatments (USA) Inc. at Los Angeles, CA[a]

Calculated Value	Initial (mg/kg)	Final (mg/kg)	Percent Removal
Mean	2222	191	91

[a] This information is from vendor-published literature [7]; therefore, quality assurance has not been evaluated.

Table 10-4. Demonstration Test Results for Volatiles Removed by Toxic Treatments (USA) Inc. [4]

12-Block Test Area

Block Number	Pre-Treatment (μg/g)	Post-Treatment (μg/g)	Percent Removal
A-25-e	54	14	73
A-26-e	28	12	56
A-27-e	642	29	96
A-28-e	444	34	92
A-29-e	850	82	90
A-30-e	421	145	65
A-31-e	788[a]	61	92
A-32-e	479	64	87
A-33-e	1133	104	91
A-34-e	431	196	54
A-35-e	283	60	79
A-36-e	153	56	64

[a] Only analyses from two of the three sample cores taken were available.

In 1987, Toxic Treatments' mobile steam extraction system was selected as the remedial action to clean up approximately 8,700 cu. yds. of soil contaminated with VOCs and SVOCs at the GATX Annex Terminal site in San Pedro, California [11, p. I-1]. Treatability testing of the technology at the site has been underway to validate its performance prior to full site remediation. This system also has been evaluated under the SITE program at the site in San Pedro, California. Toxic Treatments expects to have a second generation Detoxifier™ available soon, which will be capable of operating on grades up to 5 percent.

For the mobile technology, the most significant factor influencing cost is the time of treatment or treatment rate. Treatment rate is influenced primarily by the soil type (soils with higher clay content require longer treatment times), the waste type, and the on-line efficiency. Cost estimates for this technology are strongly dependent on the treatment rate and range. A SITE demo indicated costs of $111–317/cu. yd. (for 10 and 3 cu. yd. treatment rates, respectively). These costs are based on a 70% on-line efficiency [4, p. 28].

Solvent Service, Inc. is using and testing its first full-scale stationary Steam Injection Vapor Extraction (SIVE) system at its San Jose, California, facility for remediation to a depth of 20 feet of up to 41,000 cu. yds. of soil contaminated with numerous organic solvents [5, p. 3; 10]. Solvent Service hopes to make the SIVE system available for other applications in the future. The system consists of injection and extraction wells and a gas and liquid treatment process. Equipment for steam generation and extraction and contaminated gas/liquid treatment are trailer mounted.

Hydro-Fluent, Inc. is designing and constructing its first full-scale stationary steam extraction system to be used in Huntington Beach, California for recovery of 135,000 gallons of diesel fuel in soil to a depth of 40 feet at the Rainbow Disposal, Nichols Avenue site [12]. Bench and pilot-scale studies have been conducted.

For the stationary steam extraction system, the most significant factor influencing cost is the number of wells required per unit area, which is related to the depth of contamination and soil perme-

Table 10-5. Demonstration Test Results for Semivolatiles Removed by Toxic Treatments (USA) Inc. [4]

12-Block Test Area

Block Number	Pre-Treatment ($\mu g/g$)	Post-Treatment ($\mu g/g$)	Percent Removal
A-25-e	595	82	86
A-26-e	1117	172	85
A-27-e	1403	439	69
A-28-e	1040	576	45
A-29-e	1310	726	45
A-30-e	1073	818	24
A-31-e	781	610	22
A-32-e	994	49	95
A-33-e	896	763	15
A-34-e	698	163	77
A-35-e	577	192	67
A-36-e	336	314	7

ability. Shallow contamination requires lower operating pressures to prevent soil fracturing, and wells are placed closer together. Deeper contamination allows higher operating pressures and greater well spacing; therefore, fewer wells and lower capital cost. Cost estimates for this technology range from about $50 to $300/cu. yd., depending on site characteristics [10].

EPA CONTACT

Technology-specific questions regarding in situ steam extraction may be directed to Michelle Simon, RREL-Cincinnati, (513) 569-7469 (see Preface for mailing address).

ACKNOWLEDGMENTS

This chapter was prepared for the U.S. Environmental Protection Agency, Office of Research and Development (ORD), Risk Reduction Engineering Laboratory (RREL), Cincinnati, Ohio, by Science Applications International Corporation (SAIC) under Contract No. 68-C8-0062. Mr. Eugene Harris served as the EPA Technical Project Monitor. Mr. Gary Baker was SAIC's Work Assignment Manager. This chapter was authored by Mr. Kyle Cook of SAIC. The project team included Mr. Jim Rawe and Mr. Joe Tillman of SAIC. The author is especially grateful to Mr. Bob Hillger and Dr. John Brugger of EPA, RREL, who have contributed significantly by serving as technical consultants during the development of this chapter.

The following other Agency and contractor personnel have contributed their time and comments by participating in the expert review meetings and/or peer-reviewing the chapter:

Mr. Clyde Dial, SAIC
Mr. Vic Engleman, SAIC
Mr. Trevor Jackson, SAIC
Mr. Lyle Johnson, Western Research Institute
Dr. Kent Udell, Udell Technologies

REFERENCES

1. ROD Annual Report, FY 1989. EPA/540/8-90/006, 1990.
2. Personal Communications with the Regional Project Manager, April, 1991.
3. Udell, K.S., and L.D. Stewart. Combined Steam Injection and Vacuum Extraction for Aquifer Cleanup. Presented at Conference of the International Association of Hydrogeologists, Calgary, Alberta, Canada, 1990.

4. Applications Analysis Report—Toxic Treatments' In Situ Steam/Hot-Air Stripping Technology, San Diego, CA. EPA/540/A5-90/008, June 1991. (SITE Report.)
5. Udell, Kent S., and L.D. Stewart. Field Study of In Situ Steam Injection and Vacuum Extraction for Recovery of Volatile Organic Solvents. University of California Berkeley-SEEHRL Report No. 89-2, June 1989.
6. Udell, K.S., J.R. Hunt, and N. Sitar. Nonaqueous Phase Liquid Transport and Cleanup 2. Experimental Studies. Water Resources Research, 24(8):1259–1269, 1988.
7. La Mori, P.N. and M. Ridosh. In Situ Treatment Process for Removal of Volatile Hydrocarbons from Soils: Results of Prototype Test. EPA/600/9-87/018F, 1987.
8. Lord, A.E., Jr., R.M. Koerner, D.E. Hullings, and J.E. Brugger. Laboratory Studies of Vacuum-Assisted Steam Stripping of Organic Contaminants from Soil. In: 15th Annual Research Symposium: Remedial Action, Treatment, and Disposal of Hazardous Waste. EPA/600/9-90/006, 1990.
9. Technology Screening Guide for Treatment of CERCLA Soils and Sludges. EPA/540/2-88/004 (NTIS PB89-132674), 1988.
10. Udell, Kent S. Personal Communication. July 23, 1990.
11. Harding Lawson Associates, Remedial Design, Annex Terminal Site, San Pedro, California. Prepared for GATX Terminals Corporation, 1987.
12. Toxic Cleanup Going Underground. The Orange County Register, June 25, 1990, pp. A1 and A14.

ADDITIONAL SITE PROGRAM REFERENCES (see Appendix A for information on how to obtain SITE program publications)[4]

In Situ Steam Enhanced Recovery Process, Hughes Environmental Systems, Inc.:

Demonstration Bulletin (EPA/540/MR-94/508).
SITE Technology Capsule (EPA/540/R-94/510a), August 1995.
Innovation Technology Evaluation Report (EPA/540/R-94/510) July 1995.

HRUBETZ Environmental Service In-Situ Thermal Oxidative Process:

Demonstration Bulletin (EPA/540/MR-93/524), September 1993.

Radio Frequency Heating, KAI Technologies, Inc.:

Demonstration Bulletin (EPA/540/MR-94/528), November 1994.
Site Technology Capsule (EPA/540/R-94/528a), January 1995.
Innovative Technology Evaluation Report (EPA/540/R-94/528), April 1995.

Radio Frequency Heating, IIT Research Institute:

Demonstration Bulletin (EPA/540/MR-94/527), November 1994.
Site Technology Capsule (EPA/540/R-94/528a), March 1995.
Innovative Technology Evaluation Report (EPA/540/R-94/527), June 1995.

---

[4] See also, SITE Technology Profiles (Seventh Edition, EPA/540/R-94/526) for description of Berkeley Environmental Restoration Center's (formerly Udell Technologies, Inc.) In Situ Steam Enhanced Extraction Process completed demonstration, and Praxis Environmental Technologies, Inc. In Situ Thermal Extraction Process ongoing project.

Chapter 11

In Situ Biodegradation Treatment[1]

Jim Rawe and Evelyn Meagher-Hartzell, Science Applications International Corporation (SAIC), Cincinnati, OH

ABSTRACT

In situ biodegradation may be used to treat low-to-intermediate concentrations of organic contaminants in-place without disturbing or displacing the contaminated media. Although this technology has been used to degrade a limited number of inorganics, specifically cyanide and nitrate, in situ biodegradation is not generally employed to degrade inorganics or to treat media contaminated with heavy metals.

During in situ biodegradation, electron acceptors (e.g., oxygen and nitrate), nutrients, and other amendments may be introduced into the soil and ground water to encourage the growth of an indigenous population capable of degrading the contaminants of concern. These supplements are used to control or modify site-specific conditions that impede microbial activity and, thus, the rate and extent of contaminant degradation. Depending on site-specific cleanup goals, in situ biodegradation can be used as the sole treatment technology or in conjunction with other biological, chemical, and physical technologies in a treatment train. In the past, in situ biodegradation has often been used to enhance traditional pump-and-treat technologies by reducing the time needed to achieve aquifer cleanup standards.

One of the advantages of employing an in situ technology is that media transport and excavation requirements are minimized, resulting in both reduced potential for volatile releases and minimized material handling costs. Biological technologies that require the physical displacement of media during treatment (e.g., "land treatment" applications involving excavation for treatment in lined beds or tilling of nonexcavated soils) assume many of the risks and costs associated with ex situ technologies and cannot strictly be considered in situ applications.

As of Fall 1993, in situ biodegradation was being considered or implemented as a component of the remedy at 21 Superfund sites and 38 Resource Conservation and Recovery Act (RCRA), Underground Storage Tank (UST), Toxic Substances Control Act (TSCA), and federal sites with soil, sludge, sediment, or ground-water contamination [1, p. 13[2]; 2; 3]. This chapter provides information on the technology's applicability, the types of residuals produced, the latest performance data, the site requirements, the status of the technology, and sources for further information.

TECHNOLOGY APPLICABILITY

In situ biodegradation has been shown to be potentially effective at degrading or transforming a large number of organic compounds to environmentally-acceptable or less mobile compounds [4, p. 54; 5, p. 103; 6–9]. Soluble organic contaminants are particularly amenable to biodegradation; however, relatively insoluble contaminants may be degraded if they are accessible to microbial degraders. Classes of compounds considered amenable to biodegradation include petroleum hydrocarbons (e.g., gasoline and diesel fuel), nonchlorinated solvents (e.g., acetone, ketones, and alcohols), wood-treating wastes (e.g., creosote and pentachlorophenol), some chlorinated aromatic compounds (e.g., chlorobenzenes and biphenyls with fewer than five chlorines per molecule), and some chlorinated

---

[1] EPA/540/S-94/502.
[2] [Reference number, page number.]

aliphatic compounds (e.g., trichloroethene and dichloroethene). As advances in anaerobic biodegradation continue, many compounds traditionally considered resistant to aerobic biodegradation may eventually be degraded, either wholly or partially, under anaerobic conditions. Although not normally used to treat inorganics (e.g., acids, bases, salts, heavy metals, etc.), in situ biodegradation has been used to treat water contaminated with nitrate, phosphate, and other inorganic compounds.

Although in situ biodegradation may be used to remediate a specific site, this does not ensure that it will be effective at all sites or that the treatment efficiency achieved will be acceptable at other sites. The complex contaminant mixtures found at many Superfund sites frequently result in chemical interactions or inhibitory effects that limit contaminant biodegradability. Elevated concentrations of pesticides, highly chlorinated organics, and some inorganic salts have been known to inhibit microbial activity and thus system performance during in situ biodegradation. Treatability studies should be performed to determine the effectiveness of a given in situ biological technology at each site. Experts based out of EPA's Risk Reduction Engineering Laboratory (RREL) in Cincinnati, Ohio and the Robert S. Kerr Environmental Research Laboratory (RSKERL) in Ada, Oklahoma may be able to provide useful guidance during the treatability study and design phases. Other sources of general observations and average removal efficiencies for different treatability groups are contained in the Superfund Land Disposal Restrictions (LDR) Guide #6A, Obtaining a Soil and Debris Treatability Variance for Remedial Actions, (OSWER Directive 9347.3-06FS, September 1990) [10] and Superfund LDR Guide #6B, Obtaining a Soil and Debris Treatability Variance for Removal Actions, (OSWER Directive 9347.3-06BFS, September 1990) [11].

LIMITATIONS

Site- and contaminant-specific factors impacting contaminant availability, microbial activity, and chemical reaction rates may limit the application of in situ biodegradation. Variations in media composition and contaminant concentrations can lead to variations in biological activity and, ultimately, inconsistent degradation rates. Soil characteristics (e.g., nonuniform particle size, soil type, moisture content, hydraulic conductivity, and permeability) and the amount, location, and extent of contamination can also have a profound impact on bioremediation. The following text expands upon these factors.

The biological availability, or bioavailability, of a contaminant is a function of the contaminant's solubility in water and its tendency to sorb on the surface of the soil. Contaminants with low solubility are less likely to be distributed in an aqueous phase and may be more difficult to degrade biologically. Conversely, highly soluble compounds may leach from the soil before being degraded. In general, however, poor bioavailability can be attributed to contaminant sorption on the soil rather than a low or high contaminant solubility. The tendency of organic molecules to sorb on the soil is determined by the physical and chemical characteristics of the contaminant and soil. In general, the leaching potential of a chemical is proportional to the magnitude of its adsorption (partitioning) coefficient in the soil. Hydrophobic (i.e., "water fearing") contaminants, in particular, routinely partition from the soil water and concentrate in the soil organic matter, thus limiting bioavailability. Additionally, contaminant weathering may lead to binding in soil pores, which can limit availability even of soluble compounds. Important contaminant properties that affect sorption include: chemical structure, contaminant acidity or basicity (pKa or pKb), water solubility, permanent charge, polarity, and molecule size. In some situations, surfactants (e.g., "surface acting agents") may be used to increase the bioavailability of "bound" or insoluble contaminants. However, it may be difficult to identify a surfactant that is both nontoxic and not a preferred substrate for microbial growth.

Soil solids, which are comprised of organic and inorganic components, may contain highly reactive charged surfaces that play an important role in immobilizing organic constituents, and thus limiting their bioavailability. Certain types of inorganic clays possess especially high negative charges, thus exhibiting a high cation exchange capacity. Alternatively, clays may also contain positively charged surfaces, causing these particles to exhibit a high anion exchange capacity. Soil organic matter also has many highly reactive charged surfaces which can limit bioavailability [12].

Bioavailability is also a function of the biodegradability of the target chemical, i.e., whether it acts as a substrate, cosubstrate, or is recalcitrant. When the target chemical cannot serve as a metabolic substrate (source of carbon and energy) for microorganisms, but is oxidized in the presence of a substrate already present or added to the subsurface, the process is referred to as cooxidation and the target chemical is defined as the co-substrate [12; 13, p. 4]. Cometabolism occurs when an enzyme produced by an organism to degrade a substrate that supports microbial growth also degrades another nongrowth substrate that is neither essential for, nor sufficient to support microbial growth. Cooxidation processes are important for the biodegradation of high molecular weight polycyclic aromatic hydrocarbons (PAHs), and some chlorinated solvents, including trichloroethene (TCE). However, like surfactants, cometabolites (e.g., acetate and phenol) may be more readily mineralized by the indigenous microorganisms than the target organics [13, p. 4].

Microbial activity can be reduced by nutrient, moisture, and oxygen deficiencies, significantly decreasing biodegradation rates. Extreme soil temperatures, soil alkalinity, or soil acidity can limit the diversity of the microbial population and may suppress specific contaminant degraders. Spatial variation of soil conditions (e.g., moisture, oxygen availability, pH, and nutrient levels) may result in inconsistent biodegradation due to variations in biological activity. While these conditions may be controlled to favor biodegradation, the success of in situ biodegradation depends in a large part on whether required supplements can be delivered to areas where they are needed. Low hydraulic conductivity can hinder the movement of water, nutrients, aqueous-phase electron acceptors (e.g., hydrogen peroxide and nitrate), and, to a lesser extent, free oxygen through the contamination zone [14, p. 155]. Restrictive layers (e.g., clay lenses), although more resistant to contamination, are also more difficult to remediate due to poor permeability and low rates of diffusion [13, p. 4]. Low percolation rates may cause amendments to be assimilated by soils immediately surrounding application points, preventing them from reaching areas that are more remote, either vertically or horizontally. During the simultaneous addition of electron acceptors and donors through injection wells, excessive microbial growth or high iron or manganese concentrations may cause clogging in the wellscreen or in the soil pores near the wellscreen [15]. Variable hydraulic conductivities in different soil strata within a contaminated area can also complicate the design of flow control; minor heterogeneities in lithology can, in some cases, impede the transfer of supplements to specific subsurface locations.

Microbial activity may also be influenced by contaminant concentrations. Each contaminant has a range of concentrations at which the potential for biodegradation is maximized. Below this range microbial activity may not occur without the addition of co-substrate. Above this range microbial activity may be inhibited and, once toxic concentrations are reached, eventually arrested. During inhibition, contaminant degradation generally occurs at a reduced rate. In contrast, at toxic concentrations, contaminant degradation does not occur. The concentrations at which microbial growth is either supported, inhibited, or arrested vary with the contaminant, medium, and microbial species. Given long-term exposure, microbes have been known to acclimate to very high contaminant concentrations and other conditions inhibiting microbial activity. However, if prompt treatment is a primary goal, as is the case during most remedial activities, toxic conditions may need to be addressed by pH control, metals control (e.g., immobilization), sequential treatment, or by introducing microbial strains resistant to toxicants.

Numerous biological and nonbiological mechanisms (e.g., volatilization, sorption, chemical degradation, migration, and photodecomposition) occur during biological treatment. Since some amendments may react with the soil, site geochemistry can limit both the form and concentration of any supplements added to the soil. Thus, care must be employed when using amendments to "enhance" biological degradation. For example, ozone and hydrogen peroxide, which can be added to enhance dissolved oxygen levels in soil or ground-water systems, may react violently with other compounds present in the soil, reduce the sorptive capacity of the soil being treated, produce gas bubbles that block the pores in the soil matrix, or damage the bacterial population in the soil [4, p. 43]. Nitrogen and phosphorus (phosphate) must also be applied cautiously to avoid excessive nitrate formation [4, p. 47] and the precipitation of calcium and iron phosphates, respectively. Excessive nitrate levels in the ground water can cause health problems in humans, especially children. If calcium concentrations are high, the added phosphate can be tied up by the calcium and, therefore, may not be available to the microorganisms [16, p. 23]. Lime treatment for soil pH adjustment is dependent on several soil factors including soil texture, type of clay, organic matter content, and aluminum concentrations [4, p. 45]. Since changes in soil pH may also affect the dissolution or precipitation of materials within the soil and may increase the mobility of hazardous materials, pH amendments (acid or base) should be added cautiously and should be based on the soil's ability to resist changes in pH, otherwise known as the soil's "buffering capacity" [4, p. 46]. Since the buffering capacity varies between soils, lime and acidification requirements should be determined on a site-specific basis.

Finally, high concentrations of metals can have a detrimental effect on the biological treatment of organic contaminants in the same medium. A number of metals can be oxidized, reduced, methylated (i.e., mercury), demethylated, or otherwise transformed by various organisms to produce new contaminants. The solubility, volatility, and sorption potential of the original soil contaminants can be greatly changed in the process [17, p. 144], leading to potential significant toxicological effects, as is the case during the methylation of mercury. To avoid these complications, it is sometimes possible to pretreat or complex the metals into a less toxic or leachable form.

TECHNOLOGY DESCRIPTION

During in situ biodegradation, site-specific characteristics are modified to encourage the growth of a microbial population capable of biologically degrading the contaminants of concern. Presently, two major types of in situ systems are being employed to biodegrade organic compounds present in soils, sludges, sediments, and ground water: bioventing systems and "traditional" in situ biodegradation systems, which usually employ infiltration galleries/wells and recovery wells to deliver required amendments to the subsurface. In general, bioventing has been used to treat contaminants present in the unsaturated zone. Traditional in situ biodegradation, on the other hand, has mostly been used to treat saturated soils and ground water. The occasional treatment of unsaturated soil using traditional in situ biodegradation techniques has been generally limited to fairly shallow regions over ground water that is already contaminated.

Traditional In Situ Biodegradation

Traditional in situ biodegradation is generally used in conjunction with ground-water pumping and soil-flushing systems to circulate nutrients and oxygen through a contaminated aquifer and associated soil. The process usually involves introducing aerated, nutrient-enriched water into the contaminated zone through a series of injection wells or infiltration trenches and recovering the water downgradient. Depending upon local regulations and engineering concerns, the recovered water can then be treated and, if necessary, reintroduced to the soil onsite, discharged to the surface, or discharged to a publicly-owned treatment works (POTW). A permit may be required for the reinjection of treated water. Note that a variety of techniques can be used to introduce and distribute amendments in the subsurface. For example, a lower horizontal well is being used at the Savannah River site near Aiken, North Carolina to deliver air and methane to the subsurface. A vacuum has been applied to an upper well (in the vadose zone) located at this site to encourage the distribution of air and methane within the upper saturated zone and lower vadose zone [18; 19].

Figure 11-1 is a general schematic of a traditional in situ biodegradation system [20, p. 113; 16, p. 13]. The first step in the treatment process involves pretreating the infiltration water, as needed, to remove metals (1). Treated or contaminated ground water, drinking water, or alternative water sources (e.g., trucked water) may be used as the water source. If ground water is used, iron dissolved in the ground water may bind phosphates needed for biological growth. Excess phosphate may be added to the infiltration water at this point in the treatment process in order to complex the iron [20, p. 111]. The presence of iron will also cause a more rapid depletion of hydrogen peroxide, which is sometimes used as an oxygen source. Surface active agents may also be added at this point in the treatment process to increase the bioavailability of contaminants, especially hydrophobic or sorbed pollutants, while methane or other substances may be added to induce the cometabolic biodegradation of certain contaminants. In continuous recycle systems, toxic metals originally located in the contaminated medium may have to be removed from the recycled infiltration water to prevent inhibition of bacterial growth. The exact type of pretreatment will vary with the water source, contamination problem, and treatment system used.

Following infiltration water pretreatment, a biological inoculum can be added to the infiltration water to enhance the natural microbial population (2). A site-specific inoculum enriched from site samples may be used; commercially available cultures reported to degrade the contaminants of concern can also be used (e.g., during the remediation of "effectively sterile soils"). Project managers are cautioned against employing microbial supplements without first assessing the relative advantages associated with their use and potential competition that may occur between the indigenous and

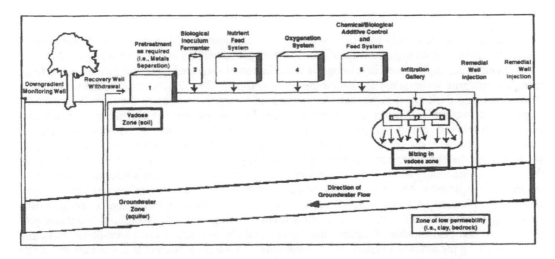

Figure 11-1. Schematic diagram of traditional in situ biodegradation of soil and ground water.

introduced organisms. The ability of microbes to survive in a foreign and possibly hostile (i.e., toxic) environment, as well as the ability to metabolize a wide range of substrates should be evaluated. The health effects of commercial inocula must also be carefully evaluated, since many products on the market are not carefully screened or processed for pathogens. It is essential that independently-reviewed data be examined before employing a commercially-marketed microbial supplement [21].

Nutrient addition can then be employed to provide nitrogen and phosphorus, two elements essential to the biological activity of both indigenous and introduced organisms (3). Optimum nutrient conditions are site-specific. Trace elements may be added at this stage, but are normally available in adequate supply in the soil or ground water.

During contaminant oxidation, energy is released as electrons are removed. Since oxygen acts as the terminal electron acceptor during aerobic biodegradation, oxygen concentrations in the subsurface may become depleted. To avoid this complication, air, oxygen, and other oxygen sources (hydrogen peroxide and ozone) can be added to the infiltration water (4). To prevent gas binding in the subsurface, and a subsequent reduction in the effective soil permeability, oxygen amendment/supplementation methods must be carefully selected. During anaerobic degradation, alternative electron acceptors (nitrate, carbonate, or sulfate) may be added to the infiltration water in place of oxygen. Alternatively, during the cooxidation of a target substrate, a cosubstrate (methanol or acetate) may be added to the infiltration water [22].

Just before the water is added to the soil or ground water, chemical additives may be used to adjust the pH (neutral is recommended for most systems) and other parameters that impact biodegradation (5). Care should be taken when making adjustments to the pH, since contaminant mobility (especially of metals) can be increased by changing the pH [4, p. 45]. Site managers are also cautioned against employing chemical additives that are persistent in the environment. The potential toxicity of additives and any synergistic effects on contaminant toxicity should also be evaluated.

During in situ bioremediation, amendment concentrations and application frequencies can be adjusted to compensate for physical/chemical depletion and high microbial demand. If these modifications fail to compensate for microbial demand, remediation may occur by a sequential deepening and widening of the active treatment layer (e.g., as the contaminant is degraded in areas near the amendment addition points, and microbial activity decreases due to the reduced substrate, the amendments move farther, increasing microbial activity in those areas). Additionally, hydraulic fracturing may be employed to improve amendment circulation within the subsurface.

The importance of using a well-designed hydraulic delivery system and thoroughly evaluating the compatibility of chemical supplements was demonstrated at sites in Park City, Kansas; Kelly AFB, Texas; and Eglin AFB, Florida. Air entrainment and iron precipitation resulted in a continued loss of injection capacity during treatment at the Park City site [23; 24], and calcium phosphate and iron precipitation resulted in the failure of the two field tests at Kelly and Eglin AFBs, respectively [25].

Bioventing

Bioventing uses relatively low-flow soil aeration techniques to enhance the biodegradation of soils contaminated with organic contaminants. Although bioventing is predominantly used to treat unsaturated soils, applications involving the remediation of saturated soils and ground water (e.g., using air sparging techniques) are becoming more common [26; 27]. Aeration systems similar to those employed during soil vapor extraction are used to supply oxygen to the soil (Figure 11-2). Typically, a vacuum extraction, air injection, or combination vacuum extraction and air injection system is employed [28]. An air pump, one or more air injection or vacuum extraction probes, and emissions monitors at the ground surface are commonly used. Although some systems utilize higher air flow rates, thereby combining bioventing with soil vapor extraction, low air pressures and low air flow rates are generally used to maximize vapor retention times in the soil while minimizing contaminant volatilization. An interesting modification to traditional aeration techniques has been proposed at the Picatinny Arsenal in New Jersey. Here researchers and project managers have proposed collecting TCE vapors at the surface, amending them with degradable hydrocarbons (methane, propane, or natural gas) capable of stimulating the cometabolic degradation of vapor-phase TCE, and then reinjecting the amended vapors into the unsaturated zone in an attempt to encourage the in situ bioremediation of the TCE remaining in the subsurface [29–32; 27].

Off-gas treatment (e.g., through biofiltration or carbon adsorption) will be needed during most bioventing applications to ensure compliance with emission standards and to control fugitive emissions. Off-gas treatment systems similar to those employed during soil vapor extraction may be used. These systems must be capable of effectively collecting and treating a vapor stream consisting of the original contaminants and/or any volatile degradation products generated during treatment. Although similar vapor treatment systems may be employed during soil vapor extraction and bioventing, less concentrated off-gases would be expected from a bioventing system than from a soil vapor extraction system employed at the same site. This difference in concentration is attributed to enhanced biological degradation within the subsurface.

Nutrient addition may be employed during bioventing to enhance biodegradation. Nutrient addition can be accomplished by surface application, incorporation by tilling into surface soil, and transport to deeper layers through applied irrigation water. However, in some field applications to date, nutrient additions have been found to provide no additional benefits [33]. Increasing the soil temperature may also enhance bioremediation, although in general high temperatures should be avoided since they can decrease microbial population and activity. Heated air, heated water, and low-level radio-frequency heating are some of the techniques which can be used to modify soil temperature. Soil core analyses can be performed periodically to assess system performance as determined by contaminant removal. A control plot located near the bioventing system, but not biovented, may also be used to obtain additional information to assess system performance.

PROCESS RESIDUALS

During in situ biodegradation, limited but potentially significant process residuals may be generated. Although the majority of wastes requiring disposal are generated as part of pre- and post-treatment activities, process residuals directly arising from in situ biological activities may also be generated. These process residuals may include: 1) partially degraded metabolic by-products, 2) residual contamination, 3) wastes produced during groundwater pre- and post-treatment activities, and 4) volatile contaminants that are either directly released into the atmosphere or collected within add-on emission control/treatment systems. The following text expands upon the specific types of process residuals, their control, and their impact on disposal requirements.

Ultimately biological technologies seek to mineralize hazardous contaminants into relatively innocuous by-products, specifically carbon dioxide, water, and inorganic salts. However, a number of site- and contaminant-specific factors may cause the partial degradation or "biotransformation" of a contaminant and the generation of an intermediate by-product. These metabolic by-products may be located in either the saturated or unsaturated zones. The identity, toxicity, and mobility of these partially degraded compounds should be determined, since intermediate degradation products can be as toxic or more toxic than the parent compound. Since metabolic by-products can accumulate in the soil and ground water, future remedial actions may be necessary.

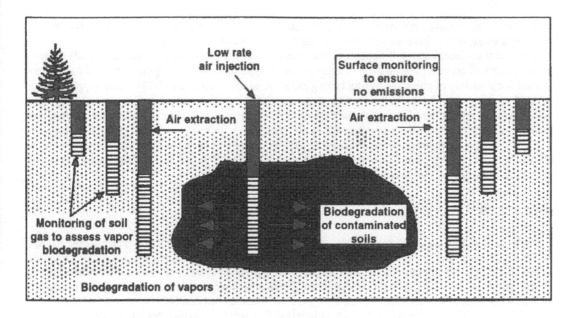

Figure 11-2. Bioventing.

In addition to intermediate degradation by-products, residual contamination may persist in the soil following treatment. Microbes are capable of degrading only that fraction of the contamination that is readily available for microbial incorporation. As a result, biologically resistant contaminants and contaminants that remain sorbed to the soil and sediment during the remedial action cannot be degraded. Depending on the nature of the contaminants and media, the "bound" fraction may slowly desorb over long periods of times (months to years), potentially recontaminating "treated" media near the residual contamination [34; 35]. Additionally, fluctuations in the water table may result in the recontamination of previously remediated soils if ground-water contamination, specifically contamination associated with the presence of a light nonaqueous phase layer (LNAPL), has not been effectively addressed.

Aboveground activities taken to ensure that the remedial action complies with regulatory requirements and adequately guards against cross-contamination and uncontrolled releases may result in the generation of a significant volume of waste requiring disposal. For example, when ground water is used to deliver amendments to the subsurface, it may be necessary to pretreat the water before it can be re-introduced to the subsurface. Additionally, in order to protect water quality outside of the treatment zone from contaminant or amendment migration, a downgradient ground water recovery and treatment system designed to collect and treat amendment- and contaminant-laden ground water may be needed. The residuals produced by these add-on treatment processes will eventually require disposal.

Significant volatile emissions may also be produced during in situ biodegradation (e.g., bioventing). Depending on their concentration, toxicity, and total volume, these emissions, which may consist of the original contaminant or any volatile degradation products produced during treatment, may need to be controlled, collected, or treated. Ultimately, the by-products of an emissions treatment/control system will require disposal.

SITE REQUIREMENTS

In situ biodegradation normally requires the installation of wells or infiltration trenches; therefore, adequate access roads are required for heavy equipment such as well-drilling rigs and backhoes. Soil-bearing capacity, traction, and soil stickiness can limit vehicular traffic [17, p. 61].

In general, the area required to set up mixing equipment is not significant. However, space requirements increase as the complexity of the various pre- and post-treatment systems increases. During the installation of infiltration galleries and wells, several hundred up to several thousand square feet of clear surface area will be required. Climate can also influence site requirements. If periods of heavy rainfall or extremely cold conditions are expected, a cover may be required.

Electrical requirements will depend on the type of technology employed. Standard 220V, three-phase electrical service may be used to supply power to pumps and mixing equipment. Since water is used for a variety of purposes during biological treatment, a readily available water supply will be needed at most sites. Municipal water or clean ground water may be used. Contaminated ground water may be used if permitted by the appropriate regulatory agency. The quantity of water needed is site- and process-specific. Waste storage is not normally required for in situ biodegradation.

Onsite analytical equipment for conducting pH and nutrient analyses will help improve operation efficiency and provide better information for process control. During bioventing applications, air emissions monitors at the ground surface are commonly used.

REGULATORY CONSIDERATIONS AND RESPONSE ACTIONS

Federal mandates can have a significant impact on the application of in situ biodegradation. RCRA LDRs that require treatment of wastes to best demonstrated available technology (BDAT) levels prior to land disposal may sometimes be determined to be applicable or relevant and appropriate requirements (ARARs) for CERCLA response actions. The in situ biodegradation technology can produce a treated waste that meets treatment levels set by BDAT, but may not reach these treatment levels in all cases. The ability to meet required treatment levels is dependent upon the specific waste constituents and the waste matrix. In cases where in situ biodegradation does not meet these levels, it still may, in certain situations, be selected for use at the site if a treatability variance establishing alternative treatment levels is obtained. Treatability variances are justified for handling complex soil and debris matrices. The following guides describe when and how to seek a treatability variance for soil and debris: Superfund LDR Guide #6A, Obtaining a Soil and Debris Treatability Variance for Remedial Actions (OSWER Directive 9347.06FS, September 1990) [10], and Superfund LDR Guide #6B, Obtaining a Soil and Debris Treatability Variance for Removal Actions (OSWER Directive 9347.06BFS, September 1990) [11]. Another approach could be to use other treatment techniques with in situ biodegradation to obtain desired treatment levels; for example, carbon treatment of recovered ground-water prior to reinfiltration into the subsurface.

When determining performance relative to ARARs and BDATs, emphasis should be placed on assessing the risk presented by a bioremediation technology. As part of this effort, risk assessment schemes, major metabolic pathways of selected hazardous pollutants, human health protocols for metabolite and pathogenicity tests, and fate protocols and issues for microorganisms and metabolites must be assessed [36]. A detailed summary of the findings of the June 17–18, 1993 EPA/Environment Canada Workshop in Duluth, Minnesota addressing Bioremediation Risk Assessment should be available in early 1994.

PERFORMANCE DATA

Performance data for Superfund sites are limited. The first record of decision (ROD) selecting in situ biodegradation as a component of the remedy was in FY87. Since then, in situ biodegradation of soil or ground-water contaminants has either been considered or selected at 22 Superfund sites and 30 RCRA, UST, TSCA, and federal sites [1][2][3]. The following two subsections address traditional in situ and bioventing applications, respectively; a third subsection has been included to briefly address information sources and data concerns related to remedial efforts performed in the private sector.

Traditional In Situ Bioremediation

Methane and phenol were employed during a series of stimulus-response studies investigating the cometabolic degradation of TCE, cis-dichloroethene (c-DCE), transdichloroethene (t-DCE), and vinyl chloride (VC) at the Moffet Field site in California. Both sets of experiments used indigenous bacteria and were performed under the induced gradient conditions of injection and extraction. During the first set of experiments, methane, oxygen, and TCE (from 50 to 100 µg/L), c-DCE, t-DCE, and VC were added to the soil to stimulate methanotrophic degradation of the injected chlorinated aliphatic compounds. Approximately 20 percent of the TCE added to the system was degraded within the 2-meter hydraulically-controlled biostimulated zone. Approximately 50 percent of the c-DCE, 90 percent of the t-DCE, and 95 percent of the VC were also degraded. During the second set of tests, methane was replaced with phenol in order to stimulate growth of an indigenous phenol-utilizing population. During 4 weeks of testing, the concentration of TCE injected into the subsurface was raised from an initial concentration of 62 µg/L to a final concentration of 1000 µg/L. A bromide tracer was used to determine transformation extent. Up to 90 percent of the TCE in the 2-meter biostimulated zone was degraded, demonstrating that even at relatively high TCE concentrations significant removal efficiencies can be achieved in situ through phenol and dissolved oxygen (DO) addition. During the course of the project, transformation yields (i.e., grams of TCE per grams of phenol) ranging from 0.0044 to 0.062 were obtained for varying concentrations of phenol and TCE. Future studies at the site will determine whether a compound more environmentally acceptable than methane or phenol can be used to induce an indigenous population that effectively degrades TCE [37; 7; 8].

A 40- by 120-foot test zone in an aquifer that receives leachate from an industrial landfill at the Du Pont Plant near Victoria, Texas was used to demonstrate the in situ biotransformation of tetrachloroethene (PCE), TCE, DCE, chloroethane, and VC to ethane and ethylene using microbial reductive dehalogenation under sulfate-reducing conditions. Ground water from this zone was alternately amended with either benzoate or sulfate and circulated through the aquifer. Initially PCE and TCE concentrations were approximately 10 and 1 micro-mole (µM), respectively. After a year of treatment the halogenated compounds were reduced to concentrations near or below 0.1 µM. PCE and TCE degraded to DCE rapidly following the introduction of benzoate. A decrease in sulfate concentrations led to increases in the vinyl chloride concentrations. Therefore, sulfate concentrations were kept above 10 mg/L until the DCE was further biodegraded. After approximately 6 months of treatment, most of the DCE, chloroethane, and VC biodegraded to produce ethane and ethylene [38].

A field-scale in situ bioremediation system, consisting of downgradient ground-water extraction wells and an upgradient infiltration system, was installed at a gasoline contaminated site owned by the San Diego Gas and Electric Company. [Note: extracted ground water was amended with nutrients (nitrate and phosphate) prior to reinfiltration into the subsurface]. Due to the relatively low rate of ground-water extraction (approximately 800 to 900 gallons per day) and the low hydraulic gradient at the site (0.004), it took nearly 2 years (until June/July 1991) for the added nitrate to reach the downgradient well and overtake the xylene (BTX) plume. BTX concentrations, which ranged from 25 to 50 mg/L for the preceding 2-year period, dropped markedly as nitrate levels in the ground water increased. By late August 1991, benzene and toluene concentrations had dropped below the detection limit (0.01 mg/L), and total xylene concentrations had dropped to 0.02 mg/L. The coincident occurrence of nitrate appearance and BTX loss in the aquifer, as well as an eight-fold increase in the percentage of denitrifiers present in the ground water (from 1 to 8 percent), points to a potential stimulatory effect nitrate may have on BTX loss in situ [5].

An in situ bioremediation system consisting of four injection and three recovery wells was employed to treat gasoline contamination present in the saturated zone at a former service station in Southern California. During treatment, recovered ground water was amended with hydrogen peroxide (from 500 to 1,000 mg/L) and nutrients and reinjected into the aquifer. Prior to treatment, total fuel hydrocarbons in the saturated clay soils ranged from below detection limits to 32 mg/kg as BTX. Maximum groundwater concentrations were 2,700 µg/L for benzene; 6,600 µg/L for toluene; 4,100 µg/L for xylene; and 45,000 µg/L for TPH [4]. After 10 months, BTX and TPH levels in the ground water and saturated soils had dropped below the detection limits. Roughly 1,350 kilograms of hydrogen peroxide were introduced to the aquifer over 10 months, roughly two times the estimated requirements based on the estimated mass of hydrocarbon in the saturated zone (i.e., 110 kg of fuel hydrocarbon and 2 to 3 kg of dissolved hydrocarbons). After 34 months of treatment, soil hydrocarbon concentrations ranged from below the detection limit to 321 ppm as TPH; benzene was not detected in any samples [39].

Following successful laboratory treatability testing, General Electric performed a 10-1/2-week field study to investigate the biodegradation of polychlorinated biphenyls (PCBs) in the Hudson River sediment. Initial PCB concentrations in the sediment ranged between 20 and 40 ppm. The study attempted to enhance the aerobic bacteria native to the upper Hudson River. Six caissons were

Table 11-1. Rates of Biodegradation, Averaged Over Depth, at Three Wells at Hill AFB

| Well | Depth (ft) | Rate (mg/kg/day) | |
		September 1991	September 1992[a]
CW-1	20–90	0.97	0.30
CW-2	60–90	0.59	0.36
CW-3	10–90	0.56	0.32

[a] Since bioventing is being performed on a sandy soil, with little to no naturally occurring organic matter, a biodegradation rate approaching zero would indicate that biodegradation had finished.

installed at the Hudson River Research Station (HRRS) to isolate sections of the river bottom for this field study. Because of extensive, naturally occurring dechlorination, approximately 80 percent of the total PCBs encountered in the sediments were mono-, di-, and trichlorobiphenyls. Biodegradation was stimulated using oxygen and nutrient addition. Mixing was employed to enhance the dispersal of oxygen and nutrients within the sediment. Between 38 and 55 percent of the PCBs present in the sediment were removed by aerobic degradation during the study. This corresponds to the percentage of biologically available PCBs [9].

Bioventing

In May 1992, the U.S. Air Force began a Bioventing Initiative to examine bioventing as a remedial technique at contaminated sites across the country. The Air Force's decision to examine bioventing on such a large scale was prompted by a successful demonstration of the technology at Tyndall AFB, Florida, where bioventing coupled with moisture addition removed one-third of the TPH and nearly all of the BTEX in JP-4 contaminated soils during 7 months of treatment. The Bioventing Initiative targets 138 sites with diesel fuel, jet fuel, or fuel oil in soil. In selecting sites for the initiative, the Air Force looked for characteristics appropriate for bioventing, such as deep vadose soil, heavy hydrocarbon contamination, and high air permeability. The chosen sites represent a wide range of depths to ground water, hydrocarbon concentrations, and soil textures. Preliminary testing has been completed and 33 systems have been installed at Battle Creek Air National Guard Base and the following AFBs: Beale, Eglin, Eielson, F.E. Warren, Galena, Hanscom, Hill, K.I. Sawyer, McGuire, Newark, Offutt, Plattsburgh, Robins, Vandenberg, and Westover. According to the Air Force, initial results are very promising, with degradation rates measured as high as 5,000 mg/kg per year [40; 41].

The EPA RREL, in collaboration with the U.S. Air Force, initiated two 3-year pilot-scale bioventing field studies in mid-1991 at JP-4 contaminated fuel sites located at Eielson AFB near Fairbanks, Alaska and at Hill AFB near Salt Lake City, Utah. Four soil plots are being used to evaluate passive, active, and buried heat tape soil-warming methods during the Eielson study. The fourth plot was vented with injected air, but not artificially heated. Roughly 1 acre of soil is contaminated from a depth of 2 feet to the water table at 6 to 7 feet. At the Hill site, a series of soil gas cluster wells capable of obtaining samples up to 90 feet deep is being used with a single air injection well and two ground-water wells to remediate JP-4 contamination found at depths ranging from 35 feet to perched water at approximately 95 feet. Inert gas tracer studies, regular soil gas measurements at several locations and depths, and periodic in situ respirometry tests to measure in situ oxygen uptake rates are being performed. Final soil hydrocarbon analyses will be conducted at both sites in mid-1994 and compared with the initial soil data. In situ respirometry data from the Hill site (Table 11-1) indicate that petroleum hydrocarbons are being removed at a significant rate. Intermediate respirometry data from the test and control plots at the Eielson site indicate that higher biodegradation rates are being obtained at higher soil temperatures [42; 43].

In November 1991, a pilot-scale bioventing system originally used to treat gasoline-contaminated vadose soils at the U.S. Coast Guard Air Station in Traverse City, Michigan was converted into a ground-water biosparging process. Eight 2-inch diameter sparge wells were installed to a depth of 10 feet below the water table. A control plot located in the vicinity of the contaminated plume, but not biosparged, was established to help assess the system's performance. After 12 months of biosparging, one-third of the oily phase residue below the water table, as well as almost all the BTEX

Table 11-2. Groundwater Quality After Seven Months of Biosparging at the U.S. Coast Guard Air Station in Traverse City, Michigan

Well Depth (ft)	Benzene (µg/L)	Xylenes (µg/L)	Total Fuel Carbon (µg/L)
Control			
16	9.9	19	2,880
17.5	228	992	4,490
20.5	70	38	956
22	57	7.7	783
Sparge Plot			
15	1.9	5.3	559
18	<1	5.0	<6
19.5	<1	<1	<6
21	<1	<1	<6

initially present within the ground-water plume, was removed. (See Table 11-2 for ground-water quality data after 7 months of biosparging.) The globular nature of the oily residue limited the surface area in contact with the introduced air, thus restricting the biodegradation and vaporization of the oily-phase contaminants [44; 45].

Non-Superfund Sites

In situ biodegradation has been applied at many sites in the private sector. Those interested in accessing information generated in the private sector may want to refer to the following EPA Publications:

- U.S. Environmental Protection Agency. Bioremediation Case Studies: Abstracts. EPA/600/R-92/044, March 1992.
- U.S. Environmental Protection Agency. Bioremediation Case Studies: An Analysis of Vendor Supplied Data. EPA/600/9-92/043, March 1992.

Most of the data contained in these resources were directly supplied by the vendor and have not been technically reviewed by the EPA. Since independently-reviewed data are not always available from privately sponsored remedial efforts, in part due to proprietary issues [46, p. 1-1], readers should use these data cautiously. Often the quality of the data used to determine system effectiveness has not been substantiated by the scientific community. Thus, many vendor claims of effectiveness, specifically regarding introduced organisms and surface-active agents, are not supported within the scientific literature. Furthermore, many bioremediation firms have only limited experience working with the complex wastes normally associated with Superfund sites. Typically, these firms deal only with gasoline and petroleum product leaks and spills. Additionally, many of the systems currently on the market involve the use of in situ biodegradation in combination with other aboveground treatment technologies such as carbon adsorption, air stripping, and biological reactors. In situ biodegradation is believed to enhance the total removal efficiency of the system. However, in many cases, it is unclear how much of the degradation occurred as a result of biological or nonbiological mechanisms (volatilization, chemical destruction, etc.). How much biodegradation actually takes place in the soil or ground water, in contrast to ex situ biodegradation, is not always clear.

TECHNOLOGY STATUS

In situ biodegradation either has been considered or selected as the remedial technology at 21 Superfund sites, as well as 38 RCRA, UST, TSCA, and federal sites [1; 23; 3]. Table 11-3 lists the location, primary contaminants, treatment employed, and status of these sites. Information has also been included on three in situ biotechnology demonstrations presently being performed under the U.S. EPA Superfund Innovative Technology Evaluation (SITE) Program and seven sites selected for performance evaluations under the U.S. EPA Bioremediation Field Initiative. The data obtained during the SITE demonstrations and Bioremediation Field Initiative performance evaluations will be used to develop reliable cost and performance information on biotreatment technologies and applications.

The majority of the information found in Table 11-3 was obtained from the August 1993 version of "Bioremediation in the Field" [1]. These sites have been sorted numerically by Region and then alphabetically by site name. Sites employing "in situ land treatment" were not included in this list since these applications typically involve a significant amount of material handling. Additionally, some of the information was modified based on phone calls made to the various site project managers. This resulted in the removal of the American Creosote Works site in Florida and four pesticides sites (i.e., the Joliet Weed Control District site in the Joliet, Montana; the Lake County Weed Control site in Ronan, Montana; the Miles Airport site in Miles City, Montana; and the Richey Airport site in Richey, Montana) [47], which are no longer considering in situ treatment. Quarterly updates of this information can be obtained from subsequent versions of "Bioremediation in the Field" (Editor's Note: the latest edition, as of early 1996, was Issue No. 12, August 1995, EPA/540/N-95/500).

Most of the hardware components of in situ biodegradation systems are available off-the-shelf and present no significant availability problems. Selected cultures, nutrients, and chemical/biological additives are also readily obtainable.

Bioremediation, particularly in situ applications, which avoid excavation and emissions control costs, are generally considered cost-effective. This can be attributed in part to low operation and maintenance requirements. During setup and operation, material handling requirements are minimal, resulting in lowered worker exposures and reduced health impacts. Although in situ technologies are generally slow and somewhat difficult to control, a large volume of soil may be treated at one time.

It is difficult to generalize about treatment costs since site-specific characteristics can significantly impact costs. Typically, the greater the number of variables requiring control during biological treatment, the more problematic the implementation and the higher the cost. For example, it is less problematic to implement a technology in which only one parameter (e.g., oxygen availability) requires modification than to implement a remedy that requires modification of multiple factors (e.g., pH, oxygen levels, nutrients, microbes, buffering agents, etc.). Initial concentrations and volumes, pre- and post-treatment requirements, and air emissions and control systems will impact final treatment costs. The types of amendments employed (e.g., hydrogen peroxide) can also impact capital cost and costs associated with equipment and manpower required during their application.

In general, however, in situ bioremediation is considered to be a relatively low-cost technology, with costs as low as 10 percent of excavation or pump-and-treat costs [7, pp. 6-16]. The cost of soil venting using a field-scale system has been reported to be approximately $50 per ton as compared to incineration, which was estimated to be more than ten times this amount. A cost estimate of about $15 per cubic yard for bioventing sandy soil at a JP-4 jet fuel contaminated site has been reported by Vogel [48]. Exclusive of site characterization, the biological remediation of JP-4 contaminated soils at the Kelly Air Force Base site was estimated to be $160 to $230 per gallon of residual fuel removed from the aquifer [9]. At the French Limited site in Texas, the cost of bioremediation is projected to be almost three times less expensive than incineration. Because of the large amount of material requiring treatment at this site, it has been projected that cleanup goals will be achieved in less time by using bioremediation rather than incineration.

EPA CONTACT

Technology-specific questions regarding in situ biodegradation may be directed to Carolyn Acheson, RREL-Cincinnati, (513) 569-7190; or Guy Sewell, RSKERL-Ada, (405) 436-8166 (see Preface for mailing address).

ACKNOWLEDGMENTS

This chapter was prepared for the U.S. Environmental Protection Agency, Office of Research and Development (ORD), Risk Reduction Engineering Laboratory (RREL), Cincinnati, Ohio, by Sci-

Table 11-3. Superfund, RCRA, UST, TSCA, and Federal Sites

Site Location (Regions)	Primary Contaminants	Status/Cost	Treatment
Charlestown Navy Yard Boston, MA (1)	Sediments: wood preserving (PAHs).	Design: pilot-scale TS underway. Aerobic and anaerobic.	In situ treatment. Ex situ treatment.
General Electric, Woods Pond, Pittsfield, MA (1)	Sediments: PCBs. Volume: 250 gallons.	Design: lab-scale TS underway.	Anaerobic treatment, confined treatment facility, nutrient addition.
FAA Technical Center, Area D Atlanta County, NJ (2)	Soil (saturated sand)/groundwater: petroleum (jet fuel, NAPLs). Volume: 33K cubic yards.	Design: pilot-scale TS completed 8/92. Expected cost: capital, $286K; O&M, $200K.	Nutrient addition (soil, water). Ground water reinjection.
General Electric, Hudson River, NY (2)	Sediments: PCBs, cadmium, chromium, lead. Volume: 150 cubic feet.	Predesign: lab-scale TS completed. Incurred cost: $2.6M.	Aerobic treatment. Less than 1% of site underwent bioremediation.
Knispel Construction Site Horseheads, NY (2)	Soil/ground water: petroleum.	Completed: full-scale 10/89 Start date: 01/89. Incurred cost: O&M, $25K.	Aerobic treatment, hydrogen peroxide, nutrient addition (water). 100% of site underwent bioremediation [25].
Picatinny Arsenal, NJ (2)	Soil (vadose)/soil vapors: solvents (TCE).	Design: lab-scale studies completed.	Aerobic treatment, bioventing. Cometabolic degradation (methane, propane, or natural gas) [27].
Plattsburgh AFB Plattsburgh, NY (2)	Ground water: petroleum.	Design: pilot-scale. Start date (est.): 3/94.	Aerobic treatment, bioventing.
ARC, Gainesville, VA (3)	Soil: solvent (chlorobenzene). Volume: 2,000 cubic yards.	Completed: full-scale 6/91. Start date: 10/89.	Aerobic treatment, bioventing. Exogenous organisms. 5% of the site underwent bioremediation.
Dover AFB, Dover, DE (3)	Soil (vadose sand and silt)/ground water: petroleum, PAHs, TCE, solvents, metals (lead, iron, manganese). Volume: 365K cubic yards.	Four separate processes are planned. Field and lab TS results are expected 2/94 and 11/94.	Aerobic treatment, bioventing, air sparging. Ex situ land treatment.
L.A. Clarke & Son Fredericksburg, VA (3)	Sediments/soil: wood preserving. Volume: 119K cubic yards.	Design: pilot-scale TS started 7/92. Expected cost: $23M.	In situ treatment, creosote recovery. 5% of site will undergo bioremediation.

Table 11-3. Continued

Site Location (Regions)	Primary Contaminants	Status/Cost	Treatment
Charleston AFB, Charleston, SC (4)	Soil (vadose sand): petroleum (jet fuel), solvents (1,1-DCE; 1,1,1-TCA; TCE; VC; trans-1,2-DCE; PCE; and dichloromethane), lead. Volume: 25 cubic yards.	Pilot-scale TS started 11/92. Expected completion 12/93.	Aerobic treatment, bioventing. Less than 10% of the site under bioremediation.
Eglin AFB, FL (4)	Soil (vadose): petroleum (jet fuel).	Completed field-scale study.	Aerobic treatment, bioventing. Nutrient and hydrogen peroxide addition [27].
Savannah River Site Aiken, NC (4)	Soil (vadose)/ground water/sediments: chlorinated solvents (TCE and PCE).	Operational: pilot-scale research study.	Aerobic treatment, horizontal wells, methane addition [18; 19].
Stallworth Timber Beatrice, AL (4)	Soil (sand, silt)/ground water: wood preserving (PCP).	Pre-design.	In situ aerobic treatment, nutrient addition. Ex situ treatment, activated sludge, continuous flow. Exogenous and indigenous organisms. 100% of site will undergo bioremediation.
Allied Chemical Ironton, OH (5)	Sediments (coal and coke fines): PAHs, arsenic. Volume: 500K cubic yards. Soil (saturated)/ground water: BTEX.	Design: pilot-scale TS study completed. Expected cost: $26M. Pilot-scale TS completed.	Aerobic treatment. 50% of site will undergo bioremediation.
Amoco Production Co. Kalkaska, MI (5)			Aerobic treatment, air sparging [49].
B&G Trucking Company Rochester, MN (5)	Soil (vadose and saturated)/ground water: petroleum (lube oil). Volume: 700 cubic yards.	Operational: full-scale. Start data: 4/91. Incurred cost: $341K.	In situ treatment. Ex situ treatment, sequencing batch reactor, continuous flow. Aerobic conditions. 75% of site under bioremediation.
Bendix Corp./Allied[a] Automotive Site St. Joseph, MI (5)	Ground water: solvents (TCE, DCE, DCA, VC).	Pre-design: lab-scale TS underway.	Aerobic and anaerobic treatment.
Unnamed site[b] Buchanan, MI (5)	Ground water: BTEX, PCE, TCE, DCE.	Pilot field study started 3/93. Expected completion 3/94.	Aerobic treatment.
Galesburg/Kopper Galesburg, IL (5)	Soil: phenols, chlorophenol, PNAs, PCP, PAHs.	Pre-design. Start date (est): 12/92.	Nutrient addition. 100% of site under bioremediation.
Hentchells Traverse City, MI (5)	Soil/ground water: petroleum.	Operational: full-scale. Start date: 9/85.	Aerobic treatment, biosparging.

Site	Matrix/Contaminants	Status	Treatment
Chillicothe, OH (5)	[BTEX, acetone, THF].	Full-scale system being installed.	peroxide, nutrient addition (nitrogen, phosphorus). Ex situ treatment, GAC bioreactor. 100% of site will undergo bioremediation.
K.I. Sawyer AFB, Marquette, MI (5)	Soil (vadose sand): petroleum.	Field TS report expected 10/93.	Aerobic treatment, bioventing.
Mayville Fire Department, Mayville, MI (5)	Ground water: petroleum.	Operational: full-scale since 5/90. Completion date (est): 1/94.	Aerobic treatment, air sparging. 100% of site will undergo bioremediation.
Michigan Air National Guard, Battle Creek, MI (5)	Soil (vadose: sand, silt): petroleum, heavy metals.	Design: pilot-scale TS started 9/92. Start date (est): 9/93. Expected cost: capital, $3,000; O&M, $1,268.	Aerobic treatment, bioventing. 100% of site will undergo bioremediation.
Newark AFB, Newark, OH (5)	Soil (vadose: silt, clay): petroleum (gasoline). Volume: 60 cubic yards.	Design: pilot-scale TS started 8/92. Expected completion 8/94. Expected cost: capital, $35K; O&M, $2K.	Aerobic treatment, bioventing. 40% of site under bioremediation.
Onalaska Municipal Landfill, Lacrosse County, WI (5)	Soil (vadose and saturated sand): solvents (TCE), petroleum (total hydrocarbons), wood preserving (naphthalene). Volume: 5,000 cubic yards.	Design: lab-scale TS completed 3/92. Expected cost: capital, $400K; O&M, $20K.	Aerobic treatment, bioventing. 20% of site will undergo bioremediation.
Parke-Davis, Holland, MI (5)	Soil/ground water: petroleum, solvents, arsenic, chloride, zinc.	Pre-design.	In situ treatment. Ex situ treatment, fixed film.
Reilly Tar & Chemical[a,b], St. Louis Park, MN (5)	Soil (vadose loam): wood preserving (PAHs).	Design: pilot-scale TS started 11/92. Expected completion 11/95. Incurred cost: $25K. Expected cost: $70K.	Aerobic treatment, bioventing, nutrient addition [50].
Sheboygan River and Harbor, Sheboygan, IL (5)	Sediments (sand, silt, clay): PCBs. Volume: 2,500 cubic yards.	Lab- and pilot-scale TS are being conducted.	In situ treatment, capping of sediments. Ex situ treatment, confined treatment facility (tank). Aerobic and anaerobic conditions.
West K&L Avenue Landfill[a], Kalamazoo, MI (5)	Ground water: solvents (acetone; TCE; trans-1,2-DCE; 1,2-DCA; 1,1-DCA; BTEX; VC; methyl isobutyl ketone; MEK).	Design: pilot- and lab-scale TS ongoing.	Anaerobic treatment under sulfate reducing conditions.

Table 11-3. Continued

Site Location (Regions)	Primary Contaminants	Status/Cost	Treatment
Wright-Patterson AFB Dayton, OH (5)	Soil (vadose: sand, silt, clay): petroleum (jet fuel). Volume: 7.5K cubic yards.	Pre-design: pilot-scale studies planned. Expected completion 3/94.	Aerobic treatment, bioventing. 100% of site will undergo bioremediation.
Dow Chemical Company Plaquemine, LA (6)	Ground water: solvents (1,1-DCA; 1,2-DCA; 1,1,1-TCA; 1,1-DCE, chloroethane). Volume: 90K cubic yards.	Design: pilot-scale started 3/93. Expected cost: capital, $1M; O&M, $50K. Incurred cost: capital, $250K; O&M, $10K.	Anaerobic treatment, nutrient addition. Less than 1% of the site under bioremediation. Experiencing nutrient dispersion problems [46].
French Limited Crosby, TX (6)	Sediments (sand, silt)/sludge/soil (sand, silt, clay)/ground water: PCBs, arsenic, petroleum, BAP, VOCs.	Operational: full-scale since 1/92. Expected cost: $90M.	Aerobic treatment, pure oxygen dissolution system, nutrient addition (soil, water, sediments). 100% of site under bioremediation.
Kelly AFB San Antonio, TX (6)	Soil (vadose clay): petroleum (jet fuel), solvents (PCE, TCE, VC, DCE).	Operational: full-scale since 2/93. Completion date (est): 9/94.	Aerobic treatment, bioventing.
Fairfield Coal & Gas Fairfield, IA (7)	Soil (saturated: sand, silt, clay)/ground water: coal tar (BTEX, PAHs).	Design: pilot-scale TS started 12/91. Expected completion 12/93. Expected cost: $1.6M.	Aerobic treatment, injection and extraction wells, hydrogen peroxide, nitrate addition.
Offutt AFB LaPlatte, NE (7)	Soil (vadose: sand, silt): petroleum (TPH), arsenic, barium, lead, zinc. Volume: 700 cubic yards.	Design: pilot-scale TS started 8/92.	Aerobic treatment, bioventing. 10% of site under bioremediation.
Park City[a] Park City, KS (7)	Ground water: petroleum (lube oil), benzene. Volume: 700K cubic feet.	Design: pilot-scale TS completed. Incurred cost: $275K. Expected cost: $650K.	Ground water: in situ treatment. Possible bioventing for soils. Anaerobic and aerobic conditions [23; 24].
Burlington Northern Tie Plant Somers, MT (8)	Soil/ground water: wood preserving (PAHs). Volume: 82K cubic yards.	Installed: full-scale. Start date (est): 7/92. Expected cost: $11M.	In situ treatment. Ex situ land treatment. Aerobic conditions. 80% of site will undergo bioremediation.
Geraldine Airport Geraldine, MT (8)	Soil (vadose: sand, silt, loam, clay): pesticides (aldrin; dieldrin; endrin; chlordane; toxaphene; β-BHC; 4,4'-DDE; 4,4'-DDT; 4,4'-DDD); herbicides (2,4-D).	Pre-design.	In situ treatment. Ex situ treatment. Aerobic and anaerobic conditions.

Site	Waste/Media	Status/Cost	Treatment
Bozeman, MT (8)	PAHs, dioxins/furans.		nutrient addition. Ex situ treatment, fixed film, slurry reactor. Aerobic conditions.
Hill AFB[a] Salt Lake City, UT (8)	Soil: petroleum (JP-4 jet fuel).	Operational: full-scale since 9/91. Completion date (est): 9/94.	Aerobic treatment, bioventing. 100% of site under bioremediation [40].
Libby Groundwater Site[a] Libby, MT (8)	Soil (vadose and saturated)/ground water; wood preserving (PAHs, pyrene, PCP, dioxin, naphthalene, phenanthrene, benzene, arsenic). Volume: 45K cubic yards.	Operational: three pilot-scale efforts ongoing. Incurred cost: $4M. TS results available (est): 8/93 and 4/94.	In situ treatment (ground water), ex situ land treatment (soil), nutrient addition (soil, water). Also, treatment of ground water in bioreactor. Aerobic conditions. 75% of site under bioremediation.
Public Service Company[a] Denver, CO (8)	Ground water: petroleum. Volume: 12M gallon.	Completed: full-scale 3/92. Start date: 06/89. Incurred cost: $500K.	Aerobic treatment, hydrogen peroxide, nutrient addition, combined bioprocess.
Beale AFB Marysville, CA (9)	Soil (vadose silty clay): petroleum (gasoline, diesel), solvents (TCE), lead. Volume: 163K cubic yards.	Seven processes are planned. 4 are in design (pilot-scale), 2 are in pre-design (full-scale), and 1 is presently operating (completion date 7/95). Expected cost: capital $500K; O&M, $136K.	In situ aerobic treatment, bioventing. Ex situ aerobic treatment, pile. Results of 4 bioventing TS expected 2/94.
Converse/Montabello Corporation Yard Montabello, CA (9)	Soil (vadose silt): petroleum (gas, diesel).	Design: pilot-scale TS started 5/93. Expected completion: 12/93.	Aerobic treatment, bioventing, nutrient addition. 10% of site under bioremediation.
Former Service Station Los Angeles, CA (9)	Soil/ground water: petroleum. Volume: 3,000 cubic yards.	Completed: full-scale 3/91. Start date: 11/88. Incurred cost: $1.6M.	Aerobic treatment, hydrogen peroxide, nutrient addition (water), closed loop system. 65% of site underwent bioremediation.
Koppers Company, Inc. Orville, CA (9)	Soil (vadose: sand, clay, gravel, cobbles): wood preserving (PCP, PAHs, dioxins/furans), arsenic, chromium. Volume: 110K cubic yards.	Pre-design: pilot-scale TS planned. Expected completion 11/94. Expected cost: capital, $4.5M; O&M, $7.7M.	Aerobic treatment, nutrient addition. 30% of site will undergo bioremediation. (20 year remedial effort.)
Marine Corps Air/Ground Combat Center Twenty-Nine Palms, CA (9)	Soil: petroleum (jet fuel, gasoline, diesel, aviation fluid, transmission fluid).	Design: full-scale.	Aerobic treatment, bioventing.

Table 11-3. Continued

Site Location (Regions)	Primary Contaminants	Status/Cost	Treatment
Naval Air Station Fallon Fallon, NV (9)	Soil (vadose and saturated silt)/ ground water: petroleum (jet fuel, p-xylene, naphthalene, 1-methyl naphthalene, n-butylbenzene), arsenic.	Design: pilot-scale TS started 10/92.	Aerobic treatment, bioventing, nutrient addition (soil), oil/water separation.
Naval Weapons Station Seal Beach, CA (9)	Ground water: petroleum.	TS conducted or in progress: laboratory-scale.	Aerobic and anaerobic treatment.
Oakland Chinatown Oakland, CA (9)	Soil (saturated sand): ground water: petroleum.	Volume: 10K cubic yards. Completed: full-scale 8/90. Start date: 3/89.	Aerobic treatment, hydrogen peroxide and nutrient addition.
San Diego Gas and Electric San Diego, CA (9)	Soil (sand): petroleum (gasoline). Volume: 1,200 cubic yards.	Completed: full-scale 4/93. Start date: 10/89.	Aerobic treatment. 100% of the site underwent bioremediation [5].
Williams AFB[b] Phoenix, AZ (9)	Soil (vadose): petroleum (JP-4 jet fuel).	Pilot field testing started 5/92. Test completion 6/93.	In situ treatment, bacterial supplementation (non-indigenous micro aerofilic bacteria).
East 15th Street Service Station Anchorage, AK (10)	Soil: petroleum (TPH diesel). Volume: 1,500 cubic yards.	Operational: full-scale since 6/92. Incurred cost: $75K. Expected cost: $200K.	Aerobic treatment, bioventing. 20% of site under bioremediation.
Eielson AFB[a] Fairbanks, AK (10)	Soil (sand/silt): petroleum (JP-4 jet fuel).	Operational: pilot full-scale. Start date: 9/91. Completion date (est): 9/94.	Aerobic treatment, bioventing, soil warming [42].
Fairchild AFB Spokane, WA (10)	Soil (vadose and saturated silt)/ground water: petroleum, solvents (TCE).	3 separated processes are planned. The first process is in pre-design; a pilot scale TS should start 1/95. The remaining two started pilot-scale TSs in 4/93.	Aerobic treatment, bioventing, nutrient addition.

TS - Treatability Study; [a] Bioremediation Field Initiative; [b] Superfund Innovative Technology Evaluation (SITE) Demonstration.

ence Applications International Corporation (SAIC) under Contract No. 68-C8-0062 and Contract No. 68-C0-0048. Mr. Eugene Harris served as the EPA Technical Project Monitor. Mr. Jim Rawe served as SAIC's Work Assignment Manager. This bulletin was authored by Mr. Rawe and Ms. Evelyn Meagher-Hartzell of SAIC.

The following other Agency and contractor personnel have contributed their time and comments by participating in the expert review meetings or independently reviewing the chapter:

Mr. Hugh Russell, EPA-RSKERL
Ms. Tish Zimmerman, EPA-OERR
Mr. Al Venosa, EPA-RREL
Dr. Robert Irvine, University of Notre Dame
Dr. Ralph Portier, Louisiana State University
Mr. Clyde Dial, SAIC

REFERENCES

1.   U.S. Environmental Protection Agency. Bioremediation in the Field. EPA/540/N-93/002, August 1993. [Editor's Note: most recent issue is No. 12, August 1995 (EPA/540/N-95/500).]

2.   U.S. Environmental Protection Agency. Superfund Innovative Technology Evaluation Program: Technology Profiles, Sixth Edition. EPA/540/R-93/526, November 1993. [Seventh Edition, EPA/540/R-94/526, November, 1994.]

3.   U.S. Environmental Protection Agency. Bioremediation Field Initiative. EPA/540/F-93/510, September 1993.

4.   U.S. Environmental Protection Agency. Handbook on In Situ Treatment of Hazardous Waste-Contaminated Soils. EPA/540/2-90/002 (NTIS PB90-155607), January 1990.

5.   Gersberg, R.M., W.J. Dawsey, and H. Ridgeway. Draft Final Report: In-Situ Microbial Degradation of Gasoline. EPRI Research Report Number RP 2795-2.

6.   Norris, R.D., K. Dowd, and C. Maudlin. The Use of Multiple Oxygen Sources and Nutrient Delivery Systems to Effect In Situ Bioremediation of Saturated and Unsaturated Soils. In: Symposium on Bioremediation of Hazardous Wastes: Research, Development, and Field Evaluations. EPA/600/R-93/054, May 1993.

7.   Hopkins, G.D., L. Semprini, and P. McCarty. Field Evaluation of Phenol for Co-Metabolism of Chlorinated Solvents. In: Symposium on Bioremediation of Hazardous Wastes: Research, Development, and Field Evaluations. EPA/600/R-93/054, May 1993.

8.   Hopkins G.D., L. Semprini, and P.L. McCarty. Evaluation of Enhanced In Situ Aerobic Biodegradation of cis- and trans-1-Trichloroethylene and cis- and trans-1,2-Dichloroethylene by Phenol-Utilizing Bacteria. In: Bioremediation of Hazardous Wastes. EPA/600/R-92/126, August 1992.

9.   Abramowicz, et al. 1991. In Situ Hudson River Research Study: A Field Study on Biodegradation of PCBs in Hudson River Sediments—Final Report. February 1992.

10. U.S. Environmental Protection Agency. Superfund LDR Guide #6A: Obtaining a Soil and Debris Treatability Variance for Remedial Actions. OSWER Directive 9347.306FS (NTIS PB91-921327), 1990.

11. U.S. Environmental Protection Agency. Superfund LDR Guide #6B: Obtaining a Soil and Debris Treatability Variance for Removal Actions. OSWER Directive 9347.306BFS (NTIS PB91-921310), 1990.

12. Sims, J., R. Sims, and J. Matthews. Bioremediation of Contaminated Surface Soils. EPA/600/9-89/073 (NTIS PB90-164047), August 1989.

13. Sims, J., R. Sims, R. Dupont, J. Matthews, and H. Russell. Engineering Issue—In Situ Bioremediation of Contaminated Unsaturated Subsurface Soils. EPA/540/S-93/501, May 1993. [See Chapter 12.]

14. Piotrowski, M.R. Bioremediation: Testing the Waters. Civil Engineering, August 1989. pp. 51–53.

15. U.S. Environmental Protection Agency. International Evaluation of In-Situ Biorestoration of Contaminated Soil and Groundwater. EPA/540/2-90/012, September 1990.

16. U.S. Environmental Protection Agency. Handbook: Remedial Actions at Waste Disposal Sites (Revised). EPA/625/6-85/006 (NTIS PB87-201034), October 1985.

17. U.S. Environmental Protection Agency. Review of In-Place Treatment Techniques for Contaminated Surface Soils. Volume 2: Background Information for In Situ Treatment. EPA/540/2-84/003b (NTIS PB85-124899), November 1984.

18.   Hazen, T.C. Test Plan for In Situ Bioremediation Demonstration of the Savannah River Integrated Demonstration Project DOE/OTD TTP No.: SR 0566-01 (U). U.S. Department of Energy, WSRC-RD-91-23, Revision 3, April 23, 1992.

19.   U.S. Department of Energy. Cleanup of VOCs in Non-Arid Soils—The Savannah River Integrated Demonstration. WSRC-MS-91-290, Rev. 1, p. 6.

20.   U.S. Environmental Protection Agency. Technology Screening Guide for Treatment of CERCLA Soils and Sludges. EPA/540/2-88/004 (NTIS PB89-132674), September 1988.

21.   U.S. Environmental Protection Agency. SITE Demonstration Bulletin—Augmented In Situ Subsurface Bioremediation Process, BIO-REM, Inc., EPA/540/MR-93/527, November 1993.

22.   McCarty P. and J. Wilson. Natural Anaerobic Treatment of a TCE Plume, St. Joseph, Michigan, NPL Site. In: Bioremediation of Hazardous Wastes. EPA/600/R-92/126, August 1992.

23.   Hutchins, S.R. and J.T. Wilson. Nitrate-Based Bioremediation of Petroleum-Contaminated Aquifer at Park City, Kansas: Site Characterization and Treatability Study. In: Hydrocarbon Bioremediation. R.E. Hinchee, B.C. Alleman, R.E. Hoeppel, and R.N. Miller, (eds.), CRC Press, Boca Raton, Florida, 1994.

24.   Kennedy, L.G. and S.R. Hutchins. Applied Geologic, Microbiological, and Engineering Constraints of In-Situ BTEX Bioremediation. Remediation. Winter 1992/93.

25.   New York State Department of Environmental Conservation. Final Report, Knispel Construction Company, Horseheads, New York, October 1990.

26.   Marley, M.C., et al. The Application of the In Situ Air Sparging as an Innovative Soils and Groundwater Remediation Technology. Groundwater Monitoring Review, pp. 137–144, Spring 1992.

27.   Federal Remediation Technologies Roundtable. Synopses of Federal Demonstrations of Innovative Site Remediation Technologies. EPA/542/B-92/003, August 1992.

28.   U.S. Environmental Protection Agency. Vendor Information System for Innovative Treatment Technologies (VISITT). EPA/540/2-91/001, June 1991. [Editor's Note: Latest software is VISITT Version 3.0; see User Manual EPA/542-R-94-003, available from VISITT Hotline 800/245-4505.]

29.   Environmental Protection Agency. A Citizen's Guide to Bioventing. EPA/542/F-92/008, March 1992.

30.   U.S. Environmental Protection Agency. Bioremediation in the Field. EPA/540/2-91/018 (NTIS PB92-224708), August 1991.

31.   U.S. Environmental Protection Agency. Bioremediation in the Field. EPA/540/N-91/001 (NTIS PB93-126175), March 1992.

32.   U.S. Environmental Protection Agency. The Superfund Innovative Technology Evaluation Program: Technology Profiles, Fifth Edition. EPA/540/R-92/077, December 1992.

33.   Hinchee, R.E., D.C. Downey, R.R. DuPont, P.K. Aggarwal, and R.N. Miller. Enhancing Biodegradation of Petroleum Hydrocarbons through Soil Venting. Journal of Hazardous Materials, Vol. 27, 1991.

34.   Nelson, C.H. A Natural Cleanup. Civil Engineering, March 1993.

35.   Wilson, J.T. and D.H. Kampbell. Retrospective Performance Evaluation on In Situ Bioremediation: Site Characterization. In: Symposium on Bioremediation of Hazardous Wastes: Research, Development, and Field Evaluations. EPA/600/R-93/054, May 1993.

36.   Day, S., K. Malchowsky, T. Schultz, P. Sayre, and G. Saylor. Draft Issue Paper: Potential Risk, Environmental and Ecological Effects Resulting from the Use of GEMS for Bioremediation. U.S. Environmental Protection Agency, Office of Pollution Prevention, April 1993.

37.   Hopkins, G.D., L. Semprini, and P.L. McCarty. Field Study of In Situ Trichloroethylene Degradation in Groundwater by Phenol-Oxidizing Microorganisms. In: Abstract Proceedings Nineteenth Annual RREL Hazardous Waste Research Symposium. EPA/600/R-93/040, April 1993.

38.   Beeman, R., S. Shoemaker, J. Howell, E. Salazar, and J. Buttram. A Field Evaluation of In Situ Microbial Reductive Dehalogenation by the Biotransformation of Chlorinated Ethylenes. In: Bioremediation of Chlorinated Polycyclic Aromatic Hydrocarbon Compounds, R.E. Hinchee, A. Leeson, L. Sempini, and S.K. Ong (eds.), CRC Press, Boca Raton, Florida, 1994.

39.   Norris, R.D., K. Dowd, C. Maudlin, and W.W. Irwin. The Use of Multiple Oxygen Sources and Nutrient Delivery Systems to Effect in Situ Bioremediation of Saturated and Unsaturated Soils. In: Hydrocarbon Bioremediation, R.E. Hinchee, B.C. Alleman, R.E. Hoeppel, and R.N. Miller (eds.), CRC Press, Boca Raton, Florida, 1994.

40.   U.S. Air Force Correspondence. Subject: Air Force's Bioventing Initiative. July 1993.

41.   U.S. Environmental Protection Agency. Bioremediation in the Field. EPA/540/N-92/004, October 1992.

42.   Sayles, G.D., R.E. Hinchee, C.M. Vogel, R.C. Brenner, and R.N. Miller. An Evaluation of Concurrent Bioventing of Jet Fuel and Several Soil Warming Methods: A Field Study at Eielson Air Force Base,

Alaska. In: Symposium on Bioremediation of Hazardous Wastes: Research, Development, and Field Evaluations. EPA/600/R-93/0-54, May 1993.

43. Sayles, G.D., R.E. Hinchee, R.C. Brenner, and R. Elliot. Documenting Bioventing of Jet Fuel to Great Depths: A Field Study at Hill Air Force Base, Utah. In: Symposium on Bioremediation of Hazardous Wastes: Research, Development, and Field Evaluations. EPA/600/R-93/0-54, May 1993.

44. Kampbell, D.H., J.T. Wilson, and C.J. Griffin. Performance of Bioventing at Traverse City, Michigan. In: Bioremediation of Hazardous Wastes. EPA/600/R-92/126, August 1992.

45. Kampbell, D.H., G.J. Griffin, and F.A. Blaha. Comparison of Bioventing and Air Sparging for In Situ Bioremediation of Fuels. In: Symposium on Bioremediation of Hazardous Wastes: Research, Development, and Field Evaluations. EPA/600/R93/054, May 1993.

46. U.S. Environmental Protection Agency. Summary Report on the EPA-Industry Meeting on Environmental Applications of Biotechnology. February 1990.

47. Remediation Technologies, Inc. Results of Biotreatability Testing of Pesticide- and Herbicide-Contaminated Soils from Richey Airport in Richey, Montana. For: Montana Department of Health and Environmental Sciences. October 1993.

48. Vogel, C. Enhanced In Situ Biodegradation of Petroleum Hydrocarbons Through Soil Venting. Tech data RDV91-7, Air Force E&S Center, Tyndall AFB, Florida. 1991.

49. Barker, G.W., Y.J. Beausoleil, J.S. Huber, and S.N. Neumann. Application of In Situ Air Sparging at a Hydrocarbon Contaminated Groundwater Site. In: Speaker Abstracts: In Situ and On-Site Bioreclamation. The Second International Symposium. San Diego, California, April 5–8, 1993.

50. Minnesota Pollution Control Agency. Reilly Tar Bioventing Demonstration—Superfund Fact Sheet. January 1993.

SELECTED ADDITIONAL REFERENCES

Methodologies for Evaluating In-Situ Bioremediation of Chlorinated Solvents. EPA/600/R-92/042 (NTIS PB92-146943), March 1992.

Bioremediation Resource Guide. EPA/542-B-93-004, September 1993.

Emerging Technology Summary: Pilot-Scale Demonstration of a Two-Stage Methanotrophic Bioreactor for Biodegradation of Contaminated Unsaturated Subsurface Soils. EPA/540/S-93/505, 1993. (SITE report.)

J.R. Simplot Ex-Situ Bioremediation Technology for Treatment of Dinoseb-Contaminated Soils, Innovative Technology Evaluation Report. EPA/540/R-94/508, September 1995. (SITE report.)

J.R. Simplot Ex-Situ Bioremediation Technology for Treatment of TNT-Contaminated Soils, Innovative Technology Evaluation Report. EPA/540/R-95/529, September 1995.

Bioremediation in the Field Search System (BFSS) User Documentation. EPA/540/R-95/508a&b, 1995.

U.S. Environmental Protection Agency (EPA). 1988-1994. Annual Symposium on Bioremediation of Hazardous Wastes: Research, Development, and Field Evaluations: First (1988), Second (1989), Third (1990, EPA/600/R-90/041), Fourth (1991, EPA/600/9-91/036, 72 pp.), Fifth (1992, EPA/600/R-92/126, 119 pp.), Sixth (1993, EPA/600/R-93/054), Seventh (1994, EPA/600/R-94/161).

Chapter 12

In Situ Bioremediation of Contaminated Unsaturated Subsurface Soils[1]

J.L. Sims, **R.C. Sims**, and **R.R. Dupont**, Utah State University, Logan, UT
J.E. Matthews, U.S. EPA, Robert S. Kerr Environmental Research Laboratory, Ada, OK (Retired)
H.H. Russell, U.S. Army Corps of Engineers, Tulsa, OK

INTRODUCTION

An emerging technology for the remediation of unsaturated subsurface soils involves the use of microorganisms to degrade contaminants which are present in such soils. Although in situ bioremediation has been used for a number of years in the restoration of ground water contaminated by petroleum hydrocarbons, it has only been in recent years that in situ systems have been directed toward contaminants in unsaturated subsurface soils. Research has contributed greatly to understanding the biotic, chemical, and hydrologic parameters which contribute to or restrict the application of in situ bioremediation, and has been successful at a number of locations in demonstrating its effectiveness at field scale.

This chapter is based on findings from the research community in concert with experience gained at sites undergoing remediation. The intent of the chapter is to provide an overview of the factors involved in in situ bioremediation, outline the types of information required in the application of such systems, and point out the advantages and limitations of this technology.

BACKGROUND

Bioremediation of contaminated surface soils using in situ systems, prepared bed, and aboveground bioreactors, has been previously addressed with regard to characterization, environmental processes and variables, and field-scale applications (Sims et al., 1989; see also Chapter 13). This chapter will address processes which are currently being utilized or are in development of treating contaminated unsaturated subsurface soils in place.

In situ remediation of subsurface soils contaminated with organic chemicals is an alternative treatment technology that, in certain cases, can meet the goal of achieving a permanent cleanup at hazardous waste sites. Use of such alternatives is encouraged by the U.S. Environmental Protection Agency (U.S. EPA) for implementing the requirements of the Superfund Amendments and Reauthorization Act (SARA) of 1986. Bioremediation of subsurface soils is consistent with the philosophical thrust of SARA, for it involves use of naturally occurring microorganisms to degrade and/or detoxify hazardous constituents in the soil to protect public health and the environment. Use of in situ subsurface bioremediation techniques in conjunction with chemical and physical treatment processes, i.e., "treatment trains," is an effective means for comprehensive site-specific remediation (Ross et al., 1988; Sims, 1990). For instance, bioremediation may be utilized to lower the concentration of organic contaminants in a soil matrix before stabilization or solidification is used as a remedial alternative for metals.

Bioremediation has been shown effective in reducing the overall mass of a variety of organic contaminants. Full-scale systems have been utilized to remediate soil contaminated with both crude and refined petroleum hydrocarbons (i.e., diesel fuel, gasoline), creosote, and pentachlorophenol. To

---

[1] EPA/540/S-93/501.

date, it has not been shown effective at removing highly structured, highly insoluble compounds such as polychlorinated biphenyls and dioxins.

For the purposes of this chapter, subsurface soil refers to unsaturated soil within the vadose zone at depths greater than three feet below the land surface. The vadose zone extends from the ground surface to the upper surface of the principal water-bearing formation (Everett et al., 1982). The vadose zone usually consists of three to six feet of topsoil (weathered geological materials) which gradually merges with deeper underlying earth materials such as deposition or transported clays or sands. In this zone, water primarily coexists with air, though saturated regions may occur. Perched water tables may develop at interfaces of layers (soils having different textures) of soil having less hydraulic conductivity. Prolonged infiltration also may result in transient saturated conditions. In some regions, the entire vadose zone may be hundreds of feet thick and the travel time of constituents to ground water can be hundreds or thousands of years. Other regions may be underlain by shallow potable aquifers that are especially susceptible to contamination due to short transport times and reduced potential for pollutant attenuation by soil materials and processes.

This chapter addresses specific environmental processes, factors, and data requirements for characterizing and evaluating the application of subsurface in situ bioremediation, and describes selected field-scale applications of recovery and delivery systems to enhance in situ subsurface soil bioremediation.

OVERVIEW: IN SITU SUBSURFACE MICROBIAL PROCESSES AND CONTROLLING ENVIRONMENTAL FACTORS

The rate and extent of biodegradation of organic chemicals during subsurface in situ bioremediation are influenced by several site-specific factors. These include type and activity of microbial populations; chemical environmental factors; bioavailability of the target chemical(s) and other substrates required for cometabolism, i.e., electron donor; mass transport of moisture, nutrients, and oxygen (the terminal electron acceptor in aerobic metabolism); toxicity; and stratigraphy, heterogeneity, and geochemistry of the surface or subsurface environment. A detailed discussion of the impact of these and other factors on bioremediation can be found in Transport and Fate of Contaminants in the Subsurface (EPA/625/4-89/019) and Bioremediation of Contaminated Surface Soils (EPA/600/9-89/073-Sims et al., 1989).

Microbial Populations

Successful in situ bioremediation depends on the presence of appropriate microbial populations which can be stimulated to degrade contaminants of concern by modifying or otherwise managing environmental conditions at a site. Results of microbial characterizations of deep subsurface material have indicated that: (1) microorganisms are present at populations sufficient to change the chemistry of the environment when stimulated; (2) the microbial communities are diverse and carry out a wide range of chemical transformations; (3) a majority (>95%) of the microbes are chemotrophic bacteria that degrade organic chemicals to obtain energy; and (4) environmental characteristics identified previously (oxygen concentration, nutrient status, moisture content) are important in influencing microbial activity and degradation patterns (Fliermans and Hazen, 1990).

Microbial communities in the subsurface are diverse and adaptable. Microbial populations at older sites are usually acclimated to the contaminants of concern. Therefore, levels of critical nutrients or electron acceptors, toxicity, and adverse environmental conditions most often are the major factors which limit the extent and rate of in situ bioremediation.

Critical Environmental Conditions

There are several environmental conditions that affect activity of soil microorganisms. These factors, along with individual soil and waste characteristics, all interact to affect microbial activity at specific contaminated sites. Many of these conditions can be managed to enhance biodegradation of organic constituents in subsurface soils. Optimum ranges for the most critical of these factors are presented in Table 12-1.

Table 12-1. Critical Environmental Factors for Microbial Activity (Sims et al., 1984; Huddleston et al., 1986; Rochkind and Blackburn, 1986; Paul and Clark, 1989)

Environmental Factor	Optimum Levels
Available soil water	25%–85% of water holding capacity; -0.01 MPa
Oxygen	Aerobic metabolism: Greater than 0.2 mg/L dissolved oxygen, minimum air-filled pore space of 10%; Anaerobic metabolism: O_2 concentrations <1%
Redox potential	Aerobes and facultative anaerobes: greater than 50 millivolts; Anaerobes: less than 50 millivolts
pH	5.5–8.5
Nutrients	Sufficient nitrogen, phosphorus, and other nutrients so not limiting to microbial growth; suggested C:N:P ratio of 100:10:1
Temperature	15°C–45°C (Mesophiles)

Water content of soil is an important factor which regulates microbial activity. Soil water serves as the transport medium through which many nutrients and organic constituents diffuse to the microbial cell, and through which metabolic waste products are removed. Soil water also affects soil aeration status, nature and amount of soluble materials, osmotic pressure, pH of the soil solution, and unsaturated hydraulic conductivity of the soil (Paul and Clark, 1989). The water content of deeper subsurface soils may vary greatly. Unsaturated soil samples have been obtained even from cores collected below the water table in deep subsurface environments, and the low water content was shown to adversely affect microbial activity (Kieft et al., 1990).

Biodegradation rates often depend on the rate at which terminal electron acceptors can be supplied. A large fraction of the microbial population within soils are aerobes which use oxygen as the terminal electron acceptor. Oxygen can be easily depleted in subsurface soils where there is an oxygen demand due to plant root respiration or due to normal microbial activity throughout the depth of the unsaturated zone. Oxygen levels tend to decrease in soils having high clay and organic matter content. Clayey soils tend to retain higher moisture content, which restricts oxygen diffusion, while organic matter may increase microbial activity and deplete available oxygen. Under these circumstances, oxygen may be consumed faster than it can be replaced by diffusion from the atmosphere, and the soil may become anoxic.

Facultative anaerobic organisms (which can use oxygen or alternative electron acceptors such as nitrate or sulfate in the absence of oxygen) and obligate anaerobic organisms then become the dominant populations under such conditions. The sequence of use of various electron acceptors is determined by the redox potential and the electron affinity of the electron acceptors present (Zehnder and Stumm, 1988). The potential of alternative electron acceptors has been evaluated with nitrate at field scale for contaminants (including benzene, toluene, and xylene) in an aquifer environment (Hutchins et al., 1991).

Redox potential also affects metabolic processes in subsurface microbial populations (Paul and Clark, 1989). Redox potential provides a measurement of electron density and is related to the oxygen status of a subsurface soil. As oxygen is removed and a system becomes more reduced, there is a corresponding increase in electron density, resulting progressively in an increased negative potential.

Soil pH affects growth and activity of subsurface soil microorganisms. Fungi are generally more tolerant of acidic soil conditions (below pH 5) than bacteria. Solubility of phosphorus, a critical nutrient in biological systems, is maximized at a pH value of 6.5. A specific contaminated soil system may require management of soil pH to achieve levels that maximize microbial activity. Control of pH to enhance microbial activity may also aid in the immobilization of hazardous metals in a subsurface soil system (a pH level greater than 6 is recommended to minimize metal transport). Subsurface soil pH may be managed through addition of an aqueous phase containing pH adjusting chemicals through gravity delivery systems such as infiltration galleries or surface irrigation systems.

Microbial metabolism and growth depends upon adequate supplies of essential macro- and micronutrients. Critical nutrients such as nitrogen and phosphorous must be present and available to microorganisms in: (1) usable form; (2) appropriate concentrations; and (3) proper ratios (Dragun, 1988). If wastes are high in carbon (C), and low in nitrogen (N) and phosphorus (P), biodegradation will cease when available N and P are depleted. Therefore, fertilization of subsurface soils may be required as a management technique to enhance microbial degradation.

Soil temperature affects microbial growth and metabolic activity. Biodegradation rates decrease as temperature drops and essentially cease at temperatures below 0°C. While surface soils exhibit both diurnal and seasonal variations in temperature, changes of temperature decrease with depth. Generally, only the top 30 feet of the subsurface profile are affected by seasonal variations in temperature; temperature is generally constant and corresponds to the mean annual air temperature of the locality (Kuznetsov et al., 1963; Matthess, 1982). In the United States, temperatures in this zone range from 3°C to 25°C (Dunlap and McNabb, 1973). Due to the high specific heat of water, wet soils are less subject to larger diurnal changes than dry soils (Paul and Clark, 1989).

Bioavailability is a general term which refers to the accessibility of contaminants by degrading populations. There are two major components involved: (1) a physical aspect related to phase distribution and mass transfer limitations of the contaminant, and (2) a physiological aspect related to the suitability of the contaminant as a substrate.

Major factors which affect bioavailability include water solubility and sorption. Target chemicals may occur in one or more of the four phases comprising the subsurface soil environment: (1) soil solids, including organic matter and inorganic sand, silt, and clay particles; (2) soil water; (3) soil gas; and (4) often a nonaqueous phase liquid (NAPL). In general, chemicals that distribute to the water phase (more soluble) are more bioavailable than chemicals that either sorb strongly to solid phases or occur in a NAPL phase. NAPLs are generally degraded from the water:NAPL interface inward since the aqueous phase contains nutrients, oxygen, and moisture required for microbial life processes. The bioavailability of a NAPL phase may be increased by increasing the surface area to volume ratio of NAPL elements. This increases mass transfer of nutrients, moisture, and oxygen; and decreases toxicity by decreasing interfacial concentrations (Symons and Sims, 1988). Substrate chemicals in the gas phase have also been found to be bioavailable (Dupont et al., 1991; Miller et al., 1991). Generally, chemicals that are highly sorbed, such as high molecular weight PAHs present in creosote, petroleum, and manufactured town gas plant wastes, are found to be degraded at slower rates than chemicals that are only slightly sorbed. Since the majority of the mass of target constituents at many contaminated sites is associated with NAPL and/or solid phases, these represent the greatest challenge with regard to in situ bioremediation.

Bioavailability is also a function of the biodegradability of the target chemical, i.e., whether it acts as a substrate, cosubstrate, or is recalcitrant. The target chemical may be physically available (i.e., water soluble and/or not sorbed to solids) but not useful as a metabolic substrate. Contaminants of concern may not be the dominant organic substrate in a system. When the target chemical cannot serve as a substrate (source of carbon and energy) for microorganisms, but is oxidized in the presence of a substrate already present or added to the subsurface, the process is referred to as cooxidation and the target chemical is defined as the cosubstrate (Keck et al., 1989; Sims et al., 1989). Cooxidation processes are important for the biodegradation of high molecular weight polycyclic aromatic hydrocarbons (PAHs), and some chlorinated solvents, including trichloroethylene (TCE). Contaminants with complex molecular structures or high degrees of toxicity may not be degradable, and may persist or be recalcitrant under aerobic conditions. Examples of recalcitrant compounds include highly oxidized halogenated compounds such as polychlorinated biphenyls (PCBs), pesticides such as toxaphene, and dioxin contaminants present in wood-preserving wastes.

The toxicity of the environment may be reduced by decreasing the concentration of a toxic waste (e.g., creosote) or chemical (e.g., pentachlorophenol) within one or more subsurface phases. Concentrations of toxic chemicals in the gas phase may be reduced through soil vacuum extraction; in the water phase through soil flushing; in the NAPL phase through soil flushing with water containing viscosifiers, or with solvents or surfactants; and in the soil solid phase by inducing partitioning of contaminants from solid to fluid phases. All mobile phases in the subsurface have potential for escape; therefore, containment strategies are often necessary while the constituents with the phase are biodegraded.

Heterogeneity of the subsurface environment limits the rate and extent of in situ bioremediation. Restrictive layers (e.g., clay lenses), although more resistant to contamination, are also more difficult to remediate due to poor permeability and low rates of diffusion. Clay soils have larger porosities than silty or sandy soils and therefore larger storage capacities for contaminants, but have greater resistance to fluid flow including aqueous, gas, and NAPL phases. Also clay layers with poor hydraulic conductivity are less permeable to nutrients and oxygen. In sites that have substantial clay and silt deposits, more permeable soils will become preferential conduits for remedial fluids, and the clay/silt deposits will require much longer time frames for remediation. For example, heterogeneity of the subsurface with respect to soil layering and chemical parameters at a gas works site in the United Kingdom presented constraints on the feasibility of utilizing in situ biomediation (Thomas et al., 1991).

ENHANCEMENT OF IN SITU SUBSURFACE BIOREMEDIATION

The method of enhancing in situ bioremediation efforts depends on the four phases in which contaminants can occur, heterogeneity of subsurface matrix, and the types of delivery and recovery systems utilized. Removing limiting or controlling factors and establishing favorable conditions are the primary goals of recovery and/or delivery systems. Enhancement may be achieved by increasing bioavailability; reducing toxicity; increasing delivery of moisture, nutrients, and oxygen; and/or by introducing substrates that stimulate indigenous microbial degradative activity.

A variety of strategies may be implemented to maximize biodegradation activity in contaminated subsurface soils. The success of in situ bioremediation efforts is often determined by the effectiveness of the recovery and delivery systems used to remove major sources of contaminants and to transport nutrients and electron acceptors to the location of the remaining contaminants. Establishing optimum levels of essential nutrients and electron acceptors at specific subsurface locations is often driven by physical limitations of the subsurface matrix on transport of fluids (liquids and gases) used to delivery these amendments. Overcoming these limitations is the primary goal of a delivery system, and the development of adequate delivery technologies continues to be the major challenge of in situ bioremediation. A summary of delivery and recovery techniques commonly used to manage subsurface remediation is provided in Table 12-2.

MAKING THE SATURATED ZONE UNSATURATED

Advantages of Unsaturated Systems

Because hydraulic conductivity is a function of soil moisture content, changing a saturated soil into an unsaturated soil greatly reduces the hydraulic conductivity and therefore the downward transport of chemicals in the water phase to the ground water. Also, because oxygen diffuses through air 10,000 times faster than through water, an unsaturated environment may be maintained in an aerobic condition more easily than a saturated environment in the presence of oxygen-demanding chemicals (Table 12-3). Soil pore space that contains a gas phase also allows removal of volatile contaminants (via soil vacuum extraction) in a direction that is away from the ground water. Therefore, management of a site to change the saturated zone to an unsaturated condition may reduce potential for ground-water contamination as well as enhanced oxygen delivery to stimulate in situ biodegradation.

Physical Containment

There are a variety of approaches to establishing and maintaining dewatered conditions. In order to adequately dewater the subsurface, it is often necessary to physically isolate the treatment zone. Impermeable subsurface barriers can prevent the migration of ground water by preventing uncontaminated water from entering the contaminated site and stopping contaminated water from leaving. Extraction systems or drains must then be used to remove the ground water to create an unsaturated zone.

Commonly used barriers include slurry walls, grout curtains and sheet piling cutoff walls to retard the flow of water under and through a site (Devinny et al., 1990).

Table 12-2. Management Strategies for Addressing Factors Limiting In Situ Bioremediation of Subsurface Soils

Limiting Factor	Management Response	Delivery or Recovery Technique
Bioavailability limited due to NAPL	Reduce NAPL mass	Gravity or forced delivery; Soil flushing, Steam stripping, Hydraulic fracturing
Bioavailability limited by sorption or slow mass transport through soil matrix	Reduce sorption, increase mass transport	Soil flushing, Steam stripping, Hydraulic fracturing
Moisture	Add water or water saturated air	Gravity or forced delivery; Bioventing, Cyclic pumping
Nutrients	Add nutrients in water or as ammonia gas	Gravity or forced delivery; Bioventing, Cyclic pumping
Oxygen/Redox	Add air	Bioventing, Hydraulic fracturing, Cyclic pumping, Radial drilling, Kerfing
Toxicity	Remove chemicals	Soil vacuum extraction, Soil flushing, Steam stripping
pH	Adjust soil pH	Gravity or forced delivery
Temperature	Increase temperature	Radio frequency heating, Steam stripping
Substrate Addition	Add in water or air	Gravity or forced delivery; Bioventing, Hydraulic fracturing
Heterogeneity	Add or withdraw material in more restrictive layers	Cyclic pumping, Hydraulic fracturing, Radial drilling, Kerfing

Table 12-3. Carrier Fluid Oxygen Supply Requirements (Dupont et al., 1991)

Carrier	g Carrier/g O_2
Water	
Air Saturated	110,000
Pure O_2 Saturated	22,000
500 mg/L H_2O_2 (100% Utilization)	2,000
Air (20.0% O_2)	13

Ground-Water Removal

Ground-water removal can be accomplished by hydraulic pumping and/or drainage trenches. Hydraulic pumping using a well-point system is one such technique (Devinny et al., 1990) using short lengths of plastic or Teflon wellscreen placed in the saturated zone.

Ground water can also be removed using subsurface drains or drainage ditches. Subsurface drains are constructed by excavating a trench to the desired depth, partially backfilling the trench with highly permeable sand or gravel, placing a plastic or ceramic drain tile in the sand and gravel bed, and completing the backfilling (Devinny et al., 1990).

Drainage ditches or surface drains are similar to subsurface drains except that no collection pipes or tiles and backfills are used. They may be used at sites underlain by poorly permeable soils (Devinny et al., 1990).

RECOVERY AND DELIVERY TECHNOLOGIES FOR SUBSURFACE BIOREMEDIATION

Recovery and delivery technologies are those that facilitate transport of materials either out of or into the subsurface (Murdoch et al., 1990). Recovery technologies are primarily utilized for contaminant source reduction. High levels of contamination present as either trapped residuals or NAPLs can severely limit success of bioremediation attempts. Therefore, removal of as much of this initial contaminant mass as possible is a prerequisite to in situ bioremediation efforts.

Specific recovery and delivery technologies for enhancing in situ bioremediation of subsurface soils are identified in Table 12-2. Each identified technology is discussed below with regard to its applications and limitations, and current status.

RECOVERY TECHNOLOGIES

The principal recovery technologies used for subsurface remediation depend on the ability to move fluids. Also involved is the ability to move contaminants by altering their solubility or sorption characteristics (Murdoch et al., 1990). These techniques are used to move materials from the subsurface soil environment in order to enhance in situ bioremediation by addressing one or more limiting factors identified in Tables 12-1 and 12-2, including: soil vacuum extraction, soil flushing, steam stripping, and radio frequency heating.

Soil Vacuum Extraction

Soil vacuum extraction (SVE) (also referred to as subsurface or forced air venting, in situ air stripping, or soil vapor extraction) involves the removal or contaminants carried in the soil gas phase by reduction of the vapor pressure within the soil pores by applying a vacuum. As clean air is drawn through the soil, the contaminants are removed. This process is driven by concentration differences between solid, aqueous, and NAPL phases and the clean air that is introduced through the soil vacuum extraction process.

Vacuum extraction is most applicable to sites contaminated with highly volatile compounds, such as those associated with gasoline and solvents (e.g., perchloroethylene, trichloroethylene, dichloroethylene, trichloroethane, benzene, toluene, ethylbenzene, and xylene).

Important soil characteristics that should be measured or estimated to determine the feasibility of vacuum extraction at a specific site include physical factors that control the rate and extent of air flow through contaminated soil, and chemical factors that determine the amount of contaminant that partitions from soil to air. These factors include: bulk density (weight per volume); total porosity (void spaces between soil grains) and air-filled porosity (that portion of the total porosity filled with air); diffusivity of volatiles (amount of volatiles which move through an area over time); soil moisture content (percentage of void spaces filled with water); air phase permeability (ease with which air moves through soils); texture; structure; mineral content; surface area; temperature; organic carbon content; heterogeneity; depth of air permeable zone; and depth to water table (Metcalf & Eddy, Inc., 1991). Soils at sites where vacuum extraction is used should be fairly homogeneous and have high permeability, porosity, and uniform particle-size distributions (Metcalf & Eddy, Inc., 1991). Soil vapor transport can be severely limited in a soil with high bulk density, high soil water or high NAPL content, low porosity, and low permeability. In heterogeneous soils, air flows preferentially through more permeable zones, leaving less permeable zones untreated.

Contaminant characteristics that affect the feasibility of vacuum extraction include the extent and degree of contamination, vapor pressure, Henry's law constant, aqueous solubility, diffusivity, and partition coefficients. Due to the high solubility of many organic contaminants in NAPL phases, the presence of NAPL in subsurface soil systems may significantly affect the distribution of the compounds in various phases, and their fate in SVE systems. Specific contaminant and soil conditions that determine the feasibility of vacuum extraction are presented in Table 12-4.

The efficiency of a vacuum extraction system can be enhanced in several ways. For example, a system of air injection wells can be installed at the perimeter of a contaminated area (Metcalf & Eddy, Inc., 1991) which can be connected to air blowers to force air into the soil or remain open to the atmosphere. Use of air injection wells can result in increased soil air flow rates and a larger area through which clean air can move.

Pulsed pumping may be used to give contaminants time to desorb from solid surfaces, diffuse from restricting layers, and volatilize from residual saturation (NAPL) in the soil pore space. Using pulsed

Table 12-4. Conditions Affecting Feasibility of Use of Vacuum Extraction (U.S. EPA, 1990; Metcalf & Eddy, Inc., 1991)

Condition	Favorable	Unfavorable
Contaminant:		
Dominant form	Vapor phase	Solid or strongly sorbed to soil
Vapor pressure	>100 mm of mercury	<10 mm of mercury
Water solubility	<100 mg/L	>1,000 mg/L
Henry's law constant	>0.01 (dimensionless)	<0.01 (dimensionless)
Soil:		
Temperature	>20°C (usually will require external heating of soils)	<10°C (common in northern climates)
Air conductivity	>10^{-4} cm/s	<10^{-6} cm/s
Moisture content	<10% (by volume)	>10% (by volume)
Composition	Homogeneous	Heterogeneous
Surface area of soil matrix	<0.1 m²/g of soil	>1.0 m²/g of soil
Depth to ground water	>20 m	<1 m

pumping for recovery of contaminants allows a lower volume of air with higher concentrations of contaminants to be recovered.

If ground water is at or near the zone of soil contamination, water table rise may occur due to reduced air pressure near extraction wells (Metcalf & Eddy, Inc., 1991). Ground-water pumping may be used to counteract the water table rise, as well as to expose additional contaminated soil that can be treated by vacuum extraction.

Horizontal extraction wells (wells drilled parallel to ground surface) have been used for deep subsurface contamination at the U.S. Department of Energy Savannah River facility to access larger areas of the contaminated site (Hazen, 1992). This use of horizontal wells may be a means to reduce costs associated with deep subsurface remediation, since only a single hole may be required to access contaminated areas instead of many vertical wells.

The performance of a vacuum extraction system is monitored by system operational characteristics and by treatment efficiency characteristics (Metcalf & Eddy, Inc., 1991). System characteristics include strength of vacuum applied, air flow rate, and contaminant concentrations and moisture content in the vented gas. Wells are used to monitor pressure in the contaminated area. Efficiency of treatment is monitored by soil gas analyses and soil core analyses, to determine residual concentration of contaminants. For more detailed discussions of soil venting evaluation, see "Evaluation of Soil Venting Application" (Chapter 9).

Since soil vacuum extraction is an in situ treatment technique that requires only addition of ambient air to the subsurface, it can be applied with little disturbance to existing facilities and operations (Metcalf & Eddy, Inc., 1991). SVE can be used at sites where areas of contamination are large and deep, or when the contamination is beneath a building. The system can be easily modified, depending on additional analytical and subsurface characterization data and/or changing site conditions. Even if vacuum extraction can be implemented at a site, most of the conditions listed in Table 12-4 must be met, or the cost and time for cleanup will be prohibitive.

The use of SVE at remedial sites has been reviewed by the U.S. EPA (1989) and classified as a developed technology for remedial applications. It is currently the most commonly used in situ remedial technology (Murdoch et al., 1990). Soil vacuum extraction may be used to reduce toxic concentrations of contaminants to levels which are more conducive to bioremediation. In addition, it will also deliver oxygen to the subsurface, which is required by aerobic bacteria.

Soil Flushing

In situ soil flushing is used to accelerate movement of contaminants through unsaturated materials by solubilizing, emulsifying, or chemically modifying the contaminants. A treatment solution is applied to the soil and allowed to percolate downward and interact with contaminating chemicals. Contaminants are mobilized by the treatment solution and transported downward to a saturated zone, where they are captured in drains or wells and pumped to the surface for recovery, treatment, or disposal (Murdoch et al., 1990). In combination with bioremediation, the flushing solution may be amended with nutrients to enhance biological activity (Metcalf & Eddy, Inc., 1991).

Treatment solutions are delivered to the contaminated zone by using either gravity or forced methods. Forced delivery consists of various pumping techniques. Gravity delivery methods include surface flooding, ponding, spraying, ditching and subsurface infiltration beds and galleries (Amdurer et al., 1986). Barriers, such as slurry walls, may be required to prevent the transport of contaminants away from the site (Metcalf & Eddy, Inc., 1991). A ground-water extraction system must be used to capture the flushing solution and associated contaminants. In some cases, the flushing solution may be treated to remove the contaminants and reused, and in others it may require disposal.

Efficiency of soil flushing is related to two processes: the increase in hydraulic conductivity that accompanies an increase in water content of unsaturated soil, and selection of treatment solutions with regard to the composition of the contaminants and the contaminated medium. The hydraulic conductivity of soils decreases markedly with decreases in water content; therefore, the flow of liquids through unsaturated soils is extremely slow and the recovery of contaminants by conventional pumping techniques is not possible. With soil flushing, the water content and consequently the hydraulic conductivity of the soil is increased (Murdoch et al., 1990). However, heterogeneities in soil permeability may result in incomplete removal of contaminants.

At sites where water-soluble contaminants are present, water can be used to flush or mobilize the contaminants (Metcalf & Eddy, Inc., 1991). Surfactants can be added to increase the mobility of hydrophobic organic contaminants, such as oils and petroleum. Examples of other flushing solutions include: acidic aqueous solutions (for the removal of metals and basic organic constituents including amines, ether, and anilines), basic solutions, chelating agents, oxidizing agents, and reducing agents. Toxicity of flushing solutions to soil microorganisms should be considered when followed by bioremediation of residual contamination. The flushing solution may change physical and chemical properties of the soil environment and affect bioremediation potential.

The level of treatment that will be achieved is dependent on selection of an appropriate flushing solution, extent and time of contact between the solution and waste constituents, soil partition coefficients of the waste constituents, and the hydraulic conductivity of the soil (Metcalf & Eddy, Inc., 1991). Soil flushing is not applicable to soils with low hydraulic conductivities (e.g., less than 1 ft/ day), or for contaminants that are strongly sorbed to the soil (e.g., PCBs, dioxin).

Soil flushing has been classified by the U.S. EPA as a developed technology used for recovery in remedial applications (Murdoch et al., 1990). Although the technology has been tested at field-scale, soil flushing has not yet been used extensively in large-scale cleanup operations. As with SVE systems, soil flushing may be utilized with bioremediation as a coupled technology. Soil flushing may initially be utilized to lower toxic or extreme concentrations of contaminants to a manageable level for biological processes which may be utilized as a polishing step to remove those contaminants which were not removed through the flushing process. If biological processes are used during or after soil flushing, the compatibility of the soil flushing solution with subsurface bacteria must always be considered.

DELIVERY TECHNIQUES

The major limiting factor to the bioremediation of amenable compounds is the delivery of required nutrients, cooxidation substrates, electron acceptors or other necessary enhancers of microbial growth. Delivery techniques are used to add required materials to the subsurface environment to enhance in situ bioremediation by addressing one or more limiting factors identified in Tables 12-1 and 12-2. A variety of delivery techniques are in use or are being developed (Figures 12-1 to 12-3). These include soil venting, gravity and forced hydraulic delivery, hydraulic fracturing of low permeability zones, radial drilling, and cyclic pumping. Of these, only gravity and forced hydraulic delivery and venting systems are in common use at sites. The other three approaches are still in developmental stages.

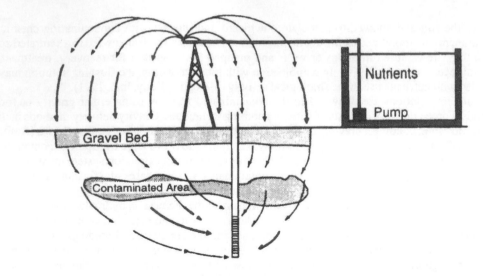

Figure 12-1. Schematic of a sprinkling system used to deliver nutrients to contaminated subsurface soil.

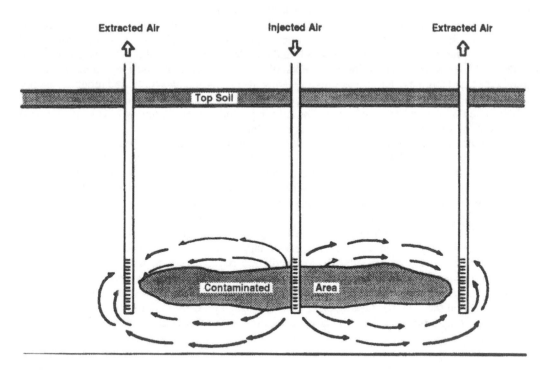

Figure 12-2. Schematic of a bioventing system designed to deliver air to contaminated subsurface soil.

IN USE: GRAVITY/FORCED HYDRAULIC DELIVERY AND BIOVENTING

Gravity and Forced Hydraulic Delivery

Irrigation technologies were among the first utilized for enhancing in situ biodegradation. Gravity methods are used to deliver water and amendments to the contaminated subsurface by applying the solutions directly over the contaminated area. Applied solutions then percolate downward through the subsurface to contaminated zones. Application methods consist of both surface and subsurface spreading (Amdurer et al., 1986).

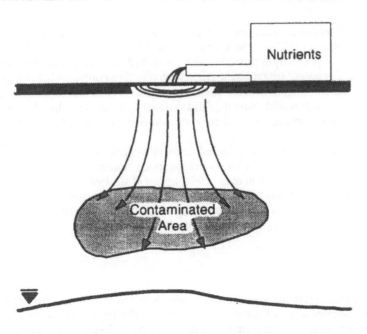

Figure 12-3. Schematic of a ponding system used to deliver nutrients to contaminated subsurface soil.

Surface application methods include flooding, ponding, ditches, and sprinkler systems. These methods are generally applicable to contamination at depths less than 15 feet. Flooding is a surface application method in which the solution is spread over the land surface in a thin sheet. Flooding is applicable to sites that are flat or gently sloped (i.e., less than 3 percent slope), uniform, without gullies or ridges, and have soils with high hydraulic conductivities (i.e., greater than 10^{-3} cm/sec; such as those found in sands, loamy sands, and sandy loams).

Ponding can be used to increase the infiltration rate of the applied solution above that achieved by flooding. Ponds are constructed by excavating into the ground or by constructing low berms. The depth of the solution in the pond becomes the driving force to increase infiltration rates. Ponding can be used in sandy or loamy soils and in flat areas.

The ditch method of surface spreading utilizes flat-bottomed, shallow, narrow ditches to transport the solution over the land surface; allowing for infiltration of the solution into the ground through both bottom and side surfaces. Gradients in the ditches are kept small to prevent erosion as well as to allow residence time for infiltration. Ditches may be constructed by excavating surface materials or by building small embankments. Ditches are used at sites where it is not desirable to completely cover an entire area with the solution.

Sprinkler systems can be used to deliver solutions uniformly and directly to the ground surface. These systems are less susceptible to topographical constraints than flooding and ponding. Sprinkler systems have been used successfully to deliver nutrients and moisture to bioventing systems where the site was contaminated to a depth of 50 feet (Dupont et al., 1991).

Subsurface gravity delivery systems include infiltration galleries (or trenches) and infiltration beds. These systems are applicable to sites where there is deep contamination or where the surface layers have low permeability. Subsurface systems consist of excavations filled with a porous medium (e.g., coarse sands or gravels) that distribute solutions to the contaminated area. An infiltration gallery consists of a pit or trench that is filled with gravel or stones. The solution fills the pores in the gallery and is distributed to the surrounding soils in both the vertical and horizontal directions. This system is most applicable to sites with sandy or loamy soils. In sites with silty soils, an infiltration gallery can be used, but application rates will be reduced. Solutions can be introduced into the gallery by injection at locations along the length of the gallery or through perforated distribution pipe. Infiltration occurs almost entirely through the bottom, with little infiltration through sidewall surfaces. This system is applicable to soils with sandy and loamy texture, but limited to sites where the topography is relatively flat (i.e., with less than 5 percent slope) and the terrain is even. Beds can saturate larger areas than a single trench and are easier to install than a multi-trench system.

Forced systems deliver fluids under pressure into a contaminated area through open end or slotted pipes that have been placed to deliver the solution to the zone requiring treatment (Amdurer et al., 1986). These systems are generally applicable to soils with hydraulic conductivities greater than 10^{-4} cm/sec (i.e., fine sandy or coarse silty materials) and high effective porosities (i.e., ranging from 25 to 55 percent). A maximum injection pressure must be established to prevent hydraulic fracturing and uplift in the subsurface, which would cause the fluid to travel upward rather than through the contaminated area. Unlike gravity systems, a forced delivery system is theoretically independent of surface topography and climate.

Design considerations for gravity and forced delivery systems are presented in Amdurer et al. (1986). Application of gravity delivery systems in subsurface bioremediation systems has been demonstrated in bioventing systems (Dupont et al., 1991; Miller et al., 1991). In Russia, methane-oxidizing bacteria grown in fermenters have been injected into lateral core holes in a coal mine (Fliermans and Hazen, 1990). This process has been shown to reduce methane concentration in the air by 50-60 percent in one month, thus reducing the risk of explosions and fire.

Soil Bioventing

Soil bioventing incorporates soil vacuum extraction processes to deliver oxygen to the subsurface to enhance in situ bioremediation of organic contaminants. The large amounts of oxygen-saturated water required for bioremediation often cannot be delivered due to hydraulic conductivity limitations. For example, benzene and hexane, which are common hydrocarbon contaminants, require more than 3 g O_2 per g of hydrocarbon for mineralization. Soil bioventing is applicable to remediation of contaminants of low volatility and can also reduce concentrations of volatile contaminants in off-gases, thus reducing the amount of contaminants requiring off-gas treatment.

To accomplish bioventing, soil vacuum extraction processes are operated at lower than usual air flow rates to reduce vapor extraction quantities and maximize vapor retention times. Soil moisture levels necessary for biological activity are usually higher than those recommended for optimum vacuum extraction operations. The addition of nutrients may also enhance bioremediation. Nutrient addition can be accomplished by surface application, incorporation by tilling into surface soil, and transport to deeper layers through applied irrigation water. Increased soil temperatures have been shown to enhance biodegradation rates in bioventing systems (Miller et al., 1991). Possible means of increasing soil temperature include the use of heated air, heated water, or low-level radio-frequency heating. High temperature should be avoided, since this can result in decreased microbial populations and/or activity.

Soil bioventing has been demonstrated in several field applications (Dupont et al., 1991; Hinchee et al., 1991; Hoeppel et al., 1991; Miller et al., 1991; van Eyk and Vreeken, 1991; Urlings et al., 1991). At Hill Air Force Base in Utah, a JP-4 jet fuel spill occurred in January 1985 that resulted in the contamination of approximately 0.4 hectares (1 acre) to a depth of approximately 50 feet with approximately 25,000 gallons of JP-4 (Dupont et al., 1991). Soil total petroleum hydrocarbon (TPH) concentrations at the site were as high as 15,000 mg/kg, with average TPH levels of 1,500 mg/kg. Site soil consists of mixed coarse sand and gravel deposits with interspersed, discontinuous clay stringers to a confined ground-water table located approximately 600 feet below ground surface. Prior to initiating a full-scale vacuum extraction project, the fuel tanks were excavated, refurbished, and installed in an above-ground concrete cradle. Excavated soil was placed in a pile and subjected to vacuum extraction.

An SVE system consisting of 15 wells in the undisturbed soil and 10 wells in the excavated soil pile and under the tanks was installed to provide access to the contaminated soil and allow flexibility in the operation of the venting system. The system was operated in a conventional mode to maximize the recovery of volatile components of the JP-4 through volatilization. Venting was initiated on December 18, 1988, at a rate of 1,270 ft³/hr (approximately 0.04 pore volumes/day) and gradually increased to approximately 74,000 ft³/hr (approximately 2.5 pore volumes/day) as the hydrocarbon levels in the vent gas decreased over time. The venting rate during the startup period was limited by the operating conditions of the catalytic incinerator used to treat the collected vent gas. This high-rate operating mode was maintained from December 18, 1988, through September 15, 1989 with approximately 340 pore volumes (245×10^6 ft³) of soil gas extracted from the site.

In situ respiration tests conducted during the high-rate SVE operating period indicated that significant respiration was occurring without nutrient or moisture addition, and that enhancement of biodegradation might be possible under modified site management conditions. Biodegradation was a significant removal mechanism during the initial high-rate venting, accounting for 15 to 25 percent of the recovered hydrocarbon. To asses the potential for enhancing biodegradation rates, a series of laboratory and field biotreatability studies were conducted to evaluate moisture and nutrient additions. The effect of SVE system operational parameters on biodegradation rates was also evaluated by decreasing air flow rates and increasing flow path length.

A number of in situ respiration tests were conducted during the field studies to asses the impact of different engineering management options on microbial activity. A total of three tests were conducted to monitor the effect of different management approaches, including: (1) flow rate and operating configuration modifications, (2) moisture addition, and (3) moisture and nutrient addition. Biodegradation reactions were estimated based on cumulative oxygen consumption and carbon dioxide production. All biodegradation calculations were normalized to background CO_2 and O_2 concentrations so that the effects of field management techniques could be isolated from changes in background respiration taking place during the study.

The results of these studies indicated that moisture addition and operational modifications significantly enhanced biodegradation rates. Based on analyses of O_2 uptake rates, moisture addition (35% to 50% field capacity) was shown to statistically accelerate in situ respiration at the site. However, nutrient addition generally did not statistically increase the degradation rates of residual JP-4 constituents. The operational modifications (reduced air flow rate, increased path length) significantly improved biodegradation rates. Fuel removal due to biodegradation increased to greater than 80 percent, resulting in an additional 12,000 lb of total petroleum hydrocarbons being degraded during the bioventing portion of the study. Initial hydrocarbon (on a carbon equivalent basis) removal rates of 70 lb/day were maintained at an average rate of greater than 100 lb/day following system operating modifications.

Soil bioventing was also investigated at Tyndall Air Force Base in Florida to remediate sandy soils contaminated by past jet fuel storage activities (Miller, 1990; Miller et al., 1991). Hydrocarbon concentrations in the soil ranged from 30 to 23,000 mg/kg. The contaminated area was dewatered prior to system installation. The impact of moisture and nutrient addition was investigated during a 7-month period. Moisture addition had no significant effect on biodegradation rate in this system. Nutrient addition also did not affect biodegradation rate, since naturally occurring nutrients were present in adequate quantities to support the amount of biodegradation observed. Biodegradation rates were shown to be affected by soil temperature and followed predicted rates based on the van't Hoff-Arrhenius equation. Fifty-five percent removal was attributed to biodegradation during the period of study, but a series of flow rate tests showed that biodegradation could be increased to 85 percent by decreasing air flow rates. The optimal air flow conditions were found to be the removal of 0.5 air flow volumes per day. The contaminated gas phase was drawn through clean soil to increase gas residence time within the soil. This augmented in situ biodegradation and eliminated the need for off-gas treatment, as well as reducing exposure to off-gas.

RESEARCH: HYDRAULIC FRACTURING, RADIAL DRILLING

Research areas are focusing on methods to increase the capacity of current systems to deliver increased concentrations of required solutions to the subsurface. Two of these systems are discussed below.

Hydraulic Fracturing

Hydraulic fracturing is a technique that involves using hydraulic pressure to induce cracking in rock or clay/silt lenses in the vicinity of a borehole, which develops a larger framework of interconnected pore space. The newly created pore space is filled with solid, granular materials, which can act as permeable channels to increase the rate and area of delivery of fluids containing nutrients or oxygen to the subsurface (Murdoch et al., 1990; Murdoch et al., 1991; Davis-Hoover et al., 1991). The hydraulic fractures may be filled with granules of slow-dissolving nutrients or oxygen-releasing chemicals, which may provide a reservoir of these compounds for the enhancement of bioremediation. This technique could also potentially be used in recovery systems, e.g., by increasing extraction of vapor phases in soils with low permeabilities, or by forming horizontal sheet-like drains to capture leachates in soil flushing systems.

Hydraulic fracturing has been successfully utilized in petroleum engineering in many types of geologic materials, ranging from granite to poorly consolidated sediments. For remedial applications, it has been demonstrated in soft clay soils at shallow depths, but has not yet been demonstrated in a wide range of soils or at waste sites. For use in remedial applications, hydraulic fracturing has been classified by the U.S. EPA as an emerging technology (i.e., research on its use is in progress) (Murdoch et al., 1990).

Radial Drilling

Radial well technology consists of drilling horizontal wells radially outward from a central borehole. This enhances access to a contaminated subsurface environment by increasing the volume serviced by each vertical well (Murdoch et al., 1990). Radial wells can be placed at the same level or on multiple levels in the same borehole. The use of horizontal wells allows access to fracture zones that are perpendicular to the ground surface and allows contaminated areas to be entered laterally rather than vertically.

Radial wells have been installed in both consolidated rock and unconsolidated materials (Murdoch et al., 1990). In unconsolidated formations, drilling rates range from 5 to 120 ft/min, while in very hard, homogeneous basalt, rates range from 0.10 to 0.50 ft/min. For use in remedial applications, radial well drilling has been classified by the U.S. EPA as an emerging technology (i.e., research on its use is in progress)(Murdoch et al., 1990).

WASTE, SOIL, AND SITE INFORMATION REQUIREMENTS FOR EVALUATION AND MANAGEMENT OF IN SITU BIOREMEDIATION

Adequate site characterization including: surface and subsurface soil characteristics, hydrogeology, and microbiological characteristics, serve as the basis for rational design of any subsurface soil bioremediation system. A thorough site characterization is necessary to determine both the three-dimensional extent of contamination as well as engineering and management constraints which may limit the rate and extent of remediation. Specific characterization information regarding waste, soil, and hydrogeology is required in order to assess the potential effectiveness of bioremediation. Specific waste characterization information required includes the relative aerobic biodegradability of waste chemicals under optimum conditions. Important hydraulic, physical, and chemical properties of soils that affect the behavior of organic constituents in the vadose zone are presented in Sims et al. (1989). Subsurface soil characterization information required includes identification of limiting soil environmental factors identified in Table 12-1. Required site characterization information includes identification of potential limiting factors with regard to relative ease of delivery and recovery listed in Table 12-2.

Based upon waste, subsurface soil, and site characterization information, appropriate containment strategies need to be considered for the mobile contaminant phases associated with the subsurface. Naturally occurring containment may be sufficient with regard to preventing escape of mobile phases under existing site conditions. However, other containment strategies may need to be considered if materials are to be added or removed from the subsurface to stimulate microbial activity. These may include volatiles removed in vacuum extraction, water used to add oxygen and nutrients, or NAPLs, if soil flushing is carried out.

For each chemical (or chemical class), the information required is summarized as: (1) characteristics related to potential leaching, e.g., water solubility, octanol/water partition coefficient, solid sorption coefficient; (2) volatilization, e.g., vapor pressure, relative volatilization index; (3) Henry's law constant; (4) potential biodegradation, e.g., half-life, degradation rate, biodegradability index; and (5) chemical reactivity, e.g., hydrolysis half-life, soil redox potential (Sims et al., 1984; Sims et al., 1989).

Information from waste and site characterization studies, and laboratory evaluations of biodegradation may be integrated by using appropriate mathematical models to predict: (1) the potential for bioremediation of and (2) the potential for cross-contaminating other media (i.e., ground water under the contaminated area, atmosphere over the site or at the site boundaries, surface waters, etc.). The models used will be highly dependent on site characteristics and contaminants of interest. These may range from "back-of-the-envelope" calculations to sophisticated fate and transport computer models.

MASS BALANCE APPROACH TO IN SITU SUBSURFACE BIOREMEDIATION

Successful subsurface bioremediation depends upon thorough characterization and management of each subsurface phase with regard to containment, stimulation of microbial activity, and monitoring strategies. The chemical mass balance approach provides a framework for evaluating, managing, and monitoring subsurface soil bioremediation (Sims, 1990). Mass balance helps obtain specific information that is needed to determine fate and behavior, evaluate and select management options for in situ bioremediation, and monitor treatment effectiveness for specific chemicals in specific subsurface phases. The information needed to construct a mass balance for subsurface contamination simultaneously addresses site characterization and biodegradation rates.

A necessary first step in mass balance requires characterizing each phase present in the subsurface with regard to location, amount, and heterogeneity of the subsurface environment to assess which chemicals are associated with which phase(s). This information allows determination of the relative bioavailability of chemicals. For example, chemicals associated with aqueous and gas phases are generally more bioavailable than chemicals associated with solid and NAPL phases. In addition, chemicals associated with aqueous and gas phases are more prone to migration. This information also allows determination of the need for containment by defining where contamination is migrating under the influence of natural processes. The problem can be defined in the context of mobility versus biodegradation for chemicals. Is the rate of biodegradation (either natural or enhanced) such that chemicals which are prone to leaching or volatilization degrade before either occurs? Using mathematical models or other tools, chemicals can be ranked in order of their relative tendencies to leach, volatilize, or remain in-place under subsurface site-specific conditions. Containment and management options can then be selected that address specific escape and attenuation pathways. For example, SVE may be appropriate as a managerial tool to remove highly volatile, biologically recalcitrant chemicals from soil before switching to a bioventing mode to remove less volatile, easily biodegraded compounds. Specific waste phases may be addressed at specific times during bioremediation. Finally, comprehensive monitoring programs can be designed to track specific chemicals in specific phases in the subsurface at specific times.

After a phase is contained through natural or managed processes, techniques to enhance microbial activity may be applied. Monitoring strategies can then be designed to ensure that the rate and extent of biodegradation within each phase, as well as transfer of chemicals between phases, are measured. Biodegradation rates of organic compounds in soil systems are generally measured by monitoring their disappearance in a soil through time. Rates of degradation are often expressed as a function of the concentration of one or more of the constituents being degraded. This is accomplished by measuring at specific time intervals the concentration of contaminants of interest (in the medium of interest, i.e., soil phase, gas phase, etc.), through a properly designed sampling and analysis plan. This sampling and analysis plan should be statistically valid and provide sufficient information to determine the rate of disappearance of contaminants of interest or appropriate surrogates, such as total petroleum hydrocarbons (TPH). Care should be taken to ensure that transfer or partitioning of contaminants from one phase to another is not misinterpreted as biodegradation within the source phase. Abiotic losses such as volatilization and leaching must be defined in order to accurately determine biodegradation rates. Identification of metabolic transformation products is also necessary since metabolites may be more mobile or toxic than the parent compounds. In addition, measuring only for parent compounds and not metabolites may tremendously overestimate extent of biodegradation. In addition, identification of metabolites is warranted when known daughter products are toxic.

RECOMMENDATIONS

There is currently a lack of information concerning some aspects of in situ bioremediation of subsurface soils. Specific areas where additional information is required include site characterization with regard to effects of physical, chemical, and hydrologic properties on microbial distribution, numbers, and activity. Field research to obtain these types of information is currently limited; however, this information is required in order to estimate the feasibility of bioremediation for subsurface contamination. Implementation of subsurface remediation is currently limited to a significant extent by the difficulty of establishing adequate systems for delivery and recovery of chemicals for augmenting biological activity. As research continues, these difficulties may be overcome as more information becomes available concerning the applicability of innovative technologies in the remediation of contaminated soil.

EPA CONTACTS

For further information, contact Joe Williams, RSKERL-Ada, (405) 436-8608 (see Preface for mailing address).

REFERENCES

Amdurer, M., R.T. Fellman, J. Roetzer, and C. Russ. 1986. Systems to Accelerate In Situ Stabilization of Waste Deposits. EPA/540/2-86/002.

Davis-Hoover, W.J., L.C. Murdoch, S.J. Vesper, H.R. Pahren, O.L. Sprockel, C.L. Chang, A. Hussain, and W.A. Ritschel. 1991. Hydraulic Fracturing to Improve Nutrient and Oxygen Delivery for In Situ Bioreclamation. pp. 67–82. In: In Situ Bioreclamation: Applications and Investigations for Hydrocarbon and Contaminated Site Remediation (R.E. Hinchee and R.F. Olfenbuttel, eds.). Butterworth-Heinemann, Boston, MA.

Devinny, J.S., L.G. Everett, J.C.S. Lu, and R.L. Stollar. 1990. Subsurface Migration of Hazardous Wastes. Van Nostrand Reinhold, New York, NY.

Dragun, J. 1988. The Soil Chemistry of Hazardous Materials. Hazardous Materials Control Research Institute, Silver Spring, MD.

Dunlap, W.J., and J.F. McNabb. 1973. Subsurface Biological Activity in Relation to Ground Water Pollution. EPA-660/2-73-014 (NTIS PB227-990).

Dupont, R.R., W.J. Doucette, and R.E. Hinchee. 1991. Assessment of In Situ Bioremediation Potential and the Application of Bioventing at a Fuel-Contaminated Site. pp. 262–282. In: In Situ Bioreclamation: Applications and Investigations for Hydrocarbon and Contaminated Site Remediation (R.E. Hinchee and R.F. Olfenbuttel, eds.). Butterworth-Heinemann, Boston, MA.

Everett, L.G., E.W. Hoylman, L.G. McMillion, and L.G. Wilson, 1982. Vadose Zone Monitoring Concepts at Landfills, Impoundments, and Land Treatment Disposal Areas. In: Management of Uncontrolled Hazardous Waste Sites. Hazardous Materials Control Research Institute, Silver Spring, MD.

Fliermans, C.B., and T.C. Hazen (eds.). 1990. Microbiology of the Deep Subsurface, Proceedings, First International Symposium. U.S. Department of Energy and Westinghouse Savannah River Company, Savannah, GA, January 15–19.

Hazen, T.C. 1992. Full Scale Underground Injection of Air, Methane, and Other Gases via Horizontal Wells for In Situ Bioremediation of Chlorinated Solvent Contaminated Ground Water and Soil. Proceedings, Bioremediation Case Studies at Federal Facilities, Oak Ridge National Laboratory, Oak Ridge, TN, August.

Hinchee, R.E., D.C. Downey, R.R. Dupont, P.K. Aggarwal and R.N. Miller. 1991. Enhancing Biodegradation of Petroleum Hydrocarbons Through Soil Venting. Journal of Hazardous Materials 27:315–325.

Hoeppel, R.E., R.E. Hinchee and M.F. Arthur. 1991. Bioventing Soils Contaminated with Petroleum Hydrocarbons. Journal of Industrial Microbiology 8:141–146.

Huddleston, R.L., C.A. Bleckmann, and J.R. Wolfe. 1986. Land Treatment Biological Degradation Processes. pp. 41–61. In: Land Treatment: A Hazardous Waste Management Alternative (R.C. Loehr and J.F. Malina, eds.). Water Resources Symposium No. 13, Center for Research in Water Resources, University of Texas at Austin, Austin, TX.

Hutchins, S.R., W.C. Downs, G.B. Smith, D.A. Kovacs, D.D. Fine, R.H. Douglas, and D.J. Hendrix. 1991. Effect of Nitrate Addition on Biorestoration of Fuel-Contaminated Aquifer: Field Demonstration. Ground Water 29:571–580.

Keck, J., R.C. Sims, M. Coover, K. Park, and B. Symons. 1989. Evidence for Cooxidation of Polynuclear Aromatic Hydrocarbons in Soil. Water Research 23:1467–1476.

Kieft, T.L., L.L. Rosacker, D. Willcox, and A.J. Franklin. 1990. Water Potential and Starvation Stress in Deep Subsurface Microorganisms. pp. 4-99 to 4-112. In: Microbiology of the Deep subsurface, Proceedings, First International Symposium (C.B. Fliermans and T.C. Hazen, eds). U.S. Department of Energy and Westinghouse Savannah River Company, Savannah, GA, January 15–19.

Kuznetsov, S.I., M.W. Ivanov, and N.N. Lyalikova. 1963. Introduction to Geological Microbiology. (C.H. Oppenheimer, ed.). McGraw-Hill, New York, NY.

Matthess, G. 1982. The Properties of Ground Water. John Wiley & Sons, New York, NY.

Metcalf & Eddy, Inc. 1991. Stabilization Technologies for RCRA Corrective Actions. EPA/625/6-91/026.

Miller, R.N. 1990. A Field Scale Investigation of Enhanced Petroleum Hydrocarbon Biodegradation in the Vadose Zone Combining Soil Venting as an Oxygen Source with Moisture and Nutrient Addition. Ph.D. Dissertation. Department of Civil and Environmental Engineering, Utah State University, Logan, UT.

Miller, R.N., C.C. Vogel, and R.E. Hinchee. 1991. A Field-Scale Investigation of Petroleum Hydrocarbon Biodegradation in the Vadose Zone Enhanced by Soil Venting at Tyndall AFB, Florida. pp. 283–302. In: In Situ Bioreclamation: Applications and Investigations for Hydrocarbon and Contaminated Site Remediation (R.E. Hinchee and R.F. Olfenbuttel, eds.). Butterworth-Heinemann, Boston, MA.

Murdoch, L., B. Patterson, G. Losonsky, and W. Harrar. 1990. Technologies of Delivery and Recovery for the Remediation of Hazardous Waste Sites. EPA/600/2-89/066, Risk Reduction Engineering Laboratory, U.S. Environmental Protection Agency, Cincinnati, OH.

Murdoch, L., G. Losonky, P. Cluxton, B. Patterson, I. Klich, and B. Braswell. 1991. Feasibility of Hydraulic Fracturing of Soil to Improve Remedial Actions. EPA/600/2-91/012.

Paul, E.A., and F.E. Clark. 1989. Soil Microbiology and Biochemistry. Academic Press, Inc., San Diego, CA.

Rochkind, M.L., and J.W. Blackburn. 1986. Microbial Decomposition of Chlorinated Aromatic Compounds. EPA/600/2-86/090.

Ross, D., T.P. Marziarz, and A.L. Bourquin. 1988. Bioremediation of Hazardous Waste Sites in the USA: Case Histories. pp. 395–397. In: Superfund '88, Proc., 9th National Conference, Hazardous Materials Control Research Institute, Silver Spring, MD.

Sims, R.C. 1990. Soil Remediation Techniques at Uncontrolled Hazardous Waste Sites: A Critical Review. Journal of the Air & Waste Management Association 40:704–732.

Sims, R.C., D.L. Sorenson, J.L. Sims, J.E. McLean, R. Mahmood, and R.R. Dupont. 1984. Review of In Place Treatment Techniques for Contaminated Surface Soils. Volume 2: Background Information for In Situ Treatment. EPA/540/2-84-003b (NTIS PB85-124899).

Sims, J.L., R.C. Sims, and J.E. Matthews. 1989. Bioremediation of Contaminated Surface Soils. EPA/600/9-89/073 (NTIS PB90-164047).

Symons, B.D., and R.C. Sims. 1988. Assessing Detoxification of a Complex Hazardous Waste, Using the Microtox Bioassay. Archives of Environmental Contamination and Toxicology 17:497–505.

Thomas, A.O., P.M. Johnston, and J.N. Lester. 1991. The Characterization of the Subsurface at Former Gasworks Sites in Respect of In Situ Microbiology, Chemistry, and Physical Structure. Hazardous Waste & Hazardous Materials 8:341–365.

Urlings, L.G.C.M., F. Spuy, S. Coffa, H.B.R.J. van Vree. 1991. Soil Vapor Extraction of Hydrocarbons: In Situ and On-Site Biological Treatment. pp. 321–336. In: In Situ Bioreclamation: Applications and Investigations for Hydrocarbon and Contaminated Site Remediation (R.E. Hinchee and R.F. Olfenbuttel, eds.). Butterworth-Heinemann, Boston, MA.

U.S. EPA. 1989. State of Technology Review: Soil Vapor Extraction Systems. EPA/600/289/024 (NTIS PB89-195184).

U.S. EPA. 1990. Assessing UST Corrective Action Technologies: Site Assessment and Selection of Unsaturated Zone Treatment Technologies. EPA/600/2-90/011.

van Eyk, J. and C. Vreeken. 1991. In Situ and On-Site Subsoil and Aquifer Restoration at a Retail Gasoline Station. pp. 303–320. In: In Situ Bioreclamation: Applications and Investigations for Hydrocarbon and Contaminated Site Remediation (R.E. Hinchee and R.F. Olfenbuttel, eds.). Butterworth-Heinemann, Boston, MA.

Wobber, F.J. 1989. Deep Microbiology Transitional Program Implementation Plan. DOE/ER-0431, Office of Energy Research, Office of Health and Environmental Research, U.S. Department of Energy, Washington, DC.

Zehnder, A.J.B., and W. Stumm. 1988. Geochemistry and Biochemistry of Anaerobic Habitats. In: Biology of Anaerobic Microorganisms (A.J.B. Zehnder, ed.). John Wiley & Sons, New York, NY.

ADDITIONAL REFERENCES (included in original paper, but not cited in text)

Balkwill, D.L., and F.J. Wobber. 1989. Deep Microbiology Transitional Program Plan. DOE/ER-0328, Office of Energy Research, Office of Health and Environmental Research, U.S. Department of Energy, Washington, DC.

Brown, R.A., R.A. Norris, and R.L. Raymond. 1984. Oxygen Transport in Contaminated Aquifers with Hydrogen Peroxide. In: Proceedings, Petroleum Hydrocarbons and Organic Chemicals in Groundwater-Prevention, Detection and Restoration Conference, National Water Well Association, Worthington, OH.

Dev, H., and D. Downey. 1988. Zapping Hazwastes. Civil Engineering (August): 43–45.

Dev, H., J.B. Condorelli, C. Rogers, and D. Downey. 1986. In Situ Frequency Heating Process for Decontamination of Soil. pp. 332–339. In: Solving Hazardous Waste Problems, Learning from Dioxins. ACS Symposium Series 338, American Chemical Society, New York, NY.

Ghiorse, W.C., and J.T. Wilson. 1988. Microbial Ecology of the Terrestrial Subsurface. Advances in Applied Microbiology 33:107–172.

Madsen, E.L., and J.-M. Bollag. 1989. Aerobic and Anaerobic Microbial Activity in Deep Subsurface Sediments from the Savannah River Plant. Geomicrobiology Journal 7:93–101.

Noble, D.G. 1963. Well Points for Dewatering Groundwater. Ground Water 1:21–36.

Rittmann, B.E., and P.L. McCarty. 1980. Model of Steady-State Film Biofilm Kinetics. Biotechnology and Bioengineering 22:23–43.

Stevens, D.K., W.J. Grenney, and Z. Yan. 1988. User's Manual: Vadose Zone Interactive Processes Model. Department of Civil and Environmental Engineering, Utah State University, Logan, UT.

Stevens, D.K., W.J. Grenney, Z. Yan, and R.C. Sims. 1989. Sensitive Parameter Evaluation for a Vadose Zone Fate and Transport Model. EPA/600/2-89/039, Robert S. Kerr Environmental Research Laboratory, U.S. Environmental Protection Agency, Ada, OK.

U.S. EPA. 1988. Interactive Simulation of the Fate of Hazardous Chemicals During Land Treatment of Oily Wastes: RITZ User's Guide. EPA/600/8-88-001 (NTIS PB88-195540).

Chapter 13

In Situ Bioremediation of Contaminated Ground Water[1]

J.L. Sims, Utah Water Research Laboratory, Utah State University, Logan, UT
J.M. Suflita, Dept. of Botany and Microbiology, University of Oklahoma, Norman, OK
H.H. Russell, U.S. Army Corps of Engineers, Tulsa, OK

Although in situ bioremediation has been used for a number of years in the restoration of ground water contaminated by petroleum hydrocarbons, it has only been in recent years that this technology has been directed toward other classes of contaminants. Research has contributed greatly to understanding the biotic, chemical, and hydrologic parameters which contribute to or restrict the application of in situ bioremediation, and has been successful at a number of locations in demonstrating its effectiveness at field scale.

This chapter is based on findings from the research community in concert with experience gained at sites undergoing remediation. The intent of the chapter is to provide an overview of the factors involved in in situ bioremediation, outline the types of information required in the application of such systems, and point out the advantages and limitations of this technology.

SUMMARY

In situ bioremediation, where applicable, appears to be a potential cost-effective and environmentally acceptable remediation technology. Suflita (1989a) identified characteristics of the ideal candidate site for successful implementation of in situ bioremediation. These characteristics included: (1) a homogeneous and permeable aquifer; (2) a contaminant originating from a single source; (3) a low ground-water gradient; (4) no free product; (5) no soil contamination; and (6) an easily degraded, extracted, or immobilized contaminant. Obviously, few sites meet these characteristics. However, development of information concerning site-specific geological and microbiological characteristics of the aquifer, combined with knowledge concerning potential chemical, physical, and biochemical fate of the wastes present, can be used to develop a bioremediation strategy for a less-than-ideal site.

INTRODUCTION

In situ bioremediation is a technology to restore aquifers contaminated with organic compounds. Organic contaminants found in aquifers can be dissolved in water, attached to the aquifer material, or as free phase or residual phase liquids referred to as NAPLs, which are liquids that do not readily dissolve in water (Palmer and Johnson, 1989c). Generally, NAPLs are subdivided into two classes: those that are lighter than water (LNAPLs density <1.0), and those with a density greater than water (DNAPLs density >1.0). LNAPLs include hydrocarbon fuels, such as gasoline, heating oil, kerosene, jet fuel, and aviation gas. DNAPLs include chlorinated hydrocarbons, such as 1,1,1-trichloroethene, carbon tetrachloride, chlorophenols, chlorobenzenes, tetrachloroethylene, PCBs, and creosote.

In this discussion, a technical approach is presented to assess the potential implementation of bioremediation at a specific site contaminated with an organic compound. The approach consists of (1) a site investigation to determine the transport and fate characteristics of organic waste constitu-

---

[1] EPA/540/S-92/003.

ents in a contaminated aquifer, (2) performance of treatability studies to determine the potential for bioremediation and to define required operating and management practices, (3) development of a bioremediation plan based on fundamental engineering principles, and (4) establishment of a monitoring program to evaluate performance of the remediation effort.

The pattern of contamination from a release of contaminants into the subsurface environment, as would occur from an underground leaking storage tank containing NAPLs, is complex (Figure 13-1) (Palmer and Johnson, 1989c; Wilson et al., 1989). As contaminants move through the unsaturated zone a portion is left behind, trapped by capillary forces. If the release contains volatile contaminants, a plume of vapors forms in the soil atmosphere in the vadose zone. If the release contains NAPLs less dense than water (LNAPLs), they may flow by gravity down to the water table and spread laterally. The oily phase can exist either as a free product, which can be reversed by pumping, or as a residual phase after the pore spaces have been drained. Contaminants associated with NAPLs can also partition into the aquifer's solid phase or in the vapor phase of the unsaturated zone. If the release contains DNAPLs, these contaminants can penetrate to the bottom of an aquifer, forming pools in depressions. In either case, when ground water comes into contact with any of these phases, the soluble components are dissolved into the water phase.

There are a number of techniques available to remediate ground water contaminated with organic compounds including: (1) physical containment such as slurry walls, grout curtains, and sheet pilings (Ehrenfield and Bass, 1984; see also Chapter 1); (2) hydrodynamic control using pumping wells to manipulate the hydraulic gradient or interceptor systems (Canter and Knox, 1985); (3) several methods of free product recovery (Lee and Ward 1986); and (4) extraction of contaminated ground water followed by treatment at the surface (see Chapter 4).

Alternatively, contaminated ground water can be treated in place, without extraction using in-situ chemical treatment or biological treatment with microorganisms (Thomas et al., 1987; see also Chapter 11). An advantage of in situ treatment strategies is that treatment can take place in multiple phases.

In situ chemical treatment techniques are similar to methods used to treat contaminated materials after withdrawal or excavation, but are directly applied to the materials in place (Ehrenfield and Bass, 1984). Chemical treatment may involve neutralizing, precipitating, oxidizing or reducing contaminants by injecting reactive materials into a contaminated leachate plume through injection wells, but may be limited by mass transport and concentration dependence. For treatment of shallow contaminated aquifers, permeable treatment beds containing reactive materials such as activated carbon or ion exchange resins may be constructed downgradient to intercept and treat the contaminated plume.

Biological in situ treatment of aquifers is usually accomplished by stimulation of indigenous microorganisms to degrade organic waste constituents present at a site (Thomas and Ward, 1989). The microorganisms are stimulated by injection of inorganic nutrients and, if required, an appropriate electron acceptor, into aquifer materials.

Most biological in situ treatment techniques in use today are variations of techniques developed by researchers at Suntech to remediate gasoline-contaminated aquifers. The Suntech process received a patent titled Reclamation of Hydrocarbon Contaminated Ground Waters (Raymond, 1974). The process involves the circulation of oxygen and nutrients through a contaminated aquifer using injection and production wells (Lee et al., 1988). Placement of the wells is dependent on the area of contamination and the porosity of the formation, but they are usually no more than 100 feet apart. The nutrient amendment consists of nitrogen, phosphorus, and other inorganic salts, as required, at concentrations ranging from 0.005 to 0.02 percent by weight. Oxygen for use as an electron acceptor in microbial metabolism is supplied by sparging air into the ground water. If the growth rate of microorganisms is 0.02 g/L per day, the process is estimated to require approximately 6 months to achieve 90 percent degradation of the hydrocarbons present. Cleanup is expected to be most efficient for ground waters contaminated with less than 40 ppm of gasoline. After termination of the process, the numbers of microbial cells are expected to return to background levels.

In addition to stimulating indigenous microbial populations to degrade organic waste constituents, another technique, which has not yet been fully demonstrated, is the addition of microorganisms with specific metabolic capabilities to a contaminated aquifer (Lee et al., 1988). Populations that are specialized in degrading specific compounds are selected by enrichment culturing or genetic manipulation. Enrichment culturing involves exposure of microorganisms to increasing concentrations of a contaminant or mixture of contaminants. The type of organism (or group of organisms) that

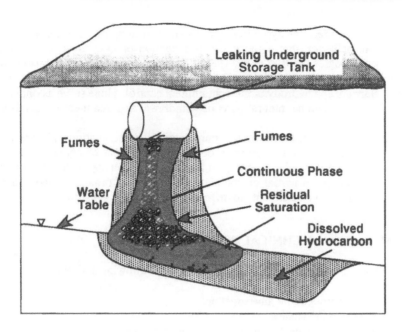

Figure 13-1. Regions of contamination in a typical release from an underground storage tank (Wilson et al., 1989).

is selected or acclimates to the contaminant depends on the source of the inoculum, the conditions used for the enrichment, and the substrate. Examples of changes that may occur during an acclimation period include an increase in population of contaminant degraders, a mutation that codes for new metabolic capabilities, and the induction or derepression of enzymes responsible for degradation of specific contaminants (Aelion et al., 1987).

It is important to note that the inoculation of a specialized microbial population into the environment may not produce the desired degree of degradation for a number of reasons (Goldstein et al., 1985; Lee et al., 1988; Suflita, 1989b). Factors that may limit the success of inoculants include contaminant concentration, pH, temperature, salinity, and osmotic or hydrostatic pressure. They may act alone or collectively to inhibit the survival of the microorganisms. The subsurface environment may also contain substances or other organisms that are toxic or inhibitory to the growth and activity of the inoculated organisms. In addition, adequate mixing to ensure contact of the organism with the specific organic constituent may be difficult to achieve at many sites. Successful inoculation of introduced organisms into simpler, more controllable environments (e.g., bioreactors such as wastewater treatment plants) to accomplish degradation has been demonstrated. However, effectiveness of inoculation into uncontrolled and poorly accessible environments such as the subsurface is much more difficult to achieve, demonstrate and assess (Thomas and Ward, 1989).

Genetic manipulation of microorganisms to produce specialized populations to degrade specific contaminants involves the acceleration and focusing of the process of evolution (Kilbane, 1986; Lee et al., 1988). Genetic manipulation can be accomplished by exposure of organisms to a mutagen, followed by enrichment culturing to isolate a population with specialized degradative capabilities, or by the use of DNA recombinant technology to change the genetic structure of a microorganism. The use of genetically engineered organisms in the environment is illegal without prior government approval (Thomas and Ward, 1989). In addition, the introduction of genetically engineered organisms into the environment would meet the same kind of barriers to success as organisms developed by enrichment culturing, or more.

Additional methods that have been suggested to enhance biodegradation include: cross acclimation, which involves the addition of a readily degradable substrate to aid in the biodegradation of more recalcitrant molecules; and analog enrichment, which involves the addition of a structural analog of a specific contaminant in order to induce degradative enzyme activity that will affect both the analog and the specific contaminant (Suflita, 1989a).

In most contaminated aquifers, the hydrogeologic system is so complex, in terms of site characteristics and contaminant behavior, that a successful remediation process must rely on the use of multiple treatment technologies (Wilson et al., 1986). The combination of several technologies, in series or in parallel, into a treatment process train may be necessary to restore ground-water quality to a required level. Barriers and hydrodynamic containment controls alone serve as only temporary plume control measures, but can be integral parts of withdrawal and treatment or in situ treatment measures.

A possible treatment train might consist of: (1) source removal by excavation and disposal; (2) free product recovery to reduce the mass in order to decrease the amount of contaminants requiring treatment; and (3) in situ treatment of remaining contamination. When applicable, biological in situ treatment offers the advantage of partial or complete destruction of organic contaminants, rather than transfer or partitioning of contaminants to different phases of the subsurface.

IN SITU BIOREMEDIATION TECHNICAL PROCESS

The in situ bioremediation technical process consists of the following activities:

1. performance of a thorough site investigation;
2. performance of treatability studies;
3. removal of source of contamination and recovery of free product;
4. design and implementation of the bioremediation technology; and
5. evaluation of performance of the technology through a monitoring program (Lee and Ward, 1986; Lee et al., 1988).

A thorough site investigation in which biological, contaminant, and aquifer characterization data are integrated, is essential for the successful implementation of a bioremediation system. Biological characterization is required to determine if a viable population of microorganisms is present which can degrade the contaminants of concern. An assessment of waste characteristics provides information for determining whether bioremediation, either alone or as part of a treatment train, is feasible for the specific contaminants at the site. Aquifer characteristics provide information on the suitability of the specific environment for biodegradative processes, as well as information required for hydraulic design and operation of the system.

Bioremediation of an aquifer contaminated with organic compounds can be accomplished by the biodegradation of those contaminants and result in the complete mineralization of constituents to carbon dioxide, water, inorganic salts, and cell mass, in the case of aerobic metabolism; or to methane, carbon dioxide, and cell mass, in the case of anaerobic metabolism. However, in the natural environment a constituent may not be completely degraded, but transformed to an intermediate product or products, which may be equally or more hazardous than the parent compound. In any event, the goal of in situ bioremediation is detoxification of a parent compound to a product or products that are no longer hazardous to human health and the environment.

In 1973 a review of ground-water microbiology was published by researchers at the U.S. EPA Robert S. Kerr Environmental Research Laboratory (RSKERL) (Dunlap and McNabb, 1973) that stimulated research into microbiology of the subsurface. Previously, biological activity in the subsurface environment below the root zone was considered unlikely and it was considered that microbial activity in the subsurface could not be of significant importance (Lee et al., 1988). However, as methods for sampling unconsolidated subsurface soils and aquifer materials without contamination from surface materials (Dunlap et al., 1977, Wilson et al., 1983, McNabb and Mallard, 1984) as well as methods to enumerate subsurface microbial organisms (Ghiorse and Wilson, 1988) were developed, evidence for microbial activity in the subsurface became convincing.

Bacteria are the predominant form of microorganisms that have been found in the subsurface, although a few higher life forms have been detected (Ghiorse and Wilson, 1988; Suflita, 1989a). The majority of microorganisms in pristine and uncontaminated aquifers are oligotrophic, because organic materials available for metabolism are likely present in low concentrations or difficult to degrade. Organic materials that enter uncontaminated subsurface environments are often refractory humic substances that resist biodegradation while moving through the unsaturated soil zone.

Many subsurface microorganisms are metabolically active and can use a wide range of compounds as carbon and energy sources, including xenobiotic compounds (Lee et al., 1988). Compounds such as acetone, ethanol, isopropanol, tertbutanol, methanol, benzene, chlorinated benzenes, chlorinated phenols, polycyclic aromatic hydrocarbons, and alkylbenzenes have been shown to degrade in samples of subsurface aquifer materials.

The rate and extent of biotransformation of organic compounds at a specific site are controlled by geochemical and hydraulic properties of the subsurface (Wilson et al., 1986). Populations of microorganisms increase until limited by a metabolic requirement, such as mineral nutrients, substrates for growth, or suitable electron acceptors. At this point, the rate of transformation of an organic material is controlled by transport processes that supply the limiting factor. Since most subsurface microorganisms are associated with the solid phase, the limiting factor must be delivered to the microbes by advection and diffusion through the mobile phases. Below the water table, all transport must be through liquid phases, and as a result, aerobic metabolism may be severely limited by the very low solubility of oxygen in water. As oxygen becomes limiting, aerobic respiration slows. However, other groups of organisms become active and continue to degrade contaminating organic materials. Under conditions of anoxia, anaerobic bacteria can use organic chemical or several inorganic anions as alternate electron acceptors (Suflita, 1989a).

Even though microorganisms may be present in a contaminated subsurface environment and have demonstrated the potential to degrade contaminants in laboratory studies, they may not be able to degrade these contaminants without a long period of acclimation. Acclimation results in development of the capability to accomplish degradation.

In summary, the rate of biological activity in the subsurface environment is generally controlled by:

1. the concentration of required nutrients in the mobile phases;
2. the advective flow of the mobile phases or the steepness of concentration gradients within the phases;
3. opportunity for colonization in the subsurface by metabolically active organisms or groups of organisms capable of degradation of the specific contaminants present;
4. presence, availability, and activity of appropriate enzymes for degradation of specific contaminants present; and
5. toxicity exhibited by the waste or co-occurring material(s) (Wilson et al., 1986; Suflita, 1989a).

METHODS TO COLLECT BIOLOGICAL SAMPLES

Traditionally, unconsolidated soils or sediments are sampled through a hollow-stem auger with a split-spoon core barrel or a conventional thin-walled sample tube (Acker, 1974, Scalf et al., 1981; Wilson et al., 1989). The hollow-stem auger acts as a temporary casing to keep the borehole open until a sample can be acquired. A borehole is drilled down to the depth to be sampled and a core barrel is inserted through the annular opening in the auger and driven or pushed while rotating the auger into the earth to collect the sample. These tools are effective in both unsaturated and saturated cohesive materials, but are not as effective in unconsolidated sands, as it is difficult to keep aquifer material out of the hollow stem auger (a phenomenon referred to as "heaving") and to keep the sample in the core barrel while the sample is being retrieved to the surface. In recent years there have been many improvements in sampling the subsurface, particularly with respect to heaving materials (Zapico et al., 1987; Leach et al., 1988).

Just as it is important to protect the integrity of samples while coring, it is as important to assure integrity while transferring sample material to containers which are to be returned to the laboratory for analysis. To prevent contamination of aquifer material samples from introduced microorganisms and to protect samples from the atmosphere to prevent injury of anaerobic microorganisms, samples are extruded inside a nitrogen-filled glove box (Figure 13-2). The glove box is prepared for sample collection by filling it with the desired number of sterile sampling jars and sterile paring devices, sealing the box, and then purging it with nitrogen gas. A slight positive pressure of nitrogen is maintained in the box by purging during extrusion and collection of the samples.

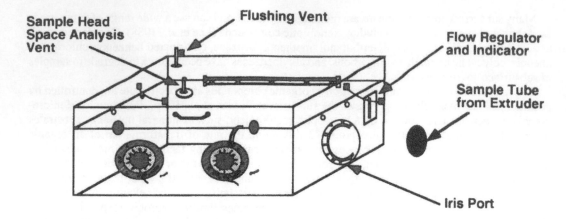

Figure 13-2. Field sampling glove box (Wilson et al., 1989).

BIOLOGICAL CHARACTERIZATION

A wide variety of methods are available to detect, enumerate and estimate biomass and metabolic activities of subsurface microorganisms. These methods include: direct light and epifluorescence microscopy, viable counts (e.g., plate counts, most probable number counts, and enrichment culture procedures), and biochemical indicators of metabolic activity such as ATP, GTP, phospholipid, and muramic acid (Ghiorse and Wilson, 1988). Levels of microorganisms ranging from 10^6 to 10^7 cells/g of dry aquifer material have been reported from uncontaminated shallow aquifers (Ghiorse and Balkwill, 1985; Lee et al., 1988). Often the distribution of microorganisms in aquifers, as it is in soils, is sporadic and nonuniform, indicating the presence of micro-environments conducive to growth and activity.

WASTE CHARACTERIZATION

The source of contamination is usually the primary object of remedial activities (Wilson et al., 1989) as the treatment of plume areas will not be effective if the source continues to release contaminants. Information concerning: (1) the areal location of the source area and contaminant plumes; (2) amounts of contaminants in the source area; and (3) amounts of contaminants released into the subsurface are required to select and apply an appropriate remediation technology and to determine cost and time requirements for completion of a remedial action. If in situ bioremediation is selected as the remedial technology, information concerning the amount and distribution of contamination is used in conjunction with hydrogeological site characteristics to locate injection and extraction wells and to optimize pumping rates and concentrations of amendments, such as nutrients and alternate electron acceptors.

The use of conventional monitoring wells can generally accurately define the geometry of the ground-water plume (Palmer and Johnson, 1989a; Wilson et al., 1989). However, there are important factors that control the quality of information collected from a network of monitoring wells, which include the amount of well purging done prior to sampling (Barcelona and Helfrich, 1986), method of sampling (Stolzenburg and Nichols, 1985), and method of well construction and installation (Keely and Boateng, 1987). Methods for ground-water sampling are presented by Scalf et al. (1981), Ford et al. (1984), and Barcelona et al. (1985). Other methods used for detecting contaminant plumes in the subsurface include geophysical techniques such as surface resistivity and electromagnetic surveys, chemical time-series sampling tests (Palmer and Johnson, 1989a), and vapor monitoring wells (Devitt et al., 1988; Palmer and Johnson, 1989b).[2]

The distribution of the source area and the extent of contamination should also be characterized by collecting cores of the solid aquifer materials. Precise information is required to define the verti-

---

[2]   Editor's Note: Other good EPA references on site characterization are U.S. EPA (1991) and Boulding (1993a, 1993b).

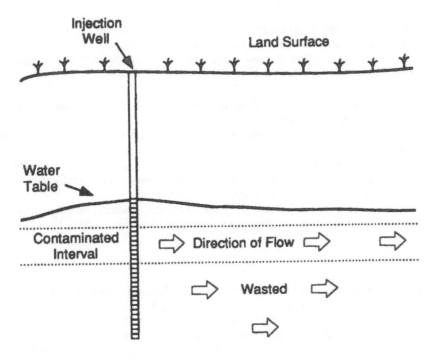

Figure 13-3. The value of accurately locating the contaminated interval (Wilson et al., 1989).

cal extent of contamination so that nutrients, oxygen and other amendments injected into the aquifer contact the contaminants. Injection into a clean part of the aquifer is a wasted effort and may give the false impression that the region of aquifer between the injection and recovery wells is clean (Figure 13-3).

Additional characteristics of waste contaminants present at a specific site that should be considered are related to their environmental fate and behavior in specific aquifer materials (Armstrong, 1987; Johnson et al., 1989). These characteristics include physical and chemical properties that determine recalcitrance, reactivity, and mobility of contaminants at the site. Information concerning partitioning of contaminants between aquifer solids and water is especially important. This information is used to evaluate the extent and rate of release of contaminants into the ground water, their mobility, and the quantity of electron acceptors and inorganic nutrients that must be supplied to support in situ bioremediation.

AQUIFER CHARACTERIZATION

Important geological characteristics of an aquifer that should be considered during a site investigation include the composition and heterogeneity of aquifer material, specific yield, hydraulic connections to other aquifers, magnitude of water table fluctuations, ground-water flow rate and direction, hydraulic conductivity distribution, permeability, bulk density, and porosity (Lee et al., 1988; Palmer and Johnson, 1989a).

Hydraulic conductivity (K) is an especially important characteristic since the aquifer must be permeable enough to allow the transport of electron acceptors and inorganic nutrients to the microorganisms in the zone of contamination. Permeable aquifer systems, i.e., aquifers with K values of 10^{-4} cm/sec or greater, are usually considered good candidates for in situ bioremediation (Thomas and Ward, 1989).

Hydraulic conductivity of an aquifer can be determined by a variety of methods (Palmer and Johnson, 1989a). Knowledge of K values at multiple locations is necessary because of the heterogeneity of aquifer materials. Laboratory methods are also available for determining hydraulic conductivity, but field-measured values represent average properties over a larger volume and utilize less disturbed materials (Palmer and Johnson, 1989a).

Aquifer characteristics play an extremely important role in determining the effectiveness of in situ bioremediation. Even in the presence of organisms acclimated to the specific waste constituents present in an aquifer, biodegradation of contaminants may be limited by unfavorable aquifer characteristics that affect microbial activity including:

1. insufficient concentrations of dissolved oxygen for aerobic metabolism of compounds susceptible to aerobic degradation;
2. excessive oxygen that inhibits anaerobic biodegradation of many halogenated compounds in the subsurface;
3. lack of a suitable alternative electron acceptor, if oxygen is unavailable or not usable;
4. insufficient inorganic nutrients, such as nitrogen, phosphorus, and trace minerals;
5. presence of toxic metals or other toxicants; and
6. other aquifer characteristics, such as pH, buffering capacity, salinity, osmotic or hydrostatic pressures, radiation, sorptive capacity, and temperature (Armstrong, 1987; Lee et al., 1988).

TREATABILITY STUDY

A treatability study is designed to determine if bioremediation is possible at a specific site, and whether there are any biological barriers to attaining cleanup goals. Even though the scientific literature may indicate that a specific chemical is likely to biodegrade in the environment, a treatability study using site-specific variables should be used to confirm that contention (Suflita, 1989a). Microcosms are generally used to conduct treatability studies. Pritchard (1981) defined a microcosm as "a calibrated laboratory simulation of a portion of a natural environment in which environmental components in as undisturbed a condition as possible, are enclosed within definable physical and chemical boundaries and studied under a set of laboratory conditions." Microcosms may range from simple batch incubation systems to large and complex flow-through devices (Suflita, 1989a).

Results of a treatability study can also provide an estimate of the rate and extent of remediation that can be attained if microorganisms are not limited by the rate of supply of an essential growth factor or by the presence of an unfavorable environmental factor.

Treatability studies to determine inorganic nutrient and electron acceptor requirements of subsurface microorganisms present at a specific site should be conducted using samples of subsurface solids as well as the ground water. Nutrient and electron acceptor requirements that will enable indigenous microorganisms to efficiently degrade organic contaminants present at a specific site can be determined by incubating contaminated subsurface materials with combinations of levels of inorganic nutrients and electron acceptors. Studies should be performed under conditions that simulate field environmental conditions. Results of the studies are used to design the bioremediation program as well as to optimize the treatment strategy.

DESIGN AND IMPLEMENTATION OF AN IN SITU BIOREMEDIATION SYSTEM

Before implementation of an in situ bioremediation system, the source of contamination in the soil and in the ground water should be removed as much as possible. In the case of a liquid fuel spill, source removal may consist of recovery of LNAPL free product from the ground water. Depending on the characteristics of the aquifer and contaminants, free product can account for as much as 91 percent of the spilled hydrocarbon, with the remaining hydrocarbon (accounting for 9–40 percent of the spill) sorbed to the soil or dissolved in the ground water (Lee et al., 1988).

Physical recovery techniques, based on the fact that LNAPL hydrocarbons are relatively insoluble in and less dense than water, are used to remove free product from a contaminated site. Physical recovery often accounts for only 30 to 60 percent of spilled hydrocarbon before yields decline. Continued pumping of recovery wells may be used to contain a spill while in-situ bioremediation is being implemented. If a spill is comprised of DNAPLs, which may sink to the bottom of the aquifer, physical recovery may be considerably more difficult to achieve.

Information from the performance of site characterization and treatability studies may be integrated with the use of comprehensive mathematical modeling to estimate the expected rates and extent of treatment at the field scale (Javandel et al., 1984; Keely, 1987). The specific model chosen should incorporate biological reaction rates, stoichiometry of waste transformation, mass-transport considerations, and spatial variability in treatment efficiency.

After assessment of site characterization and treatability studies, if results indicate that in situ bioremediation is applicable to the site and will be an effective cleanup technology, the information collected is used to design and implement the system.

When in situ bioremediation of a contaminant ground-water plume involves using methods to enhance the process, such as the addition of nutrients, additional oxygen sources, or other electron acceptors, the use of hydraulic controls to minimize migration of the plume during the in situ treatment process may be required (Thomas et al., 1987). In general, hydraulic control systems are generally less costly and time-consuming to install than physical containment structures such as slurry walls. Well systems are also more flexible, for pumping rates and well locations can be altered as the system is operated over a period of time.

Pumping-injection systems can be used to: (1) create stagnation zones at precise locations in a flow field; (2) create gradient barriers to pollution migration; (3) control the trajectory of a contaminant plume; and (4) intercept the trajectory of a contaminant plume (Shafer, 1984). The choice of a hydraulic control method depends on geological characteristics, variability of aquifer hydraulic conductivities background velocities, and sustainable pumping rates (Lee et al., 1988). Typical patterns of wells that are used to provide hydraulic controls include: (1) a pair of injection-production wells; (2) a line of downgradient pumping wells; (3) a pattern of injection-production wells around the boundary of a plume; and (4) the "double-cell" hydraulic containment system. The "double-cell" system utilizes an inner cell and an outer recirculation cell, with four cells along a line bisecting the plume in the direction of flow (Wilson, 1984).

Well systems also serve as injection points for addition of the materials used for enhancement of microbial activity and for control of circulation through the contaminated zone. The system usually includes injection and production wells and equipment for the addition and mixing of the nutrients (Lee et al., 1988). A typical system in which microbial nutrients are mixed with ground water and circulated through the contaminated portion of the aquifer through a series of injection and recovery wells is illustrated in Figure 13-4 (Raymond et al., 1978; Thomas and Ward, 1989).

Material can also be introduced to the aquifer through the use of infiltration galleries (Figure 13-5) (Brenoel and Brown, 1985; Thomas and Ward, 1989). Infiltration galleries allow movement of the injection solution through the unsaturated zone as well as the saturated zone, resulting in potential treatment of source materials that may be trapped in the pore spaces of the unsaturated zone.

Amendments to the aquifer are added to the contaminated aquifer in alternating pulses. Inorganic nutrients are usually added first through the injection system, followed by the oxygen source. Simultaneous addition of the two may result in excessive microbial growth close to the point of injection, and consequent plugging of the aquifer. High concentrations of hydrogen peroxide (greater than 10%) can be used to remove biofouling and restore the efficiency of the system.

OPERATIONS MONITORING

Both the operation and effectiveness of the system should be monitored (Lee et al., 1988). Operational factors of importance include the delivery of inorganic nutrients and electron acceptor, the point of the delivery within the aquifer in relation to the contaminated portion of the plume, and the effectiveness of containment and control of the contaminated plume.

Measurements of dissolved oxygen and nutrient levels in ground-water samples are recommended to assess whether or not bioremediation is being accomplished. Increases in microbial activities in samples of aquifer materials may be quantified relative to plume areas prior to treatment, areas within the plume that did not receive treatment, and control areas outside the plume. Carbon dioxide levels in ground-water samples may also be a useful indicator of microbial activity (Suflita, 1989b).

Measurement of contaminant levels should indicate that concentrations of contaminants are decreasing in areas receiving treatment and remaining relatively unchanged in areas that are not. If degradation pathways of specific contaminants are known, measurement of presence and concentrations of metabolic products may be used to determine whether or not bioremediation is occurring. Both soil and ground-water samples should be collected and analyzed to develop a thorough evaluation of treatment effectiveness. The use of appropriate control samples, e.g., assays of untreated

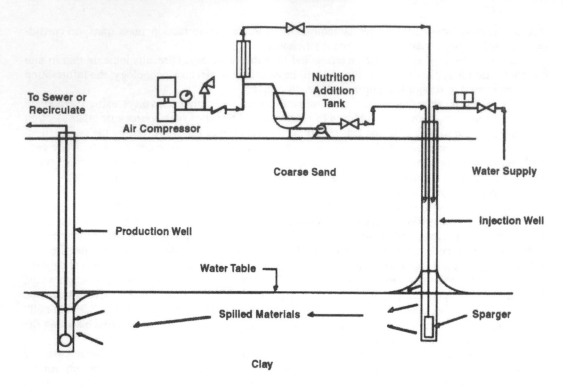

Figure 13-4. Typical schematic for aerobic subsurface bioremediation (Thomas and Ward, 1989).

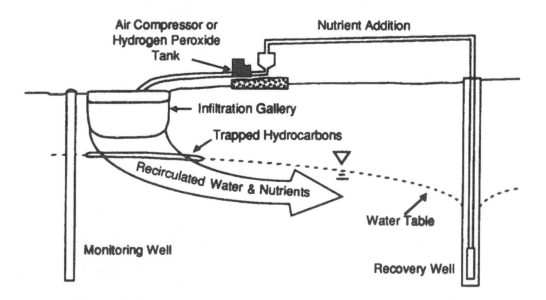

Figure 13-5. Use of infiltration gallery for recirculation of water and nutrients in in-situ bioremediation (Thomas and Ward, 1989).

areas or areas outside the plume, is highly recommended to confirm the effectiveness of the bioremediation technology (Suflita, 1989b).

The frequency of sampling should be related to the time expected for significant changes to occur along the most contaminated flow path. Important considerations include time required for water to move from injection wells to monitoring wells, seasonal variations in water table elevation or hydraulic gradient, changes in the concentration of dissolved oxygen or alternative electron acceptor, and costs of monitoring.

Advantages and Limitations in the Use of In Situ Bioremediation

There are a number of advantages and disadvantages in the use of in situ bioremediation (Lee et al., 1988). Unlike other aquifer remediation technologies, it can often be used to treat contaminants that are sorbed to aquifer materials or trapped in pore spaces. In addition to treatment of the saturated zone, organic contaminants held in the unsaturated and capillary zones can be treated when an infiltration gallery is used.

The time required to treat subsurface pollution using in situ bioremediation can often be faster than withdrawal and treatment processes. A gasoline spill was remediated in 18 months using in situ bioremediation, while pump-and-treat techniques were estimated to require 100 years to reduce the concentrations of gasoline to potable water levels (Raymond et al., 1986). In situ bioremediation often costs less than other remedial options. The areal zone of treatment using bioremediation can be larger than with other remedial technologies because the treatment moves with the plume and can reach areas that would otherwise be inaccessible.

There are also disadvantages to in situ bioremediation programs (Lee et al., 1988). Many organic compounds in the subsurface are resistant to degradation. In situ bioremediation usually requires an acclimated population of microorganisms which may not develop for recent spills or for recalcitrant compounds. Heavy metals and toxic concentrations of organic compounds may inhibit activity of indigenous microorganisms. Injection wells may become clogged from profuse microbial growth resulting from the addition of nutrients and oxygen.

In situ bioremediation is difficult to implement in low-permeability aquifers that do not permit the transport of adequate supplies of nutrients or oxygen to active microbial populations. In addition, bioremediation projects require continuous monitoring and maintenance for successful treatment.

EPA CONTACT

For further information contact Guy Sewell, RSKERL-Ada, (405) 436-8566 (see Preface for mailing address).

REFERENCES

Aelion, C.M., C.M. Swindoll, and F.K. Pfaender. 1987. Adaptation to and Biodegradation of Xenobiotic Compounds by Microbial Communities from a Pristine Aquifer. Appl. Environ. Microbiol. 53:2212–2217.

Acker, W.L., III. 1974. Bask Procedures for Soil Sampling and Core Drilling. Acker Drill Co. Scranton, PA.

Armstrong, J. 1987. Some Problems in the Engineering of Groundwater Cleanup. p. 110–120. In: N. Dee, W.F. McTernan, and E. Kaplan (eds.) Detection, Control, and Renovation of Contaminated Ground Water. Am. Soc. Civil Eng., New York, NY.

Barcelona, M.J. and J.A. Helfrich. 1986. Well Construction Purging Effects on Ground-Water Samples. Environ. Sci. Technol. 20:1179–1184.

Barcelona, M.J., J.P. Gibb, J.A. Helfrich, and E.E. Garske. 1985. Practical Guide for Ground-Water Sampling. EPA/600/2-85/104 (NTIS PB86-137304).

Boulding, J.R. 1993a. Use of Airborne, Surface, and Borehole Geophysical Techniques at Contaminated Sites: A Reference Guide. EPA/625/R-92/007.

Boulding, J.R. 1993b. Subsurface Characterization and Monitoring Techniques: A Desk Reference Guide, Vols. I and II. EPA/625/R-93/003a&b.

Brenoel, M. and R.A. Brown. 1985. Remediation of a Leaking Underground Storage Tank with Enhanced Bioreclamation. In: Proc. Fifth National Symp. on Aquifer Restoration and Ground Water Monitoring, National Water Well Association, Worthington, OH.

Canter L.W. and R.C. Knox. 1985. Ground Water Pollution Control. Lewis Publishers, Chelsea, MI.

Devitt, D.A., R.B. Evans, W.A. Jury, T.H. Starks, B. Eklund, and A. Ghalsan. 1988. Soil Gas Sensing for Detection and Mapping of Volatile Organics. EPA/600/8-87/036 (NTIS PB87-228516).

Dunlap, W.J. and J.F. McNabb. 1973. Subsurface Biological Activity in Relation to Ground Water Pollution. EPA-660/2-73/014 (NTIS PB227-990).

Dunlap, W.J., J.F. McNabb, M.R. Scalf, and R.L. Cosby. 1977. Sampling for Organic Chemicals and Microorganisms in the Subsurface. EPA-600/2-77-176 (NTIS PB272-679).

Ehrenfield, J., and J. Bass. 1984. Evaluation of Remedial Action Unit Operations of Hazardous Waste Disposal Sites. Pollution Technology Review No. 110, Noyes Publications, Park Ridge, NJ.

Ford, P.J., P.J. Turina, and D.E. Seely. 1984. Characterization of Hazardous Waste Sites—A Methods Manual, Volume II: Available Sampling Methods. EPA/600/4-84/076 (NTIS PB85-521596).

Ghiorse, W.C. and D.L. Balkwill. 1985. Microbiology of Ground Water Environments. In: G.E. Januer (ed.) Progress in Chemical Disinfection II: Problems at the Frontier. State Univ. of New York (SUNY) at Binghamton, Binghamton, NY.

Ghiorse, W.C., and J.T. Wilson. 1988. Microbial Ecology of the Terrestrial Subsurface. Adv. Appl. Microbiol. 33:107–172.

Goldstein R.M., L.M. Mallory, and M. Alexander. 1985. Reasons for Possible Failure of Inoculation to Enhance Biodegradation. Appl. Environ. Microbiol. 50:977–983.

Javandel, I., C. Doughty, and C.F. Tsang. 1984. Groundwater Transport: Handbook of Mathematical Models. Water Resources Monograph No. 10, Am. Geophysical Union, Washington, DC.

Johnson, R.L., C.D. Palmer, and W. Fish. 1989. Subsurface Chemical Processes. In: Transport and Fate of Contaminants in the Subsurface. EPA/625/4-89/019, Chapter 5.

Keely, J.F. 1987. The Use of Models in Managing Groundwater; Protection Programs. EPA/600/8-87/003 (NTIS PB87-166203).

Keely, J.F. and K. Boateng. 1987. Monitoring Well Installation, Purging, and Sampling Techniques. Part II: Case Histories. Ground Water 25:427–439.

Kilbane, J.J. 1986. Genetic Aspects of Toxic Chemical Degradation. Microbiol. Ecol. 12:135–145.

Leach, L.E., F.P. Beck, J.T. Wilson, and D.H. Kampbell. 1988. Aseptic Subsurface Sampling Techniques for Hollow-Stem Auger Drilling. In: Proc. Second Nat. Outdoor Action Conf. on Aquifer Restoration, Ground Water Monitoring and Geophysical Methods, National Water Well Association, Dublin, OH, Vol. 1, pp. 31–51.

Lee, M.D. and C.H. Ward. 1986. Ground Water Restoration. Report submitted to JACA Corporation, Fort Washington, PA.

Lee, M.D., J.M. Thomas, R.C. Borden, P.B. Bedient, J.T. Wilson, and C.H. Ward. 1988. Biorestoration of Aquifers Contaminated with Organic Compounds. CRC Critical Rev. Environ. Control 18:29–89.

McNabb, J.F. and G.E. Mallard. 1984. Microbiological Sampling in the Assessment of Groundwater Pollution. In: G. Bitton and C.P. Gerba (eds.), Groundwater Pollution Microbiology. John Wiley & Sons, New York, NY.

Palmer, C.D. and R.L. Johnson. 1989a. Determination of Physical Transport Parameters. In: Transport and Fate of Contaminants in the Subsurface. EPA/625/4-89/019, Chapter 4.

Palmer C.D. and R.L. Johnson. 1989b. Physical Processes Controlling the Transport of Contaminants in the Aqueous Phase. In: Transport and Fate of Contaminants in the Subsurface. EPA/625/4-89/019, Chapter 2.

Palmer, C.D. and R.L. Johnson. 1989c. Physical Processes Controlling the Transport of Non-Aqueous Phase Liquids in the Subsurface. In: Transport and Fate of Contaminants in the Subsurface. EPA/625/4-89/019, Chapter 3.

Pritchard, P.H. 1981. Model Ecosystems. In: R.A. Conway (ed.), Environmental Risk Analysis for Chemicals. Van Nostrand Reinhold, New York, NY.

Raymond, R.L. 1974. Reclamation of Hydrocarbon Contaminated Ground Waters. U.S. Patent Office, Washington, DC. Patent No. 3,846,290. Patented November 5, 1974.

Raymond, R.L., R.A. Brown, R.D. Norris, and E.T. O'Neill. 1986. Stimulation of Biooxidation Processes in Subterranean Formations. U.S. Patent Office, Washington, DC. Patent No. 4,588,506. Patented May 13, 1986.

Raymond, R.L., V.W. Jamison, J.O. Hudson, R.E. Mitchell, and V.E. Farmer. 1978. Field Application of Subsurface Biodegradation of Gasoline in a Sand Formation. API Publication 4430, American Petroleum Institute, Washington, DC.

Scalf, M.R., J.F. McNabb, W.J. Dunlap, R.L. Cosby, and J. Fryberger. 1981. Manual of Ground Water Sampling Procedures. EPA/600/2-81/160 (NTIS PB82-103045).

Shafer, J.M. 1984. Determining Optimum Pumping Rates for Creation of Hydraulic Barriers to Ground Water Pollutant Migration. In: Proc. Fourth National Symp. on Aquifer Restoration and Ground Water Monitoring, National Water Well Association, Worthington, OH, pp. 50-64.

Stolzenburg, T.R. and D.G. Nichols. 1985. Preliminary Results on Chemical Changes in Groundwater Samples Due to Sampling Devices. EPRI Report No. EA-4118, Electric Power Research Institute, Palo Alto, CA.

Suflita, J.M. 1989a. Microbial Ecology and Pollutant Biodegradation in Subsurface Ecosystems. In: Transport and Fate of Contaminants in the Subsurface. EPA/625/4-89/019, Chapter 7.

Suflita, J.M. 1989b. Microbiological Principles Influencing the Biorestoration of Aquifers. In: Transport and Fate of Contaminants in the Subsurface. EPA/625/4-89/019, Chapter 8.

Thomas, J.M., and C.H. Ward. 1989. In Situ Biorestoration of Organic Contaminants in the Subsurface. Environ. Sci. Technol. 23:760-766.

Thomas, J.M., M.D. Lee, P.B. Bedient, R.C. Borden, L.W. Canter, and C.H. Ward. 1987. Leaking Underground Storage Tanks: Remediation with Emphasis on In Situ Biorestoration. EPA/600/2-87/008 (NTIS PB87-168084).

U.S. Environmental Protection Agency. 1991. Site Characterization for Subsurface Remediation. EPA/625/4-91/026.

Wilson, J.L. 1984. Double-cell Hydraulic Containment of Pollutant Plumes. In: Proc. Fourth National Symp. on Aquifer Restoration and Ground Water Monitoring, National Water Well Association, Worthington, OH, pp. 65-70.

Wilson J., L. Leach, J. Michalowski, S. Vandegrift, and R. Callaway. 1989. In Situ Bioremediation of Spills from Underground Storage Tanks: New Approaches for Site Characterization, Project Design, and Evaluation of Performance. EPA/600/2-89/042 (NTIS PB89-219976).

Wilson, J.T., J.F. McNabb, D.L Balkwill, and W.C. Ghiorse. 1983. Enumeration and Characterization of Bacteria Indigenous to a Shallow Water-Table Aquifer. Ground Water 21:134-142.

Wilson, J.T., L.E. Leach, MM. Henson, and J.N. Jones. 1986. In Situ Biorestoration as a Ground Water Remediation Technique. Ground Water Monitoring Rev. 6:56-64.

Zapico, M.M., S. Vales, and J.A. Cherry. 1987. A Wireline Piston Core Barrel for Sampling Cohesionless Sand and Gravel Below the Water Table. Ground Water Monitoring Rev. 7(3):74-82.

Chapter 14

In Situ Vitrification Treatment[1]

Trevor Jackson, Science Applications International (SAIC), Cincinnati, OH

ABSTRACT

In situ vitrification (ISV) uses electrical power to heat and melt soil, sludge, mine tailings, buried wastes, and sediments contaminated with organic, inorganic, and metal-bearing hazardous wastes. The molten material cools to form a hard, monolithic, chemically inert, stable glass and crystalline product that incorporates and immobilizes the thermally stable inorganic compounds and heavy metals in the hazardous waste. The slag product material is glass-like with very low leaching characteristics.

Organic wastes are initially vaporized or pyrolyzed by the process. These contaminants migrate to the surface, where the majority are then burned within a hood covering the treatment area; the remainder are treated in an off-gas treatment system.

ISV uses a square array of four electrodes that are inserted into the surface of the ground. Electrical power is applied to the electrodes which, through a starter path or graphite and glass frit, establish an electric current in the soil. The electric current generates heat and melts the starter path and the soil; typical soil melt temperature is 2,900°F to 3,600°F. An electrode feed system (EFS) drives the electrodes in the soil as the molten mass continues to grow downward and outward until the melt zone reaches the desired depth and width. The process is repeated in square arrays until the desired volume of soil has been vitrified. The process can typically treat up to 1,000 tons of material in one melt setting.

ISV technology has been under development and testing since 1980 [1, p. 1].[2] ISV was developed originally for possible application to soils contaminated with radioactive materials. In this application, trans-uranium radionuclides are incorporated in the vitrified mass. At this time there is only one vendor of commercially available in situ vitrification systems. The technology description, status, and performance data are quoted from the published work of this vendor.

ISV is the proposed remediation technology at eight sites, six of which are EPA Superfund sites [2; 3]. Full-scale units have been constructed. Even so, the technology should be considered emerging in its full-scale application to Superfund sites. EFS mechanisms have recently been developed for pilot- and full-scale systems. This chapter provides information on the technology applicability, limitations, the types of residuals produced, the latest performance data, site requirements, the status of the technology, and sources for further information.

Site-specific treatability studies are the best means of establishing the applicability and projecting the likely performance of an ISV system. Determination of whether ISV is the best treatment alternative will be based on multiple site-specific factors, cost, and effectiveness. The EPA Contact indicated at the end of this chapter can assist in the location of other contacts and sources of information necessary for such treatability studies.

TECHNOLOGY APPLICABILITY

ISV has been reported to be effective in treating a large variety of organic and inorganic wastes based on the results of engineering- and pilot-scale tests. The technology also has proven effective-

---

[1] EPA/540/S-94/504.
[2] [Reference number, page number.]

Table 14-1. Effectiveness of ISV on General Contaminant Groups for Soil, Sludges, and Sediments

Contaminant Groups	Effectiveness		
	Soil	Sludge	Sediment
Organic			
Halogenated volatiles	■	▼	▼
Halogenated semivolatiles	■	▼	▼
Nonhalogenated volatiles	■	▼	▼
Nonhalogenated semivolatiles	■	▼	▼
Polychlorinated biphenyls (PCBs)	■	▼	■
Pesticides (halogenated)	■	■	▼
Dioxins/furans	■	▼	▼
Organic cyanides	▼	▼	▼
Organic corrosives	■	▼	▼
Inorganic			
Volatile metals	■	■	■
Nonvolatile metals	■	■	■
Asbestos	■	▼	▼
Radioactive materials	■	▼	▼
Inorganic corrosives	▼	▼	▼
Inorganic cyanides	▼	▼	▼
Reactive			
Oxidizers	▼	▼	▼
Reducers	▼	▼	▼

■ Demonstrated Effectiveness: Successful treatability test at some scale completed.

▼ Potential Effectiveness: Expert opinion that technology will work.

● No Expected Effectiveness: Expert opinion that technology will not work.

ness in treating radioactive wastes based on the results of full-scale tests. Radioactive wastes and sludges, contaminated soils and sediments, incinerator ashes, industrial wastes and sludges, medical wastes, mine tailings, and underground storage tank waste can all potentially be vitrified [4, p. 4-1].

Organic contaminants at concentrations of 5 to 10 percent by weight and inorganic contaminants at concentrations of 5 to 15 percent by weight are generally acceptable for ISV treatment [5, p 13]. The effectiveness of the ISV technology on treating various contaminants in soil, sludge, and sediments is given in Table 14-1. Examples of constituents within contaminant groups are provided in the "Technology Screening Guide for Treatment of CERCLA Soils and Sludges" [6]. Table 14-1 is based on current available information or professional judgment where no information was available. The proven effectiveness of the technology for a particular site or waste does not ensure that it will be effective at all sites or that the treatment levels achieved will be acceptable at other sites. For the ratings used for this table, demonstrated effectiveness means that at some scale, treatability tests have shown that the technology was effective for that particular contaminant and matrix. The ratings of potential effectiveness or no expected effectiveness are both based upon expert opinion. Where potential effectiveness is indicated, the technology is believed capable of successfully treating the contaminant group in a particular matrix. The technology is expected to work for all contaminant groups listed.

ISV processing required that sufficient glass-forming materials (e.g., silicon and aluminum oxides) be present within the waste materials to form and support a high-temperature melt. To form a melt, sufficient (typically 2 to 5 percent) monovalent alkali cations (e.g., sodium and potassium) must be present to provide the degree of electrical conductivity needed for the process to operate efficiently. If the natural material does not meet this requirement, fluxing materials such as sodium carbonate can be added to the base material. Typically, these conditions are met by most soils, sediments, tailings, and process sludges.

Differences in soil characteristics such as permeability and density generally do not affect overall chemical composition of the soil or the ability to use ISV. In many site locations, the soil profile may be stratified and present nonuniform characteristics that can affect the melt rate and dimensions of the vitrified mass. Before applying the ISV technology, soil stratification must be defined so that it may be factored into the remedial design.

LIMITATIONS

The ISV process can treat soils saturated with water; however, additional power is used to dry the soil prior to melting and may increase the cost of remediation by 10 percent. ISV is more economical to implement when the soil to be vitrified has a low moisture content. Progression of a melt into saturated soil enclosed in a container can result in a gaseous steam release that can cause the molten glass to spatter.

When treating a contaminated zone in an aquifer, it may be necessary to lower the water table below the zone of contamination in order to vitrify to the desired depth. Alternatively, a hydraulic barrier (e.g., slurry wall) could be placed upstream of the contamination to divert the aquifer flow around the treatment zone. Treatment in a water-saturated zone may result in movement of some of the contaminants from the treatment zone to surrounding areas, thereby reducing the amount of contaminants being destroyed, immobilized, or removed.

The maximum ISV depth obtainable is influenced by several factors, including spacing between electrodes, amount of power available, variations in soil composition and gradation between different strata, depth to ground water, soil permeability within an aquifer, surface heat loss during ISV, and waste and soil density. To date, treatment depths of only 19 feet have been demonstrated [4, 7-6].

The presence of large inclusions in the area to be treated can limit the use of the ISV process. Inclusions are highly concentrated contaminant layers, void volumes, containers, metal scrap, general refuse, demolition debris, rock, or other heterogeneous materials within the treatment volume. Figure 14-1 gives limits for inclusions within the treatment volume [7, p. 17]. If massive void spaces exist, a large subsidence could result in a very short time period. These problems, as well as those caused by other large inclusions, may be detected by ground penetrometry or other geologic investigations. Some inclusions such as void volumes, containers, and solid combustible refuse can potentially generate gases. However, the oversized hooding is intended to control and mitigate any release. If large volumes of off-gases are generated during a short time period, the off-gas treatment system may overload. Vitrification of flammable or explosive objects can result in spattering of the molten glass. Underground storage tanks can be treated only if they are filled with soil prior to the vitrification process.

Sampling and analysis of the glass matrix produced by ISV is difficult and must be carefully planned prior to conducting a treatability study or site remediation. Current EPA digestion methods for metal analyses are not designed to dissolve the glass matrix. The metal concentration measured by a standard nitric/hydrochloric acid digestion (SW-846, Method 3050) will likely be highly dependent on the particle size of the material prior to digestion. The digestion specified will not dissolve glass but will leach some metals from the exposed surfaces. Closure of mass balance for the system, therefore, can often be incomplete. However, a recently developed digestion method using hydrofluoric acid with microwave digestion has been known to improve metal analysis for this type of matrix.

TECHNOLOGY DESCRIPTION

Several methods and configurations exist for the application of ISV. At a site that has only a relatively shallow layer of contamination, the contaminated layer may be excavated and transported to a pit where the vitrification will take place. At other sites where the contamination is much deeper, thermal barriers could be placed along the site to be vitrified, and prevent the movement of heat and glass into adjacent areas. This will force the heat energy downward and melt depths will be increased.

This chapter describes the more conventional approach to using ISV; a checkerboard pattern of melts is used to encapsulate the waste and control the potential for lateral migration. The holes in the checkerboard are then vitrified to complete the remediation of the site.

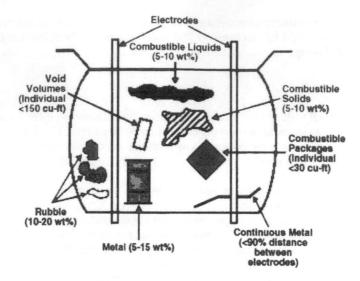

Figure 14-1. General limits for inclusion within volume to be treated.

Figure 14-2 shows a typical ISV equipment layout. ISV uses a square array of electrodes up to 18 feet apart, which is inserted to a depth of 1 to 5 feet and potentially can treat down to a depth of 20 feet to remediate a contaminated area. A full-scale system can remediate at a rate of 3 to 5 tons per hour [4, p. 3-6] until a maximum mass of 800 to 1,000 tons has been treated. Since soil is not electrically conductive once the moisture has been driven off, a conductive mixture of flaked graphite and glass frit is placed between the electrodes to act as a starter path, as shown in Figure 14-3. Power is supplied to the electrodes, which establishes an electrical current in the starter path. The resultant power heats the starter path and surrounding soil up to 3,600°F, which is well above the melting temperature of typical soils (2,000°F to 2,600°F). The graphite starter path eventually is consumed by oxidation and the current is transferred to the soil which is electrically conductive in the molten state. A typical downward growth rate is 1 to 2 inches per hour. The thermal gradient surrounding the melt is typically 300°F to 480°F per inch. As the vitrified zone grows, it incorporates metals and either vaporizes or pyrolizes organic contaminants. The pyrolyzed products migrate to the surface of the vitrified zone, where they may oxidize in the presence of oxygen. A hood placed over the processing area is used to collect combustion gases, which are treated in an off-gas treatment system.

As the melt grows downward and outward, power is maintained at sufficient levels to overcome the heat losses to the hood and surrounding soil. Generally, the melt grows outward to form a melt width approximately 50 percent wider than the electrode spacing. This growth varies as a function of electrode spacing and melt depth. The molten zone is roughly a square with slightly rounded corners, a shape that reflects the higher power density and around the electrodes. As the melt grows in size, the electrical resistance of the melt decreases; thus, the ratio between the voltage and the current must be adjusted periodically to maintain operation at an acceptable power level.

The EFS, now an integral part of all operations, enhances the ability of ISV to treat soils containing high concentrations of metal. In EFS operations, the electrodes are independently fed to the molten soil as the melt proceeds downward, instead of being placed in the soil prior to the startup of the test. The system improves processing control at sites with high concentrations of metal. For example, upon encountering a full or partial electrical short, the affected electrodes are simply raised and held above the molten metal pool at the bottom of the melt. During this time, the melt continues to grow downward. The electrodes can then be reinserted into the melt to their original depth and electrode feeding operations resumed. These advances have been incorporated into the pilot- and the full-scale ISV systems [8].

The treatment area is covered by a newly designed octagonal-shaped off-gas collection hood with a maximum distance of 60 feet between the sides. The hood has three manual viewing ports and the

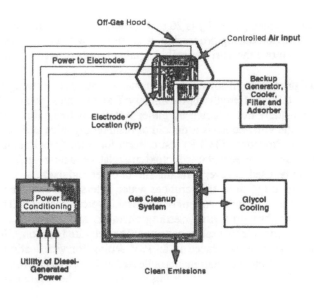

Figure 14-2. ISV equipment system.

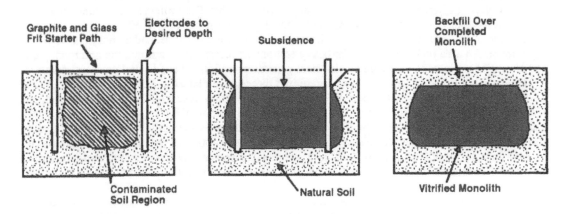

Figure 14-3. Stages of ISV processing.

provision for video monitoring or recording. The hood is connected to an off-gas treatment trailer and a backup off-gas treatment system. During the process, the off-gases are drawn by a 1,850 standard cubic feet per minute (scfm) blower into the trailer. Flow of air through the hood is controlled to maintain a vacuum of 0.5 to 2.0 inches H_2O on the system. The off-gas temperatures are typically 210°F to 750°F when they enter the treatment system. The gases are then treated by quenching, scrubbing, mist-elimination, heating, particulate filtration, and activated carbon adsorption. The backup off-gas treatment system is used in the event of a power outage, and is powered by a diesel generator. The backup system is designed to treat gases that may evolve from the melt until power is restored to the process and electrodes [9].

PROCESS RESIDUALS

The main process residual produced during operation of the ISV technology is the vitrified soil itself. The vitrified monolith is left in place after treatment, due to its nonhazardous nature. The volume of the ISV product formed generally is 20 to 45 percent less than the initial volume treated. Because of the volume reduction during processing, it is covered with clean backfill. It is possible, however, to excavate and remove the vitrified soil in smaller pieces if onsite disposal is not acceptable at a given site.

Typically, the residual product from soil applications has a compressive strength approximately 5 to 20 times greater and a tensile strength approximately 7 to 11 times greater than unreinforced concrete [4, p. 5-3]. It is usually not affected by either wet/dry or freeze/thaw cycling [10, p. 3]. Existing data indicate that the vitrified mass is devoid of residual organics and passes EPA's Toxicity Characteristic Leaching Procedure (TCLP) test criteria for priority pollutant metals. The ISV residual also has been found to have acceptable biotoxicity relative to near-surface life forms [11, p. 79]. The clean backfill can be used to revegetate the site or other end uses.

After processing for a period of time, the scrubber water, filters, and activated carbon may contain sufficient contaminants to warrant treatment or disposal. Typical treatment includes passing the contaminated scrubber water through a filter, settling chamber, and activated carbon, then either reusing the water or discharging it into a sanitary sewer. The activated carbon, filter, and the solids from the settling chamber can then be placed in an ISV setting for vitrification. In this way, the destruction/chemical incorporation of contaminants collected in the off-gas treatment system is maximized. Only residuals resulting from the last setting at a site must be treated and disposed of by means other than ISV.

SITE REQUIREMENTS

The components of the ISV system are contained in three transportable trailers: an off-gas and process control trailer; a support trailer; and an electrical trailer. The trailers are mounted on wheels sufficient for transportation to and over a compacted ground surface [12, p. 307].

The site must be prepared for the mobilization, operation, maintenance, and demobilization of the equipment. An area must be cleared for heavy equipment access roads, automobile and truck parking lots, ISV equipment, setup areas, electrical generator, equipment sheds, and workers' quarters.

The field-scale ISV equipment system required three-phase electrical power at either 12,500 or 13,800 volts, which is usually taken from a utility distribution system [13, p. 2]. At startup, the technology requires high voltage (up to 4,000 volts) to overcome the resistance of the soil, and a current of approximately 400 amps. The soil resistance decreases as the melt progresses, so that by the end of the process, the voltage decreases to approximately 400 volts and the current increases up to approximately 4,000 amps [4, p. 3-6]. Alternatively, the power may be generated onsite by means of a diesel generator. Typical applications require 800 kilowatt hour/ton (kWh/ton) to 1,000 kWh/ton.

Spent activated carbon, scrubber water, or other process waste materials may be hazardous, and the handling of these material requires that a site safety plan be developed to provide for personnel protection and special handling measures. Storage should be provided to hold these wastes until they have been tested to determine their acceptability for disposal, release, or recycling to subsequent ISV melts. Storage capacity will depend on the waste volume generated.

Site activities such as clearing vegetation, removing overburden, and acquiring backfill material are often necessary. These activities are generally advantageous from a financial point of view. For example, the cost of removal of the top portion of clean soil would generally be much less than the cost for labor and energy to vitrify the same volume of soil [4, p. 9-6].

PERFORMANCE DATA

Performance data presented in this chapter should not be considered directly applicable to other Superfund sites. A number of variables such as the specific mix and distribution of contaminants affect system performance. A thorough characterization of the site and a well-designed and conducted treatability study are highly recommended.

The performance data currently available are from the process developer. ISV has been developed through four scales of equipment: (a) bench (5 to 20 pounds); (b) engineering (50 to 2,000 pounds); (c) pilot (10 to 50 tons); and (d) full (500 to 1,000 tons). The values in parentheses are typical masses of

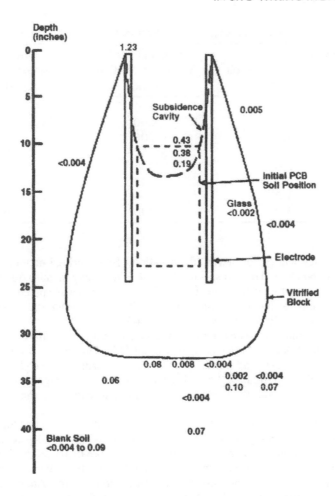

Figure 14-4. Vitrified block and surrounding soil sample positions and PCB concentrations.

vitrified products resulting from a single setting at the various scales. Several tests have been performed at each scale and on a variety of contaminated media.

An engineering-scale test was performed on a loamy–clay soil containing 500 parts per million (ppm) of PCBs. Figure 14-4 gives the final concentrations of PCBs (in ppm) in and around the vitrified block upon completion of the test [13, p. 4-3]. This figure indicates that migration of PCBs outside the vitrified block is not a significant concern. Data from off-gas emissions and soil container smears accounted for 0.05 percent by weight of the initial PCB quantity, which corresponds to a greater than 99.9 percent destruction efficiency (DE) for the ISV process. This DE does not include the removal efficiency of the off-gas treatment. Activated carbon has a 99.9 percent efficiency and can remove any of these off-gas emissions effectively. Overall, the destruction removal efficiency (DRE) range for the combined ISV and off-gas system is between 6 and 9 nines, which is greater than the 6 nines DRE required by 40 CFR 761.70 for PCB incinerators. Analysis of the off-gas also indicated the presence of small quantities of polychlorinated dibenzo-p-dioxins (PCDDs) and polychlorinated dibenzofurans (PCDFs). However, the levels reported (0.1 µg/L and 0.4 µg/L, respectively) can be removed by the off-gas treatment system. An engineering-scale test on PCB-contaminated sediments from New Bedford Harbor [4, p. 4-2] gave a similar DE (99.9999 percent) for the ISV process before additional treatment by the off-gas treatment system. During feasibility testing of PCB-contaminated soil from a Spokane, Washington site, a DE greater than 99.993 percent and a DRE greater than 99.99999 percent were obtained [14]. During engineering-scale testing of vitrification of simulated wastes from the Hanford Engineering Development Laboratory, a DRE of greater than 99.99 percent was obtained for a variety of organic contaminants [15].

Table 14-2. TCLP Extract Metal Concentrations, Idaho National Engineering Lab Soils

Metal	Maximum Allowable Leachate Concentration (mg/L)	Contaminated Soil Concentration (mg/kg)	Vitrified Product Leachate Concentration (mg/L)
Arsenic	5.0	200	<0.168
Barium	100.0	200	0.229
Cadmium	1.0	200	0.0098
Chromium	5.0	200	0.0178
Lead	5.0	2000	0.636
Mercury	0.2	200	<0.0001
Selenium	1.0	200	0.098
Silver	5.0	200	<0.023

An engineering-scale test was performed on Idaho National Engineering Laboratory spiked soil at the Pacific Northwest Laboratory. The soil was spiked with eight heavy metals (Ag, As, Ba, Cd, Cr, Hg, Pb, and Se) to 0.02 percent by weight, except for the lead, which was spiked at 0.2 percent by weight [16]. The test results for metals concentrations in the leach extract and maximum concentration limits established by the EPA are given in Table 14-2.

Feasibility testing was conducted using the bench-scale ISV equipment to treat a sample of soil from the old Jacksonville, Arkansas water treatment plant [17]. This soil was contaminated with 2,3,7,8-tetrachlorodibenzo-p-dioxin and placed in a 5-gallon can with a Pyrex-plate lid. Analytical results did not detect any dioxin or furan in the vitrified material or in the off-gas. Based on analytical detection limits, the DE was greater than 99.995 percent prior to entry into the off-gas treatment system.

Ten thousand kilograms of an industrial sludge heavily laden with zirconia and lime was vitrified successfully by the pilot-scale ISV process. The sludge contained 55 to 70 percent moisture by weight. The volume was observed to be reduced significantly (more than one-third of original volume) after the testing [18, p. 29]. Analysis of the off-gas and the scrubber water showed that the melt retained between 98 and 99 percent of the fluorides, chlorides, and sulfates. Analysis indicated that the destruction of organic carbon was good, and that ISV was effective in promoting nitrogen oxide (NO_x) destruction. This result minimizes the concern for environmental impact.

Soil from a fire training pit contaminated with fuel oils and heavy metals was bench-scale tested at the Arnold Engineering Development Center in Tennessee [19]. Results of initial testing and analyses of the soil indicated that an electrically-conducting fluxing agent (such as sodium carbonate) with a lower melting point was required as an addition to the soil for ISV processing to work effectively. The onsite pilot-scale process achieved a high destruction of organics (greater than 98 percent) and high retention of inorganics in the melt. Leach testing using Extraction Procedure Toxicity (EP-Tox) and TCLP tests showed that all metals of concern were below maximum permissible limits. The tests indicate that the fluxing agent should be distributed throughout the entire vitrification depth for optimum operation.

TECHNOLOGY STATUS

The only vendor supplying commercial systems for in situ vitrification of hazardous wastes is Geosafe Corporation. Geosafe is under a sublicense from the process developer, Battelle Memorial Institute. Four scales of units are in operation ranging from bench-scale to full-scale.

To date, only bench-, engineering-, and pilot-scale test results are available on in situ vitrification of hazardous wastes. Full-scale tests have been completed only on radioactive wastes. Table 14-3 indicates several sites where ISV has been selected as the remedial action [2].

In April 1991, a fire involving the full-scale collection ISV hooding occurred at the Geosafe Hanford, Washington test site. The vendor was testing a new, lighter hooding material. The hooding caught fire during the test when a spattering of the melt occurred. For a period of time after the incident, Geosafe suspended full-scale field operations. During this time, Geosafe completed analytical, model-

Table 14-3. Selected Sites Specifying ISV as the Remedial Action

Site	Mass/Volume to be Treated	Primary Contaminants	Status
Parsons Chemical	Soil: 2,000 cubic yards (yd³)	Biocides (pesticides), dioxins, metals (mercury)	Site preparation
Ionia City Landfill	Soil with debris: 5,000 yd³ (15 feet deep)	Volatile organic compounds (methylene chloride, TCA, styrene, toluene), metals (lead)	Treatability testing
Rocky Mountain Arsenal	Soil: 4,600 yd³ (10 feet deep) Sludge: 5,800 yd³ (10 feet deep)	Biocides (pesticides), metals (arsenic, mercury)	Remedial design
Wasatch Chemical	Soil: 3,600 yd³ (5 feet deep) sludge, solids	Semivolatile organic compounds (hexachlorobenzene, pentachlorophenol), biocides (pesticides), dioxins	Remedial design
Transformer Service Facility/TSCA Demonstration	Soil: 3,500 tons	PCBs	Site preparation
Arnold AFB, Site 10	Soil with debris: 10,000 tons	Mixed organics, heavy metals	Site preparation
Crab Orchard Wildlife Refuge	Soil: 40,000 tons	PCBs and lead	Pre-design
Anderson Development	Soil: 4,000 tons	4,4'-methylene bis (2-chloroaniline) (MBOCA)	Pre-design

ing, and engineering-scale testing to allow confident design; defined necessary process revisions; finalized design and fabrication of a new metal off-gas collection hood; and performed additional operational acceptance testing to demonstrate the capabilities of the equipment and operational procedures [20]. The new off-gas collection hood design is composed entirely of metal rather than high-temperature fabric, which was previously used. The new design is heavier than the fabric hood, but is capable of being transported by the same equipment.

Cost estimates for this technology range from $300 to $650 per ton of contaminated soil treated. The most significant factor influencing cost is the depth of the soil to be treated. High moisture content requires that additional energy be used to dry out the soil before the melting process can begin, thus increasing the cost. Other factors that influence the cost of remediation by ISV are: the amount of site preparation required; the specific properties of the contaminated soil (e.g., dry density); the required depth of processing; and the unit price of electricity.

EPA CONTACT

Technology-specific questions regarding ISV may be directed to Ed Barth, (513) 569-7669, or Teri Richardson, RREL-Cincinnati, (513) 569-7949 (see Preface for mailing address).

ACKNOWLEDGMENTS

This chapter was prepared for the U.S. Environmental Protection Agency, Office of Research and Development (ORD), Risk Reduction Engineering Laboratory (RREL), Cincinnati, Ohio, by Science Applications International Corporation (SAIC) under Contract No. 68-C0-0048. Mr. Eugene Harris served as the EPA Technical Project Monitor. Mr. Jim Rawe was SAIC's Work Assignment

Manager. Dr. Trevor Jackson (SAIC) was the primary author. The author is especially grateful to Ms. Teri Richardson of EPA-RREL, who contributed significantly by serving as a technical consultant during the development of this chapter.

The following other Agency and contractor personnel have contributed their time and comments by participating in the expert review meetings or peer reviews of the chapter:

Mr. Edward Bates, EPA-RREL
Mr. Briant Charboneau, Wastren, Inc.
Mr. Kenton Oma, Eckenfelder, Inc.
Mr. Eric Saylor, SAIC

REFERENCES

1.   Geosafe Corporation. In Situ Vitrification for Permanent Treatment of Hazardous Wastes. Presented at Advances in Separations: A Focus on Electrotechnologies for Products and Waste, Battelle, Columbus, 1989.

2.   Innovative Treatment Technologies, Semi-Annual Status Report (Fourth Edition). EPA/542/R-92/011, U.S. Environmental Protection Agency, October 1992. [Annual Report (Sixth Edition) was published as EPA 542-R-94-006, September, 1994.]

3.   Conversations with J. Hansen, of Geosafe. April 19, 1993.

4.   Vitrification Technologies for Treatment of Hazardous and Radioactive Waste. EPA/625/R-92/002, May 1992.

5.   Geosafe Corporation. Application and Evaluation Considerations for In Situ Vitrification Technology: A Treatment Process for Destruction and/or Permanent Immobilization of Hazardous Materials. April 1989.

6.   Technology Screening Guide for Treatment of CERCLA Soils and Sludges. EPA/540/2-88/004 (NTIS PB89-132674), 1988. pp. 55–60.

7.   FitzPatrick, V.F. and J.E. Hansen. In Situ Vitrification for Remediation of Hazardous Wastes. Presented at 2nd Annual HazMat Central Conference, Chicago, Illinois, 1989.

8.   Farnsworth, R.K., K.H. Oma, and C.E. Bigelow. Initial Tests on In Situ Vitrification Using Electrode Feeding Techniques. Prepared for the U.S. Department of Energy, under Contract DE-AC06-76RLO 1830, 1990.

9.   In Situ Vitrification Technology Update. Geosafe Corporation. November 1992.

10. Hansen, J.E., C.L. Timmerman, and S.C. Liikala. Status of In Situ Vitrification Technology: A Treatment Process for Destruction and/or Permanent Immobilization. In: Proceedings of Annual HazMat Management Conference International, Atlantic City, New Jersey, 1990. pp. 317–330.

11. Greene, J.C., et al. Comparison of Toxicity Results Obtained from Eluates Prepared from Non-Stabilized and Stabilized Waste Site Soils. In: Proceedings for the 5th National Conference on Hazardous Wastes and Hazardous Materials, Las Vegas, Nevada, 1988. pp. 77–80.

12. FitzPatrick, V.F., C.L. Timmerman, and J.L. Buelt. In Situ Vitrification--An Innovative Thermal Treatment Technology. Proceedings: Second International Conference on New Frontiers for Hazardous Waste Management. EPA/600/9-87/018F, U.S. Environmental Protection Agency, 1987. pp. 305–322.

13. Timmerman, C.L. In Situ Vitrification of PCB Contaminated Soils. EPRI CS-4839. Electric Power Research Institute, Palo Alto, California, 1986.

14. Timmerman, C.L. Feasibility Testing of In Situ Vitrification of PCB-Contaminated Soil from a Spokane, WA Site. Prepared for Geosafe Corporation, Kirkland, Washington, under Contract 14506, 1986.

15. Koegler, S.S. Disposal of Hazardous Wastes by In Situ Vitrification. Prepared for the U.S. Department of Energy, under Contract DE-AC06-76LO 1830, 1987.

16. Farnsworth, R.K., et al. Engineering-Scale Test No. 4: In Situ Vitrification of Toxic Metals and Volatile Organics Buried in INEL Soils. Prepared for the U.S. Department of Energy, under Contract DE-AC06-76RLO 1830, 1991.

17. Mitchell, S.J. In Situ Vitrification of Dioxin Contaminated Soils. Prepared for American Fuel and Power Corporation, Panama City, Florida, under Contract 2311211874, 1987.

18. Buelt, J.L., and S.T. Freim. Demonstration of In Situ Vitrification for Volume Reduction of Zirconia/Lime Sludges. Prepared for Teledyne Wah Chang, Albany, Oregon, under Contract 2311205327, 1986.

19.   Timmerman, C.L. Feasibility Testing of In Situ Vitrification of Arnold Engineering Development Center Contaminated Soils. Prepared for the U.S. Department of Energy, under Subcontract DE-AC05-84OR21400, 1989.
20.   Correspondence from Geosafe Corp. to Mr. Edward R. Bates (RREL), September 17, 1991.

PART II

Ex Situ Treatment Methods for Contaminated Soils, Ground Water, and Hazardous Waste

Chapter 15

Air Stripping of Aqueous Solutions[1]

Jim Rawe, Science Applications International Corporation (SAIC), Cincinnati, OH

ABSTRACT

Air stripping is a means to transfer contaminants from aqueous solutions to air. Contaminants are not destroyed by air stripping but are physically separated from the aqueous solutions. Contaminant vapors are transferred into the air stream and, if necessary, can be treated by incineration, adsorption, or oxidation. Most frequently, contaminants are collected in carbon adsorption systems and then treated or destroyed in this concentrated form. The concentrated contaminants may be recovered, incinerated for waste heat recovery, or destroyed by other treatment technologies. Generally, air stripping is used as one in a series of unit operations and can reduce the overall cost for managing a particular site. Air stripping is applicable to volatile and semivolatile organic compounds. It is not applicable for treating metals and inorganic compounds.

During 1988, air stripping was one of the selected remedies at 30 Superfund sites [1].[2] In 1989, it was a component of the selected remedy at 38 Superfund sites [2]. An estimated 1,000 air-stripping units are presently in operation at sites throughout the United States [3]. Packed-tower systems typically provide the best removal efficiencies, but other equipment configurations exist, including diffused-air basins, surface aerators, and cross-flow towers [4, p. 2; 5, p. 10–48]. In packed-tower systems, there is no clear technology leader by virtue of the type of equipment used or mode of operation. The final determination of the lowest cost alternative will be more site-specific than process equipment dominated.

This chapter provides information on the technology applicability, the technology limitations, a description of the technology, the types of residuals produced, site requirements, the latest performance data, the status of the technology, and sources of further information.

TECHNOLOGY APPLICABILITY

Air stripping has been demonstrated in treating water contaminated with volatile organic compounds (VOCs) and semivolatile compounds. Removal efficiencies of greater than 98 percent for VOCs and greater than or equal to 80 percent for semivolatile compounds have been achieved. The technology is not effective in treating low-volatility compounds, metals, or inorganics [6, p. 5-3]. Air stripping has commonly been used with pump-and-treat methods for treating contaminated ground water.

This technology has been used primarily for the treatment of VOCs in dilute aqueous waste streams. Effluent liquid quality is highly dependent on the influent contaminant concentration. Air stripping at specific design and operating conditions will yield a fixed, compound-specific percentage removal. Therefore, high influent contaminant concentrations may result in effluent concentrations above discharge standards. Enhancements, such as high temperature or rotary air stripping, will allow less-volatile organics, such as ketones, to be treated [6, p. 5-3].

Table 15-1 shows the effectiveness of air stripping on general contaminant groups present in aqueous solution. Examples of constituents within contaminant groups are provided in Reference 7, "Technology Screening Guide for Treatment of CERCLA Soils and Sludges." This table is based on the current available information or professional judgment where no information was available. The

---

[1] EPA/540/2-91/022.
[2] [Reference number, page number.]

Table 15-1. Effectiveness of Air Stripping on General Contaminant Groups from Water

Contaminant Group	Effectiveness
Organic	
Halogenated volatiles	▪
Halogenated semivolatiles[a]	▼
Nonhalogenated volatiles	▪
Nonhalogenated semivolatiles	●
PCBs	●
Pesticides	●
Dioxins/furans	●
Organic cyanides	●
Organic corrosives	●
Inorganic	
Volatile metals	●
Nonvolatile metals	●
Asbestos	●
Radioactive materials	●
Inorganic corrosives	●
Inorganic cyanides	●
Reactive	
Oxidizers	●
Reducers	●

▪ Demonstrated Effectiveness: Successful treatability test at some scale completed.
▼ Potential Effectiveness: Expert opinion that technology will work.
● No Expected Effectiveness: Expert opinion that technology will not work.
[a] Only some compounds in this category are candidates for air stripping.

proven effectiveness of the technology for a particular site or contaminant does not ensure that it will be effective at all sites or that the treatment efficiencies achieved will be acceptable at other sites. For the ratings used for this table, demonstrated effectiveness means that, at some scale, treatability testing demonstrated the technology was effective for that particular contaminant group. The ratings of potential effectiveness and no expected effectiveness are both based upon expert judgment. Where potential effectiveness is indicated, the technology is believed capable of successfully treating the contaminant group in a particular matrix. When the technology is not applicable or will probably not work for a particular contaminant group, a no-expected-effectiveness rating is given.

LIMITATIONS

Because air stripping of aqueous solutions is a means of mass transfer of contaminants from the liquid to the air stream, air pollution control devices are typically required to capture or destroy contaminants in the off-gas [8]. Even when off-gas treatment is required, air stripping usually provides significant advantages over alternatives such as direct carbon adsorption from water because the contaminants are more favorably sorbed onto activated carbon from air than from water. Moreover, contaminant destruction via catalytic oxidation or incineration may be feasible when applied to the off-gas air stream.

Aqueous solutions with high turbidity or elevated levels of iron, manganese, or carbonate may reduce removal efficiencies due to scaling and the resultant channeling effects. Influent aqueous media with pHs greater than 11 or less than 5 may corrode system components and auxiliary equipment. The air stripper may also be subject to biological fouling. The aqueous solution being air stripped may need pretreatment to neutralize the liquid, control biological fouling, or prevent scaling [6; 9].

Contaminated water with VOC or semivolatile concentrations greater than 0.01 percent generally cannot be treated by air stripping. Even at lower influent concentrations, air stripping may not be able to achieve cleanup levels required at certain sites. For example, a 99 percent removal of trichloroethene (TCE) from ground water containing 100 parts per million (ppm) would result in an effluent concentration of 1 ppm, well above drinking water standards. Without heating, only volatile organic contaminants with a dimensionless Henry's law constant greater than 10^{-2} are amenable to continuous flow air stripping in aqueous solutions [6; 5]. In certain cases, where a high removal efficiency is not required, compounds with lower Henry's law constants may be air stripped. Ashworth et al. published the Henry's law constants for 45 chemicals [10, p. 25]. Nirmalakhandan and Speece published a method for predicting Henry's law constants when published constants are unavailable [11]. Air strippers operated in a batch mode may be effective for treating water containing either high contaminant concentrations or contaminants with lower Henry's law constants. However, batch systems are normally limited to relatively low average flow rates.

Several environmental impacts are associated with air stripping. Air emissions of volatile organics are produced and must be treated. The treated wastewater may need additional treatment to remove metals and nonvolatiles. Deposits, such as metal (e.g., iron) precipitates may occur, necessitating periodic cleaning of air-stripping towers [6, p. 5-5]. In cases where heavy metals are present and additional treatment will be required, it may be beneficial to precipitate those metals prior to air stripping.

TECHNOLOGY DESCRIPTION

Air stripping is a mass transfer process used to treat ground water or surface water contaminated with volatile or semivolatile organic contaminants. At a given site, the system is designed based on the type of contaminant present, the contaminant concentration, the required effluent concentration, water temperature, and water flow rate. The major design variables are gas pressure drop, air-to-water ratio, and type of packing. Given those design variables, the gas and liquid loading (i.e., flows per cross-sectional area), tower diameter and packing height can be determined. Flexibility in the system design should allow for changes in contaminant concentration, air and water flow rates, and water temperature. Figure 15-1 is a schematic of a typical process for the air stripping of contaminated water.

In an air-stripping process, the contaminated liquid is pumped from a ground water or surface water source. Water to be processed is directed to a storage tank (1) along with any recycle from the air-stripping unit.

Air stripping is typically performed at ambient temperature. In some cases, the feed stream temperature is increased in a heat exchanger (2). Heating the influent liquid increases air-stripping efficiency and has been used to obtain a greater removal of semivolatile organics such as ketones. At temperatures close to 100°C, steam stripping may be a more practical treatment technique [8, p. 3].

The feed stream (combination of the influent and recycle) is pumped to the air stripper (3). Three basic designs are used for air strippers: surface aeration, diffused-air systems, and specially designed liquid-gas contactors [4, p. 3]. The first two of these have limited application to the treatment of contaminated water due to their lower contaminant removal efficiency. In addition, air emissions from surface-aeration and diffused-air systems are frequently more difficult to capture and control. These two types of air strippers will not be discussed further. The air stripper in Figure 15-1 is an example of a liquid-gas contactor.

The most efficient type of liquid-gas contactor is the packed tower [4, p. 3]. Within the packed tower, structures called packing provide surface area on which the contaminated water can form a thin film and come in contact with a countercurrent flow of air. Air-to-water ratios may range from 10:1 to 300:1 on a volumetric basis [14, p. 8]. Selecting packing material that will maximize the wetted surface area will enhance air stripping. Packed towers are usually cylindrical and are filled with either random or structured packing. Random packing consists of pieces of packing dumped onto a support structure within the tower. Metal, plastic, or ceramic pieces come in standard sizes and a variety of shapes. Smaller packing sizes generally increase the interfacial area for stripping and improve the mass-transfer kinetics. However, smaller packing sizes result in an increased pressure drop of the air stream and an increased potential for precipitate fouling. Tripacks®, saddles, and slotted rings are the shapes most commonly used for commercial applications. Structured packing consists of trays fitted to the inner diameter of the tower and placed at designated points along the height of the tower. These trays are made of metal gauze, sheet metal, or plastic. The choice of which type of

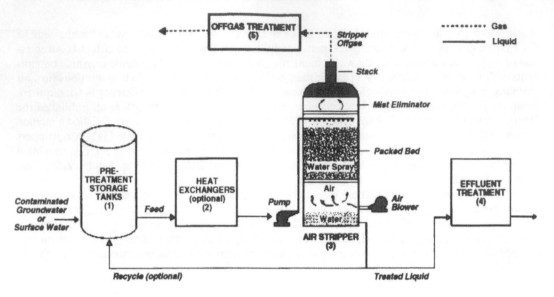

Figure 15-1. Schematic diagram of air-stripping system [8, p. 20; 13, p. 43].

packing to use depends on budget and design constraints. Random packing is generally less expensive. However, structured packing reportedly provides advantages such as lower pressure-drop and better liquid distribution characteristics [4, p. 5].

The processed liquid from the air-stripper tower may contain trace amounts of contaminants. If required, this effluent is treated (4) with carbon adsorption or other appropriate treatments.

The off-gas can be treated (5) using carbon adsorption, thermal incineration, or catalytic oxidation. Carbon adsorption is used more frequently than the other control technologies because of its ability to remove hydrocarbons cost-effectively from dilute (< 1 percent) air streams [8, p. 5].

PROCESS RESIDUALS

The primary process residual streams created with air-stripping systems are the off-gas and liquid effluent. The off-gas is released to the atmosphere after treatment; activated carbon is the treatment most frequently applied to the off-gas stream. Where activated carbon is used, it is recommended that the relative humidity of the air stream be reduced. Once spent, the carbon can be regenerated onsite or shipped to the original supplier for reactivation. If spent carbon is replaced, it may have to be handled as a hazardous waste. Catalytic oxidation and thermal incineration also may be used for off-gas treatment [15, p. 10; 8, p. 5]. Sludges, such as iron precipitates, build up within the tower and must be removed periodically [6, p. 5-5]. Spent carbon can also result if carbon filters are used to treat effluent water from the air-stripper system. Effluent water containing non-volatile contaminants may need additional treatment. Such liquids are treated onsite or stored and removed to an appropriate facility. Biological, chemical, activated carbon, or other appropriate treatment technologies may be used to treat the effluent liquid. Once satisfactorily treated, the water is sent to a sewage treatment facility, discharged to surface water, or returned to the source, such as an underground aquifer.

SITE REQUIREMENTS

Air strippers are most frequently permanent installations, although mobile systems may be available for limited use. Permanent installations may be fabricated onsite or may be shipped in modular form and constructed onsite. Packing is installed after fabrication or construction of the tower. A concrete pad will be required to support the air-stripper tower in either case. Access roads or compacted soil will be needed to transport the necessary materials.

Standard 440V, three-phase electrical service is needed. Water should be available at the site to periodically clean scale or deposits from packing materials. The quantity of water needed is site-specific. Typically, treated effluent can be used to wash scale from packing.

Contaminated liquids are hazardous, and their handling requires that a site safety plan be developed to provide for personnel protection and special handling measures. Spent activated carbon may be hazardous and require similar handling. Storage may be needed to hold the treated liquid until it has been tested to determine its acceptability for disposal or release. Depending upon the site, a method to store liquid that has been pretreated may be necessary. Storage capacity will depend on liquid volume.

Onsite analytical equipment for conducting various analyses, including gas chromatography capable of determining site-specific organic compounds for performance assessment, make the operation more efficient and provide better information for process control.

PERFORMANCE DATA

System performance is measured by comparing contaminant concentrations in the untreated liquid with those in the treated liquid. Performance data on air-stripping systems, ranging from pilot-scale to full-scale operation, have been reported by several sources, including equipment vendors. Data obtained on air strippers at Superfund sites also are discussed below. The data are presented as originally reported in the referenced documents. The quality of this information has not been determined. The key operating and design variables are provided when they were available in the reference.

An air-stripping system, which employed liquid-phase GAC to polish the effluent, was installed at the Sydney Mine site in Valrico, Florida. The air-stripping tower was 4 feet in diameter, 42 feet tall, and contained a 24-foot bed of 3.5-inch diameter polyethylene packing. The average design water flow was 150 gallons per minute (gpm) with a hydraulic loading rate of 12 gpm/ft^2 and a volumetric air-to-water ratio of approximately 200:1. The air-stripping tower was oversized for use at future treatment sites. Effluent water from the air stripper was polished in a carbon adsorption unit. Table 15-2 summarizes the performance data for the complete system; it is unclear how much removal was accomplished by the air stripper and how much by the activated carbon. Influent concentrations of total organics varied from approximately 25 parts per billion (ppb) to 700 ppb [13, p. 41].

Air stripping was used at well 12A in the city of Tacoma, Washington. Well 12A had a capacity of 3,500 gpm and was contaminated with chlorinated hydrocarbons, including 1,1,2,2-tetrachloroethane; trans-1,2-dichloroethene (DCE); TCE; and perchloroethylene. The total VOC concentration was approximately 100 ppb. Five towers were installed and began operation on July 15, 1983. Each tower was 12 feet in diameter and was packed with 1-inch polypropylene saddles to a depth of 20 feet. The water flowrate was 700 gpm for each tower, and the volumetric air-to-water ratio was 310:1. The towers consistently removed 94 to 98 percent of the influent 1,1,2,2-tetrachloroethane with an overall average of 95.5 percent removal. For the other contaminants, removal efficiencies in excess of 98 percent were achieved [16, p. 112].

Another remedial action site was Wurtsmith Air Force Base in Oscoda, Michigan. The contamination at this site was the result of a leaking underground storage tank near a maintenance facility. Two packed-tower air strippers were installed to remove TCE. Each tower was 5 feet in diameter and 30 feet tall, with 18 feet of 16 mm pall ring packing. The performance summary for the towers, presented in Table 15-3, is based on evaluations conducted in May and August 1982 and January 1983. Excessive biological growth decreased performance and required repeated removal and cleaning of the packing. Operation of the towers in series, with a volumetric air-to-water ratio of 25:1 and a water flow of 600 gpm (2,270 L/min), removed 99.9 percent of the contaminant [17, p. 119].

A 2,500 gpm air stripper was used to treat contaminated ground water during the initial remedial action at the Verona Well field site in Battle Creek, Michigan. This well field is the major source of public potable water for the city of Battle Creek. The air stripper was a 10-foot diameter tower packed to a height of 40 feet with 3.5 inch pall rings. The air stripper was operated at 2,000 gpm with a 20:1 volumetric air-to-water ratio. Initial problems with iron oxide precipitating on the packed rings were solved by recirculating sodium hypochlorite through the stripper about four times per year

Table 15-2. Performance Data for the Groundwater Treatment System at the Sydney Mine Site, FL [13, p. 42]

Contaminant	Concentration	
	Influent (µg/L)	Effluent (µg/L)
Volatile organics		
Benzene	11	ND[a]
Chlorobenzene	1	ND
1,1-Dichloroethane	39	ND
Trans-1,2-dichloropropane	1	ND
Ethylbenzene	5	ND
Methylene chloride	503	ND
Toluene	10	ND
Trichlorofluoromethane	71	ND
Meta-xylene	3	ND
Ortho-xylene	2	ND
Extractable organics		
3-(1,1-Dimethylethyl) phenol	32	ND
Pesticides		
2,4-D	4	ND
2,4,5-TP	1	ND
Inorganics		
Iron (mg/L)	11	<0.03

[a] ND = Not detected at method detection limit of 1 µg/L for volatile organics and 10 µg/L for extractable organics and pesticides

Table 15-3. Air-Stripper Performance Summary at Wurtsmith AFB [17, p. 121]

G/L (vol)	Water Flow (L/min)	Single Tower (% Removal)	Series Operation (% Removed)
10	1,135	95	99.8
10	1,700	94	99.8
10	2,270	86	96.0
18	1,135	98	99.9
18	1,700	97	99.9
18	2,270	90	99.7
25	1,135	98	99.9
25*	1,700	98	99.9
25	2,270	98	99.9

Influent TCE concentration: 50–8,000 µg/L; Water temperature: 283°K

[8, p. 8-9]. The total VOC concentration of 131 ppb was reduced by approximately 82.9 percent [15, p. 56]. The air stripper off-gas was treated via vapor phase granular activated carbon beds. The off-gas was heated prior to entering the carbon beds to reduce its humidity to 40 percent.

An air stripper is currently operating at the Hyde Park Superfund site in New York. Treatek, Inc., which operates the unit, reports the system is treating about 80,000 gallons per day (gpd) of landfill leachate. The contaminants are in the range of 4,000 ppm total organic carbon (TOC). The air stripper is reportedly able to remove about 90 percent of the TOCs [18]. A report describing the performance of the air stripper is expected to be published during 1991.

The primary VOCs at the Des Moines Superfund site were TCE; 1,2-DCE; and vinyl chloride. The TCE initial concentration was approximately 2,800 ppb and gradually declined to the 800 to 1,000 ppb range after 5 months. Initial ground-water concentrations of 1,2-DCE were unreported while the

Table 15-4. Air-Stripper Performance at Eau Claire Municipal Well Field [12, p. C-1]

Contaminant	Influent Concentration (ppb)	Removal Efficiency (%)
1,1-Dichloroethene	0.17–2.78	88
1,1-Dichloroethane	0.38–1.81	93
1,1,1-Trichloroethane	4.32–14.99	99
Trichloroethene	2.53–11.18	98

concentration of vinyl chloride ranged from 38 ppb down to 1 ppb. The water flow rate to the air stripper ranged from 500 to 1,850 gpm and averaged approximately 1,300 gpm. No other design data were provided. TCE removal efficiencies were generally above 96 percent, while the removal efficiencies for 1,2-DCE were in the 85 to 96 percent range. No detectable levels of vinyl chloride were observed in the effluent water [12, p. B-1].

VOCs were detected in the Eau Claire municipal well field in Eau Claire, Wisconsin, as part of an EPA ground-water supply survey in 1981. An air stripper was placed on-line in 1987 to protect public health and welfare until completion of the remedial investigation/feasibility study (RI/FS) and final remedy selection. Data reported on the Eau Claire site were for the period beginning August 31, 1987 and ending February 15, 1989. During this period, the average removal efficiency was greater than 88 percent for the four chlorinated organic compounds studied. The average removal efficiencies are shown in Table 15-4. The air stripper had a 12-foot diameter and was 60 feet tall, with a packed bed of 26 feet. Water feed rates were approximately 5 to 6 million gallons per day (mgd). No other design parameters were reported [12, p. C-1].

In March 1990, an EPA study reviewed the performance data from a number of Superfund sites, including the Brewster Well Field, Hicksville MEK Spill, Rockaway Township, Western Processing, and Gilson Road Sites [15].

Reported removal efficiencies at the Brewster Well Field site in New York were 98.50 percent, 93.33 percent, and 95.59 percent for tetrachloroethene (PCE); TCE; and 1,2-DCE; respectively. Initial concentrations of the three contaminants were 200 ppb (PCE), 30 ppb (TCE) and 38 ppb (1,2-DCE) [15, p. 55]. The 300 gpm air stripper had a tower diameter of 4.75 feet, packing height of 17.75 feet, air-to-water ratio of 50:1, and used 1-inch saddles for packing material [15, p. 24].

A removal efficiency of 98.41 percent was reported for methyl ethyl ketone (MEK) at the Hicksville MEK Spill site in New York. The reported influent MEK concentration was 15 ppm. The air stripper had a 100 gpm flowrate, an air-to-water ratio of 120:1, a tower diameter of 3.6 feet, a packing height of 15 feet, and used 2-inch Jaeger Tripack packing material. Water entering the air stripper was heated to approximately 180° to 195°F by heat exchangers [15, p. 38].

The Rockaway Township air stripper had a flowrate of 1,400 gpm, tower diameter of 9 feet, packing height of 25 feet, air-to-water ratio of 200:1, and used 3-inch Tellerettes packing material. The performance data are shown in Table 15-5 [15, p. 18].

The Western Processing site had two air-stripping towers treating different wells in parallel. The first tower had a 100 gpm (initial) and 200 gpm (maximum) flowrate, a tower diameter of 40 feet, a packing height of 40.5 feet, an air-to-water ratio of 160:1 (initial) and 100:1 (maximum), and used 2-inch Jaeger Tripack packing material. The second tower had a 45 gpm (initial) and 60 gpm (maximum) flowrate, a tower diameter of 2 feet, packing height of 22.5 feet, air-to-water ratio of 83.1:1 (initial) and 62.3:1 (maximum), and used 2-inch Jaeger Tripack packing material [15, p. 31]. The performance data are presented in Table 15-6.

The Gilson Road Site used a single column high-temperature air stripper (HTAS) which had a 300 gpm flowrate (heated influent), tower diameter of 4 feet, packing height of 16 feet, air-to-water ratio of 51.4:1, and used 16 Koch-type trays at 1-foot intervals [15, p. 42–45]. The performance data are provided in Table 15-7. Due to the relatively high influent concentration and the high (average) removal efficiency, this system required supplemental control of the volatiles in the off-gas.

Another EPA study, completed in August 1987, analyzed performance data from 177 air-stripping systems in the United States. The study presented data on systems design, contaminant types, and loading rates, and reported removal efficiencies for 52 sites. Table 15-8 summarizes data from 46 of those sites, illustrating experiences with a wide range of contaminants [19]. Reported efficiencies

Table 15-5. Air-Stripper Performance at Rockaway Township, NJ [15, p. 53]

Contaminant	Influent Concentration (ppb)	Removal Efficiency (%)
Trichloroethylene	28.3	99.99
Methyl-tert-butyl ether	3.2	99.99
1,1-Dichloroethylene	4.0	99.99
cis-1,2-Dichloroethylene	6.4	99.99
Chloroform	1.3	99.99
1,1,1-Trichloroethane	20.0	99.99
1,1-Dichloroethane	2.0	99.99
Total VOC	65.2	99.99

Table 15-6. Air-Stripper Performance at Western Processing, WA [15, p. 61]

Contaminant	Influent Concentration (ppb)	Removal Efficiency (%)
Benzene	73	93.15
Carbon tetrachloride	5	—
Chloroform	781	99.36
1,2-Dichloroethane	22	77.27
1,1-Dichloroethylene	89	94.38
1,1,1-Trichloroethane	1,440	99.65
Trichloroethylene	8,220	99.94
Vinyl chloride	159	99.37
Dichloromethane	8,170	99.63
Tetrachloroethylene	378	98.68
Toluene	551	99.09
1,2-Dichlorobenzene	11	54.55
Hexachlorobutadiene	250	96.00
Hexachloroethane	250	96.00
Isobutanol	10	0.00
Methyl ethyl ketone	1,480	70.27

Table 15-7. Air-Stripper Performance at the Gilson Road Site, NH [15, p. 65]

Contaminant	Influent Concentration (ppb)	Average Removal Efficiency (%)
Isopropyl alcohol	532	95.30
Acetone	473	91.93
Toluene	14,884	99.87
Dichloromethane	236	93.79
1,1,1-Trichloroethane	1,340	99.45
Trichloroethylene	1,017	99.71
Chloroform	469	99.06
Total VOC	18,951	99.41

should be interpreted with caution. Low efficiencies reported in some instances may not reflect the true potential of air stripping, but may instead reflect designs intended to achieve only modest removals from low-level contaminant sources. It is also important to recognize that, because different system designs were used for these sites, the results are not directly comparable from site to site.

Table 15-8. Summary of Reported Air-Stripper Removal Efficiencies from 46 Sites [19]

Contaminant	No. of Data Points	Influent Concentration (μg/L)		Reported Removal Efficiency[a] (%)	
		Average	Range	Average	Range
Aniline	1	226	NA[b]	58	NA
Benzene	3	3,730	200–10,000	99.6	99–100
Bromodichloromethane	1	36	NA	81	NA
Bromoform	1	8	NA	44	NA
Chloroform	1	530	1500	48	NA
Chlorobenzene	0	95	NA	ND[c]	ND
Dibromochloromethane	1	34	NA	60	NA
Dichloroethylene	7	409	2–3,000	98.6	96–100
Diisopropyl ether	2	35	20–50	97.0	95–99
Ethyl benzene	1	6,370	100–1,400	99.8	NA
Ethylene dichloride	7	173	5–1,000	99.3	79–100
Methylene chloride	1	15	9–20	100	NA
Methyl ethyl ketone	1	100	NA	99	NA
2-Methylphenol	1	160	NA	70	NA
Methyl tertiary butylether	2	90	50–130	97.0	95–99
Perchloroethylene	17	355	3–4,700	96.5	86–100
Phenol	1	198	NA	74	NA
1,1,2,2-Tetrachloroethane	1	300	NA	95	NA
Trichloroethane	8	81	5–300	95.4	70–100
Trichloroethylene	34	7,660	1–200,000	98.3	76–100
1,2,3-Trichloropropane	1	29,000	NA	99	NA
Toluene	2	6,710	30–23,000	98	96–100
Xylene	4	14,823	17–53,000	98.4	96–100
Volatile organic compounds	3	44,000	57–130,000	98.8	98–99.5
Total Volatile Organics	46	11,120	12–205,000	97.5	58.1–100

[a] Note that the averages and ranges presented in this column represent more data points than are presented in the second column of this table because the removal efficiencies were not available for all air strippers.

[b] NA = Not Applicable. Data available for only one stripper.

[c] ND = No Data. Insufficient data available.

TECHNOLOGY STATUS

Air stripping is a well-developed technology with wide application. During 1988, air stripping of aqueous solutions was a part of the selected remedy at 30 Superfund sites [1]. In 1989, air stripping was a part of the selected remedy at 38 Superfund Sites [2].

The factors determining the cost of an air stripper can be categorized as those affecting design, emission controls, and operation and maintenance (O&M). Design considerations such as the size and number of towers, the materials of construction, and the desired capacity influence the capital costs. Equipment cost components associated with a typical packed-tower air stripper include tower shell, packing support, water distributor, mist eliminator, packing, blower and motor, engineering, and contractor overhead and profit. The addition of an air treatment system roughly doubles the cost of an air-stripping system [3; 6, p. 5-5]. Onsite regeneration or incineration of carbon may increase the cost associated with emission controls. The primary O&M cost components are operating labor, repair and upkeep, and energy requirements of blower motor and pumps [12].

Adams et al. made cost estimates based on flows from 0.1 to 10 mgd assuming a removal efficiency of 99 percent. The process was optimized for packed tower volume and energy consumption. Figure 15-2 presents general cost curves for three flow rates based on their work. Air emissions controls were not included in the costs. Within the range of Henry's law coefficients of 0.01 to 1.0,

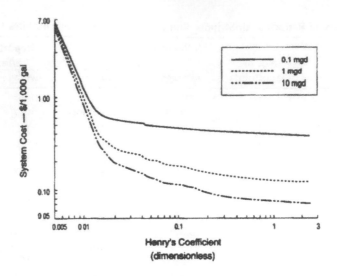

Figure 15-2. Cost estimates for air stripping without air emission controls as a function of the Henry's law coefficient.

the cost ranged from $0.07/1,000 gallons to $0.70/1,000 gallons. As the Henry's law coefficient approached 0.005, the costs rapidly rose to $7.00/1,000 gallons [20, p. 52].

According to Hydro Group, Inc., the cost of air stripping may range from $0.04 to $0.17 per 1,000 gallons [21, p. 7]. The Des Moines Superfund site unit cost for ground-water treatment is estimated to be about $0.45/1,000 gallons based on a 1,250 gpm treatment rate and an average O&M cost of $200,000/year for 10 years at 10 percent interest. The Eau Claire site had a unit cost of roughly $0.14/1,000 gallons, assuming a 5-year operation period and an average treatment rate of 7 million gpd [12, p. C-6].

Recent developments in this technology include high temperature air stripping (HTAS) and rotary air stripping. A full-scale HTAS system was demonstrated at McClellan AFB to treat ground water contaminated with fuel and solvents from spills and storage tank leaks. The combined recycle and makeup was heated to 65°C, and a removal efficiency of greater than 99 percent was achieved [8, p. 9]. The rotary design, marketed under the name HIGEE, was demonstrated at a U.S. Coast Guard air station in East Bay Township, Michigan. At a gas-to-liquid ratio of 30:1 and a rotor speed of 435 rpm, removal efficiencies for all contaminants, except 1,2-DCE, exceeded 99 percent. The removal efficiency for 1,2-DCE was not reported [4, p. 19].

Raising influent liquid temperature increases mass-transfer rates and the Henry's law constants. This results in improved removal efficiencies for VOCs and the capability to remove contaminants that are less volatile. Table 15-9 illustrates the influence that changes in liquid temperature have on contaminant removal efficiencies. Note that steam stripping may be the preferred treatment technology at a feed temperature approaching 100°C, because the higher temperatures associated with steam stripping allow organics to be removed more efficiently than in HTAS systems. However, steam stripping uses more fuel and therefore will have higher operating costs. Additionally, the capital costs for steam stripping may be higher than for HTAS if higher-grade construction materials are needed at the elevated temperatures used in steam stripping [8, p. 3].

Rotary air strippers use centrifugal force rather than gravity to drive aqueous solutions through the specially designed packing. This packing, consisting of thin sheets of metal wound together tightly, was developed for rotary air strippers because of the strain of high centrifugal forces. The use of centrifugal force reportedly results in high removal efficiencies due to formation of a very thin liquid film on wetted surfaces. The rotary motion also causes a high degree of turbulence in the gas phase. The turbulence results in improved liquid distribution over conventional gravity-driven air strippers. The biggest advantage of rotary strippers is the high capacity for a relatively small device. Disadvantages include the potential for mechanical failures and additional energy requirements for the drive motor. Water carryover into the air effluent stream may cause problems with certain emis-

Table 15-9. Influence of Feed Temperature on Removal of Water Soluble Compounds from Ground Water [8, p. 15]

Compound	Percent Removed at Selected Temperature		
	12°C	35°C	73°C
2-Propanol	10	23	70
Acetone	35	80	95
Tetrahydrofuran	50	92	>99

sion control devices used to treat the contaminated air. Cost and performance data on rotary air strippers are very limited [4, p. 16].

EPA CONTACT

Technology-specific questions regarding air stripping of liquids may be directed to Douglas Grosse, (513) 569-7844 or Jeff Adams, RREL-Cincinnati, (513) 569-7516 (see Preface for mailing address).

ACKNOWLEDGMENTS

This chapter was prepared for the U.S. Environmental Protection Agency, Office of Research and Development (ORD), Risk Reduction Engineering Laboratory (RREL), Cincinnati, Ohio, by Science Applications International Corporation (SAIC) under Contract No. 68-C8-0062. Mr. Eugene Harris served as the EPA Technical Project Monitor. Mr. Gary Baker was SAIC's Work Assignment Manager. This chapter was authored by Mr. Jim Rawe of SAIC. The Author is especially grateful to Mr. Ron Turner, Mr. Ken Dostal and Dr. James Heidman of EPA, RREL, who have contributed significantly by serving as technical consultants during the development of this chapter.

The following other Agency and contractor personnel have contributed their time and comments by participating in the expert review meeting and/or peer reviewing the chapter:

Mr. Ben Blaney, EPA-RREL
Dr. John Crittenden, Michigan Technological University
Mr. Clyde Dial, SAIC
Dr. James Gossett, Cornell University
Mr. George Wahl, SAIC
Ms. Tish Zimmerman, EPA-OERR

REFERENCES

1. ROD Annual Report, FY 1988. EPA/540/8-89/006, U.S. Environmental Protection Agency, 1989.
2. ROD Annual Report, FY 1989. EPA/540/8-90/006, U.S. Environmental Protection Agency, 1990.
3. Lenzo, F. and K. Sullivan. Ground Water Treatment Techniques: An Overview of the State-of-the-Art in America. Presented at the First US/USSR Conference on Hydrogeology, Moscow, July 3–5, 1989.
4. Singh, S.P. and R.M. Counce. Removal of Volatile Organic Compounds from Groundwater: A Survey of the Technologies. Prepared for the U.S. Department of Energy, under Contract DE-AC05-84OR21400, 1989.
5. Handbook; Remedial Action at Waste Disposal Sites (Revised). EPA/625/6-85/006 (NTIS PB87-201034), 1985, pp. 10-48 through 10-52.
6. Mobile Treatment Technologies for Superfund Wastes. EPA/540/2-86/003F (NTIS PB87-110656), 1986, pp. 5-3 through 5-6.
7. Technology Screening Guide for Treatment of CERCLA Soils and Sludges. EPA/540/2-88/004 (NTIS PB89-132674), 1988.
8. Blaney, B.L. and M. Branscome. Air Strippers and their Emissions Control at Superfund Sites. EPA/600/D-88/153, U.S. Environmental Protection Agency, Cincinnati, Ohio, 1988.

9. Umphres, M.D. and J.H. Van Wagner. An Evaluation of the Secondary Effects of Air Stripping. EPA/600/S2-89/005, U.S. Environmental Protection Agency, Cincinnati, Ohio, 1990.

10. Ashworth, R.A., G.B. Howe, M.E. Mullins, and T.N. Rogers. Air-Water Partitioning Coefficients of Organics in Dilute Aqueous Solutions. Journal of Hazardous Materials, 18:25–36, 1988.

11. Nirmalakhandan, N.N. and R.E. Speece. QSAR Model for Predicting Henry's Constants. Environmental Science and Technology, 22:1349–1357, 1988.

12. Young, C., et al. Innovative Operational Treatment Technologies for Application to Superfund Site - Nine Case Studies. EPA/540/2-90/006 (PB90-202656), 1990.

13. McIntyre, G.T., et al. Design and Performance of a Groundwater Treatment System for Toxic Organics Removal. Journal WPCF, 58(1):41–46,1986.

14. A Compendium of Technologies Used in the Treatment of Hazardous Wastes. EPA/625/8-87/014 (NTIS PB90-274893), 1987.

15. Air/Superfund National Technical Guidance Study Series: Comparisons of Air Stripper Simulations and Field Performance Data. EPA/450/1-90/002, U.S. Environmental Protection Agency, 1990.

16. Byers, W.D. and C.M. Morton. Removing VOC from Groundwater; Pilot, Scale-up, and Operating Experience. Environmental Progress, 4(2):112–118, 1985.

17. Gross, R.L. and S.G. TerMaath. Packed Tower Aeration Strips Trichloroethylene from Groundwater. Environmental Progress, 4(2):119–124, 1985.

18. Personal communication with vendor.

19. Air Stripping of Contaminated Water Sources - Air Emissions and Controls. EPA/450/3-87/017, U.S. Environmental Protection Agency, 1987.

20. Adams, J.Q. and R.M. Clark. Evaluating the Costs of Packed-Tower Aeration and GAC for Controlling Selected Organics. Journal AWWA, 1:49–57, 1991.

21. Lenzo, F.C. Air Stripping of VOCs from Groundwater: Decontaminating Polluted Water. Presented at the 49th Annual Conference of the Indiana Water Pollution Control Association, August 19–21, 1985.

Chapter 16

Granular Activated Carbon Treatment[1]

Margaret Groeber, Science Applications International Corporation (SAIC), Cincinnati, OH

ABSTRACT

Granular activated carbon (GAC) treatment is a physicochemical process that removes a wide variety of contaminants by adsorbing them from liquid and gas streams [1, p. 6-3].[2] This treatment is most commonly used to separate organic contaminants from water or air; however, it can be used to remove a limited number of inorganic contaminants [2, p. 5-17]. In most cases, the contaminants are collected in concentrated form on the GAC, and further treatment is required.

The contaminant (adsorbate) adsorbs to the surfaces of the microporous carbon granules until the GAC becomes exhausted. The GAC may then be either reactivated, regenerated, or discarded. The reactivation process destroys most contaminants. In some cases, spent GAC can be regenerated, typically using steam to desorb and collect concentrated contaminants for further treatment. If GAC is to be discarded, it may have to be handled as a hazardous waste.

Site-specific treatability studies are generally necessary to document the applicability and potential performance of a GAC system. This chapter provides information on the technology applicability, technology limitations, a technology description, the types of residuals produced, site requirements, latest performance data, status of the technology, and sources for further information.

TECHNOLOGY APPLICABILITY

Adsorption by activated carbon has a long history of use as a treatment for municipal, industrial, and hazardous waste streams. The concepts, theory, and engineering aspects of the technology are well developed [3]. It is a proven technology with documented performance data. GAC is a relatively nonspecific adsorbent and is effective for removing many organic and some inorganic contaminants from liquid and gaseous streams [4].

The effectiveness of GAC as an adsorbent for general contaminant groups is shown in Table 16-1. Examples of constituents within contaminant groups are provided in Technology Screening Guide for Treatment of CERCLA Soils and Sludges [5]. This table is based on current available information or professional judgment when no information was available. The proven effectiveness of the technology for a particular site or waste does not ensure that it will be effective at all sites or that the treatment efficiency achieved will be acceptable at other sites. For the ratings used for this table, demonstrated effectiveness means that, at some scale, treatability was tested to show that, for that particular contaminant and matrix, the technology was effective. The ratings of potential effectiveness and no expected effectiveness are based upon expert judgment. Where potential effectiveness is indicated, the technology is believed capable of successfully treating the contaminant group in a particular matrix. When the technology is not applicable or will probably not work for a particular combination of contaminant group and matrix, a no-expected-effectiveness rating is given.

The effectiveness of GAC is related to the chemical composition and molecular structure of the contaminant. Organic wastes that can be treated by GAC include compounds with high molecular weights and boiling points and low solubility and polarity [6]. Organic compounds treatable by GAC are listed in Table 16-2. GAC has also been used to remove low concentrations of certain types of inorganics and metals; however, it is not widely used for this application [1, p. 6-13].

---

[1] EPA/540/2-91/024.
[2] [Reference number, page number.]

Table 16-1. Effectiveness of Granular Activated Carbon on General Contaminant Groups

Contaminant Groups	Liquid/Gas
Organic	
Halogenated volatiles	■
Halogenated semivolatiles	■
Nonhalogenated volatiles[a]	■
Nonhalogenated semivolatiles	■
PCBs	■
Pesticides	■
Dioxins/furans	■
Organic cyanides[a]	▼
Organic corrosives[a]	■
Inorganic	
Volatile metals[a]	■
Nonvolatile metals[a]	■
Asbestos	●
Radioactive materials[a]	■
Inorganic corrosives	▼
Inorganic cyanides[b]	■
Reactive	
Oxidizers[b]	■
Reducers	●

■   Demonstrated Effectiveness: Successful treatability test at some scale completed.
▼   Potential Effectiveness: Expert opinion that technology will work.
●   No Expected Effectiveness: Expert opinion that technology will not work.
[a]   Technology is effective for some contaminants in the group; it may not be effective for others.
[b]   Applications to these contaminants involve both adsorption and chemical reaction.

Table 16-2. Organic Compounds Amenable to Adsorption by GAC [1]

Class	Example
Aromatic solvents	Benzene, toluene, xylene
Polynuclear aromatics	Naphthalene, biphenyl
Chlorinated aromatics	Chlorobenzene, PCBs, endrin, toxaphene, DDT
Phenolics	Phenol, cresol, resorcinol, nitrophenols, chlorophenols alkyl phenols
Aromatic amines and high molecular weight aliphatic amines	Aniline, toluene, diamine
Surfactants	Alkyl benzene sulfonates
Soluble organic dyes	Methylene blue, textile dyes
Fuels	Gasoline, kerosene, oil
Chlorinated solvents	Carbon tetrachloride, perchloroethylene
Aliphatic and aromatic acids	Tar acids, benzoic acids
Pesticides/herbicides	2,4-D, atrazine, simazine, aldicarb, alachlor, carbofuran

Almost all organic compounds can be adsorbed onto GAC to some degree [2, p. 5-17]. The process is frequently used when the chemical composition of the stream is not fully analyzed [1, p. 6-3]. Because of its wide-scale use, GAC has probably been inappropriately selected when an alternative technology may have been more effective [7]. GAC can be used in conjunction with other treatment technologies. For example, GAC can be used to remove contaminants from the off-gas from air stripper and soil vapor extraction operations [7; 8, p. 73; 9].

LIMITATIONS

Compounds that have low molecular weight and high polarity are not recommended for GAC treatment. Streams with high suspended solids (50 mg/L) and oil and grease (10 mg/L) may cause fouling of the carbon and require frequent backwashing. In such cases, pretreatment prior to GAC is generally required. High levels of organic matter (e.g., 1,000 mg/L) may result in rapid exhaustion of the carbon. Even lower levels of background organic matter (e.g., 10–100 mg/L) such as fulvic and humic acids may cause interferences in the adsorption of specifically targeted organic contaminants which are present in lower concentrations. In such cases, GAC may be most effectively employed as a polishing step in conjunction with other treatments.

The amount of carbon required, regeneration/reactivation frequency, and the potential need to handle the discarded GAC as a hazardous waste are among the important economic considerations. Compounds not well adsorbed often require large quantities of GAC, and this will increase the costs. In some cases the spent GAC may be a hazardous waste, which can significantly add to the cost of treatment.

TECHNOLOGY DESCRIPTION

Carbon is an excellent adsorbent because of its large surface area, which can range from 500–2000 m^2/g, and because its diverse surfaces are highly attractive to many different types of contaminants [3]. To maximize the amount of surface available for adsorption, an activation process which increases the surface-to-volume ratio of the carbon is used to produce an extensive network of internal pores. In this process, carbonaceous materials are converted to mixtures of gas, tars, and ash. The tar is then burned off and the gases are allowed to escape to produce a series of internal micropores [1, p. 6-6]. Additional processing of the GAC may be used to render it more suitable for certain applications (e.g., impregnation for mercury or sulfur removal).

The process of adsorption takes place in three steps [3]. First the contaminant migrates to the external surface of the GAC granules. It then diffuses into the GAC pore structure. Finally, a physical or chemical bond forms between the contaminant and the internal carbon surface.

The two most common reactor configurations for GAC adsorption systems are the fixed bed and the pulsed or moving bed [3]. The fixed-bed configuration is the most widely used for adsorption from liquids, particularly for low to moderate concentrations of contaminants. GAC treatment of contaminated gas streams is done almost exclusively in fixed-bed reactors. The following technical discussion applies to both gas and liquid streams.

Figure 16-1 is a schematic diagram of a typical single-stage, fixed-bed GAC system for use on a liquid stream. The contaminant stream enters the top of the column (1). As the waste stream flows through the column, the contaminants are adsorbed. The treated stream (effluent) exits out the bottom (2). Spent carbon is reactivated, regenerated, or replaced once the effluent no longer meets the treatment objective (3). Although Figure 16-1 depicts a downward flow, the flow direction can be upward, depending on design considerations.

Suspended solids in a liquid stream or particulate matter in a gaseous stream accumulate in the column, causing an increase in pressure drop. When the pressure drop becomes too high, the accumulated solids must be removed; for example, by backwashing. The solids removal process necessitates adsorber downtime, and may result in carbon loss and disruption of the mass transfer zone. Pretreatment for removal of solids from streams to be treated by GAC is, therefore, an important design consideration.

As a GAC system continues to operate, the mass-transfer zone moves down the column. Figure 16-2 shows the adsorption pattern and the corresponding effluent breakthrough curve [3]. The breakthrough curve is a plot of the ratio of effluent concentration (C_e) to influent concentration (C_o) as a function of water volume or air volume treated per unit time. When a predetermined concentration appears in the effluent (C_B), breakthrough has occurred. At this point, the effluent quality no longer meets treatment objectives. When the carbon becomes so saturated with the contaminants that they can no longer be adsorbed, the carbon is said to be spent ($C_e=C_o$). Alternative design arrangements may allow individual adsorbers in multi-adsorber systems to be operated beyond the breakpoint as far as complete exhaustion. This condition of operation is defined as the operating limit ($C_e=C_L$) of the adsorber.

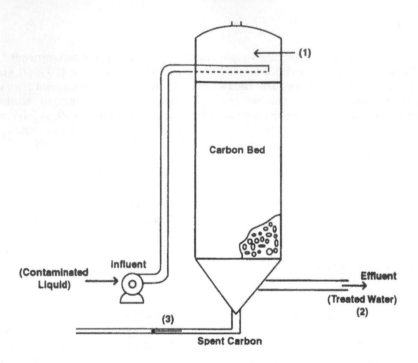

Figure 16-1. Schematic diagram of fixed-bed GAC system.

The major design variables for liquid phase applications of GAC are empty bed contact time (EBCT), GAC usage rate, and system configuration. Particle size and hydraulic loading are often chosen to minimize pressure drop and reduce or eliminate backwashing. System configuration and EBCT have an impact on GAC usage rate. When the bed life is longer than 6 months and the treatment objective is stringent ($C_e/C_o < 0.05$), a single adsorber or a combination of single beds operating in parallel is preferred. For a single adsorber, the EBCT is normally chosen to be large enough to minimize GAC usage rate. When less stringent objectives are required ($C_e/C_o > 0.3$), blending of effluents from partially saturated adsorbers can be used to reduce GAC usage rate. When stringent treatment objectives are required ($C_e/C_o < 0.05$) and GAC bed life is short (less than 6 months) multiple beds in series may be used to decrease GAC usage rate.

For gas-phase applications, the mass transfer zone is usually very short if the relative humidity is low enough to prevent water from filling the GAC pores. The adsorption zone (Figure 16-2) for gas-phase applications is small relative to bed depth, and the GAC is nearly saturated at the breakpoint. Accordingly, EBCT and system configuration have little impact on GAC usage rate, and a single bed or single beds operated in parallel are commonly used.

GAC can be reactivated either onsite or offsite. The choice is usually dictated by costs which are dependent on the site and on the proximity of offsite facilities that reactivate carbon. Generally, onsite reactivation is not economical unless more than 2,000 pounds per day of GAC are required to be reactivated. Even so, an offsite reactivation service may be more cost-effective [10].

The basic evaluation technique for initial assessment of the feasibility of GAC treatment is the adsorption isotherm test. This test determines if a compound is amenable to GAC adsorption and can be used to estimate minimum GAC usage rates. More detailed testing, such as small-scale column tests and pilot tests, should be conducted if the isotherms indicate GAC can produce an effluent of acceptable quality at a reasonable carbon usage rate [10].

PROCESS RESIDUALS

The main process residual produced from a GAC system is the spent carbon containing the hazardous contaminants. When the carbon is regenerated, the desorbed contaminants must be treated

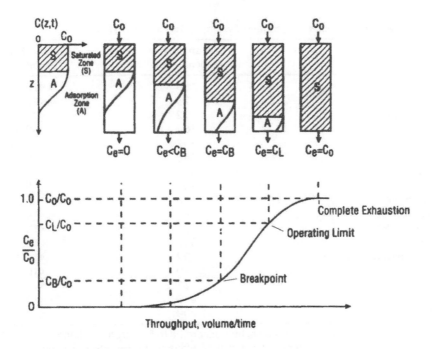

Figure 16-2. Breakthrough characteristics of fixed-bed GAC adsorber [3].

or reclaimed. Reactivation of carbon is typically accomplished by thermal processes. Elevated temperatures are employed in the furnace and afterburners to destroy the accumulated contaminants. If the carbon cannot be economically reactivated, the carbon must be discarded and may have to be treated and disposed of as a hazardous waste. In some cases, the influent to GAC treatment must be pretreated to prevent excessive head loss. Residues from pretreatment (e.g., filtered suspended solids) must be treated or disposed of. Solids collected from backwashing may need to be treated and disposed of as a hazardous waste.

SITE REQUIREMENTS

GAC equipment generally has small space requirements and sometimes can be incorporated in mobile units. The rapidity of startup and shutdown also makes GAC amenable to mobile treatment. Carbon beds or columns can be skid-mounted and transported by truck or rail [2, p. 5-19].

As previously stated, spent carbon from the treatment of streams containing hazardous substances is generally considered hazardous, and its transportation and handling requires that a site safety plan be developed to provide for personnel protection and special handling measures. Storage may have to be provided to hold the GAC-treated liquid until its acceptability for release has been determined. If additional treatment is required, adequate space must be provided for these systems.

PERFORMANCE DATA

Performance data on full-scale GAC systems have been reported by several sources including equipment vendors. Data on GAC systems at several Superfund sites and other cleanup sites are discussed in this section. The data presented for specific contaminant removal effectiveness were obtained from publications developed by the respective GAC system vendors. The quality of this information has not been determined; however, it does give an indication of the efficiency of GAC.

A GAC system was employed for leachate treatment at the Love Canal Superfund site in Niagara Falls, New York. The results of this operation are listed in Tables 16-3 and 16-4 [11].

Table 16-5 summarizes a number of experiences by Calgon Corporation in treating contaminated ground water at many other non-Superfund sites. Table 16-5 identifies the sources of contamination

Table 16-3. Love Canal Leachate Treatment System[a] (March 1979) [11]

Priority Pollutant Compounds Identified	Carbon System Influent µg/L	Carbon System Effluent µg/L
Hexachlorobutadiene	109	<20
1,2,4-Trichlorobenzene	23	<20
Hexachlorobenzene	32	<20
α-BHC	184	<0.01
γ-BHC	392	0.12
β-BHC	548	<0.01
Heptachlor	573	<0.01
Phenol	4,700[b]	<5[b]
2,4-Dichlorophenol	10	<5
Methylene chloride	180	<10
1,1-Dichloroethylene	28	<10
Chloroform	540	<10
Carbon tetrachloride	92	<10
Trichloroethylene	240	<10
Dibromochloromethane	21	<10
1,1,2,2-Tetrachloroethylene	270	<10
Chlorobenzene	1,200	<10

[a] Samples were analyzed by Recra Research, Inc., according to EPA protocol dated April 1977 (sampling and analysis procedures of screening for industrial effluents for priority pollutants).

[b] The data represent phenol analysis conducted by Calgon in June 1979, as earlier results were suspect.

along with operating parameters and results [12]. While these sites were not regulated under CERCLA, the type and concentration of contaminants are typical of those encountered at a Superfund site.

The Verona Well Field Superfund site in Battle Creek, Michigan used GAC as a pretreatment for the air stripper. This arrangement reduced the influent concentrations which allowed the air stripper to comply with the National Pollution Discharge Elimination System (NPDES) permit. The system had two parallel trains: a single unit and two units in series. Approximately one-third of the total flow was directed to the first train, while the remaining flow went to the other train. Performance data for removal of total volatile organic compounds (TVOC) on selected operating days are given in Table 16-6 [13].

A remediation action at the U.S. Coast Guard Air Station in Traverse City, Michigan, resulted in GAC being used to treat contaminated ground water. The ground water was pumped from the extraction well system to the GAC system. The treated water was then discharged to the municipal sewer system. Concentrations of toluene in the monitoring wells were reduced from 10,329 parts per billion (ppb) to less than 10 ppb in approximately 100 days [14].

TECHNOLOGY STATUS

GAC is a well-proven technology. It has been used in the treatment of contaminated ground water at a number of Superfund sites. Carbon adsorption has also been used as a polishing step following other treatment units at many sites. In 1988, the number of sites where activated carbon was listed in the Record of Decision was 28; in 1989, that number was 38.

Costs associated with GAC are dependent on waste stream flow rates, type of contaminant, concentrations, and site and timing requirements. Costs are lower with lower concentration levels of a contaminant of a given type. Costs are also lower at higher flow rates. At liquid flow rates of 100-million gallons per day (mgd), costs range from $0.10 to 1.50/1,000 gallons treated. At flow rates of 0.1 mgd, costs increase to $1.20 to 6.30/1,000 gallons treated [12].

EPA CONTACT

Technology-specific questions regarding GAC treatment may be directed to Douglas Grosse, (513) 569-7844, or Ronald Turner, RREL-Cincinnati, (513) 569-7775 (see Preface for mailing address).

Table 16-4. Love Canal Leachate Treatment System[a] (June 1979) [11]

Priority Pollutant Compounds Identified	Raw Leachate µg/L	Carbon System Effluent µg/L
2,4,6-Trichlorophenol	85	<10
2,4-Dichlorophenol	5,100	ND
Phenol	2,400	<10
1,2,3-Trichlorobenzene	870	ND
Hexachlorobenzene	110	ND
2-Chloronaphthalene	510	ND
1,2-Dichlorobenzene	1,300	ND
1,3 and 1,4-Dichlorobenzene	960	ND
Hexachlorobutadiene	1,500	ND
Anthracene and phenanthrene	29	ND
Benzene	28,000	<10
Carbon tetrachloride	61,000	<10
Chlorobenzene	50,000	12
1,2-Dichloroethane	52	ND
1,1,1-Trichloroethane	23	ND
1,1-Dichloroethane	66	ND
1,1,2-Trichloroethane	780	<10
1,1,2,2-Tetrachloroethane	80,000	<10
Chloroform	44,000	<10
1,1-Dichloroethylene	16	ND
1,2-Trans-dichloroethylene	3,200	<10
1,2-Dichloropropane	130	ND
Ethyl benzene	590	<10
Methylene chloride	140	46
Methyl chloride	370	ND
Chlorodibromomethane	29	ND
Tetrachloroethylene	44,000	12
Toluene	25,000	<10
Trichloroethylene	5,000	ND

[a] Samples were analyzed by Carborundum Corporation according to EPA protocol dated April 1977 (sampling and analysis procedures for screening of industrial effluents for priority pollutants).

ND = nondetectable.

ACKNOWLEDGMENTS

This chapter was prepared for the U.S. Environmental Protection Agency, Office of Research and Development (ORD), Risk Reduction Engineering Laboratory (RREL), Cincinnati, Ohio, by Science Applications International Corporation (SAIC) under Contract No. 68-C8-0062. Mr. Eugene Harris served as the EPA Technical Project Monitor. Mr. Gary Baker was SAIC's Work Assignment Manager. This bulletin was authored by Ms. Margaret M. Groeber of SAIC. The author is especially grateful to Mr. Ken Dostal and Dr. James Heidman of EPA, RREL, who have contributed significantly by serving as a technical consultant during the development of this chapter.

The following other Agency and contractor personnel have contributed their time and comments by participating in the expert review meetings and/or peer reviewing the chapter:

Dr. John C. Crittenden, Michigan Technological University
Mr. Clyde Dial, SAIC
Mr. James Rawe, SAIC
Dr. Walter J. Weber, Jr., University of Michigan
Ms. Tish Zimmerman, EPA-OERR

Table 16-5. Performance Data at Selected Sites [12]

Source of Contaminants	Typical Influent Conc. (mg/L)	Typical Effluent Conc. (µg/L)	Carbon Usage Rate (lb/1000 gal)	Total Contact Time (min.)
Truck spill				
Methylene chloride	21	<1.0	3.9	534
1,1,1-Trichloroethane	25	<1.0	3.9	534
Rail car spills				
Phenol	63	<1.0	5.8	201
Orthochlorophenol	100	<1.0	5.8	201
Vinylidine chloride	2–4	<10.0	2.1	60
Ethyl acrylate	200	<1.0	13.3	52
Chloroform	0.020	<1.0	7.7	160
Chemical spills				
Chloroform	3.4	<1.0	11.6	262
Carbon tetrachloride	130–135	<1.0	11.6	262
Trichloroethylene	2–3	<1.0	11.6	262
Tetrachloroethylene	70	<1.0	11.6	262
Dichloroethyl ether	1.1	<1.0	0.45	16
Dichloroisopropyl ether	0.8	<1.0	0.45	16
Benzene	0.4	<1.0	1.9	112
DBCP	2.5	<1.0	0.7–3.0	21
1,1,1-Trichloroethane	0.42	<10	1.5	53
Trichlorotrifluoroethane	5.977	<10	1.5	53
cis-1,2-Dichloroethylene	.005	<1.0	0.25	121
Onsite storage tanks				
cis-1,2-Dichloroethylene	0.5	<1.0	0.8	64
Tetrachloroethylene	7.0	<1.0	0.8	64
Methylene chloride	1.5	<100	4.0	526
Chloroform	0.30–0.50	<100	1.19	26
Trichloroethylene	3–8	<1.0	1.54	36
Isopropyl alcohol	0.2	<10.0	1.54	36
Acetone	0.1	<10.0	1.54	36

1,1,1-Trichloroethane	12	5.0	1.0	52
1,2-Dichloroethylene	0.5	<1.0	1.0	52
Xylene	8.0	<1.0	1.0	52
Landfill site				
TOC	20	<5000	1.15	41
Chloroform	1.4	<1.0	1.15	41
Carbon tetrachloride	1.0	<1.0	1.15	41
Gasoline spills, tank leakage				
Benzene	9–11	<100 Total	<1.01	214
Toluene	5–7		<1.01	214
Xylene	6–10		<1.01	214
Methyl t-butyl ether	0.030–0.035	<5.0	0.62	12
Di-isopropyl ether	0.020–0.040	<1.0	0.10–0.62	12
Trichloroethylene	0.050–0.060	<1.0	0.62	12
Chemical by-products				
Di-isopropyl methyl phosphonate	1.25	<50	0.7	30
Dichloropentadiene	0.45	<10	0.7	30
Manufacturing residues				
DDT	0.004	<0.5	1.1	31
TOC	9.0		1.1	31
1,3-Dichloropropene	0.01	<1.0	1.1	31
Chemical landfill				
1,1,1-Trichloroethane	0.060–0.080	<1.0	<0.45	30
1,1-Dichloroethylene	0.005–0.015	0.005	<0.45	30

Table 16-6. TVOC Removal with GAC at Verona Well Superfund Site [13]

| Operating Day | Influent Feed Concentration (ppb) | Effluent | |
		Train (1) Concentration (ppb)	Train (2) Concentration (ppb)
1	18,812	NA	25
9	12,850	11	7
16	9,290	41	17
27	6,361	260	426
35	7,850	484	575
42	7,643	412	551
49	7,577	405	524
57	5,591	452	558
69	10,065	377	475
92	6,000	444	509
106	3,689	13	702
238	4,671	246	263

NA - not available

REFERENCES

1. Voice, T.C. Activated-Carbon Adsorption. In: Standard Handbook of Hazardous Waste Treatment and Disposal, H.M. Freeman, ed. McGraw-Hill, New York, NY, 1989.

2. Mobile Treatment Technologies for Superfund Wastes. EPA/540/2-86/003F (NTIS PB87-110656), 1986.

3. Weber, W.J., Jr., Evolution of a Technology. Journal of the Environmental Engineering Division, American Society of Civil Engineers, 110(5):899–917, 1984.

4. Sontheimer, H., et al. Activated Carbon for Water Treatment. DVGW-Forschungsstelle, Karlsruhe, Germany. Distributed in the U.S. by AWWA Research Foundation, Denver, CO. 1988.

5. Technology Screening Guide for Treatment of CERCLA Soils and Sludges. EPA/540/2-88/1004 (NTIS PB89-132674), 1988.

6. A Compendium of Technologies Used in the Treatment of Hazardous Wastes. EPA/625/8-87/014 (NTIS PB90-274093), 1987.

7. Lenzo, F. and K. Sullivan. Ground Water Treatment Techniques—An Overview of the State-of-the-Art in America. Paper presented at First US/USSR Conference on Hydrology. Moscow, U.S.S.R. July 3–5, 1989.

8. Crittenden, J.C., et al. Using GAC to Remove VOC's from Air Stripper Off-Gas. Journal AWWA, 80(5):73–84, May 1988.

9. Stenzel, M.H. and U.S. Gupta. Treatment of Contaminated Groundwaters with Granular Activated Carbon and Air Stripping. Journal of the Air Pollution Control Association, 35(12):1304–1309, 1985.

10. Stenzel, M.H. and J.G. Rabosky. Granular Activated Carbon—Attacks Groundwater Contaminants. Marketing Brochure for Calgon Carbon Corporation, Pittsburgh, Pennsylvania.

11. McDougal, W.J., et al., Containment and Treatment of the Love Canal Landfill Leachate, Journal WPCF, 52(12):2914–2923, 1980.

12. O'Brien, R.P. There is an Answer to Groundwater Contamination. Water/Engineering & Management, May 1983.

13. CH2M Hill. Thomas Solvent-Raymond Road Groundwater Extraction Well Treatment System Monitoring Report. June 1988.

14. Sammons, J.H. and J.M. Armstrong. Use of Low Flow Interdiction Wells to Control Hydrocarbon Plumes in Groundwater. In: Proceedings of the Natural Conference on Hazardous Wastes and Hazardous Materials Control Research Institute. Silver Spring, Maryland, 1986.

15. Adams, J.Q. and R.M. Clark. Evaluating the Costs of Packed Tower Aeration and GAC for Controlling Selected Organics. Journal AWWA, 83(1):49–57, January 1991.

Chapter 17

Soil Washing Treatment[1]

U.S. Environmental Protection Agency, Risk Reduction Engineering Laboratory, Cincinnati, OH

ABSTRACT

Soil washing is a water-based process for mechanically scrubbing soils ex situ to remove undesirable contaminants. The process removes contaminants from soils in one of two ways: by dissolving or suspending them in the wash solution (which is later treated by conventional wastewater treatment methods) or by concentrating them into a smaller volume of soil through simple particle size separation techniques (similar to those used in sand and gravel operations). Soil washing systems incorporating both removal techniques offer the greatest promise for application to soils contaminated with a wide variety of heavy metal and organic contaminants.

The concept of reducing soil contamination through the use of particle size separation is based on the finding that most organic and inorganic contaminants tend to bind, either chemically or physically, to clay and silt soil particles. The silt and clay, in turn, are attached to sand and gravel particles by physical processes, primarily compaction and adhesion. Washing processes that separate the fine (small) clay and silt particles from the coarser sand and gravel soil particles effectively separate and concentrate the contaminants into a smaller volume of soil that can be further treated or disposed. The clean, larger fraction can be returned to the site for continued use. This set of assumptions forms the basis for the volume-reduction concept upon which most soil washing technology applications are being developed.

At the present time, soil washing is used extensively in Europe and has had limited use in the United States. During 1986–1989, the technology was one of the selected source control remedies at eight Superfund sites.

The final determination of the lowest cost alternative will be more site-specific than process equipment dominated. Vendors should be contacted to determine the availability of a unit for a particular site. This chapter provides information on the technology applicability, the types of residuals resulting from the use of the technology, the latest performance data, site requirements, the status of the technology, and where to go for further information.

TECHNOLOGY APPLICABILITY

Soil washing can be used either as a stand-alone technology or in combination with other treatment technologies. In some cases, the process can deliver the performance needed to reduce contaminant concentrations to acceptable levels and, thus, serve as a stand-alone technology. In other cases, soil washing is most successful when combined with other technologies. It can be cost-effective as a preprocessing step in reducing the quantity of material to be processed by another technology such as incineration; it also can be used effectively to transform the soil feedstock into a more homogeneous condition to augment operations in the subsequent treatment system. In general, soil washing is effective on coarse sand and gravel contaminated with a wide range of organic, inorganic, and reactive contaminants. Soils containing a large amount of clay and silt typically do not respond well to soil washing, especially if it is applied as a stand-alone technology.

---

[1] EPA/540/2-90/017. Available publications from EPA's Superfund Innovative Technology Evaluation (SITE) program soil washing demonstrations are listed at the end of the Chapter.

A wide variety of chemical contaminants can be removed from soils through soil washing applications. Removal efficiencies depend on the type of contaminant as well as the type of soil. Volatile organic contaminants often are easily removed from soil by washing; experience shows that volatiles can be removed with 90–99 percent efficiency or more. Semivolatile organics may be removed to a lesser extent (40–90 percent) by selection of the proper surfactant. Metals and pesticides, which are more insoluble in water, often require acids or chelating agents for successful soil washing. The process can be applicable for the treatment of soils contaminated with specific listed Resources Conservation and Recovery Act (RCRA) wastes and other hazardous wastes including wood-preserving chemicals (pentachlorophenol, creosote), organic solvents, electroplating residues (cyanides, heavy metals), paint sludges (heavy metals), organic chemicals production residues, pesticides and pesticides production residues, and petroleum/oil residues [1, p. 659; 2, p. 15; 4; 7 through 13].[2]

The effectiveness of soil washing for general contaminant groups and soil types is shown in Table 17-1 [1, p. 659; 3, p. 13; 15, p. 1]. Examples of constituents within contaminant groups are provided in Reference 3, Technology Screening Guide for Treatment of CERCLA Soils and Sludges. This table is based on currently available information or professional judgment where definitive information is currently inadequate or unavailable. The proven effectiveness of the technology for a particular site or waste does not ensure that it will be effective at all sites or that the treatment efficiency achieved will be acceptable at other sites. For the ratings used in this table, good to excellent applicability means the probability is high that soil washing will be effective for that particular contaminant and matrix. Moderate to marginal applicability indicates situations where care needs to be exercised in choosing the soil washing technology. When Not Applicable is shown, the technology will probably not work for that particular combination of contaminant group and matrix. Other sources of general observations and average removal efficiencies for different treatability groups are the Superfund LDR Guide #6A, Obtaining a Soil and Debris Treatability Variance for Remedial Actions (OSWER Directive 9347.3-06FS), [16] and Superfund LDR Guide #6B, Obtaining a Soil and Debris Treatability Variance for Removal Actions (OSWER Directive 9347.3-07FS) [17].

Information on cleanup objectives as well as the physical and chemical characteristics of the site soil and its contaminants is necessary to determine the potential performance of this technology and the requirements for waste preparation and pretreatment. Treatability tests are also required at the laboratory screening, bench-scale and/or pilot-scale level(s) to determine the feasibility of the specific soil washing process being considered and to understand waste preparation and pretreatment steps needed at a particular site. If bench-test results are promising, pilot-scale demonstrations should normally be conducted before final commitment to full-scale implementation. Treatability study procedures are explained in the EPA's forthcoming document entitled Superfund Treatability Study Protocol: Bench-Scale Level of Soils Washing for Contaminated Soils [14].

Table 17-2 contains physical and chemical soil characterization parameters that must be established before a treatability test is conducted on a specific soil washing process. The parameters are defined as either "key" or "other" and should be evaluated on a site-specific basis. Key parameters represent soil characteristics that have a direct impact on the soil washing process. Other parameters should also be determined, but they can be adjusted prior to the soil washing step based on specific process requirements. The table contains comments relating to the purpose of the specific parameter to be characterized and its impact on the process [6, p. 90; 14, p. 35].

Particle size distribution is the key physical parameter for determining the feasibility of using a soil washing process. Although particle size distribution should not become the sole reason for choosing or eliminating soil washing as a candidate technology for remediation, it can provide an initial means of screening for the potential use of soil washing. Figure 17-1 presents a simplistic particle size distribution range of curves that illustrate a general screening definition for soil washing technology.

In its simplest application, soil washing is a particle size separation process that can be used to segregate the fine fractions from the coarse fractions. In Regime I of Figure 17-1, where coarse soils are found, the matrix is very amenable to soil washing using simple particle size separation.

Most contaminated soils will have a distribution that falls within Regime II of Figure 17-1. The types of contaminants found in the matrix will govern the composition of the washing fluid and the overall efficiency of the soil washing process.

---

[2] [Reference number, page number.]

Table 17-1. Applicability of Soil Washing on General Contaminant Groups for Various Soils

Contaminant Groups	Matrix	
	Sandy/Gravelly Soils	Silty/Clay Soils
Organic		
Halogenated volatiles	■	▼
Halogenated semivolatiles	▼	▼
Nonhalogenated volatiles	▼	▼
Nonhalogenated semivolatiles	▼	▼
PCBs	▼	▼
Pesticides (halogenated)	▼	▼
Dioxins/furans	▼	▼
Organic cyanides	▼	▼
Organic corrosives	▼	▼
Inorganic		
Volatile metals	■	▼
Nonvolatile metals	■	▼
Asbestos	●	●
Radioactive materials	▼	▼
Inorganic corrosives	▼	▼
Inorganic cyanides	▼	▼
Reactive		
Oxidizers	▼	▼
Reducers	▼	▼

■ Good to excellent applicability: High probability that technology will be successful.

▼ Moderate to marginal applicability: exercise care in choosing technology.

● Not applicable: expert opinion that technology will not work.

In Regime III of Figure 17-1, soils consisting largely of finer sand, silt, and clay fractions, and those with high humic content, tend to contain strongly adsorbed organics that generally do not respond favorably to systems that work by only dissolving or suspending contaminants in the wash solution. However, they may respond to soil washing systems that also incorporate a particle size separation step whereby contaminants can be concentrated into a smaller volume.

LIMITATIONS

Contaminants in soils containing a high percentage of silt- and clay-sized particles typically are strongly adsorbed and difficult to remove. In such cases, soil washing generally should not be considered as a stand-alone technology.

Hydrophobic contaminants generally require surfactants or organic solvents for their removal from soil. Complex mixtures of contaminants in the soil (such as a mixture of metals, nonvolatile organics, and semivolatile organics) and frequent changes in the contaminant composition in the soil matrix make it difficult to formulate a single suitable washing fluid that will consistently and reliably remove all of the different types of contaminants from the particles. Sequential washing steps may be needed. Frequent changes in the wash formulation and/or the soil/wash fluid ratio may be required [3, p. 76; 14, p. 7].

While wastewater additives such as surfactants and chelants may enhance some contaminant removal efficiencies in the soil washing portion of the process, they also tend to interfere with the downstream wastewater treatment segments of the process. The presence of the additives in the washed soil and in the wastewater treatment sludge may cause some difficulty in their disposal [14, p. 7; 15, p. 1]. Costs associated with handling the additives and managing them as part of the residuals/wastewater streams must be carefully weighed against the incremental improvements in soil washing performance that they may provide.

Table 17-2. Waste Soil Characterization Parameters

Parameter	Purpose and Comment
Key Physical	
Particle size distribution:	
>2 mm	Oversize pretreatment requirements
0.25–2 mm	Effective soil washing
0.063–0.25 mm	Limited soil washing
<0.063 mm	Clay and silt fraction—difficult soil washing
Other Physical	
Type, physical form, handling properties	Affects pretreatment and transfer requirement
Moisture content	Affects pretreatment and transfer requirements
Key Chemical	
Organics	
Concentration	
Volatility	
Partition coefficient	Determine contaminants and assess separation and washing efficiency, hydrophobic interaction, washing fluid compatibility, changes in washing fluid with changes in contaminants. May require preblending for consistent feed. Use the jar test protocol to determine contaminant partitioning.
Metals	Concentration and species of constituents (specific jar test) will determine washing fluid compatibility, mobility of metals, post-treatment.
Humic acid	Organic content will affect adsorption characteristics of contaminants on soil. Important in marine/wetland sites.
Other Chemical	
pH, buffering capacity	May affect pretreatment requirements, compatibility with equipment materials of construction, wash fluid compatibility.

TECHNOLOGY DESCRIPTION

Figure 17-2 is a general schematic of the soil washing process [1, p. 657; 3, p. 72; 15, p. 1].

Soil preparation (1) includes the excavation and/or moving of contaminated soil to the process where it is normally screened to remove debris and large objects. Depending upon the technology and whether the process is semibatch or continuous, the soil may be made pumpable by the addition of water.

A number of unit processes occur in the soil washing process (2). Soil is mixed with washwater and possibly extraction agent(s) to remove contaminants from soil and transfer them to the extraction fluid. The soil and washwater are then separated, and the soil is rinsed with clean water. Clean soil is then removed from the process as product. Suspended soil particles are recovered directly from the spent washwater, as sludge, by gravity means, or they may be removed by flocculation with a selected polymer or chemical, and then separated by gravity. These solids will most likely be a smaller quantity but carry higher levels of contamination than the original soil and, therefore, should be targeted for either further treatment or secure disposal. Residual solids from recycle water cleanup may require post-treatment to ensure safe disposal or release. Water used in the soil washing process is treated by conventional wastewater treatment processes to enable it to be recycled for further use.

Wastewater treatment (3) processes the blowdown or discharge water to meet regulatory requirements for heavy metal content, organics, total suspended solids, and other parameters. Whenever possible, treated water should be recycled to the soil washing process. Residual solids, such as spent

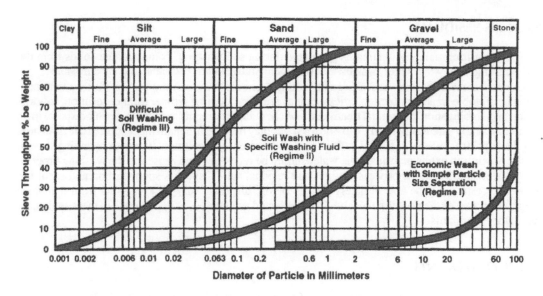

Figure 17-1. Soil washing applicable particle size range.

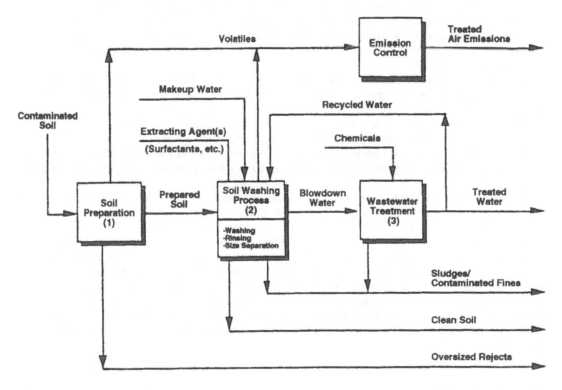

Figure 17-2. Aqueous soil washing process.

ion exchange resin and carbon, and sludges from biological treatment may require post-treatment to ensure safe disposal or release.

Vapor treatment may be needed to control air emissions from excavation, feed preparation, and extraction; these emissions are collected and treated, normally by carbon adsorption or incineration, before being released to the atmosphere.

PROCESS RESIDUALS

There are four main waste streams generated during soil washing: contaminated solids from the soil washing unit, wastewater, wastewater treatment sludges and residuals, and air emissions.

Contaminated clay fines and sludges resulting from the process may require further treatment using acceptable treatment technologies (such as incineration, low temperature desorption, solidification and stabilization, biological treatment, and chemical treatment) in order to permit disposal in an environmentally safe manner [16]. Blowdown water may need treatment to meet appropriate discharge standards prior to release to a local, publicly owned wastewater treatment works or receiving stream. To the maximum extent practical, this water should be recovered and reused in the washing process. The wastewater treatment process sludges and residual solids, such as spent carbon and spent ion exchange resin, must be appropriately treated before disposal. Any air emissions from the waste preparation area or the washing unit should be collected and treated, as appropriate to meet applicable regulatory standards.

SITE REQUIREMENTS

Access roads are required for transport of vehicles to and from the site. Typically, mobile soil washing process systems are located onsite and may occupy up to 4 acres for a 20 ton/hour unit; the exact area will depend on the vendor system selected, the amount of soil storage space, and/or the number of tanks or ponds needed for washwater preparation and wastewater treatment.

Typical utilities required are water, electricity, steam, and compressed air. An estimate of the net (consumed) quantity of local water required for soil washing, assuming water cleanup and recirculation, is 130,000–800,000 gallons per 1,000 cubic yards (2,500,000 lbs.) of soil (approximately 0.05 to 0.3 gallons per pound).

Because contaminated soils are usually considered hazardous, their handling requires that a site safety plan be developed to provide for personnel protection and special handling measures during soil washing operations.

Moisture content of soil must be controlled for consistent handling and treatment; this can be accomplished, in part, by covering excavation, storage, and treatment areas.

Fire hazard and explosion consideration should be minimal, since the soil washing fluid is predominantly water. Generally, soil washing does not require storing explosive, highly reactive materials.

Climatic conditions, such as annual or seasonal precipitation, cause surface runoff and water infiltration. Berms, dikes, or other runoff control methods may be required. Cold weather freezing must also be considered for aqueous systems and soil excavation operations.

Proximity to a residential neighborhood will affect plant noise requirements and emissions permitted in order to minimize their impact on the population and meet existing rules and regulations.

If all or part of the processed soil is to be redeposited at the site, storage areas must be provided until analytical data are obtained that verifies that treatment standards have been achieved. Onsite analytical capability could expedite the storage/final disposition process. However, soil washing might be applied to many different contaminant groups. Therefore, the analytes that would have to be determined are site specific, and the analytical equipment that must be available will vary from site to site.

PERFORMANCE DATA

The performances of soil washing processes currently shown to be effective in specific application are listed in Table 17-3 [1; 2; 4; 7 through 13]. Also listed are the range of particle size treated, contaminants successfully extracted, by-product wastes generated, extraction agents used, major extraction equipment for each system, and general process comments.

The data presented for specific contaminant removal effectiveness were obtained from publications developed by the respective soil washing system vendors. The quality of this information has not been determined.

RCRA Land Disposal Restrictions (LDRs) that require treatment of wastes to best demonstrate available technology (BDAT) levels prior to land disposal may sometimes be determined to be applicable or relevant and appropriate requirements (ARARs) for CERCLA response actions. The soil washing technology can produce a treated waste that meets treatment levels set by BDAT, but may not reach these treatment levels in all cases. The ability to meet required treatment levels is dependent upon the specific waste constituents and the waste matrix. In cases where soil washing does not meet these levels, it still may, in certain situations, be selected for use at the site if a treatability variance establishing alternative treatment levels is obtained. The EPA has made the treatability variance process available in order to ensure that LDRs do not unnecessarily restrict the use of alternative and innovative treatment technologies. Treatability variances may be justified for handling complex soil and debris matrices. The following guides describe when and how to seek a treatability variance for soil and debris: Superfund LDR Guide #6A, Obtaining a Soil and Debris Treatability Variance for Remedial Actions (OSWER Directive 9347.3-06FS) [16], and Superfund LDR Guide #6B, Obtaining a Soil and Debris Treatability Variance for Removal Actions (OSWER Directive 9347.3-07FS) [17]. Another approach could be to use other treatment techniques in series with soil washing to obtain desired treatment levels.

TECHNOLOGY STATUS

During 1986–1989, soil washing technology was selected as one of the source control remedies at eight Superfund sites: Vineland Chemical, New Jersey; Koppers Oroville Plant, California; Cape Fear Wood Preserving, North Carolina; Ewan Property, New Jersey; Tinkam Garage, New Hampshire; United Scrap, Ohio; Koppers/Texarkana, Texas; and South Cavalcade, Texas [18].

A large number of vendors provide a soil washing technology. Table 17-3 shows the current status of the technology for 14 vendors. The front portion of the table indicates the scale of equipment available from the vendor and gives some indication of the vendor's experience by showing the year it began operation.

Processes evaluated or used for site cleanups by the EPA are identified separately by footnote in Table 17-3.

The following soil washing processes that are under development have not been evaluated by the EPA or included in Table 17-3. Environmental Group, Inc. of Webster, Texas, has a process that reportedly removes metals and oil from soil. Process efficiency is stated as greater than 99 percent for lead removal from soils cleaned in Concord, California; greater than 99 percent for copper, lead, and zinc at a site in Racine, Wisconsin; and 94 percent for PCB removal on a Morrison-Knudsen Company project. The process does not appear to separate soil into different size fractions. Detailed information on the process is not available. Consolidated Sludge Company of Cleveland, Ohio, has a soil washing system planned that incorporates their Mega-sludge Press at the end of the process for dewatering solids. The system has not yet been built.

Vendor-supplied treatment costs of the processes reviewed ranged from $50 to $205 per ton of feed soil. The upper end of the cost range includes costs for soil residue disposal.

EPA CONTACT

Technology-specific questions regarding soil washing may be directed to Dennis Timberlake, RREL-Cincinnati, (513) 569-7547 (see Preface for mailing address).

REFERENCES

1. Assink, J.W. Extractive Methods for Soil Decontamination; a General Survey and Review of Operational Treatment Installations. In: Proceedings from the First International TNO Conference on Contaminated Soil, Utrecht, Netherlands, 1985.
2. Raghavan, R., D.H. Dietz, and E. Coles. Cleaning Excavated Soil Using Extraction Agents: A State-of-the-Art Review. EPA 600/2-89/034 (NTIS PB89-212757), 1988.

Table 17-3. Summary of Performance Data and Technology Status - Part I

Proprietary Vendor Process/EPA	Highest Scale of Operation	Year Operation Began	Range of Particle Size Treated	Contaminants Extracted from Soil	Extraction Agent(s)
U.S. Processes					
(1) Soil Cleaning Company of America [5; 15, p. 2]	Full-scale 15 tons/hr	1988	Bulk soil	Oil and grease	Hot water with surfactant
(2)ᵃ Biotrol Soil Treatment System (BSTS) [4, p. 6; 12]	Pilot-scale 500 lbs/hr	Fall, 1987	Above clay size and below 0.5 in. Some cleaning of fine particles in bio-reactor	Organics - pentachloro-phenol, creosote, naphthalene, pyrene, fluorene, etc.	Proprietary conditioning chemicals
(3) EPA's Mobile Countercurrent Extractor [9; 5, p. 5]	Pilot-scale 4.1 tons/hr	Modified with drum washer and shake-down - 1982 Full-scale - 1986	2–25 mm in drum washer <2 mm in four-stage extractor	Soluble organics (phenol, etc.) Heavy metals (Pb, etc.)	Various solvents, additives, surfactants, redox acids and bases. Chelating agent (EDTA)
(4)ᵃ EPA's First Generation Pilot Drum Screen Washer [10, p. 8]	Pilot-scale	1988	Oversize (>2 mm) removed prior to treatment	Petroleum hydrocarbons	Biodegradable surfactant (aqueous slurry)
(5)ᵃ MTA Remedial Resources (MTARRI) Froth Flotation [11; 15, p. 2]	Bench-scale	N/A	Oversize removed prior to treatment	Organics (oil) Heavy metals (inorganics) removed using counter-current decantation with leaching	Surfactants and alkaline chemicals added upstream of froth flotation cells. Acid for leaching.
Non-U.S. Processes					
(6) Ecotechniek BV [2, p. 17]	Commercial 100 ton/hr mass	1982	Sandy soil	Crude oil	None. Water-sand slurry heated to 90°C max. with steam.
(7) Bodemsanering Nederland BV (BSN) [2, p. 17]	Commercial 20 ton/hr	1982	>100 mm removed No more than 20% <63 μm Sludge <30 μm not cleaned	Oil from sandy soil	None. Uses high pressure water jet for soils washing.

Process	Scale/Throughput	Date	Pretreatment/Feed Size	Contaminants	Additives
[2, p. 20; 7, p. 5]	15–20 ton/hr	Commercial - 1986 With fines removal - 1987	treatment: coarse screens, electromagnet blade washer	Limited heavy metals removal experience	oscillation/vibration Surfactants Acid/base
(9) HWZ Bodemsanering BV [2, p. 17]	Commercial 20–25 ton/hr	1984	<10 mm and >63 μm	Cyanide, chlorinated HC, some heavy metals, PNA	Sodium hydroxide to adjust pH Surfactants
(10) Heijman Melieutechniek BV [2, p. 17; 7, p. 6]	Pilot-scale 10–15 ton/hr	1985	<10 mm and no more than 30% <63 μm	Cyanide, heavy metals, mineral oil (water immiscible hydrocarbons)	Proprietary extraction agents. Hydrogen peroxide (H_2O_2) added to react with extracted CN to form CO_2 and NH_3
(11) Heidemij Froth Flotation [7, p. 8]	Full-scale	N/A	<4 mm and no more than 20% <50 μm	Cyanide, heavy metals, chlorinated HCs, oil, toluene, benzene, pesticides, etc.	Proprietary surfactants and other proprietary chemicals
(12) EWH Alsen-Breitenburg Dekomat System [2, p. 20]	Pilot-scale 8–10 cu. m/hr	N/A	<80 mm Clays treated offsite	Oil from sandy soil	Proprietary
(13) TBSG Industrieveitietungen Oil Crep I System [7, p. 7]	Pilot-scale 1986	1986	Sand <50 mm Particles <100 μm treated offsite	Hydrocarbon and oil	Proprietary combination of surfactants, solvents, and aromatic hydrocarbons
(14) Klockner Umweltechnik Jet-Modified BSN [2, p. 20]	Pilot-scale	N/A	No more than 20% <63 μm	Aliphatics and aromatics with densities < water, volatile organics, some other hydrocarbons	None. Soil blasted with a water jet (at 5,075 psi)

[a] Process evaluated or used for site cleanup by the EPA. N/A = Not Available.

Table 17-3. Summary of Performance Data and Technology Status - Part II

Proprietary Vendor Process/EPA	By-Product Wastes Generated	Extraction Equipment	Efficiency of Contaminant Removal			Additional Process Comments
			Contaminant	% Removal	Residual ppm	
U.S. Processes						
(1) Soil Cleaning of America	Wet oil	Screw conveyors	Oil and grease	50–83	250–600	Three screw conveyors operated in series, hot water with surfactant injected into each stage. Final soil rinse on a fourth screw conveyor.
(2)[a] Biotrol Soil Treatment System (BSTS)	Oil and grease; Sludge from biological treatment	Agitated conditioning tank; Froth flotation; Slurry bioreactor	For the case presented: 90–95% for pentachlorophenol; to residuals <115 ppm. 85–95% for most other organics; to residuals <1 ppm.			Dewatered clays and organics to be treated offsite by incineration, solidification, etc. Washed soil was approx. 78% of feed. Therefore, significant volume reduction was achieved.
(3) EPA's Mobile Counter-Current Extractor	Clay fraction; Recovered organics (extractor skimmings); Spent carbon (oversize)	Drum screen; Water knife; Soil scrubber; 4-stage counter-current chemical extractor	Phenol; As_2O_3	90 from in. soil; 80 from or. soil; 50–80	1; 96; 0.5–1.3	Clay fraction treated elsewhere.
(4)[a] EPA's First Generation Pilot Drum Screen Washer (PDSW)	Sludge; Flocculated fines	Drum screen washer	Oil and grease; Soil size fraction: 0.25–2; <0.25	99; 90	<5; 2400	Process removal efficiency increases if extracting medium is heated. Install wet classifiers beneath the PDSW to remove waste-water from treated soil. Auger classifiers are required to discharge particles effectively.

Process	Feed	Equipment	Contaminant / % removal	Value	Comments
(5)* MTA Remedial Resources (MTARRI) Froth Flotation	Flocculation from	Reagent blend tank, Flotation cells, Counter-current decantation	Volatile organics 98-99+ Semivolatile organics 98-99+ Most fuel products 98-99+	<50 <250 <2200	Flotation cells mixed by underflow weir gates. Induced air blown down a center shaft in each cell. Continuous flow operation. Froth contains 5-10% wt% of feed soil.
Non-U.S. Processes (6) Ecotechniek BV	Wet oil	Jacketed, agitated tank	About 90% 20,000 ppm residual oil		Effectiveness of process dependent on soil particle size and type of oil to be separated.
(7) Bodemsanering Nederland BV (BSN)	Oil/organics recovered from wastewater fines	Water jet	Selected results: Aromatics >81 PNAs 85 Crude oil 97	>45 15 2300	No comments.
(8) Harbauer of America	Carbon which may contain contaminants	Conditioning tank, Low frequency vibration unit	Organic-Cl 96 Tot. organics 86-94 Tot. phenol 86-90 PAH 84-88 PCB	ND 159-201 7-22.5 91.4-97.5 0.5-1.3	Vibrating screw conveyor used. Cleaned soil separated from extractant liquor in stages; coarse soil by sedimentation, medium fraction in hydroclone, fines (15-20 μm) by vacuum filter press.
(9) HWZ Bodemsanering BV	Fines Sludge containing iron cyanide Large particles—carbon, wood, grass	Scrubber (for caustic addition), Upflow classifier	CN 95 PNAs 98 Chlorinated-HC 98 Heavy metals 75	5-15 15-20 <1 75-125	When the fines fraction (<63 μm) is greater than 20%, the process is not economical. HWZ has had some problems in extracting PNAs and oily material.

Table 17-3. Continued

Proprietary Vendor Process/EPA	By-Product Wastes Generated	Extraction Equipment	Efficiency of Contaminant Removal		Additional Process Comments
(10) Heijman Milieutechniek BV	Flocculated fines sludge Oil (if any) and silt	Mix tank followed by soils fraction equipment—hydroclones, sieves, tilt plate separators	Cyanide 93–99 Heavy metal cations approx. 70	<15 <200	Process works best on sandy soils with a minimum of humus-like compounds. Because no sand or charcoal filters are employed by Heijman, the system does not remove contaminants such as chlorinated hydrocarbons.
(11) Heidemij Froth Flotation	Contaminated float	Conditioning tank Froth flotation tanks	Cyanide >95 Heavy metals >90 avg Chlorinated-HC >99 Oil >99	5 >150 0.5 20	Process has broad application for removing hazardous materials from soil. Most experience has been on a laboratory scale.
(12) EWH Alsen-Breitenburg Dekomat System	Recovered oil Flocculated fines (sludge)	High-shear stirred tank	About 95% oil removed		Cleaned soil from high shear stirred tank is separated into fractions using vibrating screens, screw classifiers, hydroclones, and sedimentation tanks.
(13) TBSG Industrieveitiet-Ungen Oil Crep I System	Oil phase containing Oil Crep I	Screw mixer followed by a rotating separation drum for oil recovery	>95% removal of hydrocarbons has been achieved. Results are influenced by other contaminants present.		Oil Crep system was used successfully in Flansburg, FRG (in 1986) to remove PCBs, PAHs, and other hydrocarbons.
(14) Klockner Umwelttechnik High Pressure Water Jet-Modified BSN	Oil/organics recovered from wastewater fines Sludge	Water jet - circular nozzle arrangement	Selected results: HC 96.3 Chlorinated-HC >75 Aromatics 99.8 PAHs 95.4 Phenol >99.8	82.05 <0.01 <0.02 15.48 <0.01	No comments

a Process evaluated or used for site cleanup by the EPA.
N/A = Not Available.

3. Technology Screening Guide for Treatment of CERCLA Soils and Sludges. EPA 540/2-88/004 (NTIS PB89-132674), 1988.
4. M.K. Stinson, et al. Workshop on the Extractive Treatment of Excavated Soil. U.S. Environmental Protection Agency, Edison, New Jersey, 1988.
5. Smarkel, K.L. Technology Demonstration Report—Soil Washing of Low Volatility Petroleum Hydrocarbons. California Department of Health Services, 1988.
6. Guide for Conducting Treatability Studies Under CERCLA, Interim Final. EPA/540/2-89/058, U.S. Environmental Protection Agency, 1989.
7. Nunno, T.J., J.A. Hyman, and T. Pheiffer. Development of Site Remediation Technologies in European Countries. Presented at Workshop on the Extractive Treatment of Excavated Soil. U.S. Environmental Protection Agency, Edison, New Jersey, 1988.
8. Nunno, T.J. and J.A. Hyman. Assessment of International Technologies for Superfund Applications. EPA/540/2-88/003 (NTIS PB90-106428), 1988.
9. Scholz, R. and J. Milanowski. Mobile System for Extracting Spilled Hazardous Materials from Excavated Soils, Project Summary. EPA/600/S2-83/100, 1983.
10. Nash, J. Field Application of Pilot Scale Soils Washing System. Presented at Workshop on the Extracting Treatment of Excavated Soil. U.S. Environmental Protection Agency, Edison, New Jersey, 1988.
11. Trost, P.B. and R.S. Richard. On-Site Soil Washing—A Low Cost Alternative. Presented at ADPA. Los Angeles, California, 1987.
12. Pflug, A.D. Abstract of Treatment Technologies, Biotrol, Inc., Chaska, Minnesota (no date).
13. Biotrol Technical Bulletin, No. 87-1A, Presented at Workshop on the Extraction Treatment of Excavated Soil, U.S. Environmental Protection Agency, Edison, New Jersey, 1988.
14. Superfund Treatability Study Protocol: Bench-Scale Level of Soils Washing for Contaminated Soils, Interim Report. U.S. Environmental Protection Agency, 1989. [See also, Guide to Conducting Treatability Studies Under CERCLA: Soil Washing—Interim Guidance, EPA/540/2-91/020A (NTIS PB92-170570).]
15. Innovative Technology: Soil Washing. OSWER Directive 9200.5-250FS (NTIS PB90-274184), 1989.
16. Superfund LDR Guide #6A: Obtaining a Soil and Debris Treatability Variance for Remedial Actions. OSWER Directive 9347.3-06FS (NTIS PB91-921327), 1990.
17. Superfund LDR Guide #6B: Obtaining a Soil and Debris Treatability Variance for Removal Actions. OSWER Directive 9347.3-07FS (NTIS PB91-921310), 1990.
18. ROD Annual Report, FY1989. EPA/540/8-90/006, U.S. Environmental Protection Agency, 1990.

OTHER REFERENCES

Overview-Soils Washing Technologies for: Comprehensive Environmental Response, Compensation, and Liability Act, Resource Conservation and Recovery Act, Leaking Underground Storage Tanks, Site Remediation, U.S. Environmental Protection Agency, 1989.

SITE PROGRAM SOIL WASHING DEMONSTRATION PUBLICATIONS (See Appendix A for information on how to obtain SITE Reports)

BioTrol, Inc., Soil Washing System (Wood Preserving Site):

Demonstration Bulletin (EPA/540/M5-91/003)
Technology Demonstration Summary (EPA/540/S5-91/003)
Applications Analysis Report (EPA/540/A5-91/003)
Technology Evaluation, Vol. I (EPA/540/5-91/003a, NTIS PB92-115310), Vol. II Part A (EPA/540/5-91/003b, NTIS PB92-115328), Vol. Part B (EPA/540/5-91/003c, NTIS PB92-115336).

Biogenesis™ Soil Washing Technology:

Demonstration Bulletin (EPA/540/MR-93/510)
SITE Technology Capsule (EPA/540/SR-93/510)
Innovative Technology Evaluation Report (EPA/540/R-93/510)

BESCORP Soil Washing System for Lead Battery Site Treatment:

Demonstration Bulletin (EPA/540/MR-93/503)
Applications Analysis Report (EPA/540/AR-93/503)

Bergmann USA Soil/Sediment Washing System:

Demonstration Bulletin (EPA/540/MR-92/075)
Applications Analysis Report (EPA/540/AR-92/075)

Chapter 18

Solvent Extraction[1]

Jim Rawe and George Wahl, Science Applications International Corporation (SAIC), Cincinnati, OH

ABSTRACT

Solvent extraction does not destroy hazardous contaminants, but is a means of separating those contaminants from soils, sludges, and sediments, thereby reducing the volume of the hazardous material that must be treated. Generally it is used as one in a series of unit operations, and can reduce the overall cost for managing a particular site. It is applicable to organic contaminants and is generally not used for treating inorganic compounds and metals [1, p. 64].[2] The technology generally uses an organic chemical as a solvent [2, p. 30], and differs from soil washing, which generally uses water or water with wash improving additives. Commercial-scale units are in operation. There is no clear solvent extraction technology leader because of the solvent employed, type of equipment used, or mode of operation. The final determination of the lowest cost/best performance alternative will be more site-specific than process dominated. Vendors should be contacted to determine the availability of a unit for a particular site. This chapter provides information on the technology applicability, the types of residuals produced, the latest performance data, site requirements, the status of the technology, and sources for further information.

TECHNOLOGY APPLICABILITY

Solvent extraction has been shown to be effective in treating sediments, sludges, and soils containing primarily organic contaminants such as polychlorinated biphenyls (PCBs), volatile organic compounds (VOCs), halogenated solvents, and petroleum wastes. The technology is generally not used for extracting inorganics (i.e., acids, bases, salts, heavy metals). Inorganics usually do not have a detrimental effect on the extraction of the organic components, and sometimes metals that pass through the process experience a beneficial effect by changing to a less toxic or leachable form. The process has been shown to be applicable for the separation of the organic contaminants in paint wastes, synthetic rubber process wastes, coal tar wastes, drilling muds, wood treating wastes, separation sludges, pesticide/insecticide wastes, and petroleum refinery oily wastes [3].

Table 18-1 lists the codes for the specific Resources Conservation and Recovery Act (RCRA) wastes that have been treated by the technology [3; 4, p. 11]. The effectiveness of solvent extraction on general contaminant groups for various matrices is shown in Table 18-2 [5, p. 1; 1, p. 10]. Examples of constituents within contaminant groups are provided in Reference 1, Technology Screening Guide for Treatment of CERCLA Soils and Sludges. This table is based on the current available information or professional judgment where no information was available. The proven effectiveness of the technology for a particular site or waste does not ensure that it will be effective at all sites or that the treatment efficiencies achieved will be acceptable at other sites. For the ratings used for this table, demonstrated effectiveness means that at some scale treatability was tested to show the technology was effective for that particular contaminant and matrix. The ratings of potential effectiveness or no expected effectiveness are both based upon expert judgment. Where potential effectiveness is indicated, the technology is believed capable of successfully treating the contaminated group in a particular matrix. When the technology is not applicable or will probably not work for a particular combination of contaminant group and matrix, a no-expected-effectiveness rating is given.

---

[1] EPA/540/S-94/503.

[2] [Reference number, page number.]

Table 18-1. RCRA Codes for Wastes Treated by Solvent Extraction

Wood Treating Wastes	K001
Water Treatment Sludges	K044
Dissolved Air Flotation (DAF) Float	K048
Slop Oil Emulsion Solids	K049
Heat Exchanger Bundles Cleaning Sludge	K050
American Petroleum Institute (API) Separator Sludge	K051
Tank Bottoms (leaded)	K052
Ammonia Still Sludge	K060
Pharmaceutical Sludge	K084
Decanter Tar Sludge	K089
Distillation Residues	K101

Table 18-2. Effectiveness of Solvent Extraction on General Contaminant Groups for Soil, Sludges, and Sediments

	Effectiveness		
Contaminant Groups	Soil	Sludge	Sediments
Organic			
Halogenated volatiles	▼	▼	▼
Halogenated semivolatiles	■	■	■
Nonhalogenated volatiles	■	■	▼
Nonhalogenated semivolatiles	■	■	■
PCBs	■	■	■
Pesticides	■	▼	▼
Dioxins/furans	▼	▼	▼
Organic cyanides	▼	▼	▼
Organic corrosives	▼	▼	▼
Inorganic			
Volatile metals	●	●	●
Nonvolatile metals	●	●	●
Asbestos	●	●	●
Radioactive materials	●	●	●
Inorganic corrosives	●	●	●
Inorganic cyanides	●	●	●
Reactive			
Oxidizers	●	●	●
Reducers	●	●	●

■ Demonstrated Effectiveness: Successful treatability test at some scale completed.
▼ Potential Effectiveness: Expert opinion that technology will work.
● No Expected Effectiveness: Expert opinion that technology will not work.

LIMITATIONS

Organically bound metals can co-extract with the target organic pollutants and become a constituent of the concentrated organic waste stream. This is an unfavorable occurrence because the presence of metals can restrict both disposal and recycle options.

The presence of detergents and emulsifiers can unfavorably influence extraction performance and material throughput. Water soluble detergents found in some raw wastes (particularly municipal) will dissolve and retain organic pollutants in competition with the extraction solvent. This can impede a system's ability to achieve low concentration treatment levels. Detergents and emulsifiers can promote the evolution of foam, which hinders separation and settling characteristics and generally decreases materials throughput. Although methods exist to combat these problems, they will add to the process cost.

When treated solids leave the extraction subsystem, traces of extraction solvent are present [6, p. 125]. The typical extraction solvents used in currently available systems either volatilize quickly from the treated solids or biodegrade easily. Ambient air monitoring can be employed to determine if the volatilizing solvents present a problem.

The types of organic pollutants that can be extracted successfully depend, in part, on the nature of the extraction solvent. Treatability tests should be conducted to determine which solvent or combination of solvents is best suited to the site-specific matrix and contaminants. In general, solvent extraction is least effective on very high molecular weight organics and very hydrophilic (having an affinity for water) substances.

Some commercially available extraction systems use solvents that are flammable, toxic, or both [7, p. 2]. However, there are standard procedures used by chemical companies, service stations, etc. that can be used to greatly reduce the potential for accidents. The National Fire Protection Association (NFPA) Solvent Extraction Plants Standard (No. 36) has specific guidelines for the use of flammable solvents [8, p. 4-60].

TECHNOLOGY DESCRIPTION

Some type of pretreatment is necessary. This may involve physical processing and, if needed, chemical conditioning after the contaminated medium has been removed from its original location. Soils and sediments can be removed by excavation or dredging. Liquids and pumpable sludges can be removed and transported using diaphragm or positive displacement pumps.

Any combination of material classifiers, shredders, and crushers can be used to reduce the size of particles being fed into a solvent extraction process. Size reduction of particles increases the exposed surface area, thereby increasing extraction efficiency. Caution must be applied to ensure that an overabundance of fines does not lead to problems with phase separation between the solvent and treated solids. The optimum particle size varies with the type of extraction equipment used.

Moisture content may affect the performance of a solvent extraction process, depending on the specific system design. If the system is designed to treat pumpable sludges or slurries, it may be necessary to add water to solids or sediments to form a pumpable slurry. Other systems may require reduction of the moisture content in order to treat contaminated media effectively.

Chemical conditioning may be necessary for some wastes or solvent extraction systems. For example, pH adjustment may be necessary for some systems to ensure solvent stability or to protect process equipment from corrosion.

Depending on the nature of the solvent used, solvent extraction processes may be divided into three general types. These include processes using the following types of solvents: standard, liquefied gas (LG), and critical solution temperature (CST) solvents. Standard solvent processes use alkanes, alcohols, ketones, or similar liquid solvents at or near ambient temperature and pressure. These types of solvents are used to treat contaminated solids in much the same way as they are commonly used by analytical laboratories to extract organic contaminants from environmental samples. LG processes use propane, butane, carbon dioxide, or other gases which have been pressurized at or near ambient temperature. Systems incorporating CST solvents utilize the unique solubility properties of those solvents. Contaminants are extracted at one temperature where the solvent and water are miscible and then the concentrated contaminants are separated from the decanted liquid fraction at another temperature where the solvent has minimal solubility in water. Triethylamine is an example of a CST solvent. Triethylamine is miscible in water at temperatures less than 18°C and only slightly miscible above this temperature.

A general schematic diagram of a standard solvent extraction process is given in Figure 18-1 [9, p. 5]. These systems are operated in either batch or continuous mode and consist of four basic process steps: (1) extraction, (2) separation, (3) desorption, and (4) solvent recovery.

In the first step, solids are loaded into an extraction vessel and the vessel is purged with an inert gas. Solvent is then added and mixed with the solids. Designs of vessels used for the extraction stage vary from countercurrent, continuous-flow systems to batch mixers. The ratio of solvent-to-solids also varies, but normally remains within a range from 2:1 to 5:1. Solvent selection may also be a consideration. Ideally, a hydrophilic (having an affinity for water) solvent or mixture of hydrophilic/

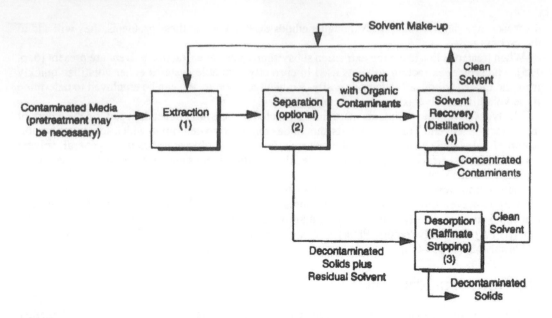

Figure 18-1. General schematic of a standard solvent extraction process.

hydrophobic (lacking an affinity for water) solvents is mixed with the solids. This hydrophilic solvent or solvent mixture will dewater the solids and solubilize organic materials. Subsequent extractions may use only hydrophobic solvents. The contact time and type of solvent used are contaminant-specific and are usually selected during treatability studies.

Depending on the type of contaminated medium being treated, three phases may exist in the extractor: solid, liquid, and vapor. Separation of solids from liquids can be achieved by allowing solids to settle and pumping the contaminant-containing solvent to the solvent recovery system. If gravity separation is not sufficient, filtration or centrifugation may be necessary. Residual solids will normally go through additional solvent washes within the same vessel (for batch systems) or in duplicate reaction vessels until cleanup goals are achieved. The settled solids retain some solvent which must be removed. This is often accomplished by thermal desorption.

Solvent recovery occurs in the final process step. Contaminant-laden solvent, along with the solvent vapors removed during the desorption or raffinate stripping stage, are transferred to a distillation system. To facilitate separation through volatilization and condensation, low boiling point solvents are used for extraction. Condensed solvents are normally recycled to the extractor; this conserves solvent and reduces costs. Water may be evaporated or discharged from the system, and still bottoms, which contain high boiling point contaminants, are recovered for future treatment.

In Figure 18-2, a general schematic diagram of an LG extraction process is shown [9, p. 7]. The same basic steps associated with standard solvent processes are used with LG systems; however, operating conditions are different. Increased pressure and temperature are required in order for the solvent to take on LG characteristics.

Pumps or screw augers move the contaminated feed through the process. In the extractor, the slurry is vigorously mixed with the hydrophobic solvent. The extraction step can involve multiple stages, with feed and solvent moving in countercurrent directions.

The solvent/solids slurry is pumped to a decanting tank where phase separation occurs. Solids settle to the bottom of the decanter and are pumped to a desorber. Here, a reduction in pressure vaporizes the solvent, which is recycled, and the decontaminated slurry is discharged.

Contaminated solvent is removed from the top of the decanter and is directed to a solvent recovery unit. A reduction of pressure results in separating organic contaminants from the solvent. The organic contaminants remain in the liquid phase and the solvent is vaporized and removed. The solvent is then compressed and recycled to the extractor. Concentrated contaminants are removed for future treatment.

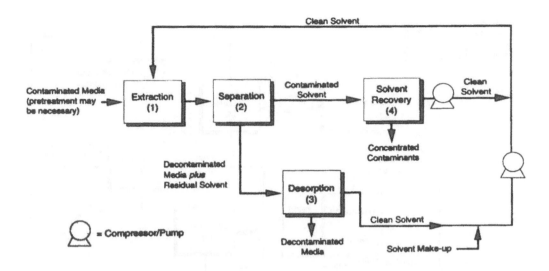

Figure 18-2. General schematic of an LG solvent extraction process.

CST processes use extraction solvents for which solubility characteristics can be manipulated by changing the temperature of the fluid. Such solvents include those binary (liquid-liquid) systems that exhibit an upper CST (sometimes referred to as upper consolute temperature), a lower CST (sometimes referred to as lower consolute temperature), or both. For such systems, mutual solubilities of the two liquids increase while approaching the CST. At or beyond the CST, the two liquids are completely miscible in each other. Figure 18-3 is a general schematic of a typical lower CST solvent extraction process. Again, the same four basic process steps are used; however, the solvent recovery step consists of numerous unit operations [9, p. 8].

PROCESS RESIDUALS

Three main product streams are produced from solvent extraction processes. These include treated solids, concentrated contaminants (usually the oil fraction), and separated water. Each of these streams should be analyzed to determine its suitability for recycle, reuse, or further treatment before disposal. Treatment options include: incineration, dehalogenation, pyrolysis, etc.

Depending on the system used, the treated solids may need to be dewatered, forming a dry solid and a separate water stream. The volume of product water depends on the inherent dewatering capability of the individual process, as well as the process-specific requirements for feed slurrying. Some residual solvent may remain in the soil matrix. This can be mitigated by solvent selection, and if necessary, an additional separation stage. Depending on the types and concentrations of metal or other inorganic contaminants present, post-treatment of the treated solids by some other technique (e.g., solidification/stabilization) may be necessary. Since the organic component has been separated, additional solids treatment should be simplified.

The organic solvents used for extraction of contaminants normally will have a limited effect on mobilizing and removing inorganic contaminants such as metals. In most cases, inorganic constituents will be concentrated and remain with the treated solids. If these remain below cleanup levels, no further treatment may be required. Alternatively, if high levels of leachable inorganic contaminants are present in the product solids, further treatment such as solidification/stabilization, soil washing, or disposal in a secured landfill may be required. The exception here is organically bound metals. Such metals can be extracted and recovered with the concentrated contaminant (oil) fraction. High concentrations of specific metals, such as lead, arsenic, and mercury, within the oil fraction can restrict disposal and recycle options.

Concentrated contaminants normally include organic contaminants, oils and grease (O&G), naturally occurring organic substances found in the feed solids, and some extraction fluid. Concentration

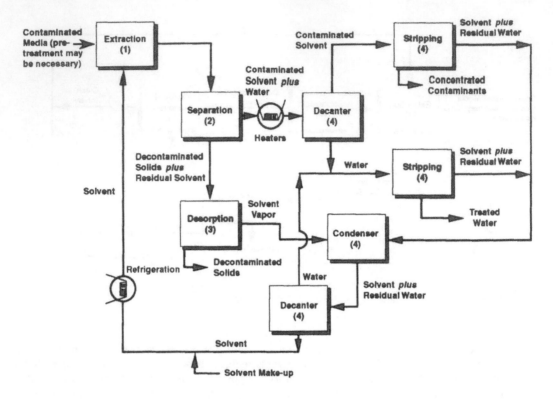

Figure 18-3. General schematic of a CST solvent extraction process.

factors may reduce the overall volume of contaminated material to 1/10,000 of the original waste volume, depending on the volume of the total extractable fraction. The highly-concentrated waste stream which results is either destroyed or collected for reuse. Incineration has been used for destruction of this fraction. Dechlorination of contaminants such as PCBs remains untried, but is a possible treatment. Resource recovery may also be a possibility for waste streams which contain useful organic compounds.

Use of hydrophilic solvents with moisture-containing solids produces a solvent/water mixture and clean solids. The solvent and water mixture are separated from the solids by physical means such as decanting. Some fine solids may be carried into the liquid stream. The solvent is normally separated from the water by distillation [10]. The water produced via distillation will contain water-soluble contaminants from the feed solids, as well as trace amounts of residual solvent and fines which passed through the separation stage. If the feed solids were contaminated with emulsifying agents, some organic contaminants may also remain with the water fraction. Furthermore, the volume of the water fraction can vary significantly from one site to another, and with the use of dewatering as a pretreatment. Hence, treatment of this fraction is dependent upon the concentration of contaminants present in the water and the flowrate and volume of residual water. In some cases, direct discharge to a publicly owned treatment works (POTW) or stream may be acceptable; alternatively, onsite aqueous treatment systems may be used to treat this fraction prior to discharge.

Solvent extraction units are designed to operate without air emissions. Nevertheless, during a recent SITE Demonstration Test, solvent concentrations were detected in 2 of 23 samples taken from the off-gas vent system [11]. Corrective measures were taken to remedy this. In addition, emissions of dust and fugitive contaminants could occur during excavation and materials handling operations.

SITE REQUIREMENTS

Solvent extraction units are transported by trailers. Therefore, adequate access roads are required to get the units to the site. Typical commercial-scale units of 25 to 125 tons per day (tpd) require a setup area of 1,500 to 10,000 square feet [12]. NFPA recommends an exclusion zone of 50 feet around solvent extraction systems operating with flammable solvents [8, p. 4-61].

Standard 440V three-phase electrical service is needed. Depending on the type of system used, between 50 and 10,000 gallons per day (gpd) of water must be available at the site [12]. The quantity of water needed is vendor- and site-specific.

Contaminated soils or other waste materials are hazardous and their handling requires that a site safety plan be developed to provide for personnel protection and special handling measures. Storage should be provided to hold the process product streams until they have been tested to determine their acceptability for disposal or release. Depending upon the site, a method to store waste that has been prepared for treatment may be necessary. Storage capacity requirements will depend on waste volume.

Onsite analytical equipment for conducting O&G analyses and a gas chromatograph capable of determining site-specific organic compounds for performance assessment will shorten analytical turnaround time and provide better information for process control.

PERFORMANCE DATA

Full-scale and pilot-scale performance data are currently available from only a few vendors: CF Systems, Resources Conservation Company (RCC), Terra-Kleen Corporation, and Dehydro-Tech Corporation. Lab-scale performance data are also available from these and other vendors. Data from Superfund Innovative Technology Evaluation (SITE) demonstrations are peer-reviewed and have been acquired in independently verified tests with stringent quality standards. Likewise, performance data from remedial actions at Superfund sites or EPA-sponsored treatability tests are assumed to be valid. The quality of other data has not been determined.

The CF Systems' 25-tpd commercial unit treated refinery sludge at Port Arthur, Texas, and operated with an online availability of greater than 90 percent. Extraction efficiencies for BTX and polynuclear aromatic hydrocarbon (PAH) compounds were greater than 99 percent. As demonstrated by Table 18-3, the typical level of organics in the treated solids met or exceeded the EPA Best Demonstrated Available Technology (BDAT) standards required for these listed refinery wastes [13].

Pilot-scale activities include the United Creosoting Superfund Site treatability study and the SITE demonstration at New Bedford Harbor, Massachusetts. During the spring of 1989, CF Systems conducted a pilot-scale treatability study for EPA Region VI and the Texas Water Commission at the United Creosoting Superfund Site in Conroe, Texas. The treatability study's objective was to evaluate the effectiveness of the CF Systems process for treating soils contaminated with pentachlorophenol (PCP), dioxins, and creosote-derived organic contaminants, such as PAHs. Treatment data from the field demonstration (Table 18-4) show that the total PAH concentration in the soil was reduced by more than 95 percent. Untreated soil had total PAH concentrations ranging from 2,879 to 2,124 mg/kg [13].

The SITE demonstration was conducted during the fall of 1988 to obtain specific operating and cost information for making technology evaluations for use at other Superfund sites. Under the SITE Program, CF Systems demonstrated an overall PCB reduction of more than 90 percent (see Table 18-5) for harbor sediments with inlet concentrations up to 2,575 ppm [14, p. 6]. An extraction solvent blend of propane and butane was used in this demonstration.

The ability of the RCC full-scale B.E.S.T.® process to separate oil feedstock into product fractions was evaluated by the EPA at the General Refining Superfund Site near Savannah, Georgia, in February 1987. The test was conducted with the assistance of EPA's Region X Environmental Services Division in cooperation with EPA's Region IV Emergency Response and Control Branch [15, p. 1]. The site was operated as a waste oil reclamation and re-refining facility from the early 1950s until 1975. As a result of those activities, four acidic oily sludge ponds with high levels of heavy metals (Pb = 200 to 10,000 ppm, Cu = 83 to 190 ppm) and detectable levels of PCBs (2.9 to 5 ppm) were produced. The average composition of the sludge from the four lagoons was 10 percent oil, 20 percent solids, and 70 percent water by weight [15, p. 13]. The transportable 70-tpd B.E.S.T.® unit processed approximately 3,700 tons of sludge at the General Refining Site. The treated solids from this unit were backfilled to the site, product oil was recycled as a fuel oil blend, and the recovered water was pH adjusted and transported to a local industrial wastewater treatment facility. Test results (Table 18-6) showed that the heavy metals were mostly concentrated in the solids product fraction. Toxicity Characteristic

Table 18-3. Contaminant Concentrations in Typical Solids Treated by CF Systems' Process at Port Arthur, Texas Refinery

Compound	mg/kg (ppm)	BDAT
Benzene	BDL	14
Ethyl benzene	BDL	14
Toluene	BDL	14
Xylenes	1.5	22
Naphthalene	2.2	42
Phenanthrene	3.4	34
2-Methylphenol	BDL	6.2
Anthracene	BDL	28
Benzo(a)anthracene	BDL	28
Pyrene	1.6	36
Chrysene	BDL	15
Benzo(a)pyrene	BDL	12
Phenol	BDL	3.6
4-Methyl phenol	BDL	6.2
Bis(2-E.H.)phthalate	BDL	7.3
Di-n-butyl phthalate	BDL	3.6

BDL below detection limits.

Table 18-4. CF Systems' Performance Data at United Creosote Superfund Site

Compound	Feed Soil (mg/kg)	Treated Soil (mg/kg)	Reduction (percent)
PAHs			
Acenaphthene	360	3.4	99
Acenaphthylene	15	3.0	80
Anthracene	330	8.9	97
Benzo(a)anthracene	100	7.9	92
Benzo(a)pyrene	48	12	75
Benzo(b)fluoranthene	51	9.7	81
Benzo(g,h,i)perylene	20	12	40
Benzo(k)fluoranthene	50	17	66
Chrysene	110	9.1	92
Dibenzo(a,h)anthracene	ND	4.3	NA
Fluoranthene	360	11	97
Fluorene	380	3.8	99
Indeno(1,2,3-cd)pyrene	19	11	58
Naphthalene	140	1.5	99
Phenanthrene	590	13	98
Pyrene	360	11	97
Total PAH concentration	2879	122.6	96

Notes: mg/kg on a dry weight basis. ND indicates not detected. NA indicates not applicable.

Leaching Procedure (TCLP) test results showed heavy metals to be in stable forms that resisted leaching, illustrating a potential beneficial side effect when metals are treated by the process [4, p. 13].

During the summer of 1992, a SITE demonstration was conducted to test the ability of the B.E.S.T.® system to remove PAHs and PCBs from contaminated sediments obtained from the Grand Calumet River. The pilot-scale B.E.S.T.® system was primarily contained on two skids and had an average daily capacity of 90 pounds of contaminated sediments. As Table 18-7 demonstrates, more than 96

Table 18-5. Extraction of New Bedford Harbor Sediments Using CF Systems' Process

Test #	Initial PCB Concentration (ppm)	Final PCB Concentration (ppm)	Reduction (percent)	Number of Passes Through Extractor
1	350	8	98	9
2	288	47	84	1
3	2,575	200	92	6

Table 18-6. B.E.S.T.™ Process Data from the General Refining Superfund Site

Metals	Initial Concentration (mg/kg)	Product Solids Metal (ppm)	TCLP Levels (ppm)
As	<0.6	<0.5	<0.0
Ba	239	410	<0.03
Cr	6.2	21	<0.05
Pb	3,200	23,000	5.2
Se	<4.0	<5.0	0.008

percent of the PAHs and greater than 99 percent of the PCBs initially present in the sediments collected from Transect 6 and Transect 28 of the Grand Calumet River were removed [16].

Terra-Kleen Corporation has compiled remedial results for its solvent extraction system at three sites; Treband Superfund site, in Tulsa, Oklahoma; Sand Springs Substation site, Sand Springs, Oklahoma; and Pinette's Salvage Yard Superfund site, Washburn, Maine. PCBs were the primary contaminant at each of these sites. Table 18-8 summarizes the performance at the Treband site. Preliminary results from the Pinette's Salvage Yard site are given in Table 18-9 [17].

The Carver-Greenfield (C-G) Process®, developed by Dehydro-Tech Corporation, was evaluated during a SITE demonstration at an EPA research facility in Edison, New Jersey. During the August 1991 test, about 640 pounds of drilling mud contaminated with indigenous oil and elevated levels of heavy metals were shipped to the EPA in Edison, New Jersey from the PAB Oil Site in Abbeville, Louisiana. The pilot-scale unit was trailer-mounted and capable of treating about 100 lb/hr of contaminated drilling mud. The process removed about 90 percent of the indigenous oil (as measured by solids/oil/water analysis). The indigenous total petroleum hydrocarbon (TPH) removals were essentially 100 percent for both runs [18, p. 1].

E.S. Fox Limited has determined performance data for the Extraksol® Process developed by Sanivan Group of Montreal, Quebec, Canada. Performance data on contaminated soils and refinery wastes for the 1 ton per hour (tph) mobile unit are shown in Table 18-10 [19]. The process uses a proprietary solvent that reportedly achieved removal efficiencies up to 99 percent (depending on the number of extraction cycles and the type of soil) on solids with contaminants such as PCBs, O&G, PAHs, and PCP.

RCRA Land Disposal Restrictions (LDRs) that require treatment of wastes to BDAT levels prior to land disposal may sometimes be determined to be applicable or relevant and appropriate requirements (ARARs) for CERCLA response actions. The solvent extraction technology can produce a treated waste that meets treatment levels set by BDAT, but may not reach these treatment levels in all cases. The ability to meet required treatment levels is dependent upon the specific waste constituents and the waste matrix. In cases where solvent extraction does not meet these levels, it still may, in certain situations, be selected for use at the site if a treatability variance establishing alternative treatment levels is obtained. The EPA has made the treatability variance process available in order to ensure that LDRs do not unnecessarily restrict the use of alternative and innovative treatment technologies. Treatability variances may be justified for handling complex soil and debris matrices. The following guides describe when and how to seek a treatability variance for soil and debris: Superfund LDR Guide #6A, Obtaining a Soil and Debris Treatability Variance for Remedial Actions (OSWER Directive

Table 18-7. Summary of Results from the SITE Demonstration of the RCC B.E.S.T.® Process (Averages from Three Runs)

Parameter	Transect 28 Sediment			Transect 6 Sediment		
	PCBs	PAHs	Triethylamine	PCBs	PAHs	Triethylamine
Concentration in untreated sediment, mg/kg	12.1	550	NA	425	70,900	NA
Concentration in treated solids, mg/kg	0.04	22	45.1	1.8	510	103
Removal from sediment percent	99.7	96.0	NA	99.6	99.3	NA
Concentration in oil product, mg/kg	NA[a]	NA[a]	NA[a]	2,030	390,000	73.3[b]
Concentration in water product, mg/L	<0.003	<0.01	1.0	<0.001	<0.01	2.2

NA = Not applicable.

[a] The Transect 28 oil product was sampled at the end of the last run conducted on Transect 28 material. When the oil was sampled, there was not sufficient oil present for oil polishing (using the solvent evaporator to remove virtually all of the triethylamine for the oil). Excess triethylamine was therefore left in the oil.

[b] This oil product was sampled following oil polishing.

Table 18-8. Terra-Kleen Soil Restoration Unit PCB Removal at Treband Superfund Site[a]

Initial Level (ppm)	Final Level (ppm)	Site Goal (ppm)	Reduction (percent)
740	77	<100	89.6
810	3	<100	99.6
2,500	93	<100	96.3

[a] Soil type: sand and concrete dust.

Table 18-9. Terra-Kleen Soil Restoration Unit PCB Removal at Pinette's Salvage Yard NPL Site[a]

Initial Level (ppm)	Final Level (ppm)	Site Goal (ppm)	Reduction (percent)
41.8	2.7	<5.0	93.5
76.9	4.31	<5.0	94.4
381	3.59	<5.0	99.1

[a] Full-scale data. Soil type: glacial till (gravel, sand, silt, and grey marine clay).

Table 18-10. Summary of 1-tph Extrakol® Process Performance Data

Contaminant	Matrix	In (ppm)	Out (ppm)	Reduction (percent)
O&G	Clayey Soil	1,800	182	89.9
O&G	Oily Sludge	72,000	2,000	97.2
O&G	Fuller's Earth	313,000	3,700	98.8
PAH	Clayey Soil	332	55	83.4
PAH	Oily Sludge	240	10	95.8
PCB	Clayey Soil	150	14	90.7
PCB	Clayey Soil	54	4.4	91.8
PCP	Porous Gravel	81.4	<0.21	99.7
PCP	Activated Carbon	744	83	88.8

Note: Treated concentrations are based on criteria to be met and not process efficiency.

9347.3-06FS, September 1990) [20], and Superfund LDR Guide #6B, Obtaining a Soil and Debris Treatability Variance for Removal Actions (OSWER Directive 9347.3-06BFS, September 1990) [21]. Another approach would be to use other treatment techniques in series with solvent extraction to obtain desired treatment levels.

TECHNOLOGY STATUS

As of October 1992, solvent extraction has been chosen as the selected remedy at eight Superfund sites. Two of these, General Refining, Georgia and Treband Warehouse, Oklahoma were emergency responses that have been completed. The other sites include Norwood PCBs, Massachusetts; O'Conner, Maine; Pinette's Salvage Yard, Maine; Ewan Property, New Jersey; Carolina Transformer, North Carolina; and United Creosoting, Texas [22, p. 51].

Solvent extraction systems are at various stages of development. The following is a brief discussion of several systems that have been identified.

CF Systems uses liquefied hydrocarbon gases such as propane and butane as solvents for separating organic contaminants from soils, sludges, and sediments. To date, the CF Systems process has been used in the field at three Superfund sites; nine petrochemical facilities and remediation sites; and a centralized treatment, storage, and disposal (TSD) facility. The CF Systems solvent extraction technology is available in several commercial sizes and the Mobile Demonstration Unit is available for onsite treatability studies. CF Systems has supplied three commercial-scale extraction units for the treatment of a variety of wastes [23, p. 3-12]. A 60-tpd treatment system was designed to extract organic liquids from a broad range of hazardous waste feeds at ENSCO's El Dorado, Arkansas, incinerator facility. A commercial-scale extraction unit is installed at a facility in Baltimore, Maryland, to remove organic contaminants from a 20 gallons-per-minute (gpm) wastewater stream. A PCU-200 extraction unit was installed and successfully operated at the Star Enterprise (Texaco) refinery in Port Arthur, Texas. This unit was designed to treat listed refinery wastes to meet or exceed the EPA's BDAT standards. A 220 tpd extraction unit is currently being designed for use at the United Creosoting Superfund site in Conroe, Texas.

RCC's B.E.S.T.® system uses aliphatic amines (typically triethylamine) as the solvent to separate and recover contaminants in either batch or continuous operation [4, p. 2]. It can extract contaminants from soils, sludges, and sediments. In batch mode of operation, a pumpable waste is not required. RCC has a transportable B.E.S.T.® pilot-scale unit available to treat soils and sludges. This pilot-scale equipment was used at a Gulf Coast refinery treating various refinery waste streams and treated PCB-contaminated soils at an industrial site in Ohio during November of 1989. A full-scale unit with a nominal capacity of 70 tpd was used to clean 3,700 tons of PCB-contaminated petroleum sludge at the General Refining Superfund Site in Savannah, Georgia, in 1987 [16].

Terra-Kleen Corporation's Soil Restoration Unit was developed for remedial actions involving soil, debris, and sediments contaminated with organic compounds. The Soil Restoration Unit is a mobile system which uses various combinations of up to 14 patented solvents, depending upon target contaminants present. These solvents are nontoxic and not listed hazardous wastes [17].

Dehydro-Tech Corporation's C-G Process is designed for the cleanup of Superfund sites with sludges, soils, or other water-bearing wastes containing hazardous compounds, including PCBs, polycyclic aromatics, and dioxins. A transportable pilot-scale system capable of treating 30 to 50 lb/hr of solids is available. Over 80 commercial C-G Process facilities have been licensed in the past 30 years to solve industrial waste disposal problems. More than half of these plants were designed to dry and remove oil from slaughterhouse waste (rendering plants) [12].

NuKEM Development Company/ENSR developed a technique to remove PCBs from soils and mud several years ago. Their solvent extraction method involves acidic conditions, commercially available reagents to prepare the soil matrix for exposure to the solvent, and ambient temperatures and pressures [24]. NuKEM Development Company/ENSR is not currently marketing this technology for the treatment of contaminated soils and sludges. Another application being reviewed is the treatment of refinery sludges (K wastes and F wastes). The Solvent Extraction Process (SXP) system developed for treating these wastes has six steps; acidification, dispersion, extraction, raffinate solvent recovery, stabilization/filtration, and distillation. A pilot-scale SXP system has performed tests on over 20 different sludges. According to the vendor, preliminary cost estimates for treating 5,000 tons per year of a feed with 10 percent solids and 10 percent oil appear to be less than $300 per ton [25].

The Extraksol® process was developed in 1984 by Sanivan Group, Montreal, Quebec, Canada [26, p. 35]. It is applicable to treatment of contaminated soils, sludges, and sediments [26, p. 45]. The 1-tph unit is suitable for small projects with a maximum of 300 tons of material to be treated. A transportable commercial scale unit, capable of processing up to 8 tph, was constructed by E.S. Fox Ltd. At present, the assembled unit is available for inspection at the fabricator's facility in Welland, Ontario, Canada [19].

The Low Energy Extraction Process (LEEP), developed by ART International, Inc., is a patented solvent extraction process that can be used onsite for decontaminating soils, sludges and sediments. LEEP uses common organic solvents to extract and concentrate organic pollutants such as PCB, PAH, PCP, creosotes, and tar-derived chemicals [27, p. 250]. Bench-scale studies were conducted on PCB contaminated soils and sediments, base neutral contaminated soils and oil refinery sludges. ART has designed and constructed a LEEP Pilot Plant with a nominal solids throughput of 200 lb/hr [12]. The pilot plant has been operational since March 1992. Recently, a 13 tph (dry basis) commercial facility capable of treating soil contaminated with up to 5 percent tar was completed for a former manufactured gas plant site.

Phønix Miljø, Denmark has developed the Soil Regeneration Plant, a 10 tph transportable solvent extraction process. This process consists of a combined liquid extraction and steam stripping process operating in a closed loop. A series of screw conveyors is used to transfer the contaminated soil through the process. Contaminants are removed from soil in a countercurrent extraction process. A drainage screw separates the soil from the extraction liquid. The extraction liquid is distilled to remove contaminants and is then recycled. The soil is steam heated to remove residual contaminants before exiting the process [28].

Cost estimates for solvent extraction range from $50 to $900 per ton [12]. The most significant factors influencing costs are the waste volume, the number of extraction stages, and operating parameters such as labor, maintenance, setup, decontamination, demobilization, and lost time resulting from equipment operating delays. Extraction efficiency can be influenced by process parameters such as solvent used, solvent/waste ratio, throughput rate, extractor residence time, and number of extraction stages. Thus, variation of these parameters, in particular hardware design and/or configuration, will influence the treatment unit cost component but should not be a significant contributor to the overall site costs.

EPA CONTACT

Technology-specific questions regarding solvent extraction may be directed to Mark Meckes, (513) 569-7348, or Ronald Turner, RREL-Cincinnati, (513) 569-7775 (see Preface for mailing address).

ACKNOWLEDGMENTS

This updated chapter was prepared for the U.S. EPA, Office of Research and Development (ORD), Risk Reduction Engineering Laboratory (RREL), Cincinnati, Ohio, by Science Applications International Corporation (SAIC) under EPA Contract No. 68-C0-0048. Mr. Eugene Harris served as the EPA Technical Project Monitor. Mr. Jim Rawe (SAIC) was the Work Assignment Manager. He and Mr. George Wahl (SAIC) co-authored the revised bulletin. The authors are especially grateful to Mr. Mark Meckes of EPA-RREL, who contributed significantly by serving as a technical consultant during the development of this chapter. The authors also want to acknowledge the contributions of those who participated in the development of the original bulletin.

The following other Agency and contractor personnel have contributed their time and comments by peer-reviewing the chapter:

Dr. Ben Blaney, EPA-RREL
Mr. John Moses, CF Technologies, Inc.
Dr. Ronald Dennis, Lafayette College

REFERENCES

1. Technology Screening Guide for Treatment of CERCLA Soils and Sludges. EPA/540/2-88/004 (NTIS PB89-132674), 1988.
2. Raghavan, R., D.H. Dietz, and E. Coles. Cleaning Excavated Soil Using Extraction Agents: A State-of-the-Art Review. EPA 600/2-89/034 (NTIS PB89-212757), 1988.
3. CF Systems Corporation, marketing brochures (no dates).
4. Austin, D.A. The B.E.S.T.® Process—An Innovative and Demonstrated Process for Treating Hazardous Sludges and Contaminated Soils. Presented at 81st Annual Meeting of APCA, Preprint 88-6B.7, Dallas, Texas, 1988.
5. Innovative Technology B.E.S.T.® Solvent Extraction Process. OSWER Directive 9200.5-253FS (NTIS PB90-274218), 1989.
6. Reilly, T.R., S. Sundaresan, and J.H. Highland. Cleanup of PCB Contaminated Soils and Sludges by a Solvent Extraction Process: A Case Study. Studies in Environmental Science, 29:125–139, 1986.
7. Weimer, L.D. The B.E.S.T.® Solvent Extraction Process Applications with Hazardous Sludges, Soils and Sediments. Presented at the Third International Conference, New Frontiers for Hazardous Waste Management, Pittsburgh, Pennsylvania, 1989.
8. Fire Protection Handbook. Fourteenth Ed. National Fire Protection Association, 1976.
9. Guide for Conducting Treatability Studies under CERCLA: Solvent Extraction. EPA/540/R-92/016a, U.S. Environmental Protection Agency, 1992.

10.   Blank, Z. and W. Steiner. Low Energy Extraction Process-LEEP: A New Technology to Decontaminate Soils, Sediments, and Sludges. Presented at HazTech International 90, Houston Waste Conference, Houston, Texas, May 1990.

11.   Technology Evaluation Report—Resources Conservation Company, Inc. B.E.S.T.® Solvent Extraction Technology. EPA/540/R-92/079a, 1993.*

12.   Vendor Information System for Innovative Treatment Technologies (VISITT) Database, Version 1.0. U.S. Environmental Protection Agency. [Editor's Note: VISITT Version 4.0 was released in 1995; call VISITT help line (800/245-4505 or 703/883-8448 for information on how to obtain the software and user manual.]

13.   Site Remediation of Contaminated Soil and Sediments: The CF Systems Solvent Extraction Technology. CF Systems, marketing brochure (no date).

14.   Technology Evaluation Report—CF Systems Organics Extraction System, New Bedford, MA, Volume I. EPA/540/5-90/002, 1990.*

15.   Evaluation of the B.E.S.T.® Solvent Extraction Sludge Treatment Technology Twenty-Four Hour Test. EPA/600/2-88/051, U.S. Environmental Protection Agency, 1988.

16.   Applications Analysis Report—Resources Conservation Company, Inc. B.E.S.T.® Solvent Extraction Technology. EPA/540/AR-92/079, 1993.*

17.   Cash, A.B. Full Scale Solvent Extraction Remedial Results. Presented at American Chemical Society, I&EC Division Special Symposium; Emerging Technologies for Hazardous Waste Management, Atlanta, GA, September 21–23, 1992.

18.   Applications Analysis Report—The Carver-Greenfield Process˚; Dehydro-Tech Corporation. EPA/540/AR-92/002, 1992.*

19.   Non-Thermal Extraction of Hazardous Wastes from Soil Matrices. ES Fox Limited, marketing brochure (no date).

20.   Superfund LDR Guide #6A: (2nd Edition) Obtaining a Soil and Debris Treatability Variance for Remedial Actions. OSWER Directive 9347.3-06FS (NTIS PB91-921327), 1990.

21.   Superfund LDR Guide #6B: Obtaining a Soil and Debris Treatability Variance for Removal Actions. OSWER Directive 9347.3-06BFS (NTIS PB91-921310), 1990.

22.   Innovative Treatment Technologies: Semi-Annual Status Report (Fourth Edition). EPA 542-R-92-011, U.S. Environmental Protection Agency, 1992. [Editor's Note: Sixth Edition, EPA 542-R-95-005 was published September, 1994.]

23.   Applications Analysis Report—CF Systems Organics Extraction System, New Bedford, MA. EPA/540/A5-90/002, 1990.*

24.   Massey, M.J. and S. Darian. ENSR Process for the Extractive Decontamination of Soils and Sludges. Presented at the PCB Forum, International Conference for the Remediation of PCB Contamination, Houston, Texas, 1989.

25.   Chelemer, M.J. ENSR's SXP System for Treating Refinery K-Wastes: An Update. ENSR Consulting and Engineering, marketing brochure (no date).

26.   Paquin, J. and D. Mourato. Soil Decontamination with Extraksol. Sanivan Group, Montreal, Canada (no date), pp. 35–47.

27.   The Superfund Innovative Technology Evaluation Program: Technology Profiles Fifth Edition. EPA/540/R-92/077, U.S. Environmental Protection Agency, 1992. [Editor's Note: Seventh Edition, EPA/540/R-94/526 was published November, 1994.]

28.   Phønix Miljø Cleans Contaminated Soil On-Site: In Mobile Extraction Plant. Phønix Miljø Marketing Information (no date).

ADDITIONAL SITE PROGRAM REFERENCES

Terra-Kleen Solvent Extraction Technology:
    Demonstration Bulletin (EPA/540/MR-94/521), September 1994.
    SITE Technology Capsule (EPA/540/R-94/521a), February 1995.

* Superfund Innovative Technology Evaluation (SITE) program report. See Appendix A for information on how to obtain SITE reports.

Chapter 19

Chemical Oxidation Treatment[1]

Margaret Groeber, Science Applications International Corporation (SAIC), Cincinnati, OH

ABSTRACT

Oxidation destroys hazardous contaminants by chemically converting them to nonhazardous or less toxic compounds that are ideally more stable, less mobile, and/or inert. However, under some conditions, other hazardous compounds may be formed. The oxidizing agents most commonly used for the treatment of hazardous contaminants are ozone, hydrogen peroxide, hypochlorites, chlorine, and chlorine dioxide. Current research has shown the combination of these reagents or ultraviolet (UV) light and an oxidizing agent(s) makes the process more effective [1; 2; 3, p. 11].[2] Treatability studies are necessary to document the applicability and performance of chemical oxidation systems technology for a specific site.

Chemical oxidation is a developed technology commonly used to treat liquid mixtures containing amines, chlorophenols, cyanides, halogenated aliphatic compounds, mercaptans, phenols, and certain pesticides [4, p. 7.76; 5, p. 7.42]. In lab-scale tests, chemical oxidation has been shown to be effective for chlorinated organics [6, p. 229].

This chapter provides information on the technology applicability, limitations, a technology description, the types of residuals produced, site requirements, current performance data, status of technology, and sources of further information.

TECHNOLOGY APPLICABILITY

Chemical oxidation effectively treats liquids that contain oxidizable contaminants; however, it can be used on slurried soils and sludges. Because it is a nonselective treatment, it is most suited to media with low concentrations of contaminants.

The effectiveness of chemical oxidation technology on general contaminant groups is shown in Table 19-1. Examples of constituents within contaminant groups are provided in Technology Screening Guide for Treatment of CERCLA Soils and Sludges [7]. This table is based on the current available information or professional judgment when no information was available. The proven effectiveness of the technology for a particular site or waste does not ensure that it will be effective at all sites or that the treatment efficiency achieved will be acceptable at other sites. For the ratings used for this table, demonstrated effectiveness means that, at some scale, treatability was tested to show that, for that particular contaminant and matrix, the technology was effective. The ratings of potential effectiveness and no-expected-effectiveness are based upon expert judgment. Where potential effectiveness is indicated, the technology is believed capable of successfully treating the contaminant group in a particular matrix. When the technology is not applicable or will probably not work for a particular combination of contaminant group and matrix, a no-expected-effectiveness rating is given.

Chemical oxidation depends on the chemistry of the oxidizing agent(s) and the chemical contaminants. Table 19-2 lists selected organic compounds by their relative ability to be oxidized. Chemical oxidation has also been used as part of a treatment process for cyanide-bearing wastes and metals such as arsenic, iron, and manganese [8, p. 4.4]. Metal oxides formed in the oxidation process more readily precipitate out of the treated medium.

Table 19-1. Effectiveness of Chemical Oxidation on General Contaminant Groups for Liquids, Soils, and Sludges[a]

Contaminant Groups	Liquids	Soils, Sludges
Organic		
Halogenated volatiles	■	▼
Halogenated semivolatiles	■	▼
Nonhalogenated volatiles	■	▼
Nonhalogenated semivolatiles	▼	■
PCBs	▼	●
Pesticides	■	▼
Dioxins/furans	▼	●
Organic cyanides	■	■
Organic corrosives	▼	▼
Inorganic		
Volatile metals	■	▼
Nonvolatile metals	■	▼
Asbestos	●	●
Radioactive materials	●	●
Inorganic corrosives	●	●
Inorganic cyanides	■	■
Reactive		
Oxidizers	●	●
Reducers	■	▼

■ Demonstrated Effectiveness: Successful treatability test at some scale completed.
▼ Potential Effectiveness: Expert opinion that technology will work.
● No Expected Effectiveness: Expert opinion that technology will not work.
[a] Enhancement of the chemical oxidation process is required for the less easily oxidizable compounds for some contaminant groups.

Table 19-2. Selected Organic Compounds by Relative Ability To Be Oxidized

Ability To Be Oxidized	Examples
High	phenols, aldehydes, amines, some sulfur compounds
Medium	alcohols, ketones, organic acids, esters, alkyl-substituted aromatics, nitro-substituted aromatics, carbohydrates
Low	halogenated hydrocarbons, saturated aliphatics, benzene

The oxidation of some compounds will require a combination of oxidizing agents or the use of UV light with an oxidizing agent(s) [1; 2; 3, p. 10]. An example of such a situation is polychlorinated biphenyls (PCBs), which do not react with ozone alone, but have been destroyed by combined UV-ozone treatment [5, p. 7.48]. Enhanced chemical oxidation has been used at several Superfund sites [3; 9].

LIMITATIONS

If oxidation reactions are not complete, residual hazardous compounds may remain in the contaminant stream. In addition, intermediate hazardous compounds may be formed (e.g., trihalomethanes, epoxides, and nitrosamines) [10; 11, p. 190]. Incomplete oxidation may be caused by insufficient quantity of the oxidizing agent(s), inhibition of oxidation reactions by low or high pH, the strength of the oxidizing agent(s), the presence of interfering compounds that consume reagent, or inadequate mixing or contact time between contaminant and oxidizing agent(s) [12, p. 10.52]. It is important to monitor the concentrations of residual oxidizing agent(s), contaminants, and products to ensure a complete reaction has occurred. It may be necessary to monitor reaction conditions such as pH, the strength of the oxidizing agent(s), the presence of interfering compounds that consume reagent, or inadequate mixing or contact time between contaminant and oxidizing agent(s) [12, p. 10.52]. It is important to monitor the concentrations of residual oxidizing agent(s), contaminants, and products to ensure a complete reaction has occurred. It may be necessary to monitor reaction conditions such as pH, temperature, and contact time to optimize the reaction. Determination of potential reactions and rates may be critical to prevent explosions or formation of unwanted compounds.

Oil and grease in the media should be minimized to optimize the efficiency of the oxidation process. Oxidation is not cost-effective for highly concentrated wastes because of the large amounts of oxidizing agent(s) required.

Chemical oxidation can be used on soils and sludges if there is complete mixing of the oxidizing agent(s) and the oxidizable hazardous component in the matrix.

Ozonation systems generally have higher capital costs than those using other oxidizing agents because an ozone generator must be used. They must also have an ozone decomposition unit to prevent emission of excess ozone into the ambient air, which further adds to the cost.

Although hydrogen peroxide is considered a relatively safe oxidant, proper storage and handling is required [5, p. 7.44]. The hydrogen peroxide reaction may be explosive when introduced into high-organic materials [11, p. 190].

The cost of generating UV light and the problem of scaling or coating on the lamps are two of the biggest drawbacks to UV-enhanced chemical oxidation systems. They do not perform as well in turbid waters and slurries because the reduced light transmission lowers the effectiveness [13].

TECHNOLOGY DESCRIPTION

Chemical oxidation is a process in which the oxidation state of a contaminant is increased while the oxidation state of the reactant is lowered. The electrons gained by the oxidizing agent are lost by the contaminant. An example of a common oxidation reaction is:

$$NaCN + H_2O_2 \rightarrow NaCNO + H_2O$$

NaCN	+	H_2O_2	$\rightarrow$	NaCNO	+	H_2O
(sodium cyanide)		(hydrogen peroxide)		(sodium cyanate)		(water)

In this reaction, the oxidation state of carbon in the sodium cyanide is increased while the oxidation state of each oxygen in the hydrogen peroxide is decreased.

Chemical oxidation is used when hazardous contaminants can be destroyed by converting them to nontoxic or less hazardous compounds. Contaminants are detoxified by actually changing their chemical forms. The process is nonselective; therefore, any oxidizable material reacts. The oxidizing agent(s) must be well mixed with the contaminants in a reactor to produce effective oxidation. In order for the oxidation reaction to occur, the pH must be maintained at a proper level; therefore, pH adjustment may be necessary [10; 14].

Figure 19-1 shows a process flow diagram for a chemical oxidation system. The main component is the process reactor. Oxidant is fed into the mixing unit (1), then the reactor (2). Reaction products and excess oxidant are scrubbed prior to venting to the ambient air. The pH and the temperature in the reactor are controlled to ensure the reaction goes to completion. The reaction can be enhanced with the addition of UV light.

Common commercially available oxidants include ozone, hydrogen peroxide, hypochlorites, chlorine and chlorine dioxide. Treatment of hazardous contaminants requires a strong oxidizing agent(s), such as ozone or hydrogen peroxide. Ozone and combinations of ozone and hydrogen peroxide react rapidly with a large number of contaminants [3, p. 11]. Ozone has a half-life of 20 to 30 minutes at 20°C (68°F); therefore, it must be produced onsite. This requirement eliminates storage and handling problems associated with other oxidants.

Systems that use ozone in combination with hydrogen peroxide or UV radiation are catalytic ozonation processes. They accelerate ozone decomposition, thereby increasing the hydroxyl radical concentration and promoting the oxidation rate of the compounds of interest [3, p. 10]. Specifically,

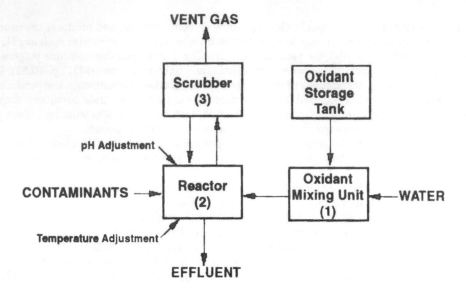

Figure 19-1. Process flow diagram for chemical oxidation system.

hydrogen peroxide, hydrogen ion, and UV radiation have been found to initiate ozone decomposition and accelerate the oxidation of refractory organics via the free radical reaction pathway [6, p. 228]. Reaction times can be 100 to 1000 times faster in the presence of UV light [11, p. 195]. Minimal emissions result from the UV-enhanced systems [15, p. 35].

PROCESS RESIDUALS

Residuals produced from chemical oxidation systems can include partially oxidized products (if the reaction does not go to completion) which may require further treatment. In some cases, inorganic salts may be formed [10]. Depending on the oxidizing agent used and the chlorine content of the contaminant, oxidation of organic compounds may result in the formation of HCl and NO_2. Ozone and hydrogen peroxide have an advantage over oxidants containing chlorine because potentially hazardous chlorinated compounds are not formed [11, p. 187].

Acid gas control is required for reactions that produce HCl. Any precipitate formed has to be filtered out and may require additional treatment to comply with the appropriate regulations [10].

SITE REQUIREMENTS

Equipment requirements for oxidation processes include storage vessels, metering equipment, and reactor vessels with some type of agitation device. UV light may also be required. All the equipment is readily available and can be skid-mounted and sent to the site.

Ozone must be generated onsite because it is not practical to store. Other oxidizing agents require onsite storage and handling. A site safety plan would have to be developed to provide for personnel protection and special handling measures. Standard 440V, three-phase electrical service may be required, depending on the reactor configuration. Water must be available onsite for cleaning and descaling operations, although the treated effluent might be used for this purpose. Water would also be needed for slurrying soils and sludges. The quantity of water needed is vendor- and site-specific.

Onsite analytical equipment may be needed to conduct pH, oil, and grease analyses. Liquid and gas chromatographs capable of determining site-specific organic compounds may be required for the operation to be more efficient and to provide better information for process control.

Table 19-3. Lorentz Barrel and Drum SITE Testing Parameters [3]

Run	pH	Time (min)	Ozone Dose (mg/L)	H_2O_2 Dose (mg/L)	UV Lamps
1	7.2	40	75	25	all on
2	6.2	40	75	25	all on
3	5.2	40	75	25	all on
4	7.2	60	75	25	all on
5	7.2	20	75	25	all on
6	7.2	40	110	25	all on
7	7.2	40	38	25	all on
8	7.2	40	110	38	all on
9	7.2	40	110	13	all on
10	7.2	40	110	13	1/2 on
11	7.2	40	110	13	1/2 on
12	7.2	40	110	13	all on
13	7.2	40	110	13	all on

PERFORMANCE DATA

Performance of full-scale chemical oxidation systems has been reported by several sources, including equipment vendors. Some of the data presented for specific contaminant removal effectiveness were obtained from publications developed by the respective chemical oxidation system vendors. The quality of this information has not been determined; however, it does give an indication of the efficiency of chemical oxidation. Data on chemical oxidation systems at Superfund sites are discussed in the following paragraphs.

Ultrox International installed its system at the Lorentz Barrel and Drum Superfund site in San Jose, California. The system uses ozone and hydrogen peroxide with UV radiation to treat contaminated ground water whose main contaminants were 1,1,1-trichloroethane (TCA), trichloroethylene (TCE), and 1,1-dichloroethane (DCA). Demonstration of this system at the Lorentz site was also part of the Superfund Innovative Technology Evaluation (SITE) program. During the SITE testing, hydraulic retention time (reaction time), ozone dose, hydrogen peroxide dose, UV radiation intensity, and pH level were varied, as shown in Table 19-3, to assess the system's performance. The results of the testing are listed in Table 19-4 [3].

The system destruction efficiency averaged more than 90 percent of the TCE in the contaminated ground water over the range of operating parameters. Destruction efficiencies for 1,1,1-TCA and 1,1-DCA increased when the ozone dosage was increased. During these runs, the destruction efficiency for 1,1,1-TCA was over 80 percent and almost 60 percent for 1,1-DCA. For a more detailed discussion, the reader should consult Reference 3.

The Ultrox® system was also used to treat contaminated ground water in Muskegon, Michigan. Before treatment, the TCE concentration was reported to be as high as 7 parts per million (ppm). The Ultrox® system has reduced effluent levels to under 2 parts per billion (ppb) [13, p. 90].

Solarchem Environmental Systems installed its Rayox™ enhanced oxidation unit at the Oswego, New York, Superfund site. This demonstration system, which uses UV radiation enhancement with ozone and hydrogen peroxide, treated collected leachate from a landfill site. Results of the testing are listed in Table 19-5 [9].

Peroxidation Systems' **perox-pure**™ Organic Destruction process uses hydrogen peroxide and UV light to destroy dissolved organic contaminants. It has been used at a number of sites to reduce contaminants up to 90 percent. The **perox-pure**™ has much lower effectiveness on aliphatic compounds, such as TCA, because they are not as reactive [15]. Table 19-6 is a partial list of contaminants treated and applications where the **perox-pure**™ process has been used [16].

Table 19-7 lists performance data for several sites using the full-scale **perox-pure**™ system [17; 18]. Most organics were reduced to extremely low levels by the **perox-pure**™ treatment systems at every site. At Site 1, the **perox-pure**™ system, followed by an air stripper, was able to destroy 4 of the 6 organics below detection limits. It also eliminated over 90 percent of the air emissions as compared to

Table 19-4. Lorentz Barrel and Drum SITE Test Results (Contaminated Ground Water) [3]

Run	1,1,1-TCA Influent* (μg/L)	Effluent* (μg/L)	% Removed	TCE Influent* (μg/L)	Effluent* (μg/L)	% Removed	1,1-DCA Influent* (μg/L)	Effluent* (μg/L)	% Removed
1	4.0	1.2	70	86.0	4.6	95	11.5	6.2	46
2	3.7	0.6	83	55.0	2.4	96	10.0	3.2	69
3	3.8	1.3	65	64.0	3.6	94	10.0	6.7	35
4	3.9	1.8	53	56.0	3.4	94	12.0	7.8	32
5	4.1	1.4	66	50.0	6.2	88	10.0	6.4	36
6	3.9	1.0	73	73.0	1.0	98	11.0	5.2	54
7	4.7	3.0	37	70.0	17.0	76	13.0	9.2	30
8	3.5	0.7	80	59.0	0.7	99	9.8	4.7	52
9	4.3	0.8	83	65.0	1.2	98	11.0	5.3	54
10	3.4	0.6	82	57.0	1.6	97	10.0	3.9	62
11	3.8	0.8	80	57.0	1.3	98	11.0	5.4	50
12	3.3	0.4	87	52.0	0.6	99	11.0	3.8	65
13	3.2	0.5	85	49.0	0.6	99	10.0	4.2	60

* Mean value

Table 19-5. Oswego Leachate Test Results [9]

Volatile Organic Compounds (VOCs)	Inlet (ppb)	Outlet (ppb)	% Removed
Methylene chloride (MeCl)	204	1	99.5
1,1-Dichloroethylene (DCE)	118	0	100
1,1-DCA	401	15.7	96
t-1,2-DCE	3690	14.9	99.6
1,2-DCA	701	109	85
1,1,1-TCE	261	3.1	98.9
Benzene	469	1.8	99.6
Methyl isobutyl ketone	47	2.2	95.8
1,1,2,2-Tetrachloroethane	344	4.2	98.8
Toluene	3620	3.9	99.9
Chlorobenzene	704	0	100
Ethylbenzene	2263	1.1	99.9
M-,P-Xylene	4635	1.3	99.9
O-Xylene	6158	2.4	99.9

the previous arrangement, which used an air stripper followed by the **perox-pure**™ system. At Site 5, the system was modified to pretreat the influent to remove iron and calcium. This resulted in no organics being detected in the effluent.

The Purus Inc. enhanced oxidation system was demonstrated on contaminated ground water at Lawrence Livermore National Laboratory (LLNL). Benzene, toluene, ethylbenzene, and xylene (BTEX) levels were reduced from 5 ppm to as little as 5 ppb [19, p. 9]. The Purus system is also being used to treat air streams from air stripping of ground water and vacuum extraction of soils under the SITE emerging technology program and LLNL.

Other case studies have shown greater than 99 percent destruction of the pesticides DDT, PCP, PCB, and Malathion with ozone/UV radiation [4, p. 7.67].

TECHNOLOGY STATUS

Chemical oxidation is a well-established technology used for disinfection of drinking water and wastewater and is a common treatment for cyanide wastes. Enhanced systems are now being used more frequently to treat hazardous streams. This technology has been applied to Resource Recovery and Conservation Act (RCRA) wastes and has been used on Superfund wastes [7]. In 1988, chemical oxidation was listed in the Record of Decision at Lorentz Barrel & Drum in San Jose, California and southern Maryland Wood, in Hollywood, MD. In 1989, chemical oxidation was listed at Sullivan's Ledge in New Bedford, Massachusetts; Bog Creek Farm in Howell Twp., New Jersey; Ott/Story/Cordova Chemical in Dalton Twp., Michigan; Burlington Northern in Somers, Montana; and Sacramento Army Depot in Sacramento, California.

Operating costs can be competitive with other treatment technologies such as air stripping and activated carbon. However, oxidation is becoming a more attractive option because the contaminants are destroyed rather than transferred to another media. Operating costs for mobile chemical oxidation systems have ranged from $70 to $150 per 1,000 gallons of water treated [8, p. 4.5]. Operating costs for the Ultrox® enhanced system have varied dramatically from $0.15 to $90/1000 gallons treated, depending on the type of contaminants, their concentration, and the desired cleanup standard. The greatest expense for this system is the cost of electricity to operate the ozone generator and UV lamps [13, p. 92].

EPA CONTACT

Technology-specific questions regarding chemical oxidation may be directed to Douglas Grosse, (513) 569-7844, or Ronald Turner, RREL-Cincinnati, (513) 569-7775 (see Preface for mailing address).

Table 19-6. Applications of perox-pure™ Systems at Selected Sites [16]

Location	Type	Contaminant
CA	Ground water	Tetrahydrofuran
CA	Leachate	Mixed organic acids
CA	Ground water	TCE
CA	Ground water	TCE, TCA, CCl₄, MeCl
MA	Dredge water	PCBs
NH	Leachate	Ketones, VOCs
MD	Ground water	TCE, Perchloroethylene (PCE), TCA, DCE
MA	Ground water	MeCl, TCA, dichloromethane (DCM)
CA	Municipal water	Humic acid/color control
CA	Ground water	TCE, PCE, TCA, DCE
WA	Ground water	Pentachlorophenol
CO	Misc. wastes	Hydrazine, DIMP
CO	Ground water	Benzene, toluene, xylene (BTX)
CT	Bioeffluent	Chlorobenzene
CA	Ground water	TCE, TCA, PCE, DCE
NY	Groundwater	TCE, DCE, PCE, TCA
CA	Ground water	TCA, TCE
NY	Ground water	TCE, DCE, DCA, TCA
PA	Effluent	Phenol
CA	Ground water	BTX
PA	Effluent	Nitrated esters
NJ	Ground water	TCE, DCE, PCE, MeCl
AZ	Ground water	BTEX
TX	Effluent	Phenols, nitrophenols
MA	Ground water	BTX
CO	Waste	Hydrazine
CA	Ground water	TCE, PCE, BTX, TCA
AR	Ground water	Acrylic acid, butyl acrylate
OH	Recycle	Bacteria, phenol, formaldehyde
LA	Ground water	TCE, polynuclear aromatic hydrocarbons (PAHs)
AZ	Ground water	TCE
UT	Effluent	Isopropyl alcohol (IPA), TOC, TCA, DCE, methyl ethyl ketone (MEK)
NJ	Effluent	Phenol
CA	Ground water	TCE, PCE, DCE, TCA, MeCl, chloroform
CA	Effluent	BTX
CA	Ground water	BTX
CA	Ground water	TCE, Freon, MeCl, BTX
NC	Effluent	MeCl, phenol, PAHs

ACKNOWLEDGMENTS

This chapter was prepared for the U.S. Environmental Protection Agency, Office of Research and Development (ORD), Risk Reduction Engineering Laboratory (RREL), Cincinnati, Ohio, by Science Applications International Corporation (SAIC) under Contract No. 68-C8-0062. Mr. Eugene Harris served as the EPA Technical Project Monitor. Mr. Gary Baker was SAIC's Work Assignment Manager. This chapter was authored by Ms. Margaret M. Groeber of SAIC. The author is grateful to Mr. Ken Dostal of EPA, RREL, who has contributed significantly by serving as a technical consultant during the development of this chapter.

The following other Agency and contractor personnel have contributed their time and comments by participating in the expert review meetings and/or peer-reviewing the chapter:

Table 19-7. Full-Scale perox-pure™ Performance Data [17; 18]

Location	Contaminant	Influent (µg/L)	Effluent (µg/L)
Site 1	MeCl	30	1.5
Source of influent			
not reported	1,1-DCA	42	BDL
	1,2-DCE	2466	BDL
	1,1,1-TCA	1606	1218
	TCE	1060	BDL
	PCE	3160	BDL
Site 2			
Concentrated Wastewater	Hydrazine	1,200,000	<1
	Monomethyl Hydrazine	100,000	<10
	Unsymmetrical Dimethyl		
	Hydrazine	1,500,000	<10
	Nitrosodimethylamine	1,500	<0.02
	Chlorinated Organics	75,000	<1
	Pesticides/Herbicides	500	<1
Site 3			
Contaminated Ground Water	1,2-DCE	6.2	BDL
	TCE	66.3	BDL
	Chloroform	2.1	BDL
Site 4			
Source of influent			
not reported	MeCl	600–800	33
	1,1,1-TCA	200–400	26
	1,1-DCE	50–250	<1
Site 5			
Contaminated Ground Water	Benzene	7,600	ND[a]
	Toluene	24,000	ND[a]
	Chlorobenzene	8,800	ND[a]
	Ethylbenzene	3,300	ND[a]
	Xylenes	46,000	ND[a]
Site 6			
Contaminated Ground Water	MeCl	903	11
	1,1,1-TCA	60	6

Detection Limits not Reported
BDL = Below Detection Limit
ND = Nondetected
[a] With Pretreatment

Mr. Clyde Dial, SAIC
Mr. James Rawe, SAIC
Dr. Thomas Tiernan, Wright State University
Dr. Robert C. Wingfield, Jr., Fisk University
Ms. Tish Zimmerman, EPA-OERR

REFERENCES

1. Ku, Y. and S-C. Ho. The Effects of Oxidants on UV Destruction of Chlorophenols. Environmental Progress 9(4):218, 1990.

2. Kearny, P.C., et al. UV-Ozonation of Eleven Major Pesticides as a Waste Disposal Pretreatment. Chemosphere. 16(10–12):2321–2330, 1987.

3. U.S. Environmental Protection Agency. Technology Evaluation Report: SITE Program Demonstration of the Ultrox® International Ultraviolet Radiation/Oxidation Technology. EPA 540/5-89/012 (NTIS

PB90-198177), January 1990. [See also Applications Analysis Report (EPA/540/A5-89/012), Technology Demonstration Summary (EPA/540/S5-89/012), and Demonstration Bulletin (EPA/540/M5-89/012)—see Appendix A for information on how to obtain SITE reports.]

4.   Novak, F.C. Ozonation. In: Standard Handbook of Hazardous Waste Treatment and Disposal. Harry M. Freeman, ed. McGraw-Hill, New York, New York, 1989.

5.   Fochtman, E.G. Chemical Oxidation and Reduction. In: Standard Handbook of Hazardous Waste Treatment and Disposal. Harry Freeman, ed., McGraw-Hill, New York, NY, 1989.

6.   Glaze, W.H. Drinking-Water Treatment with Ozone. Environmental Science and Technology. 21(3):224–230, 1987.

7.   Technology Screening Guide for Treatment of CERCLA Soils and Sludges. EPA/540/2-88/004, U.S. Environmental Protection Agency, Washington, DC, 1989.

8.   Mobile Treatment Technologies for Superfund Wastes. EPA 540/2-86/003(f), U.S. Environmental Protection Agency, Washington, DC, 1986.

9.   Marketing Brochure for Rayox®. Leachate Remediation at the Oswego Superfund Site using Rayox®—A Second Generation Enhanced Oxidation Process. Solarchem Environmental Systems, Inc., Richmond Hill, Ontario.

10.   Corrective Action: Technologies and Application. Seminar Publication EPA/625/4-89/020, September 1989.

11.   Systems to Accelerate In Situ Stabilization of Waste Deposits. EPA/540/2-86/002 (NTIS PB87-112306), 1986.

12.   Handbook Remedial Action at Waste Disposal Sites (Revised). EPA/625/6-85/006 (NTIS PB87-201034), 1985.

13.   Roy, K. Researchers Use UV Light for VOC Destruction. Hazmat World, May: 82-92, 1990.

14.   A Compendium of Techniques Used in the Treatment of Hazardous Wastes. EPA/625/8-87/014 (NTIS PB90-274893), September 1987.

15.   Roy, K. UV-Oxidation Technology Shining Star or Flash in the Pan?, Hazmat World, June: 35-50, 1990.

16.   Marketing brochure for perox-pure™ organic destruction process. Peroxidation System Inc., Tucson, Arizona, September 1990. [See SITE program reports listed below.]

17.   Froelich, E. The perox-pure™ Oxidation System—A Comparative Summary. Presented at the American Institute of Chemical Engineers. 1990 Summer National Meeting, San Diego, CA, August 19–22, 1990.

18.   Froelich, E. Advanced Chemical Oxidation of Contaminated Using perox-pure™ Oxidation System. Presented at Chemical Oxidation: Technology for the 1990's. Vanderbilt University, February 20–22, 1991.

19.   New UV Lamp Said to Achieve Photolysis of Organics, HazTECH News. 6(2):9, 1991.

ADDITIONAL SITE PROGRAM REFERENCES (See Appendix A for information on how to obtain SITE reports]

perox-pure™ Chemical Oxidation Technology:

Demonstration Bulletin (EPA/540/MR-93/501)
Technology Demonstration Summary (EPA/540/SR-93/501)
Applications Analysis (EPA/540/AR-93/501)
Technology Evaluation (EPA/540/R-93/501; NITS PB93-213528)

CWM PO*WW*ER™ Evaporation-Catalytic Oxidation Technology:

Demonstration Bulletin (EPA/540/MR-93/506)
Technology Demonstration Summary (EPA/540/SR-93/506)
Applications Analysis (EPA/540/AR-93/506)
Technology Evaluation Vol. I (EPA/540/R-93/506A; NTIS PB94-160637)
Technology Evaluation Vol. II (EPA/540/R-93/506B; NTIS PB94-160660)

CAV-OX® Ultraviolet Oxidation Process Magnum Water Technology:

Demonstration Bulletin (EPA/540/MR-93/520)
Technology Demonstration Summary (EPA/540/SR-93/520)
Applications Analysis (EPA/540/AR-93/520)

Chapter 20

Chemical Dehalogenation Treatment: APEG Treatment[1]

U.S. Environmental Protection Agency, Risk Reduction Engineering Laboratory, Cincinnati, OH

ABSTRACT

The chemical dehalogenation system discussed in this report is alkaline metal hydroxide/polyethylene glycol (APEG), which is applicable to aromatic halogenated compounds. The metal hydroxide that has been most widely used for this reagent preparation is potassium hydroxide (KOH) in conjunction with polyethylene glycol (PEG) [6, p. 461][2] (typically, average molecular weight of 400 Daltons) to form a polymeric alkoxide referred to as KPEG [16, p. 835]. However, sodium hydroxide has also been used in the past and most likely will find increasing use in the future because of patent applications that have been filed for modification to this technology. This new approach will expand the technology's applicability and efficacy and should reduce chemical costs by facilitating the use of less costly sodium hydroxide [18]. A variation of this reagent is the use of potassium hydroxide or sodium hydroxide/tetraethylene glycol, referred to as ATEG, that is more effective on halogenated aliphatic compounds [21]. In some KPEG reagent formulations, dimethyl sulfoxide (DMSO) is added to enhance reaction rate kinetics, presumably by improving rates of extraction of the haloaromatic contaminants [19; 22].

Previously developed dehalogenation reagents involved dispersion of metallic sodium in oil or the use of highly reactive organosodium compounds. The reactivity of metallic sodium and these other reagents with water presented a serious limitation to treating many waste matrices; therefore, these other reagents are not discussed in this chapter and are not considered APEG processes [1, p. 1].

The reagent (APEG) dehalogenates the pollutant to form a glycol ether and/or a hydroxylated compound and an alkali metal salt, which are water soluble by-products. This treatment process chemically converts toxic materials to nontoxic materials. It is applicable to contaminants in soil [11, p. 1], sludges, sediments, and oils [2, p. 183]. It is mainly used to treat halogenated contaminants including polychlorinated biphenyls (PCBs) [4, p. 137], polychlorinated dibenzo-p-dioxins (PCDDs) [11, p. 1], polychlorinated dibenzofurans (PCDFs), polychlorinated terphenyls (PCTPs), and some halogenated pesticides [8, p. 3; 14, p. 2]. This technology has been selected as a component of the remedy for three Superfund sites. Vendors should be contacted to determine the availability of a treatment system for use at a particular site. The estimated costs of treating soils range from $200 to $500/ton. This chapter provides information on the technology applicability, the types of residuals resulting from the use of the technology, the latest performance data, site requirements, the status of the technology, and where to go for further information.

TECHNOLOGY APPLICABILITY

This technology is primarily for treating and destroying halogenated aromatic contaminants. The matrix can be soils, sludges, sediments, or oils. If a waste site has contaminants other than halogenated compounds, other alternatives should be considered.

The concentrations of PCBs that have been treated are reported to be as high as 45,000 ppm. Concentrations were reduced to less than 2 parts per million per individual PCB congener. Polychlo-

---

[1] EPA/540/2-90/015.
[2] [Reference number, page number.]

Table 20-1. Effectiveness of APEG Treatment on General Contaminant Groups for Various Matrices

Contaminant Groups	Effectiveness			
	Sediments	Oils	Soil	Sludge
Organic				
Halogenated volatiles	▼	▼	▼	▼
Halogenated semivolatiles	▼	▼	▼	▼
Nonhalogenated volatiles	●	●	●	●
Nonhalogenated semivolatiles	●	●	●	●
PCBs	■	■	■	■
Pesticides (halogenated)	▼	■	■	▼
Dioxins/furans	■	■	■	■
Organic cyanides	●	●	●	●
Organic corrosives	●	●	●	●
Inorganic				
Volatile metals	●	●	●	●
Nonvolatile metals	●	●	●	●
Asbestos	●	●	●	●
Radioactive materials	●	●	●	●
Inorganic corrosives	●	●	●	●
Inorganic cyanides	●	●	●	●
Reactive				
Oxidizers	●	●	●	●
Reducers	●	●	●	●

■ Demonstrated Effectiveness: Successful treatability test at some scale completed.
▼ Potential Effectiveness: Expert opinion that technology will work.
● No Expected Effectiveness: Expert opinion that technology will not work.

rinated dibenzo-p-dioxins (PCDDs) and polychlorinated dibenzofurans (PCDFs) have been treated to nondetectable levels at part per trillion sensitivity. The process has successfully destroyed PCDDs and PCDFs contained in contaminated pentachlorophenol oil. For a contaminated activated carbon matrix, direct treatment was less effective, and the reduction of PCDDs/PCDFs to concentrations less than 1 ppb was better achieved by first extracting the carbon matrix with a solvent and then treating the extract [15, p. 1].

All field applications of this technology to date have been in various matrices and not on specific Resource Conservation and Recovery Act (RCRA) listed wastes. The effectiveness of APEG on general contaminant groups for various matrices is shown in Table 20-1. Examples of constituents within contaminant groups are provided in Reference 23, Technology Screening Guide for Treatment of CERCLA Soils and Sludges. This table is based on the currently available information, or professional judgment when no information was available. The proven effectiveness of the technology for a particular site or waste does not ensure that it will be effective at all sites or that the treatment efficiency achieved will be acceptable at other sites. For the ratings used for this table, demonstrated effectiveness means that, at some scale, treatability was tested to show that, for that particular contaminant and matrix, the technology was effective. The ratings of potential effectiveness and no expected effectiveness are based upon expert judgment. Where potential effectiveness is indicated, the technology is believed capable of successfully treating the contaminant group in a particular matrix. When the technology is not applicable or will probably not work for a particular combination of contaminant group and matrix, a no-expected-effectiveness rating is given.

LIMITATIONS

The APEG technology is not intended as an in situ treatment. APEG will dehalogenate aliphatic compounds. If the mixture is reacted longer and at significantly higher temperatures than for aromatic compounds, it is recommended that a related reagent KTEG be considered for these contaminants. KTEG has been shown at laboratory scale to be effective on halogenated aliphatic compounds such as ethylene dibromide, carbon tetrachloride, ethylene dichloride, chloroform, and dichloromethane (methylene chloride) [18, p. 2]. The necessary treatment time and temperature for KTEG use can be determined from laboratory tests.

Treatability tests should be conducted prior to the final selection of the APEG technology to identify optimum operating factors such as quantity of reagent, temperature, and treatment time. These tests can be used to identify such things as water content, alkaline metals and high humus content in the soils, glycol extractables content, presence of multiple phases, and total organic halides that have the potential to affect processing times and costs [19].

The treated soil may contain enough residual reagent and treatment by-products that their removal could be required before final disposal. If necessary, such by-products are usually removed by washing the soil two or three times with water. The soil will have to be neutralized by lowering the pH prior to final disposal.

Specific safety aspects for the operation must be considered. Treatment of certain chlorinated aliphatics in high concentrations with APEG may produce compounds that are potentially explosive (e.g., chloroacetylenes) and/or cause a fire hazard. The use of DMSO or similar reagents may lead to formation of highly flammable volatile organics (e.g., methyl sulfide) [18, §IV C]. Severe corrosivity can be a concern when DSMO is teamed with other APEG reagents. Alkaline reactive materials such as metallic aluminum will compete with the contaminants for the reagent and may produce hydrogen gas (explosive). Vapors from heating oily soils, which are often the matrix in which PCBs are found, can also create such potential problems as fires and noxious fumes. These problems can often be solved by taking appropriate corrective actions during elevated temperature processing.

The operation must also be conducted with care because of the elevated temperatures and production of steam, the use of caustics in the process, and the presence of acids that are used for neutralization. If DMSO is used, care must be taken to prevent its coming into contact with skin, for it enhances transport of PCBs through the skin, thus increasing the risk of exposure.

TECHNOLOGY DESCRIPTION

Figure 20-1 is a schematic of the APEG treatment process.

Waste preparation includes excavation and/or moving the soil to the process where it is normally screened (1) to remove debris and large objects and to produce particles that are sufficiently small to allow treatment in the reactor without binding the mixer blades.

Typically, the reagent components are mixed with the contaminated soil in the reactor (2). The material must be well mixed with the reagent to allow effective treatment. Treatment proceeds inefficiently without mixing. This mixture is heated to between 100 and 180°C. The reaction proceeds for 1 to 5 hours, depending upon the type, quantity, and concentration of the contaminants. The treated material goes from the reactor to a separator (3) where the reagent is removed and can be recycled (4).

During the reaction, water is vaporized in the reactor, condensed (5) and collected for further treatment or recycled through the washing process, if required. Carbon filters (7) are used to trap any volatile organics that are not condensed. In the washer (6), the soil is neutralized by the additions of acid. It is then dewatered (8) before disposal.

PROCESS RESIDUALS

There are three main waste streams generated by this technology: the treated soil, the wash water, and possible air emissions. The treated soil will need to be analyzed to determine if it meets the regulatory requirements for the site before final disposition can be made. The soil's pH must be adjusted before disposal. The chemistry of this technology is specific to halogenated organics and, based upon a test conducted by the EPA in 1985, results in by-product compounds that appear to be neither toxic nor of concern. In that test the EPA checked for (1) mutagenicity, (2) toxicity, and (3) bioaccumulation/bioconcentration of the by-products of two different contaminants: tetrachlorobenzene and 2,3,7,8-TCDD that had been treated by the process [3, p. 80]. The individual by-product chemical compounds were not determined. These compounds and the residual levels of reagent or catalyst did not present a serious health or environmental problem [12, p. 2].

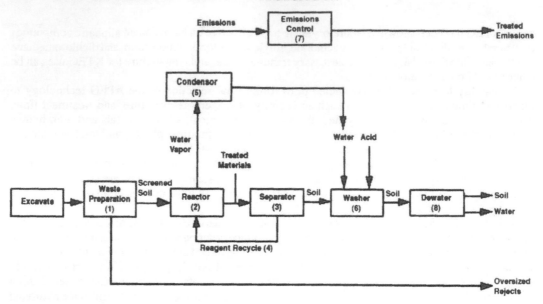

Figure 20-1. APEG treatment process.

Waste wash water contains only trace amounts of contaminants and reagents and would be expected to meet appropriate discharge standards, enabling it to be discharged to a local, publicly owned treatment works or receiving stream. Volatile air emissions can be released due to the heating and mixing that occurs with the process. They are usually captured by condensation and/or on activated carbon. The contaminated carbon is usually incinerated.

SITE REQUIREMENTS

APEG treatment units are transported by trailers [13, p. 54]. Therefore, adequate access roads are required to get the unit to the site. The system that operated in Guam, which used a 1.5-ton batch reactor, required an area of 100 feet by 100 feet.

Energy requirements involve heating the reactor and removing the water by volatilization. For the reactor used in Guam, a standard 440V, three-phase electrical service was required along with a diesel steam-generating plant rated at 600 lb/h and 80 psi [13, p. 53]. A standard municipal water supply, or equivalent, is adequate for this process.

Contaminated soils or other waste materials are hazardous and their handling requires that a site safety plan be developed to provide for personnel protection and special handling measures.

A means of containing and cleaning up accidental spills must be provided. The reagents (KOH, acids, etc.) should be stored in drums with containment beneath and provisions to pump any spills to a holding area for neutralization [19, p. 2].

The process residuals normally must be stored until their level of contaminants are verified to be below those established for the site. Depending upon the site, a method to store waste may be necessary. Storage capacity will depend on waste volume.

Onsite analytical capabilities are highly desirable. Extraction equipment and gas chromatography/mass spectrometer capabilities should be available to measure contaminants of interest and to provide information for process control.

PERFORMANCE DATA

This technology's performance has been evaluated from bench-scale tests to field tests in large reactors. Table 20-2 summarizes the results of several more important applications of the technology and their results.

Table 20-2. APEG Field Performance Data

Site/Date	Contaminant/ Waste Form	Concentration Before	Concentration After	Volume Treated
Signo Trading, NY/1982	dioxin/liquid	135 ppb	<1 ppb	15 gallons
Montana Pole Butte, MT/1986 (16, p. 838) [5, p. 1]	dioxin furans/oil	147–83,923 ppb	<1 ppb	10,000 gallons
Western Processing Kent, WA/1986 [16, p. 838]	dioxin/liquid and sludge	120 ppb	<0.3 ppb	7,550 gallons
Wide Beach Erie County, NY/1985	PCBs (Aroclor 1254)/soil	120 ppm	<2 ppm	1 ton
Guam U.S.A./1988 793 gal. reactor [13, p. 43]	PCBs/soil	2500[a] ppm with hot spots as high as 45,860 ppm	<1[a,b] ppm	22 tons soil 3.4 tons crushed rock
Bengart & Memel Buffalo, NY/1986 55 gal. drum [10, p. 13]	PCBs/soil	51 out of 52 drums, 108 ppm	<27 ppm	52 fifty-five gallon drums
Economy Products Omaha, NE/1987	TCDD, 2, 4-D,2,4,5-T/liquid	1.3 ppm 17,800 ppm 2,800 ppm	ND 334 ppm 55 ppm	20 gallons

[a] value is an average value.
[b] per resolvable PCB cogener.

RCRA Land Disposal Restrictions (LDRs) that require treatment of wastes to best demonstrated available technology (BDAT) levels prior to land disposal may sometimes be determined to be applicable or relevant and appropriate requirements (ARARs) for CERCLA response actions. The APEG treatment technology can produce a treated waste that meets treatment levels set by BDAT, but may not reach these treatment levels in all cases. The ability to meet required treatment levels is dependent upon the specific waste constituents and the waste matrix. In cases where APEG treatment does not meet these levels it still may, in certain situations, be selected for use at the site if a treatability variance establishing alternative treatment levels is obtained. The EPA has made the treatability variance process available in order to ensure that LDRs do not unnecessarily restrict the use of alternative and innovative treatment technologies. Treatability variances may be justified for handling complex soil and debris matrices. The following guides describe when and how to seek a treatability variance for soil and debris: Superfund LDR Guide #6A, Obtaining a Soil and Debris Treatability Variance for Remedial Actions, (OSWER Directive 9347.3-06FS) [20]; and Superfund LDR Guide #6B, Obtaining a Soil and Debris Treatability Variance for Removal Actions (OSWER Directive 9347.3-07FS) [17]. Another approach could be to use other treatment techniques in series with APEG treatment to obtain desired treatment levels.

TECHNOLOGY STATUS

The APEG process has been selected for cleanup of PCB contaminated soils at three Superfund sites: Wide Beach, New York (September 1985), Re-Solve, Massachusetts (September 1987), and Sol Lynn, Texas (March 1988).

This technology has received approval from the EPA's Office of Toxic Substance under the Toxic Substances Control Act for PCB treatment.

Significant advances are currently being made to the APEG technology. These advances employ water rather than costly PEG to wet the soil and require shorter reaction times and less energy. These advances should greatly enhance the economics of the process. Performance information on this modified process is not available at this time for inclusion in this chapter [18].

This technology uses standard equipment. The reaction vessel must be equipped to mix and heat the soil and reagents. A detailed engineering design for a continuous feed, full-scale system for use in Guam is currently being completed. It is estimated that a full-scale system can be fabricated and placed in operation in 6 to 12 months. Costs to use APEG treatment are expected to be in a range of $200–$500/ton.

EPA CONTACT

Technology-specific questions regarding APEG technology may be directed to Terry Lyons, (513) 569-7589, or Dennis Timberlake, RREL-Cincinnati, (513) 569-7547 (see Preface for mailing address).

REFERENCES

1. Adams, G.P. and R.L. Peterson. Non-Sodium Process for Removal of PCBs from Contaminated Transformer Oil, Presented at the APCA National Meeting in Minneapolis, 1986.
2. Brunelle, D.J. and D. Singleton. Destruction/Removal of Polychlorinated Biphenyls From Non-Polar— Media Reaction of PCB with Poly (Ethylene Glycol)/KOH. Chemosphere, 12:183–196, 1983.
3. Carpenter, B.H. PCB Sediment Decontamination Processes—Selection for Test and Evaluation, Research Triangle Institute, 1987.
4. Carpenter, B.H. and D.L. Wilson. Technical/Economic Assessment of Selected PCB Decontamination Processes. Journal of Hazardous Materials, 17:125–148, 1988.
5. des Rosiers, P.E. APEG Treatment of Dioxin- And Furan-Contaminated Oil at an Inactive Wood Treating Site in Butte, Montana, Presented at the Annual Meeting of the American Wood Preserves Institute, Washington, DC, 1986.
6. Kornel, A., C.J. Rogers, and H. Sparks. KPEG Application from the Laboratory to Guam. In: Proceedings of the Third International Conference on New Frontiers for Hazardous Waste Management. EPA/ 600/9-89/072, 1989.
7. Lauch, R., et al. Evaluation of Treatment Technologies for Contaminated Soil and Debris. In: Proceedings of the Third International Conference on New Frontiers for Hazardous Waste Management. EPA/600/ 9-89/072, 1989.
8. Locke, B. et al. Evaluation of Alternative Treatment Technologies for CERCLA Soils and Debris (Summary of Phase I and Phase II). EPA Contract No. 68-03-3389, U.S. Environmental Protection Agency, Risk Reduction Engineering Laboratory, Cincinnati, Ohio (no date).
9. NATO/CCMS. Demonstration of Remedial Action Technologies for Contaminated Land and Groundwater. In: Proceedings of the NATO/CCMS Second International Workshop, Hamburg, Federal Republic of Germany, 1988. pp. 97–99.
10. Novosad, C.F., E. Milicic, and R. Peterson. Decontamination of a Small PCB Soil Site by the Galson APEG Process, Presented before the Division of Environmental Chemistry, American Chemical Society, New Orleans, 1987.
11. Peterson, R.L, M. Edwins, and C. Rogers. Chemical Destruction/Detoxification of Chlorinated Dioxins in Soils. In: Proceedings of the Eleventh Annual Research Symposium, Incineration and Treatment of Hazardous Wastes. EPA/600/9-85/028, 1985.
12. Peterson, R.L., et al. Comparison of Laboratory and Field Test Data in the Chemical Decontamination of Dioxin Contaminated Soils, Presented at the ACS Meeting in New York, NY, 1986.
13. Taylor, M.L., et al. (PEI Associates). Comprehensive Report on the KPEG Process for Treating Chlorinated Wastes. EPA Contract No. 68-033413, U.S. Environmental Protection Agency, Risk Reduction Engineering Laboratory, Cincinnati, Ohio, 1989.
14. Tiernan, T.O., et al. Dechlorination of Organic Compounds Contained in Hazardous Wastes Using the KPEG Reagent. In: Proceedings of the Symposium on Hazardous Waste Treatment, American Chemical Society Symposium Series, 1990.
15. Tiernan, T.O., et al. Dechlorination of PCDD and PCDF Sorbed on Activated Carbon Using the KPEG Reagent. Chemosphere, 19(16):573–578, 1989.

16. Tiernan, T.O., et al. Laboratory and Field Test to Determine the Efficacy of KPEG Reagent for Detoxification of Hazardous Wastes Containing Polychlorinated Dibenzo-p-Dioxins (PCDD) and Dibenzo-furans (PCDF) and Soils Contaminated with Such Chemical Wastes. Chemosphere, 18(1-16):835–841, 1989.

17. Superfund LDR Guide #6B: Obtaining a Soil and Debris Treatability Variance for Removal Actions. OSWER Directive 9347.3-07FS (NTIS PB91-921310), 1990.

18. U.S. Patent Number 4,675,464, Chemical Destruction of Halogenated Aliphatic Hydrocarbons, June 23, 1987, with Employee Report of Invention (Charles Rogers and Alfred Kornel), Case No. WQO-485-89(E), U.S. Environmental Protection Agency, 1988.

19. Innovative Technology: Glycolate Dechlorination. OSWER Directive 9200.5-254FS (PB90-274226), 1989.

20. Superfund LDR Guide #6A: Obtaining a Soil and Debris Treatability Variance for Remedial Actions. OSWER Directive 9347.3-06FS (NTIS PB91-921327), 1990.

21. Catalytic Dehydrohalogenation: A Chemical Destruction Method for Halogenated Organics, Project Summary. EPA/600/S2-86/113, 1987.

22. Peterson, R.L., U.S. Patent Nos. 4,474,013 (3/4/86), 4,532,028 (7/30/85), and 4,447,541 (5/8/84).

23. Technology Screening, Guide for Treatment of CERCLA Soils and Sludges. EPA/540/2-88/004 (NTIS PB89-132674), 1989.

OTHER REFERENCES

Brunelle, D.J. and D.A. Singleton. Chemical Reaction of Polychlorinated Biphenyls on Soils with Poly(Ethyiene Glycol)/KOH. Chemosphere, 14(2):173–181, 1985.

Kornel, A. and C. Rogers. PCB Destruction: A Novel Dehalogenation Reagent. Journal of Hazardous Materials, 12:161–176, 1985.

Arienti, M., L. Wilk, M. Jasinski, and N. Prominski. Dioxin-Containing Wastes Treatment Technologies. Noyes Data Corporation, Park Ridge, New Jersey, 1988. pp. 156–168.

Chapter 21

Slurry Biodegradation[1]

U.S. Environmental Protection Agency, Risk Reduction Engineering Laboratory, Cincinnati, OH

ABSTRACT

In a slurry biodegradation system, an aqueous slurry is created by combining soil or sludge with water. This slurry is then biodegraded aerobically using a self-contained reactor or in a lined lagoon. Thus, slurry biodegradation can be compared to an activated sludge process or an aerated lagoon, depending on the case.

Slurry biodegradation is one of the biodegradation methods for treating high concentrations (up to 250,000 mg/kg) of soluble organic contaminants in soils and sludges. There are two main objectives for using this technology: to destroy the organic contaminant and, equally important, to reduce the volume of contaminated material. Slurry biodegradation is not effective in treating inorganics, including heavy metals. This technology is in developmental stages but appears to be a promising technology for cost-effective treatment of hazardous waste.

Slurry biodegradation can be the sole treatment technology in a complete cleanup system, or it can be used in conjunction with other biological, chemical, and physical treatment. This technology was selected as a component of the remedy for polychlorinated biphenyl (PCB)-contaminated oils at the General Motors Superfund site at Massena, New York, [11, p. 2][2] but has not been a preferred alternative in any record of decision [6, p. 6]. It may be demonstrated in the Superfund Innovative Technology Evaluation (SITE) program. Commercial-scale units are in operation. Vendors should be contacted to determine the availability of a unit for a particular site. This chapter provides information on the technology applicability, the types of residuals produced, the latest performance data, site requirements, the status of the technology, and sources for further information.

TECHNOLOGY APPLICABILITY

Biodegradation is a process that is considered to have enormous potential to reduce hazardous contaminants in a cost-effective manner. Biodegradation is not a feasible treatment method for all sites. Each vendor's process may be capable of treating only some contaminants. Treatability tests to determine the biodegradability of the contaminants and the solids/liquid separation that occurs at the end of the process are very important.

Slurry biodegradation has been shown to be effective in treating highly contaminated soils and sludges that have contaminant concentrations ranging from 2,500 mg/kg to 250,000 mg/kg. It has the potential to treat a wide range of organic contaminants such as pesticides, fuels, creosote, pentachlorophenol (PCP), PCBs, and some halogenated volatile organics. It is expected to treat coal tars, refinery wastes, hydrocarbons, wood-preserving wastes, and organic and chlorinated organic sludges. The presence of heavy metals and chlorides may inhibit the microbial metabolism and require pretreatment. Listed Resource Conservation and Recovery Act (RCRA) wastes it has treated are shown in Table 21-1 [10, p. 106].

The effectiveness of this slurry biodegradation on general contaminant groups for various matrices is shown in Table 21-2 [12, p. 13]. Examples of constituents within contaminant groups are provided in Reference 12, Technology Screening Guide for Treatment of CERCLA Soils and Sludges. This table is based on current available information or professional judgment when no informa-

---

[1] EPA/540/2-90/016.
[2] [Reference number, page number.]

Table 21-1. RCRA-Listed Hazardous Wastes Treatable Using Slurry Biodegradation

Wood Treating Wastes	K001
Dissolved Air Floatation (DAF) Float	K048
Slop Oil Emulsion Solids	K049
American Petroleum Institute (API) Separator Sludge	K051

Table 21-2. Degradability Using Slurry Biodegradation Treatment on General Contaminant Groups for Soils, Sediments, and Sludges

Contaminant Groups	Biodegradability All Matrices
Organic	
Halogenated volatiles	▼
Halogenated semivolatiles	■
Nonhalogenated volatiles	▼
Nonhalogenated semivolatiles	■
PCBs	▼
Pesticides	■
Dioxins/furans	●
Organic cyanides	▼
Organic corrosives	●
Inorganic	
Volatile metals	●
Nonvolatile metals	●
Asbestos	●
Radioactive materials	●
Inorganic corrosives	●
Inorganic cyanides	▼
Reactive	
Oxidizers	●
Reducers	●

■ Demonstrated Effectiveness: Successful treatability tests at some scale completed.
▼ Potential Effectiveness: Expert opinion that technology will work.
● No Expected Effectiveness: Expert opinion that technology will not work.

tion was available. The proven effectiveness of the technology for a particular site or waste does not ensure that it will be effective at all sites or that the treatment efficiency achieved will be acceptable at other sites. For the ratings used for this table, demonstrated biodegradability means that, at some scale, treatability was tested to show that, for that particular contaminant and matrix, the technology was effective. The ratings of potential biodegradability and no expected biodegradability are based upon expert judgment. Where potential biodegradability is indicated, the technology is believed capable of successfully treating the contaminant group. When the technology is not applicable or will probably not work for a particular contaminant group, a no-expected biodegradability rating is given. Another source of general observations and average removal efficiencies for different treatability groups is contained in the Superfund LDR Guide #6A, Obtaining a Soil and Debris Treatability Variance for Remedial Actions, (OSWER Directive 9347.3-06FS [10], and Superfund LDR Guide #6B, Obtaining a Soil and Debris Treatability Variance for Removal Actions, (OSWER Directive 9347.3-07FS [9].

LIMITATIONS

The various characteristics limiting the process feasibility, the possible reasons for these, and actions to minimize impacts of these limitations are listed in Table 21-3 [11, p. 2]. Some of these

Table 21-3. Characteristics Limiting the Slurry Biodegradation Process

Characteristics Limiting the Process Feasibility	Reasons for Potential Impact	Actions to Minimize Impacts
Variable waste composition	Inconsistent biodegradation caused by variation in biological activity	Dilution of waste stream. Increase mixing
Nonuniform particle size	Minimize the contact with microorganisms	Physical separation
Water solubility	Contaminants with low solubility are harder to biodegrade	Addition of surfactants or other emulsifiers
Biodegradability	Low rate of destruction inhibits process	Addition of microbial culture capable of degrading particularly difficult compounds or longer residence time
Temperature outside 15–35°C range	Less microbial activity outside this range	Temperature monitoring and adjustments
Nutrient deficiency	Lack of adequate nutrients for biological activity	Nutrient monitoring; adjustment of the carbon/ nitrogen/phosphorus ratio
Oxygen deficiency	Lack of oxygen is rate limiting	Oxygen monitoring and adjustments
Insufficient mixing	Inadequate microbes/solids/ organics contact	Optimize mixing characteristics
pH outside 4.5–8.8 range	Inhibition of biological activity	Sludge pH monitoring. Addition of acidic or alkaline compounds
Microbial population	Insufficient population results in low biodegradation rates	Culture test, addition of culture strains
Water and air emissions discharges	Potential environmental and/or health impacts	Post-treatment processes (e.g., air scrubbing, carbon filtration)
Presence of elevated, dissolved levels of: • Heavy metals • Highly chlorinated organics • Some pesticides, herbicides • Inorganic salts	Can be highly toxic to microorganisms	Pretreatment processes to reduce the concentration of toxic compounds in the constituents in the reactor to nontoxic range

actions could be a part of the pretreatment process. The variation of these characteristics in a particular hardware design, operation, and/or configuration for a specific site will largely determine the viability of the technology and cost-effectiveness of the process as a whole.

TECHNOLOGY DESCRIPTION

Figure 21-1 is a schematic of a slurry biodegradation process.

Waste preparation (1) includes excavation and/or moving the waste material to the process where it is normally screened to remove debris and large objects. Particle size reduction, water addition, and pH and temperature adjustment are other important waste preparation steps that may be required to achieve the optimum inlet feed characteristics for maximum contaminant reduction. The desired inlet feed characteristics [6, p. 14] are:

Organics: 0.025–25% by weight   Solids particle size: less than 1/4"
Solids: 10–40% by weight   Temperature: 15–35°C
Water: 60–90% by weight   pH: 4.5–8.8

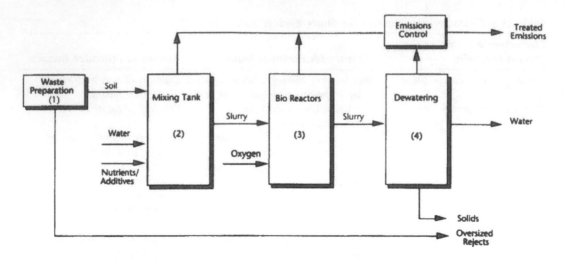

Figure 21-1. Slurry biodegradation process.

After appropriate pretreatment, the wastes are suspended in a slurry form and mixed in a tank (2) to maximize the mass transfer rates and contact between contaminants and microorganisms capable of degrading those contaminants. Aerobic treatment in batch mode has been the most common mode of operation. This process can be performed in contained reactors (3) or in lined lagoons [7, p. 9]. In the latter case, synthetic liners have to be placed in existing unlined lagoons, complicating the operation and maintenance of the system. In this case, excavation of a new lagoon or above-ground tank reactors should be considered. Aeration is provided by floating or submerged aerators or by compressors and spargers. Mixing is provided by aeration alone or by aeration and mechanical mixing. Nutrients and neutralizing agents are supplied to relieve any chemical limitations to microbial activity. Other materials, such as surfactants, dispersants, and compounds supporting growth and inducing degradation of contaminant compounds, can be used to improve the materials' handling characteristics or increase substrate availability for degradation [8, p. 5]. Microorganisms may be added initially to seed the bioreactor or added continuously to maintain the correct concentration of biomass. The residence time in the bioreactor varies with the soil or sludge matrix; physical/chemical nature of the contaminant, including concentration; and the biodegradability of the contaminants. Once biodegradation of the contaminants is completed, the treated slurry is sent to a separation/dewatering system (4). A clarifier for gravity separation, or any standard dewatering equipment, can be used to separate the solid phase and the aqueous phase of the slurry.

PROCESS RESIDUALS

There are three main waste streams generated in the slurry biodegradation system: the treated solids (sludge or soil), the process water, and possible air emissions. The solids are dewatered and may be further treated if they still contain organic contaminants. If the solids are contaminated with inorganics and/or heavy metals, they can be stabilized before disposal. The process water can be treated in an onsite treatment system prior to discharge, or some of it (as high as 90 percent by weight of solids) is usually recycled to the front end of the system for slurrying. Air emissions are possible during operation of the system (e.g., benzene, toluene, xylene [BTX] compounds); hence, depending on the waste characteristics, air pollution control, such as activated carbon, may be necessary [4, p. 29].

SITE REQUIREMENTS

Slurry biodegradation tank reactors are generally transported by trailer. Therefore, adequate access roads are required to get the unit to the site. Commercial units require a setup area of 0.5 to 1 acre per million gallons of reactor volume.

Standard 440V three-phase electrical service is required. Compressed air must be available. Water needs at the site can be high if the waste matrix must be made into slurry form. Contaminated soils or other waste materials are hazardous and their handling requires that a site safety plan be developed to provide for personnel protection and special handling measures.

Climate can influence site requirements by necessitating covers over tanks to protect against heavy rainfall or cold for long residence times.

Large quantities of wastewater that results from dewatering the slurried soil or that is released from a sludge may need to be stored prior to discharge to allow time for analytical tests to verify that the standard for the site has been met. A place to discharge this wastewater must be available.

Onsite analytical equipment for conducting dissolved oxygen, ammonia, phosphorus, pH, and microbial activity are needed for process control. High-performance liquid chromatographic and/or gas chromatographic equipment is desirable for monitoring organic biodegradation.

PERFORMANCE DATA

Performance results on slurry biodegradation systems are provided based on the information supplied by various vendors. The quality assurance for these results has not been evaluated. In most of the performances, the cleanup criteria were based on the requirements of the client; therefore, the data do not necessarily reflect the maximum degree of treatment possible.

Remediation Technologies, Inc.'s (ReTeC) full-scale slurry biodegradation system (using a lined lagoon) was used to treat wood preserving sludges (K0001) at a site in Sweetwater, Tennessee, and met the closure criteria for treatment of these sludges. The system achieved greater than 99 percent removal efficiency for most compounds and over 99 percent reduction in volume attained for PCP and polynuclear aromatic hydrocarbons (PAHs) (Table 21-4 and Table 21-5).

Data for one of these pilot-scale field demonstrations, which treated 72,000 gallons of oil refinery sludges, are shown in Figure 21-2 [8, p. 24]. In this study, the degradation of PAHs was relatively rapid and varied, depending on the nature of the waste and loading rate. The losses of carcinogenic PAHs (principally the 5- and 6-ring PAHs) ranged from 30 to 80 percent over 2 months while virtually all of the noncarcinogenic PAHs were degraded. The total PAH reduction ranged from 70 to 95 percent, with a reactor residence time of 60 days.

ECOVA's full-scale, mobile slurry biodegradation unit was used to treat more than 750 cubic yards of soil contaminated with 2,4-dichlorophenoxy acetic acid (2,4-D) and 4-chloro-2-methyl-phenoxyacetic acid (MCPA) and other pesticides such as alachlor, trifluralin, and carbofuran. To reduce 2,4-D and MCPA levels from 800 ppm in soil and 400 ppm in slurry to less than 20 ppm for both in 13 days, 26,000-gallon bioreactors capable of handling approximately 60 cubic yards of soil were used. The residuals of the process were further treated through land application [3, p. 4]. Field application of the slurry biodegradation system designed by ECOVA to treat PCP-contaminated wastes has resulted in a 99-percent decrease in PCP concentrations (both in solid and aqueous phase) over a period of 24 days [3, p. 5].

Performance data for Environmental Remediation, Inc. (ERI) is available for the treatment of American Petroleum Institute (API) separator sludge and wood-processing wastes. Two lagoons containing an olefin sludge from an API separator were treated. In one lagoon, containing, 4,000 cubic yards of sludge, a degradation time of 21 days was required to achieve 68 percent volume reduction and 62 percent mass oil and grease reduction at an operating temperature of 18°C. In the second lagoon, containing 2,590 cubic yards of sludge, a treatment time of 61 days was required to achieve 61 percent sludge reduction and 87.3 percent mass oil and grease reduction at an operating temperature of 14°C [1, p. 367].

At another site, the total wood-preserving constituents were reduced to less than 50 ppm. Each batch process was carried out with a residence time of 28 days in 24-foot diameter, 20-foot-height tank reactors handling 40 cubic yards per batch [6]. The mean concentrations of K001 constituents before treatment and the corresponding concentrations after treatment, for both settled solids and supernatant, are provided in Table 21-6 [2, p. 11]. The supernatant was discharged to a local, publicly owned wastewater treatment works.

RCRA Land Disposal Restrictions (LDRs) that require treatment of wastes to best demonstrated available technology (BDAT) levels prior to land disposal may sometimes be determined to be applicable or relevant and appropriate requirements (ARARs) for CERCLA response actions. Slurry

Table 21-4. Results Showing Reduction in Concentration for Wood Preserving Wastes

Compounds	Initial Concentration		Final Concentration		Percent Removal	
	Solids (mg/kg)	Slurry (mg/kg)	Solids (mg/kg)	Slurry (mg/kg)	Solids (mg/kg)	Slurry (mg/kg)
Phenol	14.6	1.4	0.7	<0.1	95.2[a]	92.8
Pentachlorophenol	687	64	12.3	0.8	98.2	92.8
Naphthalene	3,670	343	23	1.6	99.3[a]	99.5[a]
Phenanthrene & Anthracene	30,700	2,870	200	13.7	99.3	99.5
Fluoranthene	5,470	511	67	4.6	98.8	99.1
Carbazole	1,490	139	4.9	0.3	99.7	99.8

[a]   May be due to combined effect of volatilization and biodegradation. [Source: ReTeC, 50,000 gal. reactor].

Table 21-5. Results Showing Reduction in Volume for Wood Preserving Wastes

Compounds	Before Treatment (total pounds)	After Treatment (total pounds)	Percent Volume Reduction
Phenol	368	41.4	88.8[a]
Pentachlorophenol	141,650	193.0	99.9
Naphthalene	179,830	36.6	99.9[a]
Phenanthrene & Anthracene	2,018,060	303.1	99.9
Fluoranthene	190,440	341.7	99.8
Carbazole	114,260	93.7	99.9

[a]   May be due to combined effect of volatilization and biodegradation. [Source: ReTeC, 50,000 gal. reactor].

biodegradation can produce a treated waste that meets treatment levels set by BDAT, but may not reach these treatment levels in all cases. The ability to meet required treatment levels is dependent upon the specific waste constituents and the waste matrix. In cases where slurry biodegradation does not meet these levels, it still may, in certain situations, be selected for use at the site if a treatability variance establishing alternative treatment levels is obtained. The EPA has made the treatability variance process available in order to ensure that LDRs do not unnecessarily restrict the use of alternative and innovative treatment technologies. Treatability variances may be justified for handling complex soil and debris matrices. The following guides describe when and how to seek a treatability variance for soil and debris: Superfund LDR Guide #6A, Obtaining a Soil and Debris Treatability Variance for Remedial Actions, (OSWER Directive 9347.3-06FS) [10] and Superfund LDR Guide #6B, Obtaining a Soil and Debris Treatability Variance for Removal Actions (OSWER Directive 9347.3-07FS) [9]. Another approach could be to use other treatment techniques in series with slurry biodegradation to obtain desired treatment levels.

TECHNOLOGY STATUS

Biotrol, Inc. has a pilot-scale slurry bioreactor that consists of a feed storage tank, a reactor tank, and a dewatering system for the treated slurry. It was designed to treat the fine-particle slurry from its soil-washing system. Biotrol's process was included in the SITE program demonstration of its soil-washing system at the MacGillis and Gibbs wood-preserving site in New Brighton, Minnesota, during September and October of 1989. Performance data from the SITE demonstration are not currently available; the Demonstration and Applications Analysis Report is scheduled to be published in late 1990.[3]

---

[3]   Editor's Note: The Biotrol soil washing system and slurry biodegradation demonstration project results were published in 1991; see the Biotrol SITE reports listed at the end of Chapter 17.

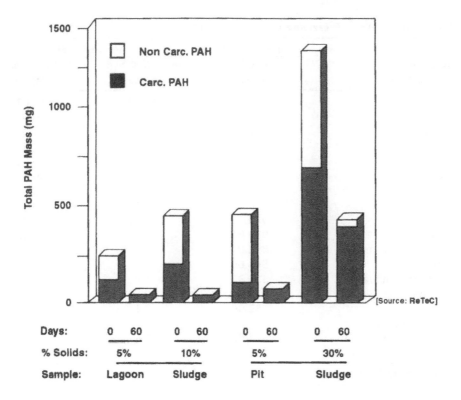

Figure 21-2. Pilot-scale results on oil refinery sludges.

ECOVA Corporation has a full-scale mobile slurry biodegradation system. This system was demonstrated in the field on soils contaminated with pesticides and PCP. ECOVA has developed an innovative treatment approach that utilizes contaminated ground water on site as the make-up water to prepare the slurry for the bioreactor.

ERI has developed a full-scale slurry biodegradation system. ERI's slurry biodegradation system was used to reduce sludge volumes and oil and grease content in two wastewater treatment lagoons at a major refinery outside of Houston, Texas, and to treat 3,000 cubic yards of wood-preserving waste (creosote-K001) over a total cleanup time of 18 months.

Environmental Solutions, Inc. reportedly has a full-scale slurry biodegradation system, with a treatment capacity of up to 100,000 cubic yards, that has been used to treat petroleum and hydrocarbon sludges.

Groundwater Technology, Inc. reportedly has a full-scale slurry biodegradation system, which employs flotation, reactor, and clarifier/sedimentation tanks in series, that has been used to treat soils contaminated with heavy oils, PAHs, and light organics.

ReTeC's full-scale slurry biodegradation system was used in two major projects: Valdosta, Georgia, and Sweetwater, Tennessee. Both projects involved closure of RCRA-regulated surface impoundments containing soils and sludges contaminated with creosote constituents and PCP. Each project used in-ground, lined slurry-phase bioreactor cells operating at 100 cubic yards per week. Residues were chemically stabilized and further treated by tillage. For final closure, the impoundment areas and slurry-phase cells were capped with clay and a heavy-duty asphalt paving [5]. ReTeC has also performed several pilot-scale field demonstrations with their system on oil refinery sludges (RCRA K048-51).

One vendor estimates the cost of full-scale operation to be $80 to $150 per cubic yard of soil or sludge, depending on the initial concentration and treatment volume. The cost to use slurry biodegradation will vary depending upon the need for additional pre- and post-treatment and the addition of air emission control equipment.

Table 21-6. Results of Wood Preserving Waste Treatment

Wood Preserving Waste Constituents	Before Treatment In Soil (mg/kg)	After Treatment	
		In Settled Soil (mg/kg)	In Supernatant (mg/L)
2-Chlorophenol	1.89	<0.01	<0.01
Phenol	3.91	<0.01	<0.01
2,4-Dimethylphenol	7.73	<0.01	<0.01
2,4,6-Trichlorophenol	6.99	<0.01	<0.01
p-Chloro-m-cresol	118.62	<0.01	<0.01
Tetrachlorophenol	11.07	<0.02	<0.02
2,4-Dinitrophenol	4.77	<0.03	<0.03
Pentachlorophenol	420.59	3.1	<0.01
Naphthalene	1078.55	<0.01	0.04
Acenaphthylene	998.80	1.4	1.60
Phenanthrene + Anthracene	6832.07	3.8	3.00
Fluoranthene	1543.06	4.9	16.00
Chrysene + Benz(a)anthracene	519.32	1.4	8.20
Benzo(b)fluoranthene	519.32	<0.03	4.50
Benzo(a)pyrene	82.96	0.1	2.50
Indeno(1,2,3-cd)pyrene + Dibenz(a,h)anthracene	84.88	0.5	1.70
Carbazole	135.40	<0.05	1.70

Source: Environmental Solutions, Inc.

EPA CONTACT

Technology-specific questions regarding slurry biodegradation may be directed to Dr. Ronald Lewis, RREL-Cincinnati, (513) 569-7856 (see Preface for mailing address).

REFERENCES

1. Christiansen, J., T. Koenig, and G. Lucas. Topic 3: Liquid/Solids Contact Case Study. In: Proceedings from the Superfund Conference, Environmental Remediation, Inc., Washington, D.C., 1989. pp. 365–374.
2. Christiansen, J., B. Irwin, E. Titcomb, and S. Morris. Protocol Development for the Biological Remediation of a Wood-Treating Site. In: Proceedings from the 1st International Conference on Physicochemical and Biological Detoxification and Biological Detoxification of Hazardous Wastes, Atlantic City, New Jersey, 1989.
3. ECOVA Corporation. Company Project Description (no date).
4. Kabrick, R., D. Sherman, M. Coover, and R. Loehr. September 1989, Biological Treatment of Petroleum Refinery Sludges. Presented at the Third International Conference on New Frontiers for Hazardous Waste Management, Remediation Technologies, Inc., Pittsburgh, Pennsylvania, 1989.
5. ReTeC Corporation. Closure of Creosote and Pentachlorophenol Impoundments. Company Literature (no date).
6. Richards., D.J. Remedy Selection at Superfund Sites on Analysis of Bioremediation, 1989 AAAS/EPA Environmental Science and Engineering Fellow, 1989.
7. Stroo, H.F. Remediation Technologies Inc. Biological Treatment of Petroleum Sludges in Liquid/Solid Contact Reactors. Environmental and Waste Management World 3(9):9–12, 1989.
8. Stroo, H.F., J. Smith, M. Torpy, M. Coover, and R. Kabrick. Bioremediation of Hydrocarbon Contaminated/Solids Using Liquid/Solids Contact Reactors, Company Report, Remediation Technologies, Inc., (no date), 27 pp.
9. Superfund LDR Guide #6B: Obtaining a Soil and Debris Treatability Variance for Removal Actions. OSWER Directive 9347.3-07FS (NTIS PB91-921310), 1990.

10. Superfund LDR Guide #6A: Obtaining a Soil and Debris Treatability Variance for Remedial Actions. OSWER Directive 9347.3-06FS (NTIS PB91-921327), 1990.
11. Innovative Technology: Slurry-Phase Biodegradation. OSWER Directive 9200.5-252FS (NTIS PB90-274200), 1989.
12. Technology Screening Guide for Treatment of CERCLA Soils and Sludges. EPA/540/2-88/004 (NTIS PB89-132674), 1988.

ADDITIONAL SITE PROGRAM REFERENCES (See Appendix A for information on how to obtain SITE program reports)

Pilot-Scale Demonstration of a Slurry-Phase Biological Reactor for Creosote-Contaminated Soil:*

Demonstration Bulletin (EPA/540/M5-91-009), February 1992.
Technology Demonstration Summary (EPA/540/S5-91-009), September 1993.
Applications Analysis Report (EPA/540/A5-91/009), January 1993.
Technology Evaluation (EPA/540/5-91/009; NITS PB93-205532).

* ECOVA Corporation's EIMCO Biolift™ Reactor.

Chapter 22

Rotating Biological Contactors[1]

Denise Scott and Evelyn Meagher-Hartzell, Science Applications International Corporation (SAIC), Cincinnati, OH

ABSTRACT

Rotating biological contactors (RBCs) employ aerobic fixed film treatment to degrade either organic and/or nitrogenous (ammonia-nitrogen) constituents present in aqueous waste streams. Treatment is achieved as the waste passes by the media, enabling fixed-film systems to acclimate biomass capable of degrading organic waste [1, p. 91].[2] Fixed-film RBC reactors provide a surface to which soil organisms can adhere; many indigenous soil organisms are effective degraders of hazardous wastes.

An RBC consists of a series of corrugated plastic discs mounted on a horizontal shaft. As the discs rotate through the aqueous waste stream, a microbial slime layer forms on the surface of the discs. The microorganisms in this slime layer degrade the waste's organic and nitrogenous constituents. Approximately 40 percent of the RBC's surface area is immersed in the waste stream as the RBC rotates through the liquid. The remainder of the surface area is exposed to the atmosphere, which provides oxygen to the attached microorganisms and facilitates oxidation of the organic and nitrogenous contaminants [2, p. 6]. In general, the large microbial population growing on the discs provides a high degree of waste treatment in a relatively short time. Although RBC systems are capable of performing organic removal and nitrification concurrently, they may be designed to primarily provide either organic removal or nitrification singly [3, p. 1-2].

RBCs were first developed in Europe in the 1950s [1, p. 6]. Commercial applications in the United States did not occur until the late 1960s. Since then, RBCs have been used in the United States to treat municipal and industrial wastewaters. Because biological treatment converts organics to innocuous products such as CO_2, investigators have begun to evaluate whether biological treatment systems like RBCs can effectively treat liquid waste streams from Superfund sites. Treatability studies have been performed at at least three Superfund sites to evaluate the effectiveness of this technology in removing organic and nitrogenous constituents from hazardous waste leachate. A full-scale RBC treatment system is presently operating in at least one Superfund site in the United States.

TECHNOLOGY APPLICABILITY

Research demonstrates that RBCs can potentially treat aqueous organic waste streams from some Superfund sites. During the treatability studies for the Stringfellow, New Lyme, and Moyer Superfund sites, RBC systems efficiently removed the major organic and nitrogenous constituents in the leachates. Because waste stream composition varies from site to site, treatability testing to determine the degree of contaminant removal is an essential element of the remedial action plan. Although recent Superfund applications have been limited to the treatment of landfill leachates, this technology may be applied to ground-water treatment [4].

In general, biological systems can degrade only the soluble fraction of the organic contamination. Thus, the applicability of RBC treatment is ultimately dependent upon the solubility of the contami-

---

[1] EPA/540/S-92/007.

[2] [Reference number, page number.]

Table 22-1. Effectiveness of RBCs on General Contaminant Groups for Liquid Waste Streams

Contaminant Groups	Effectiveness
Organic	
Halogenated volatiles	■
Halogenated semivolatiles	■
Nonhalogenated volatiles	■
Nonhalogenated semivolatiles	■
PCBs	▼
Pesticides	▼
Dioxins/furans	●
Organic cyanides	▼
Organic corrosives	▼
Inorganic	
Volatile metals	●
Nonvolatile metals	●
Asbestos	●
Radioactive materials	●
Inorganic corrosives	●
Inorganic cyanides	▼
Reactive	
Oxidizers	●
Reducers	●

■ Demonstrated Effectiveness: Successful treatability tests at some scale completed.
▼ Potential Effectiveness: Expert opinion that technology will work.
● No Expected Effectiveness: Expert opinion that technology will not work.

nant. RBCs are generally applicable to influents containing organic concentrations of up to 1 percent organics, or between 40 and 10,000 mg/L of SBOD. (Note: Soluble biochemical oxygen demand, or SBOD, measures the soluble fraction of the biodegradable organic content in terms of oxygen demand.) RBCs can be designed to reduce influent biochemical oxygen demand (BOD) concentrations below 5 mg/L SBOD and ammonia-nitrogen (NH$_3$-N) levels below 1.0 mg/L [5, p. 2; 6, p. 60]. RBCs are effective for treating solvents, halogenated organics, acetone, alcohols, phenols, phthalates, cyanides, ammonia, and petroleum products [7, p. 6; 8, p. 69]. RBCs have fully nitrified leachates containing ammonia-nitrogen concentrations up to 700 mg/L [6, p. 61].

The effectiveness of RBC treatment systems on general contaminant groups is shown in Table 22-1. Examples of constituents within contaminant groups are provided in Technology Screening Guide for Treatment of CERCLA Soils and Sludges [9]. Table 22-1 is based on the current available information or professional judgment where no information was available. The proven effectiveness of the technology for a particular site or waste does not ensure that it will be effective at all sites or that the treatment efficiencies achieved will be acceptable at other sites. For the ratings used for this table, demonstrated effectiveness means that, at some scale, treatability was tested to show the technology was effective for that particular contaminant group. The ratings of potential effectiveness or no-expected-effectiveness are based upon expert judgment. Where potential effectiveness is indicated, the technology is believed capable of successfully treating the contaminant group in a particular medium. When the technology is not applicable or will probably not work for a particular combination of contaminant group and medium, a no-expected-effectiveness rating is given.

LIMITATIONS

Although RBCs have proved effective in treating waste streams containing ammonia-nitrogen and organics, they are not effective at removing most inorganics or non-biodegradable organics. Wastes containing high concentrations of heavy metals and certain pesticides, herbicides, or highly chlorinated organics can resist RBC treatment by inhibiting microbial activity. Waste streams containing toxic concentrations of these compounds may require pretreatment to remove these materials prior to RBC treatment [10, p. 3].

RBCs are susceptible to excessive biomass growth, particularly when organic loadings are elevated. If the biomass fails to slough off and a blanket of biomass forms which is thicker than 90 to 125 mils, the resulting weight may damage the shaft and discs. When necessary, excess biofilm may be reduced by either adjusting the operational characteristics of the RBC unit (e.g., the rotational speed or direction) or by employing air or water to shear off the excess biomass [11, p. 2].

In general, care must be taken to ensure that organic pollutants do not volatilize into the atmosphere. To control their release, gaseous emissions may require off-gas treatment [12, p. 31].

All biological systems, including RBCs, are sensitive to temperature changes and experience drops in biological activity at temperatures lower than 55°F. Covers should be employed to protect the units from colder climates and extraordinary weather conditions. Covers should also be used to protect the plastic discs from degradation by ultraviolet light, to inhibit algal growth, and to control the release of volatiles [13]. In general, organic degradation is optimum at a pH between 6 and 8.5. Nitrification requires the pH be greater than 6 [6, p. 61].

Additionally, nutrient and oxygen deficiencies can reduce microbial activity, causing significant decreases in biodegradation rates [14, p. 39]. Extremes in pH can limit the diversity of the microbial population and may suppress specific microbes capable of degrading the contaminants of interest. Fortunately, these variables can be controlled by modifying the system design.

TECHNOLOGY DESCRIPTION

A typical RBC unit consists of 12-foot-diameter plastic discs mounted along a 25-foot horizontal shaft. The total disc surface area is normally 100,000 square feet for a standard unit and 150,000 square feet for a high density unit. Figure 22-1 is a diagram of a typical RBC system.

As the RBC slowly rotates through the ground water or leachate at 1.5 rpm, a microbial slime forms on the discs. These microorganisms degrade the organic and nitrogenous contaminants present in the waste stream. During rotation, approximately 40 percent of the discs' surface area is in contact with the aqueous waste while the remaining surface area is exposed to the atmosphere. The rotation of the media through the atmosphere causes the oxygenation of the attached organisms. When operated properly, the shearing motion of the discs through the aqueous waste causes excess biomass to shear off at a steady rate. Suspended biological solids are carried through the successive stages before entering the secondary clarifier [2, p. 13.101].

Primary treatment (e.g., clarifiers or screens), to remove materials that could settle in the RBC tank or plug the discs, is often essential for good operation. Influents containing high concentrations of floatables (e.g., grease, etc.) will require treatment using either a primary clarifier or an alternate removal system [11, p. 2].

The RBC treatment process may involve a variety of steps, as indicated by the block diagram in Figure 22-2. Typically, aqueous waste is transferred from a storage or equalization tank (1) to a mixing tank (2) where chemicals may be added for metals precipitation, nutrient adjustment, and pH control. The waste stream then enters a clarifier (3) where the solids are separated from the liquid. The effluent from the clarifier enters the RBC (4) where the organics and/or ammonia are converted to innocuous products. The treated waste is then pumped into a second clarifier (5) for removal of the biological solids. After secondary clarification, the effluent enters a storage tank (6) where, depending upon the contamination remaining in the effluent, the waste may be stored pending additional treatment or discharged to a sewer system or surface stream. Throughout this treatment process the off-gases from the various stages should be collected for treatment (7). The actual treatment train will, of course, depend upon the nature of the waste and will be selected after the treatability study is conducted.

Staging, which employs a number of RBCs in series, enhances the biochemical kinetics and establishes selective biological cultures acclimated to successively decreasing organic loadings. As the waste stream passes from stage to stage, progressively increasing levels of treatment occur [2, p. 13.105].

In addition to maximizing the system's efficiency, staging can improve the system's ability to handle shock loads by absorbing the impact of a shock load in the initial stages, thereby enabling subsequent stages to operate until the affected stages recover [15, p. 10.200].

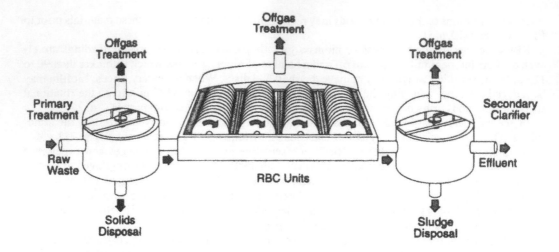

Figure 22-1. Typical RBC plant schematic [12].

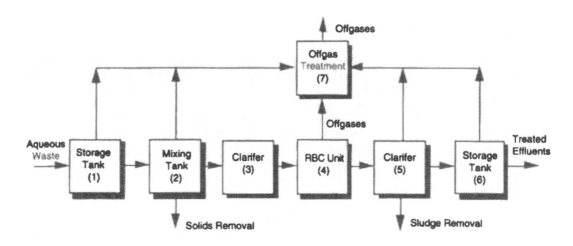

Figure 22-2. Block diagram of the RBC treatment process.

Factors affecting the removal efficiency of RBC systems include the type and concentration of organics present, hydraulic residence time, rotational speed, media surface area exposed and submerged, and pre- and post-treatment activities. Design parameters for RBC treatment systems include the organic and hydraulic load rates, design of the disc train(s), rotational velocity, tank volume, media area submerged and exposed, retention time, primary treatment and secondary clarifier capacity, and sludge production [8, p. 69].

PROCESS RESIDUALS

During primary clarification, debris, grit, grease, metals, and suspended solids (SS) are separated from the raw influent. The solids and sludges resulting from primary clarification may contain metallic and organic contaminants and may require additional treatment. Primary clarification residuals must be disposed of in an appropriate manner (e.g., land disposal, incineration, solidification, etc.).

Following RBC treatment, the effluent undergoes secondary clarification to separate the suspended biomass solids from the treated effluent. Refractory organics may contaminate both the clarified effluent and residuals. Additional treatment of the solids, sludges, and clarified effluent may be required. Clarified secondary effluents which meet the treatment standards are generally discharged to a surface stream, while residual solids and sludges must be disposed of in an appropriate manner, as outlined above for primary clarification residuals [2, p. 13.120].

Volatile organic compound (VOC)-bearing gases are often liberated as a by-product of RBC treatment. Care must be taken to ensure that off-gases do not contaminate the work space or the atmosphere. Various techniques may be employed to control these emissions, including collecting the gases for treatment [13].

SITE REQUIREMENTS

RBCs vary in size depending upon the surface area needed to treat the hazardous waste stream. A single full size unit with a walkway for access on either side of the unit takes up approximately 550 square feet [16]. The total area required for an RBC system is site-specific and depends on the number, size, and configuration of RBC units installed.

Contaminated ground water, leachates, or waste materials are often hazardous. Handling and treatment of these materials requires that a site safety plan be developed to provide for personnel protection and special handling measures. Storage should be provided to hold the process product streams until they have been tested to determine their acceptability for disposal, reuse, or release. Depending on the site, a method to store waste that has been prepared for treatment may be necessary. Storage capacity will depend on waste volume.

Onsite analytical equipment capable of determining site-specific organic compounds for performance assessment make the operation more efficient and provide better information for process control.

PERFORMANCE DATA

Limited information is available on the effectiveness of RBCs in treating waste from Superfund sites. Most of the data came from studies done on leachate from the New Lyme, Ohio; Stringfellow, California; and Moyer, Pennsylvania Superfund sites. The results of these studies are summarized below.

In order to compensate for the lack of Superfund performance data, non-Superfund applications are also discussed. The majority of the performance data for non-Superfund applications were obtained from industrial RBC operations. Theoretically, this information has a high degree of application to Superfund leachate and ground-water treatment.

The quality of the information present in this section has not been determined. The data are included as a general guidance, and may not be directly transferable to a specific Superfund site. Good characterization and treatability studies are essential in further refining and screening of RBC technology.

New Lyme Treatability Study

The EPA performed a remedy selection study on the leachate from the New Lyme Superfund site located in New Lyme Township, Ashtabula County, Ohio, to help determine the applicability of an RBC to treat hazardous waste from a Superfund site. Samples of leachate collected from various seeps surrounding the landfill showed that the leachate was highly concentrated. Results indicated that the leachate contained up to 2,000 mg/L dissolved organic carbon (DOC), 2,700 mg/L SBOD, and 5,200 mg/L soluble chemical oxygen demand (SCOD) [17, p. 12]. (Note: SCOD measures the soluble fraction of the organics amenable to chemical oxidation, as well as certain inorganics such as sulfides, sulfites, ferrous iron, chlorides, and nitrites.)

Leachate from the New Lyme site was transported from New Lyme to a demonstration-scale RBC located at the EPA's Testing and Evaluation Facility in Cincinnati, Ohio. After an adequate biomass was developed on the RBC discs using a primary effluent supplied by Mill Creek Treatment Facility (a local industrial wastewater treatment facility), the units were gradually acclimated to an influent consisting of 100 percent leachate. Results indicated that within 20 hours the RBC removed 97 percent of the gross organics, as represented by DOC, from the leachate (see Figure 22-3 and Table 22-2) [18, p. 7]. Priority pollutants were either converted and/or stripped from the leachate during treatment.

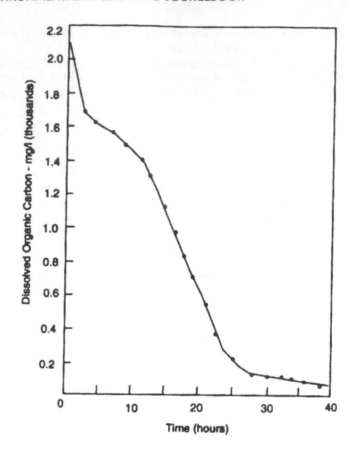

Figure 22-3. Disappearance of DOC with time [17, p. 14] Experiment 5. [The influent for Experiment 5 consisted of 100 percent leachate and the biomass on the RBCs was acclimated. Nutrient addition was also employed (at a ratio of 160/5/2 for C/N/P).]

After normal clarification, the effluent from the RBC was eligible for disposal into the sewer system leading to the Mill Creek facility.

Stringfellow Treatability Study

A remedy selection study using an RBC was conducted on leachate from the Stringfellow Superfund site located in Glen Avon, California. After the leachate from this site received lime treatment to remove metal contamination, the leachate was transported to the EPA's Testing and Evaluation Facility in Cincinnati for testing similar to the New Lyme study. The objective of this study was to determine whether the leachate from Stringfellow could be treated economically with an RBC system.

The leachate from this site was generated at a daily rate of 2,500 gallons. Compared to the New Lyme leachate, it contained moderate concentrations of gross organics with DOC values of 300 mg/L, SBOD values of 420 mg/L, and SCOD values of 800 mg/L [4, p. 44].

Results indicated that greater than 99 percent of SBOD was removed, 65 percent of DOC was removed, and 54 percent SCOD was removed within four days using the RBC laboratory scale treatment system [4, p. 44]. Table 22-3 presents pertinent information on the treatment of 100 percent leachate. Since the DOC and SCOD conversion rates were low, a significant fraction of the refractory organics remained following treatment. Activated carbon was used to reduce the DOC to limits acceptable to the Mill Creek Treatment Facility.

Table 22-2. Removal from New Lyme Leachate [17, p. 17] Experiment 5

	Influent (mg/L)	Effluent (mg/L)
SBOD	2700	4
BOD$_T$	3000	6.6
DOC	2000	17
TOC	2100	19
SCOD	5200	33
NO$_3$-N	<1	60
SS	1400	6600
VSS	240	2600
Volatile PP		
Benzene	0.28	<0.002
Toluene	4.9	<0.002
Additional Volatiles		
cis 1,2-Dichloroethene	0.94	ND
Xylenes	2.8	ND
Acetone	140	ND
Methyl Ethyl Ketone	470	ND
Total Organic Halides	—	1.2
Total Toxic Organics	<0.250	<0.010

BOD$_T$ = Total Biochemical Oxygen Demand.
NO$_3$-N = Nitrogen as Nitrate.
VSS = Volatile Suspended Solids.

Moyer Treatability Study

During a recent remedy selection study, three treatability scale RBCs were used to degrade a low-BOD (26 mg/L), high ammonia (154 mg/L) leachate from the Moyer Landfill Superfund site in Lower Providence Township near Philadelphia, Pennsylvania [19, p. 971]. The leachate has low organic strength (e.g., 26 mg/L BOD, 358 mg/L COD, and 68 mg/L TOC) which is typical of an older landfill, and it also contains mainly nonbiodegradable organic compounds [19, p. 972]. (Note: Total organic carbon, or TOC, is a measure of all organic carbon expressed as carbon.) The abundance of ammonia found in the leachate prompted investigators to attempt ammonia oxidation with an RBC system. Relatively low substrate loading rates were employed during the study (0.2, 0.4, and 0.6 gpd/square foot of disc surface area per stage). Ammonia oxidation was essentially complete (98 percent) and a maximum of 80 percent of the BOD and 38 percent of the COD in the leachate was oxidized [19, p. 980]. Runs performed using lower loading rates experienced the largest removals. A limited denitrification study was also performed using an anoxic RBC to treat an RBC effluent generated during the aerobic segment of the treatability investigation. This study demonstrated the feasibility of using denitrification to treat the nitrate produced by aerobic ammonia oxidation [19, p. 980].

Non-Superfund Applications

The Homestake Mine in Lead, South Dakota has operated an RBC wastewater treatment plant since 1984. Forty-eight RBCs treat up to 5.5 million gallons per day (MGD) (21,000 m³) of discharge water per day. The system was designed to degrade thiocyanate, free cyanide, and metal-complexed cyanides, to reduce heavy metal concentrations, and to remove ammonia, which is a by-product of cyanide degradation [20, p. 2]. Eight parallel treatment trains, utilizing five RBCs in series, were employed to degrade and nitrify the metallurgical process waters (see Table 22-4 for a characterization of the influent). The first two RBCs in each train were used to degrade the cyanides

Table 22-3. Treatment of 100% Stringfellow Leachate [4, p. 44]

	Leachate (mg/L)	RBC Effluent (mg/L)	Use APC plus Effluent (mg/L)
SBOD	420	<3.0	0.9
BOD	440		22
DOC	300	110	20
TOC	310		22
SCOD	800	360	79
COD	840		95
SS	43		23
VSS	31		14
NH$_3$-N	3.4		6.3
NO$_3$-N	44		34

APC = Activated Powered Carbon.

COD = Chemical Oxygen Demand.

and remove heavy toxic metals and particulate solids through biological adsorption. The last three RBCs employed nitrification to convert the ammonia to nitrate. Table 22-5 provides an average performance breakdown for the system. During its operation, overall performance improved significantly, as demonstrated by an 86 percent increase in the system's ability to reduce total effluent cyanide concentrations (e.g., from 0.45 to 0.06 mg/L). Concurrently, the cost per kg to treat cyanide dropped from $11.79 to $3.10, while the cost per m^3 to treat effluent decreased by 50 percent [21, p. 9]. In general, the system has responded well to any upsets or disturbances. Diesel fuels, lubricants, degreasers, biocides, dispersants, and flocculants have been periodically found in the influent wastewater, but normally only create minor upsets in the performance of the plant. During the life of the system, the number of upsets and the biomass's ability to recuperate have both improved [21, p. 6].

A significant difference between the Homestake system and the other RBC systems described within this report is that instead of removing the metals contaminating the wastewater in the pretreatment stage, metal reduction is accomplished through bioadsorption during the treatment phase. Bioadsorption of metals by biological cells is not unlike the use of activated carbon; however, the number and complexity of binding sites on the cell wall are enormous in comparison [20, p. 2].

In a study by Israel's Institute of Technology, a laboratory-scale RBC was used to treat an oil refinery wastewater. The wastewater had been pretreated using oil-water separation and dissolved air flotation. As summarized in Table 22-6, 91 percent of the hydrocarbon and 97 percent of the phenol were removed, as well as 96 percent of the ammonia-nitrogen [22, p. 4]. By gradually increasing the concentration of phenols present in the influent (e.g., over a 5 day period) from 5 mg/L to 30 mg/L, the system demonstrated that it was capable of quickly adapting to influent changes and higher phenolic loads [22, p. 6]. During this period, the RBC was able to maintain effluent COD concentrations at levels comparable to previous loadings. The system's resiliency was further demonstrated by its ability to recover from a major disturbance (e.g., such that effluent COD removal was interrupted) within 4 days [22, p. 7].

TECHNOLOGY STATUS

RBCs have been used commercially in the United States since the late 1960s to treat municipal and industrial wastes. In the past decade, studies have been performed to evaluate the effectiveness of RBCs in treating leachate from hazardous waste sites.

Treatability studies have been performed on leachate from the Stringfellow, New Lyme, and Moyer Superfund sites. Results of these studies indicate that RBCs are effective in removing organic and nitrogenous constituents from hazardous waste leachate. Additional research is needed to define the effectiveness of an RBC in treating leachates and contaminated ground water and to determine the

Table 22-4. Homestake Mine Wastewater Matrix[a]

	Decant Water (mg/L)	Mine Water (mg/L)	Influent Blend (mg/L)
Thiocyanate	110–350	1–33	35–110
Total Cyanide	5.5–65.0	0.30–2.50	0.50–11.50
WAD Cyanide	3.10–38.75	0.50–1.10	0.50–7.15
Copper	0.5–3.1	0.10–2.65	0.15–2.95
Ammonia-N	5–10	5.00–19.00	6–12
Phosphorus-P	0.10–0.20	0.10–0.15	0.10–0.15
Alkalinity	50–200	150–250	125–225
pH	7–9	7–9	7.5–8.5
Hardness	400–500	650–1400	500–850
Temperature °C	1.0–27.2	24–33	5–25

WAD = Weak Acid Dissociable.

[a] Adapted from Reference 20, p. 8.

Table 22-5. Influent, Effluent and Permit Concentrations at the Homestake Mines [20, p. 8]

	Influent (mg/L)	Effluent (mg/L)	Permit (mg/L)
Thiocyanate	62.0	<0.5	—
Total Cyanides	4.1	0.06	1.00
WAD Cyanide	2.3	<0.02	0.10
Total Copper	0.56	0.07	0.13
Total Suspended Solids	—	6.0	10.0
Ammonia-Nitrogen	5.60[a]	<0.50	1.0–3.9

[a] Ammonia peaks at 25 mg/L within the plant as a cyanide degradation by-product.

degree of organic stripping that occurs during the treatment process. RBCs are being used to treat leachate from the New Lyme Superfund site.

RBCs require a minimal amount of equipment, manpower, and space to operate. Staging of RBCs will vary from site to site, depending on the waste stream. The cost to install a single RBC unit with a protective cover and a surface area of 100,000 to 150,000 square feet ranges from $80,000 to $85,000 [16; 23]. During the Stringfellow treatability study researchers determined that by augmenting the existing carbon treatment system with RBCs, reductions in carbon costs would pay for the RBC plant within 3.3 years [4, p. 44]. The RBC plant model used to formulate this estimate was a scaled-up version of the pilot unit used during the treatability study.

EPA CONTACT

Technology-specific questions regarding rotating biological contactors may be directed to Richard Brenner, RREL-Cincinnati (513) 569-7657 (see Preface for mailing address).

ACKNOWLEDGMENTS

This chapter was prepared for the U.S. Environmental Protection Agency, Office of Research and Development (ORD), Risk Reduction Engineering Laboratory (RREL), Cincinnati, Ohio, by Science Applications International Corporation (SAIC) under Contract No. 68-C8-0062. Mr. Eugene Harris served as the EPA Technical Project Monitor. Mr. Gary Baker was SAIC's Work Assignment Manager. This chapter was written by Ms. Denise Scott and Ms. Evelyn Meagher-Hartzell of SAIC.

Table 22-6. Refinery Wastewater Quality Before and After RBC Treatment [22, p.4]

Constituent		Influent (mg/L)	Effluent (mg/L)
COD	Total	715	197
	Soluble	685	186
BOD	Total	140	8
	Soluble	128	6
Phenols		7.5	0.22
Suspended Solids	Total	32	7
	Volatile	29	6
NH$_3$-N		12.8	0.48

The following other Agency and contractor personnel have contributed their time and comments by participating in the expert review meetings and/or peer reviewing the chapter:

Dr. Robert L. Irvine, University of Notre Dame
Mr. Richard A. Sullivan, Foth & Van Dyke
Ms. Mary Boyer, SAIC
Mr. Cecil Cross, SAIC

REFERENCES

1.   Cheremisinoff, P.E. Biological Treatment of Hazardous Wastes, Sludges, and Wastewater. Pollution Engineering, May 1990.
2.   Envirex, Inc. Rex Biological Contactors: For Proven, Cost-Effective Options in Secondary Treatment. Bulletin 315-13A-51/90-3M.
3.   Design Information on Rotating Biological Contactors, EPA/600/2-84/106, June 1984.
4.   Opatken, E.J., H.K. Howard, and J.J. Bond. Stringfellow Leachate Treatment with RBC. Environmental Progress, Volume 7, No. 1, February 1988.
5.   Walker Process Corporation. EnviroDisc™ Rotating Biological Contactor. Bulletin 11-S-88.
6.   Opatken, E.J. and J.J. Bond. RBC Nitrification of High Ammonia Leachates. Environmental Progress, Volume 10, No. 4, February 1991.
7.   Guide to Treatment Technologies for Hazardous Wastes at Superfund Sites. EPA/540/2-89/052 (NTIS PB89-190821), March 1989.
8.   Data Requirements for Selecting Remedial Action Technology. EPA/600/2-87/001, January 1987.
9.   Technology Screening Guide for Treatment of CERCLA Soils and Sludges. EPA/540/2-88/004 (NTIS PB89-132674), 1988.
10. O'Shaughnessy, et al. Treatment of Oil Shale Retort Wastewater Using Rotating Biological Contactors. Presented at the Water Pollution Control Federation, 55th Annual Conference, St. Louis, Missouri, October 1982.
11. Rotating Biological Contactors: U.S. Overview. EPA/600/D-87/023, U.S. Environmental Protection Agency, January 1987.
12. Nunno, T.J. and J.A. Hyman. Assessment of International Technologies for Superfund Applications. EPA/540/2-88/003 (NTIS PB90-106428), September 1988.
13. Telephone conversation. Steve Oh, U.S. Army Corps of Engineers, September 4, 1991.
14. Corrective Action: Technologies and Applications. EPA/625/4-89/020, September 1984.
15. Lyco, Inc., Rotating Biological Surface (RBS) Wastewater Equipment: RBS Design Manual. March 1986.
16. Telephone conversation. Gerald Ornstein, Lyco Corporation, September 4, 1991.
17. Opatken, E.J., H.K. Howard, and J.J. Bond. Biological Treatment of Leachate from a Superfund Site. Environmental Progress, Volume 8, No. 1, February 1989.
18. Opatken, E.J., H.K. Howard, and J.J. Bond. Biological Treatment of Hazardous Aqueous Wastes. EPA/600/D-87/184, June 1987.

19.   Spengel, D.B. and D.A. Dzombak. Treatment of Landfill Leachate with Rotating Biological Contractors: Bench Scale Experiments. Research Journal WPCF, Vol. 63, No. 7, November/December 1991.
20.   Whitlock, J.L. The Advantages of Biodegradation of Cyanides. Journal of the Minerals, Metals and Materials Society, December 1989.
21.   Whitlock, J.L. Biological Detoxification of Precious Metal Processing Wastewaters. Homestake Mining Co., Lead, SD.
22.   Galil, N. and M. Rebhun. A Comparative Study of RBC and Activated Sludge in Biotreatment of Wastewater from an Integrated Oil Refinery. Israel Institute of Technology, Haifa, Israel.
23.   Telephone conversation. Jeff Kazmarek, Envirex Inc., September 4, 1991.

Chapter 23

Solidification/Stabilization of Organics and Inorganics[1]

Larry Fink and **George Wahl**, Science Applications International Corporation (SAIC), Cincinnati, OH

ABSTRACT

Solidification refers to techniques that encapsulate hazardous waste into a solid material of high structural integrity. Encapsulation involves either fine waste particles (microencapsulation) or a large block or container of wastes (macroencapsulation) [1, p. 2].[2] Stabilization refers to techniques that treat hazardous waste by converting it into a less soluble, mobile, or toxic form. Solidification/Stabilization (S/S) processes, as referred to in this chapter, utilize one or both of these techniques.

S/S technologies can immobilize many heavy metals, certain radionuclides, and selected organic compounds while decreasing waste surface area and permeability for many types of sludge, contaminated soils, and solid wastes. Common S/S agents include: Type 1 portland cement or cement kiln dust; lime, quicklime, or limestone; fly ash; various mixtures of these materials and various organic binders (e.g., asphalt). The mixing of the waste and the S/S agents can occur outside of the ground (ex situ) in continuous feed or batch operations, or in the ground (in situ) in a continuous feed operation. The final product can be a continuous solid mass of any size, or of a granular consistency resembling soil. During in situ operations, S/S agents are injected into and mixed with the waste and soil up to depths of 30 to 100 feet using augers.

Treatability studies are the only means of documenting the applicability and performance of a particular S/S system. Determination of the best treatment alternative will be based on multiple site-specific factors and the cost and efficacy of the treatment technology. The EPA contact identified at the end of this chapter can assist in the location of other contacts and sources of information necessary for such treatability studies.

It may be difficult to evaluate the long-term (>5 year) performance of the technology. Therefore, long-term monitoring may be needed to ensure that the technology continues to function within its design criteria.

This chapter provides information on technology applicability, the limitations of the technology, the technology description, the types of residuals produced, site requirements, the process performance data, the status of the technology, and sources for further information.

TECHNOLOGY APPLICABILITY

The U.S. EPA has established treatment standards under the Resource Conservation and Recovery Act (RCRA), Land Disposal Restrictions (LDRs) based on Best Demonstrated Available Technology (BDAT) rather than on risk-based or health-based standards. There are three types of LDR treatment standards based on the following: achieving a specified concentration level, using a specified technology prior to disposal, and "no land disposal." Achieving a specified concentration level is the most common type of treatment standard. When a concentration level to be achieved is specified for a waste, any technology that can meet the standard may be used unless that technology is otherwise prohibited [2].

---

[1] EPA/540/S-92/015.
[2] [Reference number, page number.]

The Superfund policy on use of immobilization is as follows: "Immobilization is generally appropriate as a treatment alternative only for material containing inorganics, semi-volatile and/or non-volatile organics. Based on present information, the Agency does not believe that immobilization is an appropriate treatment alternative for volatile organic compounds (VOCs). Selection of immobilization of semi-volatile compounds (SVOCs) and non-volatile organics generally requires the performance of a site-specific treatability study or non-site-specific treatability study data generated on waste which is very similar (in terms of type of contaminant, concentration, and waste matrix) to that to be treated and that demonstrates, through Total Waste Analysis (TWA), a significant reduction (e.g., a 90 to 99 percent reduction) in the concentration of chemical constituents of concern. The 90 to 99 percent reduction in contaminant concentration is a general guidance and may be varied within a reasonable range considering the effectiveness of the technology and the cleanup goals for the site. Although this policy represents EPA's strong belief that TWA should be used to demonstrate effectiveness of immobilization for organics, other leachability tests may also be appropriate in addition to TWA to evaluate the protectiveness under a specific management scenario. To measure the effectiveness on inorganics, the EPA's Toxicity Characteristic Leaching Procedure (TCLP) should be used in conjunction with other tests such as TCLP using distilled water or American Nuclear Society (ANS) 16.1" [3, p. 2].

Factors considered most important in the selection of a technology are design, implementation, and performance of S/S processes and products, including the waste characteristics (chemical and physical), processing requirements, S/S product management objectives, regulatory requirements, and economics. These and other site-specific factors (e.g., location, condition, climate, hydrology, etc.) must be taken into account in determining whether, how, where, and to what extent a particular S/S method should be used at a particular site [4, p. 7.92]. Pozzolanic S/S processes can be formulated to set under water if necessary; however, this may require different proportions of fixing and binding agents to achieve the desired immobilization and is not generally recommended [5, p. 21]. Where nonpumpable sludge or solid wastes are encountered, the site must be able to support the heavy equipment required for excavation or in situ injection and mixing. At some waste disposal sites, this may require site engineering.

A wide range of performance tests may be performed in conjunction with S/S treatability studies to evaluate short- and long-term stability of the treated material. These include total waste analysis for organics, leachability using various methods, permeability, unconfined compressive strength (UCS), treated waste and/or leachate toxicity endpoints, and freeze/thaw and wet/dry weathering cycle tests performed according to specific procedures [6, p. 4.2; 7, p. 4.1]. Treatability studies should be conducted on replicate samples from a representative set of waste batches that span the expected range of physical and chemical properties to be encountered at the site [8, p. 1].

The most common fixing and binding agents for S/S are cement, lime, natural pozzolans, and fly ash, and mixtures of these [4, p. 7.86; 6, p. 2.1]. They have been demonstrated to immobilize many heavy metals and to solidify a wide variety of wastes including spent pickle liquor, contaminated soils, incinerator ash, wastewater treatment filter cake, and waste sludge [7, p. 3.1; 9]. S/S is also effective in immobilizing many radionuclides [10]. In general, S/S is considered an established full-scale technology for nonvolatile heavy metals, although the long-term performance of S/S in Superfund applications has yet to be demonstrated [2].

Traditional cement and pozzolanic materials have yet to be shown to be consistently effective in full-scale applications treating wastes high in oil and grease, surfactants, or chelating agents without some form of pretreatment [11; 12, p. 122]. Pretreatment methods include pH adjustment, steam or thermal stripping, solvent extraction, chemical or photochemical reaction, and biodegradation. The addition of sorbents such as modified clay or powdered activated carbon may improve cement-based or pozzolanic process performance [6, p. 2.3].

Regulations promulgated pursuant to the Toxic Substances Control Act (TSCA) do not recognize S/S as an approved treatment for wastes containing polychlorinated biphenyls (PCBs) above 50 ppm. It is EPA policy that soils containing greater than 10 ppm in public/residential areas and 25 ppm in limited access/occupational areas be removed for TSCA-approved treatment/disposal. However, the policy also provides EPA regional offices with the option of requiring more restrictive levels. For example, Region 5 requires a cleanup level of 2 ppm. The proper disposition of high volume sludges, soils, and sediments is not specified in the TSCA regulations, but precedents set in the development of various records of decision (RODs) indicate that stabilization may be approved where PCBs are

Table 23-1. Effectiveness of S/S on General Contaminant Groups for Soil and Sludges

Contaminant Groups	Effectiveness Soil/Sludge
Organic	
Halogenated volatiles	●
Halogenated semivolatiles	●
Nonhalogenated volatiles	■
Nonhalogenated semivolatiles	■
PCBs	▼
Pesticides	▼
Dioxins/furans	▼
Organic cyanides	▼
Organic corrosives	▼
Inorganic	
Volatile metals	■
Nonvolatile metals	■
Asbestos	■
Radioactive materials	■
Inorganic corrosives	■
Inorganic cyanides	■
Reactive	
Oxidizers	■
Reducers	■

■ Demonstrated Effectiveness: Successful treatability tests at some scale completed.
▼ Potential Effectiveness: Expert opinion that technology will work.
● No Expected Effectiveness: Expert opinion that technology will/does not work.

effectively immobilized and/or destroyed to TSCA-equivalent levels. Some degree of immobilization of PCBs and related polychlorinated polycyclic compounds appears to occur in cement or pozzolans [15, p. 1573]. Some field observations suggest polychlorinated polycyclic organic substances such as PCBs undergo significant levels of dechlorination under the alkaline conditions encountered in pozzolanic processes. Recent tests by the EPA, however, have not confirmed these results although significant desorption and volatilization of the PCBs were documented [13, p. 41; 14, p. 3].

Table 23-1 summarizes the effectiveness of S/S on general contaminant groups for soils and sludges. Table 23-1 was prepared based on current available information or on professional judgment when no information was available. In interpreting this table, the reader is cautioned that for some primary constituents, a particular S/S technology performs adequately in some concentration ranges but inadequately in others. For example, copper, lead, and zinc are readily stabilized by cementitious materials at low to moderate concentrations, but interfere with those processes at higher concentrations [12, p. 43]. In general, S/S methods tend to be most effective for immobilizing nonvolatile heavy metals.

The proven effectiveness of the technology for a particular site or waste does not ensure that it will be effective at all sites or that treatment efficiencies achieved will be acceptable at other sites. For the ratings used in Table 23-1, demonstrated effectiveness means that at some scale, treatability tests showed that the technology was effective for that particular contaminant and matrix. The ratings of "Potential Effectiveness" and "No Expected Effectiveness" are both based upon expert judgment When potential effectiveness is indicated, the technology is believed capable of successfully treating the contaminant group in a particular matrix. When the technology is not applicable or will probably not work for a particular combination of contaminant group and matrix, a no-expected-effectiveness rating is given.

Another source of general observations and average removal efficiencies for different treatability groups is contained in the Superfund LDR Guide #6A, Obtaining a Soil and Debris Treatability Variance for Remedial Actions, (OSWER Directive 9347.3-06FS, September 1990) [16] and Superfund

LDR Guide #6B, Obtaining a Soil and Debris Treatability Variance for Removal Actions, (OSWER Directive 9347.3-06BFS, September 1990) [17]. Performance data presented in this chapter should not be considered directly applicable to other Superfund sites. A number of variables such as the specific mix and distribution of contaminants affect system performance. A thorough characterization of the site and a well-designed and conducted treatability study are highly recommended.

Other sources of information include the U.S. EPA's Risk Reduction Engineering Laboratory Treatability Database (accessible via ATTIC) and the U.S. EPA Center Hill Database (contact Patricia Erickson).

TECHNOLOGY LIMITATIONS

Tables 23-2 and 23-3 summarize factors that may interfere with stabilization and solidification processes, respectively.

Physical mechanisms that can interfere with the S/S process include incomplete mixing due to the presence of high moisture or organic chemical content resulting in only partial wetting or coating of the waste particles with the stabilizing and binding agents and the aggregation of untreated waste into lumps [6]. Wastes with a high clay content may clump, interfering with the uniform mixing with the S/S agents, or the clay surface may adsorb key reactants, interrupting the polymerization chemistry of the S/S agents. Wastes with a high hydrophilic organic content may interfere with solidification by disrupting the gel structure of the curing cement or pozzolanic mixture [11, p. 18; 18]. The potential for undermixing is greatest for dry or pasty wastes and least for freely flowing slurries [11, p. 13]. All in situ systems must provide for the introduction and mixing of the S/S agents with the waste in the proper proportions in the surface or subsurface waste site environment. Quality control is inherently more difficult with in situ products than with ex situ products [4, p. 7.95].

Chemical mechanisms that can interfere with S/S of cement-based systems include chemical adsorption, complexation, precipitation, and nucleation [1, p. 82]. Known inorganic chemical interferents in cement-based S/S processes include copper, lead, and zinc, and the sodium salts of arsenate, borate, phosphate, iodate, and sulfide [6, p. 2.13; 12, p. 11]. Sulfate interference can be mitigated by using a cement material with a low tricalcium aluminate content (e.g., Type V portland cement) [6, p. 2.13]. Problematic organic interferents include oil and grease, phenols [8, p. 19], surfactants, chelating agents [11, p. 22], and ethylene glycol [18]. For thermoplastic micro- and macro-encapsulation, stabilization of a waste containing strong oxidizing agents reactive toward rubber or asphalt must also be avoided [19, p. 10.114]. Pretreating the wastes to chemically or biochemically react or to thermally or chemically extract potential interferents should minimize these problems, but the cost advantage of S/S may be lost, depending on the characteristics and volume of the waste and the type and degree of pretreatment required. Organic polymer additives in various stages of development and field testing may significantly improve the performance of the cementitious and pozzolanic S/S agents with respect to immobilization of organic substances, even without the addition of sorbents.

Volume increases associated with the addition of S/S agents to the waste may prevent returning the waste to the landform from which it was excavated, where landfill volume is limited. Where post-closure earthmoving and landscaping are required, the treated waste must be able to support the weight of heavy equipment. The EPA recommends a minimum compressive strength of 50 to 200 psi [7, p. 4.13]; however, this should be a site-specific determination.

Environmental conditions must be considered in determining whether and when to implement an S/S technology. Extremes of heat, cold, and precipitation can adversely affect S/S applications. For example, the viscosity of one or more of the materials in the mixture may increase rapidly with falling temperatures, or the cure rate may be slowed unacceptably [20, p. 27]. In cement-based S/S processes the engineering properties of the concrete mass produced for the treatment of the waste are highly dependent on the water/cement ratio and the degree of hydration of the cement. High water/cement ratios yield large pore sizes and thus higher permeabilities [21, p. 177]. This factor may not be readily controlled in environmental applications of S/S, and pretreatment (e.g., drying) of the waste may be required.

Depending on the waste and binding agents involved, S/S processes can produce hot gases, including vapors that are potentially toxic, irritating, or noxious to workers or communities downwind from the processes [22, p. 4]. Laboratory tests demonstrate that as much as 90 percent of VOCs are

Table 23-2. Summary of Factors That May Interfere with Stabilization Processes[a]

Characteristics Affecting Processing Feasibility	Potential Interference
VOCs	Volatiles not effectively immobilized; driven off by heat of reaction. Sludges and soils containing volatile organics can be treated using a heated extruder evaporator or other means to evaporate free water and VOCs prior to mixing with stabilizing agents.
Use of acidic sorbent with metal hydroxide wastes	Solubilizes metals.
Use of acidic sorbent with cyanide wastes	Releases hydrogen cyanide.
Use of acidic sorbent with waste containing ammonium compounds	Releases ammonia gas.
Use of acidic sorbent with sulfide wastes	Releases hydrogen sulfide.
Use of alkaline sorbent (containing carbonates such as calcite or dolomite) with acid waste	May create pyrophoric waste.
Use of siliceous sorbent (soil, fly ash) with hydrofluoric acid waste	May produce soluble fluorosilicates.
Presence of anions in acidic solutions that form soluble calcium salts (e.g., calcium chloride acetate and bicarbonate)	Cation exchange reactions—leach calcium from S/S product increases permeability of concrete—increases rate of exchange reactions.
Presence of halides	Easily leached from cement and lime.

[a] Adapted from Reference 2.

volatilized during solidification and as much as 60 percent of the remaining VOCs are lost in the next 30 days of curing [23, p. 6]. In addition, if volatile substances with low flash points are involved, the potential exists for fire and explosions where the fuel-to-air ratio is favorable [22, p. 4]. Where volatilization problems are anticipated, many S/S systems now provide for vapor collection and treatment. Under dry and/or windy environmental conditions, both ex situ and in situ S/S processes are likely to generate fugitive dust with potentially harmful impacts on occupational and public health, especially for downwind communities.

Scaleup for S/S processes from bench-scale to full-scale operation involves inherent uncertainties. Variables such as ingredient flowrate control, materials mass balance, mixing, and materials handling and storage, along with the weather compared to the more controlled environment of a laboratory, all may affect the success of a field operation. These potential engineering difficulties emphasize the need for a field demonstration prior to full-scale implementation [2].

TECHNOLOGY DESCRIPTION

Waste stabilization involves the addition of a binder to a waste to immobilize waste contaminants effectively. Waste solidification involves the addition of a binding agent to the waste to form a solid material. Solidifying waste improves its material handling characteristics and reduces permeability to leaching agents such as water, brine, and inorganic and organic acids by reducing waste porosity and exposed surface area. Solidification also increases the load-bearing capacity of the treated waste, an advantage when heavy equipment is involved. Because of their dilution effect, the addition of binders must be accounted for when determining reductions in concentrations of hazardous constituents in S/S treated waste.

S/S processes are often divided into the following broad categories: inorganic processes (cement and pozzolanic) and organic processes (thermoplastic and thermosetting). Generic S/S processes involve materials that are well known and readily available. Commercial vendors have typically devel-

Table 23-3. Summary of Factors That May Interfere with Solidification Processes[a]

Characteristics Affecting Processing Feasibility	Potential Interference
Organic compounds	Organics may interfere with bonding of waste materials with inorganic binders.
Semivolatile organics or polyaromatic hydrocarbons (PAHs)	Organics may interfere with bonding of waste materials.
Oil and grease	Weaken bonds between waste particles and cement by coating the particles. Decrease in unconfined compressive strength with increased concentrations of oil and grease.
Fine particle size	Insoluble material passing through a No. 200 mesh sieve can delay setting and curing. Small particles can also coat larger particles, weakening bonds between particles and cement or other reagents. Particle size >1/4 inch in diameter not suitable.
Halides zolan	May retard setting; easily leached from cement and poz- S/S. May dehydrate thermoplastic solidification.
Soluble salts of manganese, tin, zinc, copper, and lead	Reduced physical strength of final product caused by large variations in setting time and reduced dimensional stability of the cured matrix, thereby increasing leachability potential.
Cyanides	Cyanides interfere with bonding of waste materials.
Sodium arsenate, borates, phosphates, iodates, sulfides, and carbohydrates	Retard setting and curing and weaken strength of final product.
Sulfates	Retard setting and cause swelling and spalling in cement S/S. With thermoplastic solidification, may dehydrate and rehydrate, causing splitting.
Phenols	Marked decreases in compressive strength for high phenol levels.
Presence of coal or lignite	Coals and lignites can cause problems with setting, curing, and strength of the end product.
Sodium borate, calcium sulfate, potassium dichromate, and carbohydrates	Interferes with pozzolanic reactions that depend on formation of calcium silicate and aluminate hydrates.
Nonpolar organics (oil, grease, aromatic hydrocarbons, PCBs)	May impede setting of cement, pozzolan, or organic-polymer S/S. May decrease long-term durability and allow escape of volatiles during mixing. With thermoplastic S/S, organics may vaporize from heat.
Polar organics (alcohols, phenols, organic acids, glycols)	With cement or pozzolan S/S, high concentrations of phenol may retard setting and may decrease short-term durability; all may decrease long-term durability. With thermoplastic S/S, organics may vaporize. Alcohols may retard setting of pozzolans.
Solid organics (plastics, tars, resins)	Ineffective with urea formaldehyde polymers; may retard setting of other polymers.
Oxidizers (sodium hypochlorite, potassium permanganate, nitric acid, or potassium dichromate)	May cause matrix breakdown or fire with thermoplastic or organic polymer S/S.
Metals (lead, chromium, cadmium, arsenic, mercury)	May increase setting time of cements if concentration is high.
Nitrates, cyanides	Increase setting time, decrease durability for cement-based S/S.
Soluble salts of magnesium, tin, zinc, copper, and lead	May cause swelling and cracking within inorganic matrix, exposing more surface area to leaching.

Environmental/waste conditions that lower the pH of matrix	Eventual matrix deterioration.
Flocculants (e.g., ferric chloride)	Interference with setting of cements and pozzolans.
Soluble sulfates >0.01% in soil or 150 mg/L in water	Endangerment of cement products due to sulfur attack.
Soluble sulfates >0.5% in soil or 2000 mg/L in water	Serious effects on cement products from sulfur attacks.
Oil, grease, lead, copper, zinc, and phenol	Deleterious to strength and durability of cement, lime/fly ash, fly ash/cement binders.
Aliphatic and aromatic hydrocarbons	Increase set time for cement.
Chlorinated organics	May increase set time and decrease durability of cement if concentration is high.
Metal salts and complexes	Increase set time and decrease durability for cement or clay/cement.
Inorganic acids	Decrease durability for cement (portland Type I) or clay/cement.
Inorganic bases	Decrease durability for clay/cement; KOH and NaOH decrease durability for portland cement Type III and IV.

[a] Adapted from Reference 2.

oped generic processes into proprietary processes by adding special additives to provide better control of the S/S process or to enhance specific chemical or physical properties of the treated waste. Less frequently, S/S processes combine organic binders with inorganic binders (e.g., diatomaceous earth and cement with polystyrene, polyurethane with cement, and polymer gels with silicate and lime cement) [2].

The waste can be mixed in a batch or continuous system with the binding agents after removal (ex situ) or in place (in situ). In ex situ applications, the resultant slurry can be (1) poured into containers (e.g., 55-gallon drums) or molds for curing and then off- or onsite disposal, (2) disposed in onsite waste management cells or trenches, (3) injected into the subsurface environment, or (4) reused as construction material with the appropriate regulatory approvals. In in situ applications, the S/S agents are injected into the subsurface environment in the proper proportions and mixed with the waste using backhoes for surface mixing or augers for deep mixing [5]. Liquid waste may be pretreated to separate solids from liquids. Solid wastes may also require pretreatment in the form of pH adjustment, steam or thermal stripping, solvent extraction, chemical reaction, or biodegradation to remove excessive VOCs and SVOCs that may react with the S/S process. The type and proportions of binding agents are adjusted to the specific properties of the waste to achieve the desired physical and chemical characteristics of the waste appropriate to the conditions at the site based on bench-scale tests. Although ratios of waste-to-binding agents are typically in the range of 10:1 to 2:1, ratios as low as 1:4 have been reported. However, projects utilizing low waste-to-binder ratios have high costs and large volume expansion.

Figures 23-1 and 23-2 depict generic elements of typical ex situ and in situ S/S processes, respectively. Ex situ processing involves: (1) excavation to remove the contaminated waste from the subsurface; (2) classification to remove oversize debris; (3) mixing; and (4) off-gas treatment. In situ processing has only two steps: (1) mixing; and (2) off-gas treatment. Both processes require a system for delivering water, waste, and S/S agents in proper proportions and a mixing device (e.g., rotary drum paddle or auger). Ex situ processing requires a system for delivering the treated waste to molds, surface trenches, or subsurface injection. The need for off-gas treatment using vapor collection and treatment modules is specific to the S/S project.

PROCESS RESIDUALS

Under normal operating conditions, neither ex situ nor in situ S/S technologies generate significant quantities of contaminated liquid or solid waste. Certain S/S projects require treatment of the off-gas. Prescreening collects debris and materials too large for subsequent treatment.

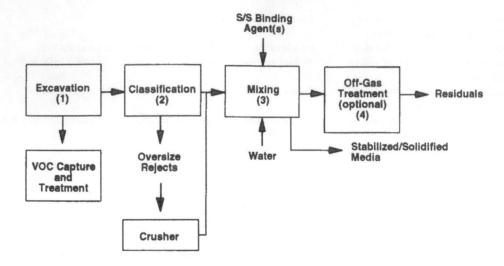

Figure 23-1. Generic elements of a typical ex situ S/S process.

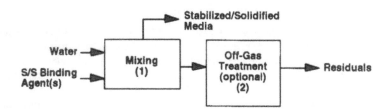

Figure 23-2. Generic elements of a typical in situ S/S process.

If the treated waste meets the specified cleanup levels, it could be considered for reuse onsite as backfill or construction material. In some instances, treated waste may have to be disposed of in an approved landfill. Hazardous residuals from some pretreatment technologies must be disposed of according to appropriate procedures.

SITE REQUIREMENTS

The site must be prepared for the construction, operation, maintenance, decontamination, and ultimate decommissioning of the equipment. An area must be cleared for heavy equipment access roads, automobile and truck parking lots, material transfer stations, the S/S process equipment, setup areas, decontamination areas, the electrical generator, equipment sheds, storage tanks, sanitary and process wastewater collection and treatment systems, workers quarters, and approved disposal facilities (if required). The size of the area required for the process equipment depends on several factors, including the type of S/S process involved, the required treatment capacity of the system, and site characteristics, especially soil topography and load-bearing capacity. A small mobile ex situ unit could occupy a space as small as that taken up by two standard flatbed trailers. An in situ system requires a larger area to accommodate a drilling rig, as well as a larger area for auger decontamination.

Process, decontamination, transfer, and storage areas should be constructed on impermeable pads with berms for spill retention and drains for the collection and treatment of stormwater runoff. Stormwater storage and treatment capacity requirements will depend on the size of the bermed area and the local climate. Standard 440V, three-phase electrical service is usually needed. The quantity and quality of process water required for pozzolanic S/S technologies are technology-specific.

S/S process quality control requires information on the range of concentrations of contaminants and potential interferents in waste batches awaiting treatment and on treated product properties such as compressive strength, permeability, leachability, and in some instances, contaminant toxicity.

PERFORMANCE DATA

Most of the data on S/S performance come from studies conducted for EPA's Risk Reduction Engineering Laboratory under the Superfund Innovative Technology Evaluation (SITE) Program. Pilot-scale demonstration studies available for review during the preparation of this chapter included: Soliditech, Inc. at Morganville, New Jersey (petroleum hydrocarbons, PCBs, other organic chemicals, and heavy metals); International Waste Technologies (IWT) process using the Geo Con, Inc. deep-soil-mixing equipment, at Hialeah, Florida (PCBs, VOCs); Chemfix Technologies, Inc., at Clackamas, Oregon (PCBs, arsenic, heavy metals); Im-Tech (formerly Hazcon) at Douglassville, Pennsylvania (oil and grease, heavy metals including lead, and low levels of VOCs and PCBs); Silicate Technology Corporation (STC), at Selma, California (arsenic, chromium, copper, pentachlorophenol and associated polychlorinated dibenzofurans and dibenzo-p-dioxins). The performance of each technology was evaluated in terms of ease of operation, processing capacity, frequency of process outages, residuals management, cost, and the characteristics of the treated product. Such characteristics included weight, density, and volume changes; UCS and moisture content of the treated product before and after freeze/thaw and wet/dry weathering cycles; permeability (or permissivity) to water; and leachability following curing and after the weathering test cycles. Leachability was measured using several different standard methods, including the EPA's TCLP. Table 23-4 summarizes the SITE performance data from these sites [20; 24–28].

A full-scale S/S operation has been implemented at the Northern Engraving Corporation (NEC) site in Sparta, Wisconsin, a manufacturing facility which produces metal nameplates and dials for the automotive industry. The following information on the site is taken from the remedial action report. Four areas at the site that have been identified as potential sources of soil, ground-water, and surface water contamination are the sludge lagoon, seepage pit, sludge dump site, and lagoon drainage ditch. The sludge lagoon was contaminated primarily with metal hydroxides consisting of nickel, copper, aluminum, fluoride, iron, and cadmium. The drainage ditch, which showed elevated concentrations of copper, aluminum, fluoride, and chromium, was used to convey effluent from the sludge lagoon to a stormwater runoff ditch. The contaminated material in the drainage ditch area and sludge dumpsite was then excavated and transported into the sludge lagoon for stabilization with the sludge present. The vendor, Geo-Con, Inc., achieved stabilization by the addition of hydrated lime to the sludge. Five samples of the solidified sludge were collected for Extraction Procedure (EP) toxicity leaching analyses. Their contaminant concentrations (in mg/L) are as follows: Arsenic (<.01); Barium (.35–1.04); Cadmium (<.005); Chromium (<.01); Lead (<.2); Mercury (<.001); Selenium (<.005); Silver (<.01); and Fluoride (2.6–4.1). All extracts were not only below the EP toxicity criteria but (with the exception of fluoride) met drinking water standards as well.

Approximately three weeks later, UCS tests on the solidified waste were taken. Test results ranged from 2.4 to 10 psi, well below the goal of 25 psi. One explanation for the low UCS could be due to shear failure along the lenses of sandy material and organic peat-like material present in the samples. It was determined that it would not be practical to add additional quantities of lime into the stabilized sludge matrix because of its high solids content. Therefore, the stabilized sludge matrix capacity will be increased to support the clay cap by installing an engineered subgrade for the cap system using a stabilization fabric and aggregate prior to cap placement [29].

The Industrial Waste Control (IWC) Site in Fort Smith, Arkansas, a closed and covered industrial landfill built in an abandoned surface coal mine, has also implemented a full-scale S/S system. Until 1978, painting wastes, solvents, industrial process wastes, and metals were disposed at the site. The primary contaminants of concern were methylene chloride, ethyl benzene, toluene, xylene, trichloroethane, chromium, and lead. Along with S/S of the onsite soils, other technologies used were: excavation, slurry wall, french drains, and a landfill cover. Soils were excavated in the contaminated region (Area C) and a total of seven lifts were stabilized with fly ash on mixing pads previously formed. A clay liner was then constructed in Area C to serve as a leachate barrier. After the lifts passed the TCLP test they were taken to Area C for in situ solidification. Portland cement was

Table 23-4. Summary of SITE Performance Data

Site	Vendor Technology	Pretreatment	Post-Treatment
Imperial Oil Co./ Champion Chemical Co. Morganville, NJ	Soliditech: Urrichem reagent, water, additives, Type II portland cement	Bulk density: 1.14 to 1.26 g/cm³; Permeability: Not determined; UCS: Not determined; Lead-TCLP Extract: 0.46 mg/L	Bulk density: 1.43 to 1.68 g/cm³; Permeability 8.9×10^{-9} to 4.5×10^{-7} cm/s; UCS: 390 to 860 psi; Lead-TCLP extract: <0.05 to <0.20 mg/L
GE Electrical Service Shop Hialeah, FL	IWT-DMS/Geo-Con; In situ injection of silicate additive	Bulk density: 1.55 g/mL; Permeability: 1.8×10^{-2} cm/s; UCS: 1.2 to 1.85 psi	Bulk density: 1.88 g/mL; Permeability: 0.24×10^{-7} to 21×10^{-7} cm/s; UCS: 300 to 500 psi
Portable Equipment Salvage Co. Clackamas, OR	Chemfix: polysilicates and dry calcium containing reagents	TCLP-Extractable (Pb, Cu, Zn): 12 to 880 mg/L	Hydraulic cond. (CSS-13): 2.4×10^{-6} to 2.7×10^{-4} cm/s; Bulk density: 2.0 to 2.6 g/cm³; TCLP-Extractable (Pb, Cu, Zn): 0.024 to 47 mg/L; Hydraulic cond. (CSS-14): 4.6×10^{-7} to 1.2×10^{-6} cm/s; Bulk density: 1.6 to 2.0 g/cm³; USC (14, 21, 28 days): 131, 136, 143 psi; Immersion UCS (30, 60, 90 days): 177, 188, 204 psi
Douglasville mL Douglasville, PA	Imtech (Hazcon): Chloranan™, water and cement	Bulk density: 1.23 g/mL; Permeability: 0.57 cm/s; TCLP-Extractable Pb: 52.6 mg/L	Bulk density (7, 28 days): 1.95, 1.99 g/mL; Permeability (7, 28 days): 1.6×10^{-9}, 2.3×10^{-9} cm/s; TCLP-Extractable Pb (7, 28 days): 0.14, 0.05 mg/L
Selma Pressure Treating Wood Preserving Site Selma, CA	Silicate Tech Corp.: alumino-silicate compounds	Arsenic-TCLP: 1.06 to 3.33 ppm; Arsenic-Distilled H₂O TCLP: 0.73 to 1.25 ppm; PCP-TWA: 1983 to 8317 ppm; Bulk density: 1.42 to 1.54 g/cm	UCS (7, 28 days): 1447, 113 psi; Arsenic-TCLP: 0.086 to 0.875 ppm; Arsenic-Distilled H_2O TCLP: <0.01 to 0.012 ppm; PCP-TWA: 14 to 158 ppm; Bulk density: 1.57 to 1.62 g/cm; Permeability: 0.8×10^{-7} to 1.7×10^{-7} cm/s; UCS: 259 to 347 psi

UCS - Unconfined Compressive Strength; TCLP - Toxicity Characteristic Leaching Procedure; TWA - Total Waste Analysis

added to solidify each lift, and they obtained the UCS goal of 125 psi. With the combination of the other technologies, the overall system appears to be functioning properly [30].

Other Superfund sites where full-scale S/S has been completed to date include Davie Landfill (82,158 yd³ of sludge containing cyanide, sulfide, and lead treated with Type I portland cement in 45 days) [31]; Pepper's Steel and Alloy (89,000 yd³ of soil containing lead, arsenic, and PCBs treated with portland cement and fly ash) [32]; and Sapp Battery and Salvage (200,000 yd³ soil fines and washings containing lead and mercury treated with portland cement and fly ash in roughly 18 months) [33], all in Region 4; and Bio-Ecology, Inc. (about 20,000 yd³ of soils, sludge, and liquid waste containing heavy metals, VOCs, and cyanide treated with cement kiln flue dust alone or with lime) in Region 6 [34]. All sites required that the waste meet the appropriate leaching test and UCS criteria. At the Sapp Battery site, the waste also met a permeability criterion of 10^{-6} cm/s [33]. Past remediation appraisals by the responsible remedial project managers indicate the S/S technologies are performing as intended.

RCRA LDRs that require treatment of wastes based on BDAT levels prior to land disposal may sometimes be determined to be Applicable or Relevant and Appropriate Requirements (ARARs) for CERCLA response actions. S/S can produce a treated waste that meets treatment levels set by BDAT but may not reach these treatment levels in all cases. The ability to meet required treatment levels is dependent upon the specific waste constituents and the waste matrix. In cases where S/S does not meet these levels, it still may in certain situations be selected for use at a site if a treatability variance establishing alternative treatment levels is obtained. Treatability variances may be justified for handling complex soil and debris matrices. The following guides describe when and how to seek a treatability variance for soil and debris: Superfund LDR Guide #6A, Obtaining a Soil and Debris Treatability Variance for Remedial Actions (OSWER Directive 9347.346FS) [16], and Superfund LDR Guide #6B, Obtaining a Soil and Debris Treatability Variance for Removal Actions (OSWER Directive 9347.346BFS) [17]. Another approach could be to use other treatment techniques in conjunction with S/S to obtain desired treatment levels.

TECHNOLOGY STATUS

In 1990, 24 RODs identified S/S as the proposed remediation technology [35]. To date, only about a dozen Superfund sites have proceeded through full-scale S/S implementation to the operation and maintenance (O&M) phase, and many of those were small pits, ponds, and lagoons. Some involved S/S for offsite disposal in RCRA-permitted facilities. Table 23-5 summarizes these sites where full-scale S/S has been implemented under CERCLA or RCRA [7, p. 34].

More than 75 percent of the vendors of S/S technologies use cement-based or pozzolanic mixtures [11, p. 2]. Organic polymers have been added to various cement-based systems to enhance performance with respect to one or more physical or chemical characteristics, but only mixed results have been achieved. For example, tests of standardized wastes treated in a standardized fashion using acrylonitrile, vinyl ester, polymer cement, and water-based epoxy yielded mixed results. Vinyl and plastic cement products achieved superior UCS and leachability to cement-only and cement-fly ash S/S, while the acrylonitrile and epoxy polymers reduced UCS and increased leachable TOC, in several instances by two or three orders of magnitude [36, p. 156].

The estimated cost of treating waste with S/S ranges from $50 to $250 per ton (1992 dollars). Costs are highly variable due to variations in site, soil, and contaminant characteristics that affect the performance of the S/S processes evaluated. Economies of scale likely to be achieved in full-scale operations are not reflected in pilot-scale data.

EPA CONTACT

Technology-specific questions regarding S/S may be directed to Ed Barth (513) 569-7669 or Ed Bates, RREL-Cincinnati, (513) 569-7774 (see Preface for mailing address).

ACKNOWLEDGMENTS

This chapter was prepared for the U.S. Environmental Protection Agency, Office of Research and Development (ORD), Risk Reduction Engineering Laboratory (RREL), Cincinnati, Ohio, by Sci-

Table 23-5. Summary of Full-Scale S/S Sites

Site	Contaminant	Physical Form	Binder	Percentage Binder(s) Added	Treatment (batch/continuous in situ)
Independent Nail, SC	Zn, Cr, Cd, Ni	Solid/soil	Portland cement	20%	Batch plant
Midwest, US Plating Company	Cu, Cr, Ni	Sludge	Portland cement	20%	In situ
Unnamed	Pb/soil 2–100 ppm	Solid/soils	Portland cement and proprietary ingredient	Cement (15–20%) Proprietary (5%)	In situ
Marathon Steel, Phoenix, AZ	Pb, Cd	Dry-landfill	Portland cement and silicates	Varied 7–15% (cement)	Concrete batch plant
Alaska Refinery	Oil/oil sludges	Sludges, variable	Portland cement and proprietary ingredient	Varied 50+	Concrete batch plant
Unnamed, KY	Vinyl chloride Ethylene dichloride	Sludges, variable	Portland cement and proprietary ingredient	Varied 25+	In situ
NE Refinery	Oil sludges, Pb, Cr, As	Sludges, variable	Kiln dust (high CaO content)	Varied, 15–30%	In situ
Velsicol Chemical	Pesticides and organics (resins, etc.) up to 45% organic	Sludges, variable	Portland cement and kiln dust, proprietary ingredient	Varied (cement 5–15%)	In situ
Amoco Wood River	Oil/solids Cd, Cr, Pb	Sludges	Proprietary ingredient	NA, proprietary	Continuous flow (proprietary)
Pepper Steel & Alloy, Miami, FL	Oil saturated soil Pb-1000 ppm PCBs-200 ppm As-1-200 ppm	Soils	Pozzolanic and proprietary ingredient	30%	Continuous feed (mixer proprietary design)
Vickery, OH	Waste acid PCBs (<500 ppm) dioxins	Sludges (viscous)	Lime and kiln dust	15% CaO 5% kiln dust	In situ
Wood Treating, Savannah, GA	Creosote wastes	Sludges	Kiln dust	20%	In situ

			...ingredient		
etc.					
API Sep. Sludge, Puerto Rico	API separator sludges	Sludges	Portland cement and proprietary ingredient	50% cement 4% proprietary	Concrete batch plant
Metaplating, WI	Al-9500 ppm Ni-750 ppm Cr-220 ppm Cu-2000 ppm	Sludges	Lime	10–25%	In situ

NA = not available.

ence Applications International Corporation (SAIC) under Contract No. 68-C84062 (WA 2-22). Mr. Eugene Harris served as the EPA Technical Project Manager. Mr. Gary Baker was SAIC's Work Assignment Manager. This chapter was written by Mr. Larry Fink and Mr. George Wahl of SAIC. The authors are especially grateful to Mr. Carlton Wiles and Mr. Edward Bates of EPA, RREL and Mr. Edwin Barth of EPA, CERI, who have contributed significantly by serving as technical consultants during the development of this chapter.

The following other EPA and contractor personnel have contributed their time and comments by participating in the expert review meetings or peer reviews of the chapter.

Dr. Paul Bishop, University of Cincinnati
Dr. Jeffrey Means, Battelle
Ms. Mary Boyer, SAIC-Raleigh
Mr. Cecil Cross, SAIC-Raleigh
Ms. Margaret Groeber, SAIC-Cincinnati
Mr. Eric Saylor, SAIC-Cincinnati

REFERENCES

1. Conner, J.R. Chemical Fixation and Solidification of Hazardous Wastes, Van Nostrand Reinhold, New York, 1990.
2. Technical Resources Document on Solidification/Stabilization and its Application to Waste Materials (Draft), Contract No. 68-C0-0003, Office of Research and Development, U.S. Environmental Protection Agency, Cincinnati, Ohio, 1991.
3. Guidance on Key Terms. Office of Solid Waste and Emergency Response. U.S. Environmental Protection Agency. Directive No. 9200.5-220, Washington, D.C., 1991.
4. Wiles, C.C. Solidification and Stabilization Technology. In: Standard Handbook of Hazardous Waste Treatment and Disposal, H.M. Freeman, Ed., McGraw Hill, New York, 1989.
5. Jasperse, B.H. Soil Mixing, Hazmat World, November 1989.
6. Handbook for Stabilization/Solidification of Hazardous Waste. EPA/540/2-86/001 (NTIS PB87-116745), June 1986.
7. Stabilization/Solidification of CERCLA and RCRA Wastes; Physical Tests, Chemical Testing Procedures, Technology, and Field Activities. EPA/625/6-89/022, May 1990.
8. Wiles, C.C. and E. Barth. Solidification/Stabilization: Is it Always Appropriate? In: Stabilization and Solidification of Hazardous, Radioactive and Mineral Wastes, 2nd Volume, ASTM STP 1123, T.M. Gilliam and C.C. Wiles (eds.), American Society for Testing and Materials, Philadelphia, Pennsylvania, December 1990, pp. 18-32.
9. Superfund Treatability Clearinghouse Abstracts. EPA/540/2-89/001 (NTIS PB90-119751), August 1989.
10. Kasten, J.L., H.W. Godbee, T.M. Gilliam, and S.C. Osborne, 1989. Round I Phase I Waste Immobilization Technology Evaluation Subtask of the Low-Level Waste Disposal Development and Demonstration Program, Prepared by Oak Ridge National Laboratories, Martin Marietta Energy Systems, Inc., Oak Ridge, Tennessee, for Office of Defense and Transportation Management under Contract DE-AC05-840R21400, May 1989.
11. JACA Corporation. Critical Characteristics and Properties of Hazardous Solidification/Stabilization. Prepared for Water Engineering Research Laboratory, Office of Research and Development, U.S. Environmental Protection Agency, Cincinnati, Ohio. Contract No. 68-03-3186, 1985.
12. Bricka, R.M. and L.W. Jones. An Evaluation of Factors Affecting the Solidification/Stabilization of Heavy Metal Sludge, Waterways Experimental Station, U.S. Army Corps of Engineers, Vicksburg, Mississippi, 1989.
13. Fate of Polychlorinated Biphenyls (PCBs) in Soil Following Stabilization with Quicklime, EPA/600/2-91/052, September 1991.
14. Convery, J. Status Report on the Interaction of PCB's and Quicklime, Risk Reduction Engineering Laboratory, Office of Research and Development, U.S. Environmental Protection Agency, Cincinnati, Ohio, June 1991.

15. Stinson, M.K. EPA SITE Demonstration of the International Waste Technologies/Geo-Con In Situ Stabilization/Solidification Process. Air and Waste Management J., 40(11):1569–1576.

16. Superfund LDR Guide #6A (2nd edition), Obtaining a Soil and Debris Treatability Variance for Removal Actions, OSWER. Directive 9347.3-06FS (NTIS PB91-921327), September 1990.

17. Superfund LDR Guide #6B, Obtaining a Soil and Debris Treatability Variance for Removal Actions, OSWER Directive 9347.3 06BFS (NTIS PB91-921310), September 1990.

18. Chasalani, D., F.K. Cartledge, H.C. Eaton, M.E. Tittlebaum, and M.B. Walsh. The Effects of Ethylene Glycol on a Cement-Based Solidification Process. Hazardous Waste and Hazardous Materials. 3(2):167–173, 1986.

19. Handbook of Remedial Action at Waste Disposal Sites. EPA/625/6-85/006 (NTIS PB87-201034), June 1985.

20. Technology Evaluation Report SITE Program Demonstration Test Soliditech, Inc. Solidification/Stabilization Process, Volume I. EPA/540/5-89/005a, U.S. Environmental Protection Agency, Cincinnati, Ohio, February 1990. [See below for complete list of Solidtech SITE reports.]

21. Kirk-Othmer. Cement. Encyclopedia of Chemical Technology, 3rd Ed., John Wiley & Sons, New York. pp. 163–193, 1981.

22. Soundararajan, R., and J.J. Gibbons, Hazards in the Quicklime Stabilization of Hazardous Waste. Unpublished paper delivered at the Gulf Coast Hazardous Substances Research Symposium, February 1990.

23. Weitzman, L., L.R. Hamel, and S. Cadmus. Volatile Emissions from Stabilized Waste, Prepared By Acurex Corporation Under Contract No. 68-02-3993 (32, 37) for the Risk Reduction Engineering Laboratory, Office of Research and Development, U.S. Environmental Protection Agency, Cincinnati, Ohio, May 1989.

24. Technology Evaluation Report SITE Program Demonstration Test International Waste Technologies In Situ Stabilization/Solidification—Hialeah, Florida, Volume I. EPA/540/5-89/004a, June 1989. [See below for complete list of IWT/Geo-Con SITE reports.]

25. Technology Evaluation Report: Chemfix Technologies, Inc. Solidification/Stabilization Process—Clackamas, Oregon, Volume I. EPA/540/5-89/011a, September 1990. [See below for complete list of Chemfix SITE reports.]

26. Technology Evaluation Report SITE Program Demonstration Test, HAZCON Solidification, Douglassville, Pennsylvania, Volume I. EPA/540/5-89/001a, February 1989. [See below for complete list of Hazcon SITE reports.]

27. Bates, E.R., P.V. Dean, and I. Klich, Chemical Stabilization of Mixed Organic and Metal Compounds: EPA SITE Program Demonstration of the Silicate Technology Corporation Process. Journal of the Air & Waste Management Association. 42(5):724–728, 1992.

28. Applications Analysis Report Silicate Technology Corporation. Solidification/Stabilization Technology for Organic and Inorganic Contaminants in Soils, EPA/540/AR-92/010, December 1992. [See below for complete list of Silicate Technology SITE reports.]

29. Eder Associates Consulting Engineers, P.C. Northern Engraving Corporation Site Remedial Action Report. Sparta, Wisconsin, 1989.

30. Remedial Construction Report. Industrial Waste Control Site. Fort Smith, Arkansas. U.S. Environmental Protection Agency, 1991.

31. Jackson, R. RPM, Davie Landfill, Florida. Personal Communication. Region 4, U.S. Environmental Protection Agency, Atlanta, Georgia, August 1991.

32. Scott, D. RPM, Pepper's Steel and Alloy. Personal Communication. Region 4, U.S. Environmental Protection Agency, Atlanta, Georgia, October 1991.

33. Berry, M. RPM, Sapp Battery and Salvage, Florida. Personal Communication. Region 4, U.S. Environmental Protection Agency, Atlanta, Georgia, August 1991.

34. Pryor, C. RPM, Bio-Ecology Systems, Texas. Personal Communication. Region 6, U.S. Environmental Protection Agency, Dallas, Texas, August 1991.

35. Rod Annual Report, FY 1990. EPA/540/8-91/067, U.S. Environmental Protection Agency, Washington D.C., July, 1991.

36. Kyles, J.H., K.C. Malinowski, J.S. Leithner, and T.F. Stanczyk. The Effect of Volatile Organic Compounds on the Ability of Solidification/Stabilization Technologies to Attenuate Mobile Pollutants. In: Proceedings of the National Conference on Hazardous Waste and Hazardous Materials. Hazardous Materials Control Research Institute, Silver Spring, MD, March 16–18, 1987.

ADDITIONAL SITE PROGRAM REFERENCES (See Appendix A for information on how to obtain SITE program reports)

Solidtech:

Demonstration Bulletin (EPA/540/M5-89/005), August 1989
Technology Demonstration Summary (EPA/540/S5-89/005), May 1990
Applications Analysis (EPA/540/A5-89/005)
Technology Evaluation Vol. I (EPA/540/5-89/005a)
Technology Evaluation Vol. II (EPA/540/5-89/005b; NTIS PB90-191768)

International Waste Technologies (IWT)/Geo-Con:

Demonstration Bulletin (EPA/540/M5-89/004), August 1989
Technology Demonstration Summary (EPA/540/S5-89/004), June 1989
Technology Demonstration Summary, Update Report (EPA/540/S5-89/004a), January 1991
Applications Analysis (EPA/540/A5-89/004)
Technology Evaluation Vol. I (EPA/540/5-89/004a)
Technology Evaluation Vol. II (EPA/540/5-89/004b; NTIS PB89-194179)
Technology Evaluation Vol. III (EPA/540/5-89/004c; NTIS PB90-269069)
Technology Evaluation Vol. IV (EPA/540/5-89/004d; NTIS PB90-269077)

Chemfix Technologies, Inc.:

Demonstration Bulletin (EPA/540/M5-89/011), October 1989
Technology Demonstration Summary (EPA/540/S5-89/011), December 1990
Applications Analysis (EPA/540/A5-89/011)
Technology Evaluation Vol. I (EPA/540/5-89/011a; NTIS PB91-127696)
Technology Evaluation Vol. II (EPA/540/5-89/011b; NTIS PB90-274127)

Imtech (formerly HAZCON, Inc.):

Demonstration Bulletin (EPA/540/M5-89/001), March 1989
Technology Demonstration Summary (EPA/540/S5-89/001), March 1989
Applications Analysis (EPA/540/A5-89/001)
Technology Evaluation Vol. I (EPA/540/5-89/001a; NTIS PB89-158810)
Technology Evaluation Vol. II (EPA/540/5-89/001b; NTIS PB89-158828)

Silicate Technology Corporation:

Demonstration Bulletin (EPA/540/MR-92/010), March 1992
Technology Demonstration Summary (EPA/540/SR-92/010), August 1995
Applications Analysis (EPA/540/AR-89/010), December 1992

Chapter 24

Mobile/Transportable Incineration Treatment[1]

U.S. Environmental Protection Agency, Risk Reduction Engineering Laboratory, Cincinnati, OH

ABSTRACT

Incineration treats organic contaminants in solids and liquids by subjecting them to temperatures typically greater than 1000°F in the presence of oxygen, which causes the volatilization, combustion, and destruction of these compounds. This chapter describes mobile/transportable incineration systems that can be moved to and subsequently removed from Superfund and other hazardous waste sites. It does not address other thermal processes that operate at lower temperatures or those that operate at very high temperatures, such as a plasma arc. It is applicable to a wide range of organic wastes and is generally not used in treating inorganics and metals. Mobile/transportable incinerators exhibit essentially the same environmental performance as their stationary counterparts. To date, 49 of the 95 records of decision (RODs) designating thermal remedies at Superfund sites have selected onsite incineration as an integral part of a preferred treatment alternative. There are commercial-scale units in operation [5].[2] This chapter provides information on the technology applicability, the types of residuals resulting from the use of the technology, the latest performance data, site requirements, the status of the technology, and where to go for further information.

TECHNOLOGY APPLICABILITY

Mobile/transportable incineration has been shown to be effective in treating soils, sediments, sludges, and liquids containing primarily organic contaminants such as halogenated and nonhalogenated volatiles and semivolatiles, polychlorinated biphenyls (PCBs), pesticides, dioxins/furans, organic cyanides, and organic corrosives. The process is applicable for the thermal treatment of a wide range of specific Resource Conservation and Recovery Act (RCRA) wastes and other hazardous waste matrices that include pesticides and herbicides, spent halogenated and nonhalogenated solvents, chlorinated phenol and chlorinated benzene manufacturing wastes, wood preservation and wastewater sludge, organic chemicals production residues, pesticides production residues, explosives manufacturing wastes, petroleum refining wastes, coke industry wastes, and organic chemicals residues [1; 2; 4; 6 through 11; 13].

Information on the physical and chemical characteristics of the waste matrix is necessary to assess the matrix's impact on waste preparation, handling, and feeding; incinerator type, performance, size, and cost; air pollution control (APC) type and size; and residue handling. Key physical parameters include wastes matrix physical characteristics (type of matrix, physical form, handling properties, and particle size), moisture content, and heating value. Key chemical parameters include the type and concentration of organic compounds including PCBs and dioxins, inorganics (metals), halogens, sulfur, and phosphorous.

The effectiveness of mobile/transportable incineration on general contaminant groups for various matrices is shown in Table 24-1 [7, p. 9]. Examples of constituents within contaminant groups are provided in Reference 7, Technology Screening Guide for Treatment of CERCLA Soils and Sludges. This table is based on current available information or professional judgment when no information was available. The proven effectiveness of the technology for a particular site or waste does not

---

[1] EPA/540/2-90/014.

[2] [Reference number, page number.]

Table 24-1. Effectiveness of Incineration on General Contaminant Groups for Soil, Sediment, Sludge, and Liquid

Contaminant Groups	Soil/Sediment	Sludge	Liquid
Organic			
Halogenated volatiles	■	■	■
Halogenated semivolatiles	■	■	■
Nonhalogenated volatiles	■	■	■
Nonhalogenated semivolatiles	■	■	■
PCBs	■	■	■
Pesticides (halogenated)	▼	■	■
Dioxins/furans	■	■	■
Organic cyanides	▼	▼	▼
Organic corrosives	▼	▼	▼
Inorganic			
Volatile metals	●	●	●
Nonvolatile metals	●	●	●
Asbestos	●	●	●
Radioactive materials	●	●	●
Inorganic corrosives	●	●	●
Inorganic cyanides	▼	▼	▼
Reactive			
Oxidizers	▼	▼	▼
Reducers	▼	▼	▼

■ Demonstrated Effectiveness: Successful treatment test at some scale completed.
▼ Potential Effectiveness: Expert opinion that technology will work.
● No Expected Effectiveness: Expert opinion that technology will not work.

ensure that it will be effective at all sites or that the treatment efficiency achieved will be acceptable at other sites. For the ratings used in this table, demonstrated effectiveness means that, at some scale, treatability was tested to show that the technology was effective for a particular contaminant and matrix. The ratings of potential effectiveness are based upon expert judgment. Where potential effectiveness is indicated, the technology is believed capable of successfully treating the contaminant group in a particular matrix. When the technology is not applicable or will probably not work for a particular combination of contaminant group and matrix, a no-expected-effectiveness rating is given. Other sources of general observations and average removal efficiencies for different treatability groups are the Superfund LDR Guide #6A, Obtaining a Soil and Debris Treatability Variance for Remedial Actions, (OSWER Directive 9347.3-06FS) [13], and Superfund LDR Guide #6B, Obtaining a Soil and Debris Treatability Variance for Removal Actions, (OSWER Directive 9347.3-06BFS) [14].

LIMITATIONS

Toxic metals such as arsenic, lead, mercury, cadmium, and chromium are not destroyed by combustion. As a result, some will be present in the ash, while others are volatilized and released into the flue gas [1, pp. 3–6].

Alkali metals, such as sodium and potassium, can cause severe refractory attack and form a sticky, low-melting-point submicron particulate, which causes APC problems. A low feed stream concentration of sodium and potassium may be achieved through feed stock blending [1, pp. 3–11].

When PCBs and dioxins are present, higher temperatures and longer residence times may be required to destroy them to levels necessary to meet regulatory criteria [7, p. 34].

Moisture/water content of waste materials can create the need to co-incinerate these materials with higher BTU streams, or to use auxiliary fuels.

The heating value (BTU content) of the feed material affects feed capacity and fuel usage of the incinerator. In general, as the heating value of the feed increases, the feed capacity and fuel usage of the incinerator will decrease. Solid materials with high calorific values also may cause transient behaviors that further limit feed capacity [9, p. 4].

The matrix characteristics of the waste affect the pretreatment required and the capacity of the incinerator, and can cause APC problems. Organic liquid wastes can be pumped to and then atomized in the incinerator combustion chamber. Aqueous liquids may be suitable for incineration if they contain a substantial amount of organic matter. However, because of the large energy demand for evaporation when treating large volumes of aqueous liquids, pretreatment to dewater the waste may be cost-effective [1, pp. 3–14]. Also, if the organic content is low, other methods of treatment may be more economical. For the infrared incinerator, only solid and solid-like materials within a specific size and moisture content range can be processed because of the unique conveyor belt feed system within the unit.

Sandy soil is relatively easy to feed and generally requires no special handling procedures. Clay, which may be in large clumps, may require size reduction. Rocky soils usually require screening to remove oversize stones and boulders. The solids can then be fed by gravity, screw feeder, or ram-type feeder into the incinerator. Some types of solid waste may also require crushing, grinding, and/or shredding prior to incineration [1, pp. 3–17].

The form and structure of the waste feed can cause periodic jams in the feed and ash handling systems. Wooden pallets, metal drum closure rings, drum shards, plastics, trash, clothing, and mud can cause blockages if poorly prepared. Muddy soils can stick to waste processing equipment and plug the feed system [9, p. 8].

The particle size distribution of the ash generated from the waste can affect the amount of particulate carry-over from the combustion chamber to the rest of the system [9, p. 16].

Incineration of halogens, such as fluorine and chlorine, generates acid gases that can affect the capacity, the water removal and replacement rates that control total dissolved solids in the process water system, and the particulate emissions [9, p. 12]. The solutions used to neutralize these acid gases add to the cost of operating this technology.

Organic phosphorous compounds form phosphorous pentoxide, which attacks refractory material, causes slagging problems and APC problems. Slagging can be controlled by feed blending or operating at lower temperatures [1, pp. 3–10].

TECHNOLOGY DESCRIPTION

Figure 24-1 is a schematic of the mobile/transportable incineration process.

Waste preparation (1) includes excavation and/or moving the waste to the site. Depending on the requirements of the incinerator type for soils and solids, various equipment is used to obtain the necessary feed size. Blending is sometimes required to achieve a uniform feed size and moisture content or to dilute troublesome components [1, pp. 3–19].

The waste feed mechanism (2), which varies with the type of the incinerator, introduces the waste into the combustion system. The feed mechanism sets the requirements for waste preparation and is a potential source of problems in the actual operation of incinerators if not carefully designed [1, pp. 3–19].

Different incinerator designs (3) use different mechanisms to obtain the temperature at which the furnace is operated, the time during which the combustible material is subject to that temperature, and the turbulence required to ensure that all the combustible material is exposed to oxygen to ensure complete combustion. Three common types of incineration systems for treating contaminated soils are rotary kiln, circulating fluidized bed, and infrared.

The rotary kiln is a slightly inclined cylinder that rotates on its horizontal axis. Waste is fed into the high end of the rotary kiln and passes through the combustion chamber by gravity. A secondary combustion chamber (afterburner) further destroys unburned organics in the flue gases [7, p. 40].

Circulating fluidized bed incinerators use high air velocity to circulate and suspend the fuel/waste particles in a combustor loop. Flue gas is separated from heavier particles in a solids separation cyclone. Circulating fluidized beds do not require an afterburner [7, p. 35].

Infrared processing systems use electrical resistance heating elements or indirect fuel-fired radiant U-tubes to generate thermal radiation [1, pp. 4–5]. Waste is fed into the combustion chamber by a conveyor belt and exposed to the radiant heat. Exhaust gases pass through a secondary combustion chamber.

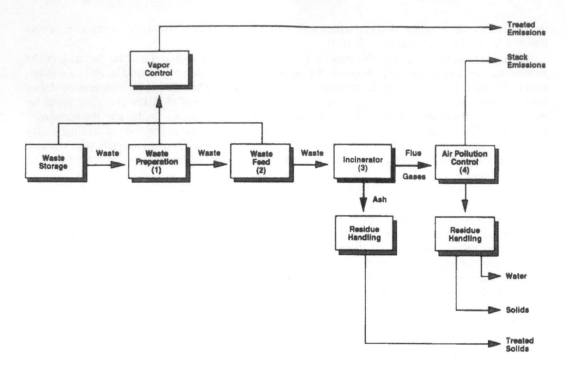

Figure 24-1. Mobile/transportable incineration process.

Off-gases from the incinerator are treated by the APC equipment to remove particulates and capture and neutralize acids (4). Rotary kilns and infrared processing systems may require both external particulate control and acid gas scrubbing systems. Circulating fluidized beds do not require scrubbing systems because limestone can be added directly into the combustor loop but may require a system to remove particulates [1, pp. 4–11; 2, p. 32]. APC equipment that can be used include venturi scrubbers, wet electrostatic precipitators, baghouses, and packed scrubbers.

PROCESS RESIDUALS

Three major waste streams are generated by this technology: solids from the incinerator and APC system, water from the APC system, and emissions from the incinerator.

Ash and treated soil/solids from the incinerator combustion chamber may be contaminated with heavy metals. APC system solids, such as fly ash, may contain high concentrations of volatile metals. If these residues fail required leachate toxicity tests, they can be treated by a process such as stabilization/solidification and disposed of onsite or in an approved landfill [7, p. 126].

Liquid waste from the APC system may contain caustic, high chlorides, volatile metals, trace organics, metal particulates, and inorganic particulates. Treatment may require neutralization, chemical precipitation, reverse osmosis, settling, evaporation, filtration, or carbon adsorption before discharge [7, p. 127].

The flue gases from the incinerator are treated by APC systems such as electrostatic precipitators or venturi scrubbers before discharge through a stack.

SITE REQUIREMENTS

The site should be accessible by truck or rail and a graded/gravel area is required for setup of the system. Concrete pads may be required for some equipment (e.g., rotary kiln). For a typical 5 tons per hour commercial-scale unit, 2 to 5 acres are required for the overall system site, including ancillary support [10, p. 25].

Standard 440V three-phase electrical service is needed. A continuous water supply must be available at the site. Auxiliary fuel for feed BTU improvement may be required.

Contaminated soils or other waste materials are hazardous and their handling requires that a site safety plan be developed to provide for personnel protection and special handling measures.

Various ancillary equipment may be required, such as liquid/sludge transfer and feed pumps, ash collection and solids handling equipment, personnel and maintenance facilities, and process-generated waste treatment equipment. In addition, a feed-materials staging area, a decontamination trailer, an ash handling area, water treatment facilities, and a parking area may be required [10, p. 24].

Proximity to a residential neighborhood will affect plant noise requirements and may result in more stringent emissions limitations on the incineration system.

Storage area and/or tanks for fuel, wastewater, and blending of waste feed materials may be needed.

No specific onsite analytical capabilities are necessary on a routine basis; however, depending on the site characteristics or a specific federal, state, or local requirement, some analytical capability may be required.

PERFORMANCE DATA

More than any other technology, incineration is subject to a series of technology-specific regulations, including the following federal requirements: the Clean Air Act 40 CFR 52.21 for air emissions; Toxic Substances Control Act (TSCA) 40 CFR 761.40 for PCB treatment and disposal; National Environmental Policy Act 40 CFR 6; RCRA 40 CFR 261/262/264/270 for hazardous waste generation, treatment performance, storage, and disposal standards; National Pollutant Discharge Elimination System 33 U.S.C. 1251 for discharge to surface waters; and the Noise Control Act P.L. 92-574. RCRA incineration standards have been proposed that address metal emissions and products of incomplete combustion. In addition, state requirements must be met if they are more stringent than the federal requirements [1, p. 6-1].

All incineration operations conducted at CERCLA sites on hazardous waste must comply with substantive and defined federal and state applicable or relevant and appropriate requirements (ARARs) at the site. A substantial body of trial burn results and other quality assured data exists to verify that incinerator operations remove and destroy organic contaminants from a variety of waste matrices to the parts per billion or even the parts per trillion level, while meeting stringent stack emission and water discharge requirements. The demonstrated treatment systems that will be discussed in the Technology Status section, therefore, can meet all the performance standards defined by the applicable federal and state regulations on waste treatment, air emissions, discharge of process waters, and residue ash disposal [1, p. A-1; 4, p. 4; 10, p. 9].

RCRA Land Disposal Restrictions (LDRs) that require treatment of wastes to best demonstrated available technology (BDAT) levels prior to land disposal may sometimes be determined to be ARARs for CERCLA response actions. The solid residuals from the incinerator may not meet required treatment levels in all cases. In cases where residues to not meet BDAT levels, mobile incineration still may be selected, in certain situations, for use at the site if a treatability variance establishing alternative treatment levels is obtained. EPA has made the treatability variance process available in order to ensure that LDRs do not unnecessarily restrict the use of alternative and innovative treatment technologies. Treatability variances may be justified for handling complex soil and debris matrices. The following guides describe when and how to seek a treatability variance for soil and debris: Superfund LDR Guide #6A, Obtaining a Soil an Debris Treatability Variance for Remedial Actions, (OSWER Directive 9347.3-06FS) [13] and Superfund LDR Guide #6B, Obtaining a Soil and Debris Treatability Variance for Removal Actions, (OSWER Directive 9347.3-06BFS) [14].

TECHNOLOGY STATUS

To date, 49 of the 95 RODs designating thermal remedies at Superfund sites have selected onsite incineration as an integral part of a preferred treatment alternative.

Table 24-2 lists the site experience of the various mobile/transportable incinerator systems. It includes information on the incinerator type/size, the site size, location, and contaminant source or waste type treated [5; 3, p. 80; 8, p. 74].

Table 24-2. Technology Status of Mobile/Transportable Incineration Systems

Treatment System/Vendor	Thermal Capacity (MM BTU/Hr)	Experience Site, Location	Waste Volume (tons)	Contaminant Source or Waste Type
Rotary Kiln Ensco	35	Sydney Mines, Valrico, FL[b]	10,000	Waste oil
		Lenz Oil NPL Site, Lemont, IL[b]	26,000	Hydrocarbon - sludge/solid/liquid
		Naval Construction Battalion Center (NCBC), Gulfport, MS	22,000	Dioxin/soil
		Union Carbide, Seadrift, TX[a]	N/A	Chemical manufacturing
		Smithville, Canada[a]	7,000	PCB transformer leaks
		Bridgeport Rental, Bridgeport, NJ[a,b]	100,000	Used oil recycling
Rotary Kiln IT	100 56	Cornhusker Army Ammunition Plant (CAAP), Grand Island, NE[b]	45,000	Munitions plant redwater pits
		Louisiana Army Ammunition Plant (LAAP), Shreveport, LA[a,b]	100,000	Munitions plant redwater lagoon
		Motco, Texas City, TX[a,b]	80,000	Styrene tar disposal pits
Rotary Kiln Vesta	8 12	Fairway Six Site, Aberdeen, NC	50	Pesticide dump
		Fort A.P. Hill, Bowling Green, VA	200	Army base
		Nyanza/Nyacol Site, Ashland, MA[b]	1,000	Dye manufacturing
		Southern Crop Services Site Delray Beach, FL	1,500	Crop dusting operation
		American Crossarm & Conduit Site, Chehalis, WA[b]	900	Wood treatment
		Rocky Boy, Havre, MT[a]	1,800	Wood treatment

NA - Not available; [a] Contracted, others completed; [b] Superfund Site; Source: References 3, 5, 8.

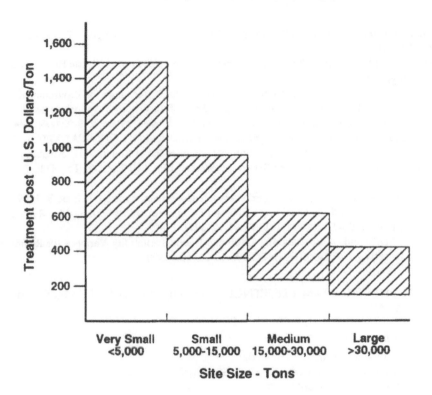

Figure 24-2. Effect of site size on incineration costs.

The cost of incineration includes fixed and operational costs. Fixed costs include site preparation, permitting, and mobilization/demobilization. Operational costs such as labor, utilities, and fuel are dependent on the type of waste treated and the size of the site. Figure 24-2 gives an estimate of the total cost for incinerator systems based on site size [12, pp. 1–3]. Superfund sites contaminated with only volatile organic compounds can have even lower costs for thermal treatment than the costs shown in Figure 24-2.

EPA CONTACT

Technology-specific questions regarding mobile/transportable incineration may be directed to Marta Richards, RREL-Cincinnati, (513) 569-7692 (see Preface for mailing address).

REFERENCES

1. High Temperature Thermal Treatment for CERCLA Waste: Evaluation and Selection of On-Site and Off-Site Systems. EPA/540/X-88/006, U.S. Environmental Protection Agency Office of Solid Waste and Emergency Response, December 1988.
2. Gupta, G., A. Sherman, and A. Gangadharan, Hazardous Waste Incineration: The Process and the Regulatory/Institutional Hurdles, Foster Wheeler Enviresponse, Inc., Livingston, NJ (no date).
3. Cudahy, J. and A. Eicher. Thermal Remediation Industry, Markets, Technology, Companies, Pollution Engineering, 1989.
4. Stumbar, J., et al. EPA Mobile Incineration Modifications, Testing and Operations, February 1986 to June 1989. EPA/600/2-90/042, 1990.
5. Cudahy, J. and W. Troxler. Thermal Remediation Industry Update II, Focus Environmental, Inc. Knoxville, TN, 1990.
6. Experience in Incineration Applicable to Superfund Site Remediation. EPA/625/9-88/008, 1988.

7. Technology Screening Guide for Treatment of CERCLA Soils and Sludges. EPA/540/2-88/004 (NTIS PB89-132674), 1988.

8. Johnson, N., and M. Cosmos. Thermal Treatment Technologies for Haz. Waste Remediation, Pollution Engineering, 1989.

9. Stumbar, J., et al. Effect of Feed Characteristics on the Performance of Environmental Protection Agency's Mobile Incineration System. In: Proceedings of the Fifteenth Annual Research Symposium, Remedial Action, Treatment and Disposal of Hazardous Wastes. EPA/600/9-90/006, 1990.

10. Shirco Infrared Incineration System, Applications Analysis Report. EPA/540/A5-89/010, U.S. Environmental Protection Agency, 1989. [See below for complete list of Shirco SITE program reports.]

11. Mobile Treatment Technologies for Superfund Wastes. EPA 540/2-86/003(F) (NTIS PB87-110656), 1986.

12. McCoy and Associates, Inc., The Hazardous Waste Consultant, Volume 7, Issue 3, 1989.

13. Superfund LDR Guide #6A: Obtaining a Soil and Debris Treatability Variance for Remedial Actions. OSWER Directive 9347.3-06FS (NTIS PB91-921327), 1990.

14. Superfund LDR Guide #6B: Obtaining a Soil and Debris Treatability Variance for Removal Actions. OSWER Directive 9347.3-06BFS (NTIS PB91-921310), 1990.

ADDITIONAL SITE PROGRAM REFERENCES (See Appendix A for information on how to obtain SITE reports)

Shirco Infrared Incineration

  Demonstration Bulletin (EPA/540/M5-88/002)
  Technology Demonstration Summary (EPA/540/S5-88/002)
  Applications Analysis (EPA/540/A5-89/010)
  Technology Evaluation-Peake Oil (EPA/540/5-88/002a)
  Technology Evaluation-Peake Oil Vol. II (EPA/540/5-88/002B; NTIS PB89-116024)
  Technology Evaluation-Rose Township (EPA/540/5-89/007a)
  Technology Evaluation-Rose Township (EPA/540/5-89/007b; NTIS PB89-167910)

Chapter 25

Issues Affecting the Applicability and Success of Remedial/Removal Incineration Projects[1]

U.S. Environmental Protection Agency, Risk Reduction Engineering Laboratory, Cincinnati, OH

INTRODUCTION

The On-Scene Coordinator (OSC) and/or Remedial Project Manager (RPM) for each Superfund site is responsible for overseeing all activities involved with the cleanup of that site. This includes oversight of Removal Actions (OSC), the Remedial Investigation/Feasibility Study (RI/FS) (RPM), Record of Decision (ROD) (RPM), and remedial design and remedial action (RD/RA) (RPM). This chapter is intended to familiarize OSCs and RPMs with issues which are important to the successful completion of incineration projects. Incineration has been a recommended method for disposing of hazardous materials, and its use in the Superfund Program is increasing rapidly. It has become one of the most often selected methods for treating hazardous constituents found at Superfund sites. Use of this chapter should assist the OSC/RPM in directing the activities of removal/remediation contractors. This report summarizes key pieces of information and lists EPA contacts that can assist the RPM/OSC in making an informed evaluation of the Remedial Design. Although the contents are based on the assumption that the reader is already somewhat familiar with incineration, a list of references is included to assist those who are less familiar with this topic.

Incineration is a proven means of destruction for many organic wastes and should be considered as a possible treatment for the cleanup of most toxic waste sites. The matrix in Figure 25-1 compares the applicability of incineration for waste treatment with that of other technologies.

An incineration system includes a number of subsystems including the following:

- Waste pretreatment

    Waste screening
    Size reduction (grinding)
    Waste mixing

- Waste feed

    Belt conveyors
    Augers
    Apron feeders
    Hoppers
    Chutes
    Pump (liquids, sludges, oils)
    Screw conveyors
    Ram feeder

- Combustion unit

    Rotary kiln/Secondary combustion chamber (SCC)
    Liquid injection
    Fluidized bed
    Infrared

---

[1] EPA/540/2-91/004.

	Aqueous Wastes										Organic Liquids							
	Metals	Highly Toxic Organics	Volatile Organics	Toxic Organics	Radioactive	Corrosive	Cyanide	Pesticide	Asbestos	Explosive	Metals	Highly Toxic Organics	Volatile Organics	Toxic Organics	Radioactive	Corrosive	Cyanide	Pesticide
In Situ Bioremediation	X	X	O	●	X	X	X	X	X	O	X	X	X	X	X	X	X	X
Activated Sludge	X	X	O	●	X	X	O	O	X	O	X	X	X	X	X	X	X	X
Filtration	●	●	X	X	●	X	X	●	●	X	●	X	X	X	O	X	X	X
Evaporation	●	●	X	X	O	X	X	●	●	O	●	X	X	X	O	X	X	X
Membrane Sep./Ion Exch.	●	●	O	●	X	X	X	●	O	●	X	X	X	X	X	X	X	X
Extraction/Soil Washing	O	O	O	O	X	X	X	O	X	O	X	X	X	X	X	X	X	X
Fixation	X	X	X	X	X	X	X	X	X	X	O	X	X	X	O	X	X	X
Phase Separation	X	O	●	●	X	X	X	O	X	●	●	O	O	O	O	O	O	O
Evaporating/Dewatering	X	X	X	X	X	X	X	X	X	O	X	X	X	X	X	X	X	X
Activated Carbon	O	●	●	●	X	X	X	●	X	●	X	O	X	X	X	X	X	X
Air Stripping/Soil Aeration	X	X	●	●	X	X	X	X	X	X	X	X	X	X	X	X	X	X
Distillation	X	●	●	●	X	X	X	●	X	X	●	●	●	●	●	●	●	●
Precipitation	●	O	O	O	X	●	●	O	●	●	●	X	X	X	O	O	X	X
Neutralization	●	O	O	O	X	●	●	O	●	X	●	X	X	X	X	●	X	X
Wet Oxidation	X	●	●	●	●	X	X	●	●	X	X	●	●	O	X	X	●	O
Pyrolysis	X	O	O	O	X	X	O	O	X	X	O	●	●	●	X	X	●	●
Incineration	O	O	O	O	X	O	O	O	X	O	O	●	●	●	O	O	●	●

Sludges/Soils

	Metals	Highly Toxic Organics	Volatile Organics	Toxic Organics	Radioactive	Corrosive	Cyanide	Pesticide	Asbestos	Explosive
	x	x	O	●	x	x	O	O	x	O
	x	O	x	x	x	x	x	x	x	x
	x	●	x	O	x	x	x	O	x	x
	O	●	x	O	x	x	x	x	x	x
	x	x	x	x	x	x	x	●	x	x
	●	O	O	O	x	x	x	●	x	O
	●	●	O	●	●	●	O	x	●	●
	x	●	O	●	●	x	x	O	x	●
	●	●	●	●	●	●	●	●	●	●
	O	●	O	●	x	x	x	●	x	●
	x	x	●	x	x	x	x	x	x	x
	x	●	●	O	x	x	O	O	x	x
	O	x	x	O	x	x	O	●	x	x
	O	x	x	x	●	x	x	x	O	
	x	●	●	●	x	x	●	O	x	x
	O	●	●	●	O	O	O	●	x	O

● Applicable; ○ Potentially Applicable; X Not Applicable

Sources: Brunner 1988a; U.S. Environmental Protection Agency 1986a.

Figure 25-1. Onsite waste treatment technology matrix.

- Heat recovery (optional - not normally applicable to onsite incineration)

- Air pollution control equipment to treat:

    Products of incomplete combustion:

    > Minimized in combustion chamber and afterburner. Afterburners can significantly reduce the toxicity of the exhaust gas from an incinerator.

    Particulate emissions:

    > Venturi scrubber
    > Wet electrostatic precipitator
    > Electrostatic precipitator
    > Quench systems
    > Fabric filter

    Acid gases:
    > Packed towers
    > Spray towers
    > Spray dryers

- Residue handling and disposal

    Ash

    > Solidification
    > Use as fill material onsite or offsite disposal

    Liquids

    > Neutralization
    > Filtration
    > Precipitation (metals)
    > Clarification

    >> Carbon adsorption or air stripping (for small amounts of organics which are sometimes recovered in scrubber water)
    >> Discharge to a POTW after successful treatment using one of the above four options.
    >> Use to cool ash from the Rotary Kiln

Figure 25-2 presents a schematic diagram of a typical incineration system.

When incineration is considered along with other possible treatment methods, the relative risks involved with the use of each of the technologies should be taken into account. Table 25-1 shows the total excess lifetime cancer risk that environmental releases from incineration pose to the most exposed individual. These values, which were developed to support the Resource Conservation and Recovery Act (RCRA) hazardous waste incineration regulations, are based on assumptions that included process upsets and covered a wide range of operating conditions. As shown in Table 25-1, the risks presented by metals are likely to be higher than those presented by Principal Organic Hazardous Constituents (POHCs) and products of incomplete combustion (PICs). The total estimated risk (including metals, POHCs, and PICs) does not exceed 1 in 100,000 and is unlikely to do so as long as all appropriate incinerator standards are met. This information should be considered in light of the other risks that are associated with a particular Superfund site, as indicated from any required risk assessments.

The information in this report was obtained through a literature survey and contacts with several EPA representatives experienced in the use of incineration for the cleanup of toxic waste sites and for the treatment of RCRA hazardous wastes (see Contact list at end of this chapter).

INCINERATION SYSTEM DESIGN, OPERATION, AND PERFORMANCE

A complete discussion of the design, operation, and performance of incineration systems is beyond the scope of this chapter. Detailed information on any of these topics can be found in the General Reference section at the end of this chapter. This information should be useful to the RPM/

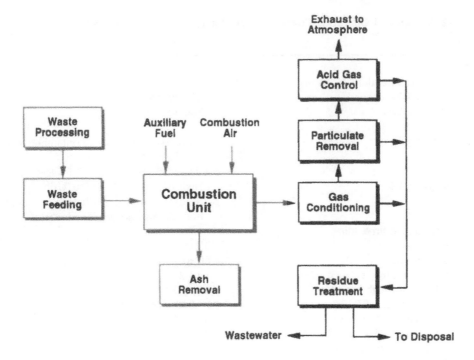

Figure 25-2. Incineration system concept flow diagram.

Table 25-1. Total Excess Lifetime Cancer Risk from Incinerator Emissions to the Maximum Exposed Individual[a]

Emission Item	Risk Range	Probability Statement
POHCs	10^{-7} to 10^{-10}	1 in 10,000,000 to 1 in 10 billion
PICs	10^{-7} to 10^{-11}	1 in 10,000,000 to 1 in 100 billion
Metals	10^{-5} to 10^{-8}	1 in 100,000 to 1 in 100,000,000
Total	10^{-5} to 10^{-8}	1 in 100,000 to 1 in 100,000,000

[a] Source: Weinberger et al., 1984.

OSC in obtaining some background and perspective on issues pertaining to the use of incineration. It is the objective of this section, and of the entire report, to provide the RPM/OSC with enough basic information, resource documents, and personal contacts to allow them to conduct technical oversight and monitoring of remedial activities. To keep the report as concise as possible, this information is presented in a series of tables, as follows:

Table 25-2. Design and Operating Characteristics of a Typical Incineration System[a]

Parameter	Typical Values
Rotary kiln	
Operating Temperature, °F	
Ashing kiln	1200 to 1800
Slagging kiln	2200 to 2600
Types of Waste	
Ashing kiln	• Low BTU waste (e.g., contaminated soils) <5000 BTU/lb
	• High BTU waste, >5000 BTU/lb
Slagging kiln	• High BTU waste >5000 BTU/lb
	• Moderate moisture & halogen content
	• Both drums and drummed wastes
Solids residence time, min	
Ashing kiln	30 to 60
Slagging kiln	60 to 100
Gas residence time, s	1 to 2
Gas velocity through kiln, ft/s	15 to 20
Heat release levels, BTU/ft^3 per h	25,000 to 40,000
Small kiln, million BTU/h	8 to 35
Large kiln, million BTU/h	35 to 100
Kiln loading, % kiln volume	
Ashing kiln	7.5 to 15
Slagging kiln	4 to 6
Kiln operating pressure, in. H$_2$O	−0.5 to −2.0
Excess air, %	75 to 200
Liquid injection unit	
Operating temperature, °F	1800 to 3100
Residence time, s	Milliseconds to 2.5
Excess air, %	10 to 60
Waste heating value, BTU/lb	4500
Secondary combustor (afterburner)	
Residence time, s	2
Operating temperature, °F	2200 typical
TSCA wastes	>2250
RCRA wastes	1600 to 2800
Excess air, %	10 to 60

[a] Sources: Tillman et al. (1990); Schaefer and Albert (1989).

These tables are excerpted from the references listed in the Cited Reference section at the end of this chapter. Some values have been updated and additional information has been added, where appropriate, to provide more complete information. These additional values were determined from discussion with various incineration experts during the development of this chapter. It is suggested that the OSC/RPM seek the advice of some of the experts listed in this chapter and of the regional RCRA incineration contacts regarding appropriate values for the incinerator to be used at their specific site.

Table 25-2 focuses on an incineration system with a rotary kiln and a liquid injection unit exhausting into a secondary combustion chamber (afterburner). Other tables in this section also rely heavily on reported information and experiences with rotary kiln incineration systems because these units have been and are scheduled to be used for the treatment of contaminated soils at most Superfund sites. Approximately 0.91 million tons of the 1.3 million tons of contaminated soils and sludges that have been treated or contracted to be treated (approximately 70 percent) by onsite thermal treatment methods been have or are projected to be treated by rotary kiln incineration. The remaining

Table 25-3. Typical Design Parameters for Air Pollution Control Equipment on Hazardous Waste Incinerators

Air Pollution Control Equipment	Typical Design Parameters
Particulate	
Electrostatic precipitators	SCA = 400–500 ft²/1000 acfm
	Gas velocity = 0.2 ft/s
Fabric filters	Pulse jet A/C = 3–4:1
	Reverse air A/C = 1.5–2:1
Venturi scrubbers	ΔP = 40–70 in. W.C.
	L/G = 8–15 gal/1000 acfm
Acid gases	
Packed towers	Superficial velocity = 6–10 ft/s
	Packing depth = 6–10 ft
	L/G = 20–40 gal/100 acfm
	Caustic scrubbing medium, maintaining pH = 6.5
	Stoichiometric ratio: 1.05
Spray dryers	Low temperature:
	Retention time 15–20 sec
	Outlet temperature 250–450°F
	Stoichiometric ratio (lime) = 2–4

SCA = specific collection area.

A/C = air-to-cloth ratio in units of ft/min.

L/G = liquid-to-gas ratio.

Source: Buonicore, 1990.

Table 25-4. Summary of Continuous Emission Monitors[a]

Pollutant	Monitor Type	Expected Concentration Range	Available Range[b]	Typical Value
O_2	Paramagnetic	3–14%	0–25%	8%
CO_2			0–21%	8%
CO	NDIR[c]	0–100 ppm	0–5000 ppm	40 ppm
NO_x	Chemiluminescent	0–4000 ppm	0–10000 ppm	200 ppm
SO2	Flame photometry	0–4000 ppm	0–5000 ppm	Varies by waste
Organic compounds (THC)	FID[d]	0–20 ppm	0–1000 ppm	<20 ppm

[a] Source: Oppelt, 1987.

[b] For available instruments only. Higher ranges are possible through dilution.

[c] Nondispersion infrared.

[d] Flame ionizing detection.

tons are fairly evenly split among low- and high-temperature desorption, circulating fluidized bed, and infrared conveyor furnaces (approximately 6 to 9 percent for each type of unit).

Table 25-3 provides an overview of design parameters for Air Pollution Control Equipment which is typically included in incineration systems. This table is useful as a reference in specifying design criteria for these systems.

Table 25-4 provides an overview of the continuous emission monitors that are typically used on incinerators. Ranges and typical values are provided. Generally, if continuous emission monitors are within the specified "typical values," the incinerator is probably operating in compliance with applicable or relevant and appropriate requirements (ARARs).

Table 25-5. Typical Automatic Waste Feed Shut Off (AWFSO) Parameters[a]

Parameter (example value)	Excess Emissions	Purpose of AWFSO Worker Safety	Equipment Protection
High CO in stack (100 ppm)[b]	X		
Low chamber temperature[b] (1400°F for rotary kiln 1700°F for SCC)	X		
High combustion gas flow (varies by size)	X		
Low pH of scrubber water (4) (e.g., not less than 6.5)	X		
Low scrubber water flow (varies by size)	X		X
Low scrubber pressure drop (20 inches W.G. for venturi)	X		X
High scrubber temp. (220°F)			X
Low sump levels (variable)			X
High chamber pressure (positive)	X	X	
High chamber temperature (2000°F for rotary kiln, 2600° for SCC)	X	X	X
Excessive fan vibration	X	X	X
Low burner air pressure (1 psig)	X		
Low burner fuel pressure (3.0 psig for natural gas)	X		
Burner flame loss		X	X
Low oxygen in stack (3 percent)[b]	X		
Loss of atomizing media	X		
High stack SO_2[b]	X		
High waste feed flow	X		
High opacity >5%	X		

[a]   Source: Oppelt (1987).

[b]   Rolling averages of these parameters can sometimes be used. (Paul Leonard comments, 10/23/90)

SCC = secondary combustion chamber.

Table 25-5 is a summary of operating parameters which are required by an operating permit to trigger an automatic cessation of feed in the event that a safe operating range is exceeded. These precautions may not always be included in incinerator designs, but do help to ensure safe operation and compliance with ARARs.

Table 25-6 is a summary of operating parameters that affect incinerator performance. This is useful general information which should assist the RPM/OSC in reviewing incinerator designs to assure the efficient performance of an incinerator at a particular site.

Table 25-7 summarizes the physical properties of solid waste which can adversely affect the performance of an incinerator. Waste streams that are difficult to treat can cause frequent shut-downs, thus significantly lengthening the time required to remediate the site. Also, some waste streams can form toxic PICs and should not be incinerated without the use of an afterburner.

Table 25-8 summarizes failure modes that can result in the incinerator failing to comply with ARARs. These conditions should be avoided.

Table 25-9 lists some of the PICs that can form from various mixtures of organic compounds. This list is particularly useful in determining what POHCs to designate during a trial burn. In addition, it provides the RPM/OSC with an indication of what organic chemicals may be emitted from an incinerator burning a particular mixture of contaminants under suboptimal conditions.

Finally, Table 25-10 lists some of the maintenance that must be done on an operating incinerator. This is useful to the RPM/OSC in determining the level of effort required to implement an incineration remedy.

Table 25-6. Example Operating Parameters and How They Affect Performance[a]

Operating Parameter	Effect
Temperature	Combustion reaction rates increase with temperature until the rates are limited by mixing. High temperatures can also elevate NO_x emissions.
Combustion gas flow rate	For a fixed chamber volume, the waste constituents remain in the chamber for a shorter time (have a lower residence time) as the flow rate increases. As the combustion gas flowrate increases, gas velocity through the chamber increases. This can result in increased entrainment of solid material (fly ash) and emission of particulates.
Waste feed rate and heat content	As waste feed rate decreases, the heat release in the combustion chamber will decrease and temperature may drop. Waste heat content can affect combustion temperature. Insufficient heat content can result in the need for auxiliary fuel which will adversely affect the economics of the process. Wide variations in heating value of the waste can cause puffing (positive pressure surges) in rotary kilns. Moisture content of the waste Moisture decreases the heat content of the waste and, as a result, reduces the combustion temperature and efficiency when high moisture waste is burned.
Air input rate	Air supplies oxygen for the combustion reactions. A minimum is needed to achieve complete combustion; however, too much air will lower the temperature (because the air must be heated) and quench combustion reactions due to excessive cooling. The additional air will increase combustion gas flow rate, which then lowers the residence times. Increased air input can increase combustion efficiency by increasing the amount of oxygen available to oxidize organic contamination.
Waste atomization	Atomizing liquid waste into smaller droplets will increase the effectiveness of fuel/air mixing and the burning rate. Waste feed and atomizing fluid (air or steam) flowrates and pressures affect atomization. Suboptimal waste feed and atomizing fluid flows will result in less efficient atomization resulting in the production of larger fuel/waste droplets.
Feed system	Consistent, reliable delivery of waste feed into the incinerator is critical to the efficient operation of an incinerator. The design of appropriate feed systems can be difficult for inconsistent or difficult feed streams.
Mixing/turbulence	A burner must be selected which induces adequate turbulence into the combustion air/fuel/waste mixture. This promotes good mixing of air and fuel which leads to efficient combustion.

[a] Source: ASME (1988).

Knowing the thermal stability of POHCs and PICs is extremely important to the design of an effective incineration system. The University of Dayton Research Institute (UDRI) has studied the thermal stability of 330 hazardous organic compounds and has ranked their thermal stability under oxidative and pyrolytic conditions. This database is available in *Environmental Science and Technology*, Volume 24, No. 3, pp. 316–328, 1990. UDRI has also determined the PICs which can be produced from various POHCs under different combustion conditions. The PICs produced from a given POHC vary depending upon whether the atmosphere is oxidative or pyrolytic. Further, mixtures

Table 25-7. Waste Properties Affecting Incineration System Performance[a]

Property	Hardware Affected	Operating Parameters Affected	Effect of Performance	Example Feeds of Concern
Heating value	Rotary kiln	Rotary kiln temperature, flue gas residence time	Feed capacity, fuel usage	Plastics, trash
Density	Rotary kiln	Weight of material held by kiln	Feed capacity	Brominated sludge (high density sludge)
Halogen and sulfur content	Quench system, air pollution control equipment design and operation	Pump cavitation, pH control, blowdown rate, particulate emissions	Feed capacity, caustic usage	Trial burn mixture, brominated sludge
Moisture	Feed system		Increased fuel usage to maintain temperature	
Particle size distribution	Cyclone, SCC, ducts, wet electrostatic precipitation (WEP), instrumentation	Kiln draft, particulate emissions, excess oxygen control, temperature control	Fouling of duct, cyclone, SCC, process water system, and instruments	Soils, brominated sludge, vermiculite
H:Cl ratio of POHCs[b]	—	Incinerator's ability to thermally destroy POHCs/PICs	As H:Cl ratio decreases thermal stability of POHCs increases and oxidation of PICs is reduced. Under oxygen starved conditions the tendency to form PICs increases as the N:Cl ratio decreases	C_2Cl_6, C_6l_6, C_2HCl_3, and similar compounds
Any fusion characteristics (determined by chemical characteristics e.g., alkalis)	Rotary kiln, cyclone, ducts, quench elbow, instrumentation	Kiln draft, temperature, excess O_2 control	Slagging of kiln, plugging of instruments and downstream equipment	Plastic, trash, brominated sludge

Sources: [a] Stumbar et al. (1989); [b] Taylor and Dellinger (1988), Tirey et al. (1990).

Table 25-8. Operating Failure Modes Leading to the Formation of Excessive Products of Incomplete Combustion (PICs) and Low Destruction and Removal Efficiency (DRE)[a]

Condition	Results
Low oxygen to fuel/waste ratio	Insufficient oxygen for complete combustion; in many cases this will reduce POHC DRE and increase propensity for PIC formation.
High air/fuel ratio	High air levels and associated gas flows lead to temperature quenching and flameouts.
Low-temperature operation	Many PICs require higher destruction temperature than parent POHCs, thus low destruction efficiency for POHCs and higher PIC emission rates.
Waste surges	Leads to overloading combustion system and incomplete combustion (starved air condition). Also can lead to fugitive emissions as a result of sudden pressurization of the system. High CO and THC levels can result.
Poor gas mixing in combustion chamber due to low turbulence	Optimum combustion of all organics not achieved. PICs can be formed from the onset of pyrolysis within the system. Localized oxygen-starved stoichiometries lower POHC DRE and increase PIC formation. CO levels increase.
Poor atomization for liquids	Droplets too large for vaporization in flame zone or droplet trajectories penetrate flame zone.
Injection waste flame impinging on cool surface such as combustion chamber wall	Can cause severe damage to the refractory. Quenches combustion reactants before combustion is complete. PICs and CO levels can increase.
Liquid waste flame impinging on cool surface such as combustion chamber wall	This can result in the release of PICs and unburned POHCs into the environment. Refractory can also be damaged.
Poorly designed or malfunctioning air pollution control (APC) device or failure of APC	PICs are absorbed on soot particles that are normally collected in the APC system. This condition will increase these particulate emissions. Dioxin formation can occur in this way.
Short residence time	Insufficient time for complete burning most critical when stable PICs are formed from POHC combustion.
High halogen content (e.g., H:Cl ratio too low)	Highly chlorinated POHCs and PICs are more difficult to oxidize than less chlorinated or unchlorinated derivatives.

[a] Sources: ASME (1988); Daniels (1989); Dellinger et al. (1989). Also, Santoleri (1989a-c).

Note: The above table is not all-inclusive, and appropriate care should be given to make certain that incinerator designs have a minimum of failure modes which could result in PIC formation. As an added precaution, secondary combustion chambers should always be used since they have been shown to reduce the toxicity of organic emissions from incinerators.

of POHCs produce different PICs than the individual POHCs would alone. Some of UDRI's results are presented in Table 25-9. Complete results can be obtained in the following references: Dellinger et al. (1984, 1989, 1990), Taylor and Dellinger (1988, 1990), Tirey et al. (1990).

INCINERATION EXPERIENCE

Incineration has been a popular method of disposing of unwanted materials for many years. Several incinerator manufacturers such as Combustion Engineering and Vulcan Iron Works have been in business for 100 years. With the advent of RCRA, the Comprehensive Environmental Re-

Table 25-9. Reaction Products Observed from Thermal Decomposition of Various Materials in UDRI Flow Reactor Studies[a]

Parent (POHC)	Product (PIC)	Condition
Carbon Tetrachloride	Tetrachloroethene Hexachloroethane Hexachlorobutadiene	Air atmosphere t_r = 2.0 s
Pentachlorobenzene Chloroform	Hexachlorobenzene CCl_4 $1,2\text{-}C_2H_2Cl_2$ C_2HCl_3 C_2Cl_4 C_2HCl_5 C_2Cl_2 $C_2H_2Cl_4$ C_3Cl_4 C_4Cl_6 C_6Cl_6	Air atmosphere t_r = 2.0 s o = 0.67, t_r = 2.0 s
Chloroform	Carbon Tetrachloride Trichloroethene Pentachloroethane Dichloroethyne Tetrachloroethene Tetrachloropropyne 1,1,2,4-Tetrachloro-1-buten-3-yne Hexachlorobutadiene	o = 0.76 and Nitrogen atmospheres
Mixture of CCl_4 53% (mole) $CHCl_3$ CH_2Cl_2 CH_3Cl	CCl_4 33% 7% 7% C_2Cl_2 $1,1\text{-}C_2H_2Cl_2$ C_2HCl_3 C_2Cl_4 C_2Cl_6 C_3Cl_4 C_4Cl_4 C_4Cl_6 C_6Cl_6 C_8Cl_8	Pyrolytic, t_r = 2.0 s $CHCl_3$ CH_2Cl_2 CH_3Cl

[a] This table was excerpted from a table appearing in a UDRI report on PIC minimization entitled Minimization and Control of Hazardous Combustion Byproducts Final Report and Project Summary prepared for U.S. EPA under cooperative agreement CR-813938-01-0 summarizing the results of flow reactor studies conducted at the University of Dayton Research Institute. The complete table can be found in the above listed reference.

sponse and Compensation and Liability Act (CERCLA), and the Superfund Amendments and Reauthorization Act of 1986 (SARA), developments in incineration have evolved with changing environmental concerns. Manufacturers have had to modify their incinerators to ensure complete destruction of all the hazardous constituents found in the variety of mixed wastes on a Superfund site. More commercial facilities were established to deal with the quantity of wastes being generated or found. The concern over transporting wastes from a hazardous waste site to a commercial facility led to the development of mobile treatment technologies, which allowed the waste to be treated onsite and thus prevented the spread of contamination. The full-scale thermal remediation projects included later in this section were all performed onsite with mobile or transportable equipment. When site

Table 25-10. Maintenance Checklist for a Rotary Kiln Incinerator[a]

Item	Procedure	Frequency
Shredder	Inspect	Daily
	Lubricate	Weekly
Kiln feeder	Inspect	Daily
	Lubricate	Weekly
Kiln burner	Check flame	Each shift
	Remove, inspect atomizer	Quarterly
Other atomizers	Remove, inspect	Quarterly
Kiln speed	Check	Daily
Kiln drive	Inspect	Daily
	Lubricate	Weekly
Kiln refractor	Inspect visually	Each shift
	Repair	As needed
Kiln seats	Inspect	Each shift
	Replace	As needed
Ash gates	Inspect	Daily
Ash conveyor	Inspect	Daily
	Lubricate	Weekly
Afterburner refractory	Inspect visually	Each shift
	Repair	As needed
Afterburner burners	Check flame	Each shift
	Remove, inspect atomizer	Quarterly
Quench	Check for leaks	Each shift
	Check outlet temperature	Each hour
	Remove, inspect atomizers	Quarterly
Waste heat boiler	Check steam pressure	Each hour
	Check pressure drop	Each shift
	Inspect tubes	Each 6 months
Particulate scrubber	Check pressure drop	Each shift
	Check water level	Each shift
	Lubricate throat drive	Monthly
Absorber	Check pressure drop	Each shift
	Inspect packing	Each 6 months
	Remove, inspect nozzles	Quarterly
Fabric filter system	Check pressure drop	Each shift
	Inspect bags	Each 6 months
	Lubricate discharge mechanism	Monthly
Main fan	Check motor amperage	Daily
	Lubricate bearings	Weekly
	Check vibration	Daily
Pumps	Check motor amperage	Weekly
	Lubricate	Weekly
	Check discharge pressure	Daily
Control instruments structions	Calibrate	Per manufacturer's in-
Analytical instruments	Calibrate	Daily
Limit controls	Test	Daily
Emergency vent	Test	Quarterly

[a] Source: Brunner (1988b).

conditions precluded the use of mobile equipment, commercial facilities were used. The Records of Decision listed in Table 25-13 all used some form of incineration or thermal treatment. Last, but not least, are the SITE Demonstrations, where new, innovative modifications such as oxygen enrichment are made to the incineration process to develop alternative systems for effectively cleaning the environment.

Onsite Mobile Treatment

When Congress authorized SARA in 1986, one of their goals was to prevent the possible spread of contamination resulting from transportation of untreated wastes. According to SARA, "The offsite transport and disposal of hazardous substances or contaminated materials... should be the least favored alternative remedial action where practicable treatment technologies are available." Because SARA also emphasizes the use of a permanent solution, incineration has become the most used method for treating hazardous waste. Using a mobile incinerator not only satisfies both of the SARA requirements, it provides a proven technology that is capable of quickly and effectively achieving a high level of waste destruction with no long-term liability. Existing technologies have demonstrated the capability of achieving >99.9999% destruction of organics while producing an organic-free ash suitable for backfilling at the site. Because onsite cleanups can be conducted without Federal, state, or local permits, the time required for startup can usually be reduced.

Even though permits may not be required for onsite cleanups, the substantive technical requirements of a permit must still be met. Offsite commercial incinerators must comply with the "offsite" policy. (OSWER Directive 9330.2-1.)

Onsite incineration includes mobile units, which are transported to a site fully operational. A unit is used to treat wastes at one site and, when the job is finished, it is moved to another site. Transportable incinerators are those which are transported to a site and are erected onsite. At some very large sites where the cleanup will require a number of years, it may be feasible to actually build an incinerator onsite. Once erected, they cannot be moved from the site without first being dismantled to some extent. Transportable incinerators are generally larger than mobile units and are best used for long-term cleanups in which a relatively large amount of material will be treated. Economic considerations are often the key factor in determining whether mobile, transportable, fixed or offsite commercial incineration will be used at a given site. Costs for onsite and offsite thermal treatment vary widely. In choosing between onsite and offsite incineration, factors which affect the economics of incineration are the type, physical form, and quantity of contaminants; applicable site cleanup criteria; and the availability of offsite incineration, including the capacity and proximity of the commercial unit, container requirements, and the method of transportation (McCormick and Duke, 1989).

Based in part on a survey conducted by McCoy and Associates, Inc. (1989), the following companies offer mobile or transportable thermal treatment of hazardous wastes:[2]

The EPA Regional Contacts listed later in this report may have more specific information concerning the capabilities of each vendor.

Chemical Waste Management

3003 Butterfield Road
Oak Brook, IL 60521
Contact: Ray Bock
Phone: (708) 218-1675
Technology: Transportable rotary kiln incinerator
Setup time: 2 months
Waste/Media: Soils, sludges, and other solids; unit
Typical cost: $200–300/ton can also burn incidental liquids
Limitations: 20–30 tons/hour; 82 million BTU/hour; minimum quantity of 10,000 tons to justify mobilization; 20,000 tons or more preferred

Environmental Systems Co. (ENSCO)

333 Executive Court
Little Rock, Arkansas 72205
Contact: Steve Hardin
Phone: (501) 223-4100

---

[2] Addresses and phone number are those included in the original bulletin and may not all be up-to-date.

Setup time: 4 to 6 weeks
Typical cost: Varies, depending on waste stream
Technology: MWP-2000 modular incinerator; rotary kiln
Waste/Media: Solids and liquids (RCRA and TSCA)
Limitations: 40 million BTU/hour; no radioactive waste or fluorinated compounds

Harmon Environmental (Williams)

1550 Pumphrey Avenue
Auburn, Alabama 36830
Contact: Bill Webster
Phone: (205) 821-9253

Setup time: 4 hours
Typical cost: $55–75/ton
Technology: Mobile rotary kiln
Waste/Media: Light fuels, diesel, gasoline
Limitations: 8 tons/hour; 24 million BTU

Haztech (Westinghouse Environmental Services)

5304 Panola Industrial Blvd., Suite E
Decatur, Georgia 30035-4013
Contact: Carol Renfroe
Phone: (404) 593-3464

Setup time: 4–6 weeks
Typical cost: $200–300/ton
Technology: Transportable infrared conveyor system
Waste/Media: Organic soils and sludges
Limitations: 100–175 tons/day; feed stream must be chopped/shredded to less than 1-in. pieces

International Technology Corporation

23456 Hawthorne Blvd.
Torrance, California 90505
Contact: Kevin R. Smith
Phone: (615) 690-3211

Setup time: 3 weeks
Typical cost: $150-450/ton
Technology: Hybrid Thermal Treatment System (HTTS); transportable rotary kiln
Waste/Media: Solids, sludges, and liquid wastes, including light contaminated materials up to
    heavy organics
Limitations: 56 million BTU/hour

Ogden Environmental Services, Inc.

P.O. Box 85178
San Diego, California 92138-5178
Contact: Robert C. Haney
Phone: (619) 455-3045

Setup time: 2–3 weeks
Typical Cost: $100–300/ton
Technology: Transportable circulating-bed combustor
Waste/Media: Soils, sludges, and liquids containing hazardous and toxic constituents includ-
    ing PCBs, hydrocarbons, oil, and munitions

O.H. Materials Corp.

16406 U.S. Route 244 East
Findlay, Ohio 45840
Contact: Greg McCartney
Phone: (419) 423-3526

Setup time: 7 days
Typical cost: $150–250/ton
Technology: Mobile infrared hazardous waste incinerator
Waste/Media: Soils, sludges, and sediments contaminated with halogenated and nonhalogenated organics
Limitations 200 tons/day; limited to solid/semisolid waste media

Thermodynamics Corporation

P.O. Box 369
Bedford Hills, New York 10507
Contact: Mark Wolstencroft
Phone: (914) 666-6066

Setup time: 2 days
Typical cost: $400/ton (depends on site and material)
Technology: Mobile rotary-kiln incinerator
Waste/Media: Handles all mediums
Limitations: 9 million BTU/hour (however, larger unit may be available in the future); solids must be crushed or shredded to 1-in. size

VESTA Technology, Ltd.

1670 West McNab Road
Ft. Lauderdale, Florida 33309
Contact: Tricia P. Jack
Phone: (305) 978-1300

Setup time: 8 hours; 24–48 hours
Typical cost: $450–750/ton; $250–600/ton
Technology: Mobile rotary-kiln incinerator (small or large unit)
Waste/Media: Liquids, solids, and sludges
Limitations: 8 million BTU/hour; 12 million BTU/hour; cannot handle heavy metals, arsenic, or mercury

Waste-Tech Services, Inc.

18400 W. 10th Avenue
Golden, Colorado 80401
Contact: John Wurster
Phone: (303) 279-9712

Setup time: 3 days
Typical cost: $700/ton
Technology: Trailer-mounted fluidized-bed incinerator
Waste/Media: Solids, liquids, sludges, slurries, soils, and gases; halogenated and nonhalogenated wastes
Limitations: 1.5 million BTU/hour; 600 pounds/hour; solid wastes with greater than 3-cubic-inch particle size require size reduction pretreatment step

Weston Services, Inc.

Weston Way
West Chester, Pennsylvania 19380
Contact: John W. Noland
Phone: (215) 430-3103

Setup time: 6 weeks
Typical cost: $250/ton
Technology: Transportable Incineration System (TIS); rotary kiln
Waste/Media: Hazardous soils, sludges, and liquids
Limitations: 7 tons/hour; 20 million BTU/hour in kiln and 20 million BTU/hour in after-burner

OFFSITE COMMERCIAL FACILITIES

Although onsite treatment is the preferred remediation method for Superfund wastes, site conditions might preclude the use of mobile or transportable incinerators. (OSWER Directive 9355.3-01) In these cases, the wastes must be transported to a commercial incinerator which is in compliance with the "offsite-policy." Currently, only 9 companies, operating 14 commercial facilities in 8 states, are capable of handling the wide spectrum of wastes that might be found at a CERCLA site. Current information regarding these facilities' compliance with the "offsite-policy" should be obtained prior to use. The following list contains the companies, incinerator location, and type of incinerator used:

Chemical Waste Management, Inc.

Incinerator location: Port Arthur, TX
Technology: Rotary kiln

Phone: 800/843-3604
        409/736-2821

Incinerator Location: Sauget, IL
Technology: Rotary kiln

Phone: 800/843-3604
        618/271-2804

Incinerator Location: Chicago, IL
Technology: Rotary kiln

Phone: 800/843-3604
        312/646-5700

ENSCO, Inc.

Incinerator Location: El Dorado, AK
Technology: Liquid injection, Rotary kiln

Phone: 501/223-4160

GSX/Thermal Oxidation Corporation

Incinerator Location: Roebuck, SC
Technology: Liquid injection

Phone: 803/576-1085

L.W.D., Inc.*

Incinerator Location: Calvert City, KY
Technology: Liquid injection, Rotary kiln

Phone: 502/395-8313

Olin Chemicals

Incinerator Location: Brandenburg, KY
Technology: Liquid injection

Phone: 800/227-7592
        502/422-2101

Rhone-Poulenc Basic Chemical Company

Incinerator Location: Baton Rouge, LA
Technology: Liquid injection

Phone: 713/688-9311

Incinerator Location: Houston, TX
Technology: Liquid injection

Phone: 713/683-3314
713/683-3315

Rollins Environmental Services, Inc.

Incinerator Location: Baton Rouge, LA
Technology: Liquid injection, Rotary kiln

Phone: 504/778-1234

Incinerator Location: Bridgeport, NJ
Technology: Liquid injection, Rotary kiln

Phone: 609/467-3105

Incinerator Location: Deer Park, TX
Technology: Liquid injection, Rotary kiln,
    Rotary reactor

Phone: 713/930-2300

Ross Incineration Services, Inc.

Incineration Location: Grafton, OH
Technology: Liquid injection, Rotary kiln

Phone: 216/748-2171

ThermalKEM, Inc.

Incinerator Location: Rock Hill, SC
Technology: Fixed hearth

Phone: 803/328-9690

* (Contact Betty Willis, EPA Region IV, regarding the permit status of this incinerator. She can be reached at FTS 257-3433)

INCINERATOR MANUFACTURERS

Incinerators can be distinguished from each other primarily by the design of their combustion chambers. Each type operates under a specific set of conditions designed to achieve maximum efficiency for the quantity and type of wastes it will handle. Many of the major incinerator manufacturers conduct extensive onsite demonstrations of their incinerator equipment to ensure maximum operating efficiency. Table 25-11 lists the manufacturers of the major incinerator types. These firms can be contacted individually for further information (see listing following table).

The locations and telephone numbers of the manufacturers listed in Table 25-11 are as follows:[3]

Basic Environmental Engineering, Inc., Glen Ellyn, IL; (312) 469-5340
Bayco Industries of California, San Leandro, CA; (415) 562-6700
Boliden Allis, Inc., Milwaukee, WI; (414) 475-2690
Brule C.E. & E., Inc., Blue Island, IL; (312) 388-7900
CE Raymond Combustion Engineering, Inc., Lisle, IL; (708) 971-2500
Cleaver-Brooks, Milwaukee, WI; (414) 962-0100
Coen Company, Burlingame, CA; (415) 697-0440
Deutsche-Babcock (Ford, Bacon & Davis), Salt Lake City, UT; (801) 583-3773
Burn-Zol Corporation, Dover, NJ; (209) 931-1297
Dorr Oliver, Inc., Stamford, CT; (203) 358-3741
Econo-Therm Energy Systems Corp., Tulsa, OK; (800) 322-7867
Environmental Elements Corp., Baltimore, MD; (301) 368-7166
EPCON Industrial Systems, Inc., The Woodlands, TX; (713) 353-2319

---

[3] Phone numbers are those given in the original bulletin and may not be up-to-date.

Table 25-11. Manufacturers of Incinerators[a]

Hearth Incinerators	Liquid Injection Incinerators	Rotary Kiln Incinerators	Fluidized Bed Incinerators
Basic Environmental Engineering	Brulé	Boliden Allis, Inc.	CE Raymond
Bayco	Burn-Zol	CE Raymond	Dorr Oliver
Burn-Zol	Coen Co.	Deutsche-Babcock	Fuller Company
Cleaver-Brooks	Hirt Combustion Engineers	Environmental Elements Corp.	Sur-Lite
Econo-Therm Energy Systems, Inc.	McGill, Inc.	Fuller Company	
Epcon Industrial Systems, Inc.	Met-Pro Corp.	Industronics, Inc.	
Int'l Waste Ind.	Peabody Int'l	Int'l Waste Energy Systems	
Kennedy Van Saun	Prenco, Inc.	Kennedy Van Saun Corp.	
	Process Combustion	ThermAll, Inc.	
	Sur-Lite	U.S. Smelting Furnace	
	Trane Thermal	vonRoll, Ltd.	
	John Zink Co.	Vulcan Iron Works	

[a] Source: U.S. Environmental Protection Agency, 1986b.

Fuller Company, Bethlehem, PA; (215) 264-6011
Hirt Combustion Engineers, Montebello, CA; (213) 728-9164
Industronics, Inc., S. Windsor, CT; (203) 289-1551
International Waste Energy Systems, Inc., St. Louis, MO; (314) 389-7275
International Waste Industries, Blue Bell, PA; (215) 643-2100
Kennedy Van Saun Corp., Danville, PA; (717) 275-3050
McGill, Inc., Tulsa, OK; (918) 445-2431
Met-Pro Corp., Harleysville, PA; (215) 723-4751
Peabody International Corporation, Stamford, CT; (203) 327-7000
Prenco, Inc., Madison Heights, MI; (313) 399-6262
Process Combustion, Pittsburgh, PA; (412) 655-0955
Sur-Lite Corporation, Santa Fe Springs, CA; (213) 693-0796
ThermAll, Inc., Peapack, NJ; (201) 234-1776
Trane Thermal Company, Conshohocken, PA; (215) 828-5400
U.S. Smelting Furnace, Belleville, IL; (618) 233-0129
vonRoll, Ltd., Cranford, NJ; (201) 272-1555
Vulcan Iron Works, Inc., Wilkes-Barre, PA; (717) 822-2161
John Zink Co., P.O. Box 702220, Tulsa, OK 74170; (918) 747-1371

FULL-SCALE, ONSITE THERMAL REMEDIATION PROJECTS

Mobile and transportable thermal treatment methods are being used at several contaminated sites throughout the United States. Table 25-12, adapted from a list developed by James Cudahy of Focus Environmental, contains information about completed, ongoing, or contracted full-scale commercial cleanups in the United States using mobile or transportable thermal equipment. In this context, a mobile thermal treatment system is defined as a truck- or skid-mounted system which takes two weeks or less for field erection and minimal foundations; a transportable system requires more than two weeks of field erection and substantial foundations. The list does not contain any pilot-scale remediation efforts or fixed-treatment methods (such as cement kilns or commercial incinerators). Of those reporting onsite problems, materials handling ranked the highest, followed by the weather. More details on each site can be obtained by contacting the responsible EPA Regional Office and the contractor.

RECORDS OF DECISION

The Superfund RODs for fiscal years (FYs) 1985 through 1988 indicate the increasing use of incineration as a remediation method. In 1984, only 8.0 percent of the total number of RODs (including action memos, enforcement decision documents, and negotiation documents) involved incineration. In 1989, 30 percent of the source control RODs that selected treatment specified incineration/thermal destruction as all or part of the remediation effort. More than half of those were for onsite treatment (U.S. EPA 1990).

The RODs listed in Table 25-13 all recommended the use of incineration/thermal destruction as part of the site remediation. More information on any of these sites can be obtained by requesting a full copy of the ROD from any EPA library or by contacting the appropriate EPA Regional Office.

SITE PROGRAM

In response to a requirement of SARA, the EPA established a program called the Superfund Innovative Technology Evaluation (SITE) Program to encourage the development and use of innovative technologies to clean up hazardous waste sites. Two of the major components of the SITE Program are the Emerging Technologies Program and the Demonstration Program. During the Emerging Technologies Program, the basic concepts of a new technology are validated through bench- and pilot-scale testing. If the technology shows promise, it may advance to the Demonstration Program. Along with other technologies selected through annual solicitation, the performance of these technologies is evaluated under field conditions. Reports discussing the procedures, sampling and ana-

Table 25-12. Full-Scale Onsite Thermal-Remediation Projects[a]

Contractor	Site Name, Location, State	Site Size, Tons	Source of Contamination/Indicator Compound	Contaminant Concentration in Treated Soil, mg/kg	Combustion Equipment	Thermal Capacity, 10^6 BTU/h	APC Equipment	Particulate Emissions, gr/dscf at 7% O_2	Project Status
Boliden Allis	Oak Creek, Oak Creek, WI	50,000	Dye manufacturing/naphthylamine	<0.5	Rotary kiln	40	Spray tower, baghouse	<0.01	Finished
Canonie	Ottati & Goss, Kingston, NH	8,000	Volatile organics	<0.2	Low-temperature direct desorber	55	Baghouse, carbon, scrubber	<0.03	Finished
Canonie	Canon Bridgewater, Bridgewater, MA	6,500	Solvent recycling/total VOC	<0.1	Low-temperature direct desorber	55	Baghouse, carbon, scrubber		Contracted
Canonie	South Kearney, South Kearney, NJ	18,000	Solvent recycling/volatile organics		Low-temperature direct desorber	55	Baghouse, carbon, scrubber		Finished
Canonie	McKin, Gray, ME	18,000	Waste treatment and disposal/trichloroethylene	<0.1	Low-temperature direct desorber	55	Baghouse, carbon, scrubber	<0.03	Finished
Chem Waste	Confidential, Northeast	35,000	PCB spills/PCBs	<2.0	High-temperature indirect desorber	35	Condensation, carbon		Contracted
ENSCO	Union Carbide, Seadrift, TX		Chemical manufacturing		Rotary kiln	35	Steam ejector scrubber		Contracted
ENSCO	Lenz Oil, Lemont, IL	26,000	Hydrocarbons	<5.0	Rotary kiln	35	Steam ejector scrubber	0.006	Finished
ENSCO	Sydney Mines, Brandon, FL	10,000	Waste-oil lagoon/hydrocarbons	<5.0	Rotary kiln	35	Steam ejector scrubber		Finished
ENSCO	NCBC, Gulfport, MS	22,000	Herbicide storage/dioxin	<15 ppt	Rotary kiln	35	Steam ejector scrubber	0.017	Finished
ENSCO	Bridgeport Rental, Bridgeport, NJ	100,000	Used-oil recycling/PCBs		Rotary kiln	100	Steam ejector scrubber		Contracted
GDC Engineering	Rubicon, Geismar, LA	52,000	Chemical manufacturing		Infrared conveyor furnace		Waterloo scrubber		Contracted
Harmon (Williams)	Bog Creek, Howell Twp., NJ	22,500			Rotary kiln	82	Cyclone, baghouse, packed bed		Contracted

Table 25-12. Continued

Contractor	Site Name, Location, State	Site Size, Tons	Source of Contamination/ Indicator Compound	Contaminant Concentration in Treated Soil, mg/kg	Combustion Equipment	Thermal Capacity, 10^6 BTU/h	APC Equipment	Particulate Emissions, g/dscf at 7% O_2	Project Status
Harmon (Williams)	Confidential, AL	600	Gasoline tank leak/petroleum hydrocarbons	<100.00	Low temperature direct desorber	21	Baghouse		Contracted
Harmon (Williams)	Prentiss Creosote, Prentiss, MS	9,200	Wood treatment/ PAHs	<2.0	Rotary kiln	82	Cyclone, baghouse, packed bed	0.011	Finished
IT Corporation	Motco, Lamarque, TX	80,000	Styrene tar disposal pits/PCBs		Rotary kiln	56	Hydrosonics tandem scrubber		Contracted
IT Corporation	Cornhusker AAP, Grand Island, NE	45,000	Munitions plant redwater pits/ trinitrotoluene	<1.34	Rotary kiln	56	Hydrosonics tandem scrubber		Ongoing
IT Corporation	Louisiana AAP, Minden, LA	100,000	Munitions plant redwater lagoon/ trinitrotoluene	<1.3	Rotary kiln	56	Hydrosonics tandem scrubber		Ongoing
IT Corporation	Sikes Pits, Crosley, TX	341,000	Chemical dumping/ hydrocarbons, metals	<100	HTDS-SK rotary kiln	56	Hydrosonics tandem scrubber	<0.08	Contracted
Kimmina	LaSalle, LaSalle, IL	69,000	PCB capacitor manufacturing/PCBs	<2.0	Rotary kiln	100	Baghouse, packed bed		Contracted
Ogden	Confidential, Sacramento, CA	22,500			Circulating fluid bed	10	Baghouse		Contracted
Ogden	Swanson River, Kenai, AK	80,000	Oil pipeline compressor oil/ PCBs	<0.1	Circulating fluid bed	10	Baghouse	<0.05	Ongoing
Ogden	Stockton, Stockton, CA	16,000	Underground tank oil leak. Total hydrocarbons	<1.0	Circulating fluid bed	10	Baghouse	<0.08	Ongoing
O.H. Materials	Gas Station, Cocoa, FL	1,000	Petroleum tank leak/benzene, toluene, xylene	<0.01	Low-temperature direct desorber	12	Venturi	0.011	Finished
O.H. Materials	Rail Yard, PA	1,500	Repetitive spills/ diesel oil	<100.0	Low-temperature direct desorber	20	Cyclone, venturi		Finished

…	New Brighton, MN	4,000	PCBs		furnace		packed bed		Finished
O.H. Materials	Rail yard, PA	1,300	Diesel tank spill/ diesel oil	<100.0	Low-temperature direct desorber	20	Cyclone, venturi		Finished
O.H. Materials	Florida Steel, Indiantown, FL	18,000	Steel mill used oils/PCBs	<2.0	Infrared conveyor furnace	30	Venturi, packed bed	0.056	Finished
O.H. Materials	Rail yards, Cleveland, OH	1,500	Petroleum hydrocarbons	<50.0	Low-temperature direct desorber	20	Cyclone, venturi	0.039	Finished
Site Recl. Systems	Koch Chemical, KS	700	Tank bottoms/ toluene, xylene		Low-temperature direct desorber	47	Baghouse		Contracted
Site Recl. Systems	Gulf Oil, multiple sites, FL	18,000	Benzene, toluene, xylene	<1.0	Low-temperature direct desorber	25	Baghouse		Contracted
Site Recl. Systems	Sun Oil, multiple sites				Low-temperature direct desorber	25	Baghouse		Contracted
Soil Remediation Co.	Multiple sites, SC	3,000	Gas and oil leaks, spills/petroleum hydrocarbons	<50.0	Low-temperature direct desorber	48	Cyclone, baghouse		Finished
Soiltech	Waukegan Harbor, Waukegan, IL	20,000	Marine motor manufacturing/PCBs		High-temperature in direct desorber	14	Baghouse, cyclone, scrubber		Contracted
TDI Services	Chevron Refinery, El Segundo, CA	30,000	API sludges	BDAT	High-temperature in direct desorber		Condensation, carbon		Contracted
Thermo-dynamics Corp.	S. Crop Services, Delray Beach, FL	1,800	Crop-dusting operation/ pentachlorophenol	0.003	Rotary kiln	7	Wet scrubber	0.035	Finished
U.S. Waste Thermal Proc.	Gas station, Temecula, CA	1,000	Petroleum tank leak/total hydrocarbons	<10.0	Infrared conveyor furnace	10	Calvert scrubber	0.008	Finished
U.S. Waste Thermal Proc.	CA	7,500	Total hydrocarbons		Infrared conveyor furnace	10	Calvert scrubber		Contracted
U.S. Waste Thermal Proc.	San Bernardino, CA	540	Total hydrocarbons	<10.0	Infrared conveyor furnace	10	Calvert scrubber		Finished
Vertac Site Contractors	Vertac, Jacksonville, AR	6,500	Chemical manufacturing/ dioxins		Rotary kiln	35	Spray dryer, baghouse, scrubber	0.08	Contracted
VESTA	Nyanza, Ashland, MA	1,000	Dye manufacturing/ nitrobenzene		Rotary kiln	8	Wet scrubber	0.02	Finished
VESTA	Rocky Boy, Havre, MT	1,800	Wood treatment/ pentachlorophenol		Rotary kiln	12	Wet scrubber		Contracted

Table 25-12. Continued

Contractor	Site Name, Location, State	Site Size, Tons	Source of Contamination/ Indicator Compound	Contaminant Concentration in Treated Soil, mg/kg	Combustion Equipment	Thermal Capacity, 10^6 BTU/h	APC Equipment	Particulate Emissions, gr/dscf at 7% O_2	Project Status
VESTA	S. Crop Services, Delray Beach, FL	1,800	Crop-dusting operation/DDT	<0.2	Rotary kiln	12	Wet scrubber	0.03	Finished
VESTA	American Crossarm, Chehalis, WA	900	Wood treatment/ dioxin	<0.001	Rotary kiln	12	Wet scrubber	0.011	Finished
VESTA	Fort A.P. Hill, Bowling Green, VA	200	Army base/dioxin	<0.001	Rotary kiln	12	Wet scrubber	0.02	Finished
Westinghouse/ Haztech	Peak Oil, Tampa, FL	7,000	Used oil recycling/ PCBs	<1.0	Infrared conveyor furnace	30	Wet scrubber	<0.08	Finished
Westinghouse/ Haztech	LaSalle, LaSalle, IL	30,000	Transformer reconditioning/ PCBs	<2.0	Infrared conveyor furnace	30	Wet scrubber	<0.08	Finished
Weston	Revenue, Springfield, IL	1,000	PAHs	<0.33	Low-temperature indirect desorber	12	Baghouse		Finished
Weston	Tinker AFB, Oklahoma City, OK	1,000	Aircraft maintenance trichloroethylene		Low-temperature indirect desorber	12	Baghouse, wet scrubber		Finished
Weston	Paxton Avenue, Chicago, IL	16,000	Waste lagoon/RCRA constituents		Rotary kiln	35	Baghouse, packed bed		Contracted
Weston	Lauder Salvage, Beardstown, IL	8,500	Metal scrap salvage/ PCBs	<2.0	Rotary kiln	35	Baghouse, packed bed	0.02	Finished

ª Source: Cudahy and Troxler, 1990.

Table 25-13. Superfund Records of Decision Recommending the Use of Incineration/Thermal Destruction for Site Remediation

Region I
- Ottati and Goss
- Re-Solve, Inc.
- Davis Liquid Waste
- Cannon Engineering Corp.
- Rose Disposal Pit
- Charles George Landfill No. 3
- Pinette's Salvage Yard
- Wells G&H
- Baird & McGuire
- O'Connor Company Site
- Norwood PCBS
- W.R. Grace

Region II
- Volney Landfill
- Williams Property
- Renora Inc.
- Brewster Wellfield
- Ewan Property
- Reich Farms
- KinBuc Landfill
- Bog Creek Farm
- Claremont Polychemical
- Fulton Terminals
- Pepe Field
- Port Washington Landfill
- Vineland State School

Region III
- Ordnance Works Disposal
- Douglassville Disposal
- Westline Site
- Wildcat Landfill
- Southern Maryland Wood
- Berks Sand Pit
- Drake Chemical Pit
- Avtex Fibers Inc.
- Tyson Dump No. 1
- MW Manufacturing Site
- Douglassville Disposal

Region IV
- Geiger (C&M Oil) Site
- Tower Chemical
- Martin Marietta Sodyeco
- Zellwood Groundwater
- Chemtronics Inc.
- Alpha Chemical Corp.
- Celanese Corp Shelby Fiber
- Amnicola Dump
- Aberdeen Pesticide Dumps
- Newsom Brothers Old Reichold
- Carolawn
- Smith's Farm Brooks

Region V
- Laskin/Poplar Oil
- Liquid Disposal
- Seymour Recycling Corp.
- Pristine Inc.
- LaSalle Electrical Utilities
- Forest Waste Disposal
- Belvidere Municipal Landfill
- Summit National Disposal Service
- Fort Wayne Reduction
- Laskin/Poplar Oil
- Wedzeb Enterprises Inc.
- Ninth Avenue Dump
- Miami County Incinerator
- Alsco Anaconda
- Cliff/Dow Dump
- Cross Brothers Pail Recycling
- Big D Campground
- Twin City Army Ammo Plant

Region VI
- Hardage/Criner
- Cleve Reber
- Bayou Bonfouca
- Brio Refinery Co. Inc.
- Koppers Co.
- South Cavalcade Street
- Gurley Pit
- Sheridan Disposal Services
- Motco Inc.
- United Creosoting Co.

Region VII
- Minker/Stout/Romaine
- Times Beach
- Hastings Groundwater

Region VIII
- Broderick Wood Products Co.
- Libby Groundwater
- Woodbury Chemical Co.
- Sand Creek Industrial

Region IX
- Lorentz Barrel and Drum Co.

Region X
- Pacific Hide & Fur
- Northwest Transformer

lytical data, results, etc., are prepared after each step. When the demonstration is completed, an Applications Analysis Report is prepared to evaluate all the information available on a particular process and to analyze the applicability of the process to other sites, waste types, and media. Also, each year the EPA publishes a document describing all the technologies that have been evaluated under the SITE Program (U.S. EPA, 1989b). Further information on the SITE Program can be obtained from:

Robert A. Olexsey, Division Director
Superfund Technology Demonstration Division
513/569-7861 FTS: 684-7861

Stephen C. James, Chief
SITE Demonstration & Evaluation Branch
513/569-7877 FTS: 684-7877

Norma M. Lewis, Chief
Emerging Technology Section
513/569-7665 FTS: 684-7665

John F. Martin, Chief
Demonstration Section
513/569-7758 FTS: 684-7758

COMPLIANCE WITH FEDERAL AND STATE ARARs

Federal Laws

Section 121 of CERCLA requires that any Superfund action that results in a hazardous substance or contaminant remaining onsite attain a level of control that is at least equivalent to any federal standard, criteria, or limitation considered applicable or relevant and appropriate (ARARs). Applicable requirements are those standards, criteria, or limitations that address a specific hazardous substance, pollutant, action, location, or other circumstance at a site. Relevant and appropriate requirements are those standards, criteria, or limitations that deal with problems or situations sufficiently similar to those encountered at the site to be considered both relevant and appropriate.

CERCLA actions conducted entirely onsite must comply only with the substantive requirements of ARARs, not the administrative requirements. Thus, CERCLA exempts any onsite action from having to obtain a federal, state, or local permit; however, the action is not exempt from complying with the substantive portions of the same laws that the permits enforce. Remedial actions that use offsite facilities during the cleanup must comply with both the substantive and the administrative portions of all legally applicable requirements. Also, these actions must be conducted only at facilities that are in compliance with all applicable federal and state requirements.

Remedial actions also must consider nonregulatory guidance manuals or advisories issued by federal or state agencies. These "to-be-considered" (TBC) materials are important because they provide interpretation and analysis of ARARs.

ARARs can be chemical-specific, location-specific, or action-specific. Chemical-specific ARARs, such as the RCRA or the Safe Drinking Water Act Maximum Contaminant Levels (MCLs), and location-specific ARARs, such as Wetlands or Wilderness area standards, are too site-specific to be dealt with here. More information on these subjects can be obtained from the document entitled CERCLA Compliance With Other Laws Manual: Interim Final, which is listed in the selected bibliography of guidance and resource documents.

Action-specific ARARs are standards or requirements related to technology- or activity-based remedial alternatives, such as incineration. Table 25-14 lists potential ARARs that are applicable to onsite incineration as a CERCLA remedial action under the EPA's HSWA omnibus authority. As new statutes are passed or regulations promulgated, other action-specific requirements will need to be added to this list. The proposed amendments to the hazardous waste incinerator regulations (55 FR 17862, April 27, 1990) and the proposed procedures and technical requirements for corrective action at waste management sites (55 FR 30798, July 27, 1990) will be important potential ARARs when promulgated.

State Laws

State regulations that are more stringent than federal standards must also be met during CERCLA actions if they are identified in a timely manner by the state and if they meet the criteria of being promulgated, generally applicable, and legally enforceable. Whether the state is the lead or the support agency, it is solely responsible for identifying potential state ARARs and documenting the particular sections that are applicable to the site under remediation. The EPA, however, always retains the responsibility for the final decision on the applicability or the possible waiver of ARARs. Examples of state laws that are potential ARARs include:

- *Siting Requirements*: Most states have locational standards that are more restrictive than the federal regulations and that are specific to a site's topographic, hydrologic, or geologic characteristics. Remedial activities, such as the use of a mobile incinerator, could be subject to siting limitations established for that type of facility or that area if those limitations are based on the protection of human health and the environment.
- *Discharge of Toxic Pollutants to Surface Waters*: The Clean Water Act required states to adopt numeric criteria for the discharge or presence of toxic pollutants applicable to the water body and sufficient to protect the designated use. A proposed discharge of incineration scrubber water into surface water could be in conflict with state regulations.
- *Cleanup Standards*: States may enact more stringent cleanup standards than those required under federal law. For example, under federal law, cleanup of releases of hazardous substances must leave no more than 25 ppm polychlorinated biphenyls (PCBs) in the area; however, under Texas law, cleanups must leave no more than 1 ppm.

Generally, CERCLA actions need not comply with local laws; however, the laws may be part of a regional plan enforceable by the state and, as such, are potential state ARARs. Table 25-14 lists potential incineration ARARs.

State standards are an integral part of determining the remediation alternatives and the level of control. The public comment period is not the time to identify conflicts between a selected remedial action and a state regulation. The document CERCLA Compliance with Other Laws Manual, Part II, contains detailed information on identifying and complying with state ARARs.

COST OF INCINERATION

Incineration costs will vary significantly from site to site. Unfortunately, costs are often sources of controversy during site remediation. The relatively high cost of incineration often eliminates it as a treatment option. This being the case, it is very important to conduct an accurate cost assessment. Since detailed cost estimation is not within the scope of this chapter, the RPM/OSC is urged to work in close coordination with the RCRA incineration contacts in each Region during the development of cost estimates for incineration projects. To provide some preliminary background information on this topic, the following information is provided.

The cost of an incineration system varies with several factors, including:

- system capacity
- types of feedstocks being fed
- regime (i.e., slagging vs. ashing)
- length-to-diameter (L/D) ratio for rotary kilns
- type of solids discharge system
- type and capacity of afterburner
- type of auxiliary fuel used
- regulatory climate

Table 25-14. Potential Incineration ARARs[a,b]

Prerequisite for Applicability	Requirement	Citation
RCRA		
RCRA hazardous waste	Analyze the waste feed to determine physical and chemical composition limits.	40 CFR 264.341
	Dispose of all hazardous waste and residues, including ash, scrubber water, and scrubber sludge, according to applicable requirements.	40 CFR 264.351
	(Note: No further requirements for wastes that are listed as hazardous solely because they exhibit one or more of the characteristics of ignitability, corrosivity, reactivity or because they fail the TCLP leaching test and a waste analysis demonstrates no Appendix VIII constituent is present that might reasonably be expected to be present.) Such wastes may also be exempted if Appendix VIII constituents are not present at significant levels.	40 CFR 264.340
	Performance standards:	
	Achieve a destruction and removal efficiency (DRE) of 99.99 percent for each principal organic hazardous constituent designated in the waste feed and 99.9999 percent for dioxins and PCB contaminated liquids.	
	Reduce hydrogen chloride emissions to 1.8 kg/hr or to 1 percent of the HCl in the stack gas before entering any pollution control device.	
	No release of particulates >180 mg/dscm (0.08 gr/dscf) corrected to 7% oxygen.	
	Emissions of CO must be <100 ppm and emissions of THC must be <20 ppm corrected to 7% oxygen.	RCRA Omnibus Authority
	Metals emissions less than those established using the tiered approach outlined in the document "Guidance on Metal and HCl Emissions for Hazardous Waste Incinerators," August 1989.	
	Trial Burn Requirements	40 CFR 270.62
	All residues must meet the RCRA Land Disposal Restrictions	40 CFR 268
	Control fugitive emissions by:	
	Keeping combustion zone sealed; or maintaining combustion-zone pressure lower than atmospheric pressure.	40 CFR 264.345
	Use automatic cutoff system to stop waste feed when operating conditions deviate or exceed established limits.	40 CFR 264.345
	Monitor various parameters during operation, including combustion temperature, waste feed rate, indication of combustion gas velocity, and carbon monoxide in stack gas.	40 CFR 264.347
CAA		
Air emissions	Remediation activities must comply with the National Ambient Air Quality Standards (NAAQS). Compliance should be determined in cooperation with the appropriate state government agency. An air permit from the state may be required.	40 CFR 50

TSCA

Liquid PCBs at concentration of 50 ppm or greater	Performance standards:	40 CFR 761.70
	2-second residence time at 1200°C (±100°C) and 3 percent excess oxygen in stack gas; or	
	1.5-second residence time at 1600°C and 2 percent excess oxygen in stack gas.	
	Combustion efficiency of at least 99.90 percent.	
	DRE>99.9999%	
	Rate and quantity of PCBs fed to the combustion system shall be measured and recorded at regular intervals of no longer than 15 minutes.	40 CFR 761.70
	Temperature of incineration shall be continuously measured and recorded.	40 CFR 761.70
	Flow of PCBs to incinerator must stop automatically whenever the combustion temperature drops below specified temperature.	40 CFR 761.70
	Monitoring must occur:	40 CFR 761.70
	When the incinerator is first used or modified; monitoring must measure for O_2, CO, CO_2, oxides of nitrogen, HCl, RCl, PCBs, total particulate matter.	40 CFR 761.70
	Whenever PCBs are being incinerated, the O_2, CO, CO_2, oxides of nitrogen and CO levels must be continuously checked; CO_2 must be periodically checked.	
	Water scrubbers must be used for HCl control.	40 CFR 761.70
Nonliquid PCBs, PCB articles, PCB equipment, and PCB containers at concentrations of 50 ppm or greater.	Mass air emissions from the incinerator shall be no greater than 0.0019 PCB per kg of the PCBs entering the incinerator (99.9999 percent DRE).	40 CFR 761.70
	Requirements as listed for liquid PCBs.	40 CFR 761.70

Table 25-14. Continued

Prerequisite for Applicability	Requirement	Citation
FIFRA		
Organic pesticides, except organic mercury, lead, cadmium, and arsenic (recommended).	Performance standards:	
	2-second residence time at 1000°C (or equivalent that will assure complete destruction)	40 CFR 165.8
	Meet requirements of CAA relating to gaseous emissions.	40 CFR 165.1
	Dispose of liquids, sludges, or solid residues in accordance with applicable federal, state, and local pollution control requirements.	40 CFR 165.8
		40 CFR 165.8
Metallo-organic pesticides, except mercury, lead, cadmium, or arsenic compounds (recommended).	Chemically or physically treat pesticides to recover heavy metals; incinerate in same manner as organic pesticides.	40 CFR 165.8
Combustible containers that formerly held organic or metallo-organic pesticides, except organic mercury, lead, arsenic, and cadmium (recommended).	Incinerate in same manner as organic pesticides.	40 CFR 165.9
OSHA		
Remediation activities	All remediation activities must comply with the policies and programs established for worker safety.	29 CFR 1910 29 CFR 1926

a Source: U.S. Environmental Protection Agency, 1988a and 1989a.

b The regulations cited herein may contain special provisions or variances applicable to the specific site under remediation. In all circumstances the actual regulations should be consulted before any decisions are formulated.

Table 25-15. Typical Costs of Incineration of Contaminated Soils[a,b]

	Incineration System Capacity (tons/h)	Unit Cost ($/ton)
Centalized rotary kiln system	Commercial unit	300 to 650
Onsite incineration		
Small site (<5,000 tons)	<5	1000 to 1500
Medium site (5,000 to 10,000 tons)	5 to 10	300 to 800
Large site (>30,000 tons)	>10	100 to 400

[a] Estimated costs are in 1988 dollars. They do not include the cost of transportation, removal of soils from the ground, or storage.

[b] Sources: Cudahy et al. (1987); Tillman et al. (1990); U.S. EPA (1988b).

These costs in turn affect the cost of waste treatment by incineration. Table 25-15 presents the estimated costs of incinerating contaminated soils in both onsite and offsite incineration systems. These costs do not include transportation, storage, or removal of the soil from the ground. The total cost of waste treatment would vary considerably from site to site, and any estimate should include the following (Evans, 1990):

- site preparation
- permitting and regulatory requirements
- capital equipment
- startup
- labor
- consumables and supplies
- utilities
- effluent treatment and disposal
- residuals/waste shipping and handling
- analytical services
- maintenance and modifications
- demobilization

GUIDANCE AND RESOURCE DOCUMENTS

EPA Hazardous Waste Incineration Guidance Series

Volume I: Guidance Manual for Hazardous Waste Incinerator Permits. SW-966, July 1983. NTIS: PB84-100577 (update expected late 1990).

Volume II: Guidance on Setting Permit Conditions and Reporting Trial Burn Results. EPA-625/6-89-019, January 1989.

Volume III: Hazardous Waste Incineration Measurement Guidance Manual. EPA-625/6-89-021, June 1989. NTIS: PB90-182759.

Volume IV: Guidance on Metals and Hydrogen Chloride Controls for Hazardous Waste Incinerators. 1989 Draft Report.

Volume V: Guidance on PIC Controls for Hazardous Waste Incinerators. 1989 Draft Report.

Volume VI: Proposed Methods for Measurement of CO, O2, THC, HCl, and Metals at Hazardous Waste Incinerators. 1989 Draft Report.

Other EPA Resource Documents

CERCLA Compliance with Other Laws Manual: Interim Final. EPA 540/G-89-006. August 1988.

CERCLA Compliance with Other Laws Manual: Part II. Clean Air Act and Other Environmental Statutes and State Requirements. EPA 540/G-89-009, August 1989.

Engineering Handbook for Hazardous Waste Incineration. SW-889, September 1981. NTIS: PB81-248163 (update expected late 1990).

Experience in Incineration Applicable to Superfund Site Remediation. EPA/625/9-88/008, December 1988.

Guidance for Conducting Remedial Investigations and Feasibility Studies Under CERCLA: Interim Final. EPA 540/G-89-004, October 1988.

Handbook: Permit Writer's Guide to Test Burn Data, Hazardous Waste Incineration. EPA 625/6-86-012, 1986.

Handbook: Quality Assurance/Quality Control (QA/QC) Procedures for Hazardous Waste Incineration. EPA-625/6-89-023, January, 1990.

The RPM Primer: An Introductory Guide to the Roles and Responsibilities of the Remedial Project Manager. EPA 540/G-87/005, 1987.

Seminar Publication: Permitting Hazardous Waste Incinerators. EPA/625/4-87/017, September, 1987.

Seminar Publication: Operational Parameters for Hazardous Waste Combustion Devices. EPA/625/R-93/008, October, 1993.

Other Resource Documents

Air Pollution Control Association (APCA). 1987. Incineration of Hazardous Waste, Critical Review Discussion Papers. Journal of the Air Pollution Control Association 37(9):1011–1024, September.

American Society of Mechanical Engineers (ASME). 1988. Hazardous Waste Incineration, A Resource Document. The ASME Research Committee on Industrial and Municipal Waste. New York City.

Brunner, C.R. 1988. Site Cleanup by Incineration. Hazardous Materials Control Research Institute, Silver Spring, MD.

Freeman, H.M., ed. 1989. Standard Handbook of Hazardous Waste Treatment and Disposal. McGraw-Hill, New York.

Oppelt, E.T. 1987. Incineration of Hazardous Wastes, A Critical Review. Journal of the Air Pollution Control Association, 27(5):558

U.S. Congress, Office of Technology Assessment. 1988. Are We Cleaning Up? 10 Superfund Case Studies-Special Report. OTA-ITE-362. U.S. Government Printing Office, Washington, DC.

U.S. Congress, Office of Technology Assessment. 1986. Ocean Incineration: Its Role Managing Hazardous Waste. OTA-O0313. U.S. Government Printing Office. Washington, DC.

EPA CONTACT

For further information on this topic please contact Laurel Staley, RREL-Cincinnati, OH, (513) 569-7863.

EPA TECHNICAL SPECIALISTS[4]

Communication between the RPM, the EPA Regional office, and the corresponding state environmental office is critical. More importantly, communication with the RCRA incineration experts and technical contacts in each Regional office who have extensive incineration expertise is vital to the success of remedial/removal activities involving incineration. Any remediation plans involving an incinerator should be sent to the Regional RCRA incinerator permit office for review. Getting this office involved early in the remediation selection process can prevent costly delays later. Each Regional office has an incinerator expert available as a technical specialist to advise and assist the RPM. Many states also have technical contacts with extensive experience in incineration. The following is a list of the EPA Headquarters and Regional incinerator experts and the corresponding state expert. If a state does not have an incinerator expert on their staff, the RPM is referred to the Regional office.

---

[4] Editor's Note: The contacts and phone numbers here are as listed in the original Engineering Bulletin, and may not be up-to-date.

Contact

Headquarters		FTS	Commercial
Sonya Sasseville, Chief Alternative Technology and Support Section		382-3132	202/382-3132
Lionel Vega, Incineration Permit Assistance		475-8988	202/475-8988

Region I.

		FTS	Commercial
Stephen Yee		833-1644	617/573-9644
John Podgurski		833-1673	617/573-9673
Connecticut	George Dews		203/566-2264
Maine	See Regional contact.		
Massachusetts	Stephen Dresszen		617/292-5832
New Hampshire	See Regional contact.		
Rhode Island	Beverly Migilore		401/277-2797
Vermont	See Regional contact.		

Region II.

		FTS	Commercial
John Brogard		264-8682	212/264-8682
Clifford Ng (Puerto Rico)		264-9579	212/264-9579
New Jersey	Thomas Sherman		609/292-1250
New York	James Dolen		518/457-6934

Region III.

		FTS	Commercial
Gary Gross		597-7940	215/597-7940
Delaware	Ken Weiss		302/736-3689
Dist. of Columbia	Angelo Tompros		202/783-3194
Maryland	Alvin Bowles		301/631-3343
Pennsylvania	Joe Hayes		717/787-7381
Virginia	Karol Akers		804/225-2496
West Virginia	Robert Weser		304/348-4022

Region IV.

		FTS	Commercial
Betty Willis		257-3433	404/347-3433
Alabama	Clyde Shearer		205/271-7700
Florida	John Griffith		904/488-0300
Georgia	Bill Mundy		404/656-2833
Kentucky	Mohammed Alauddin		502/564-6716
Mississippi	Steve Spengler		601/961-5171
N. Carolina	Bill Hamner		919/733-2178
S. Carolina	David Wilson		803/734-5200
Tennessee	Jackie Okoree-Baah		615/741-3424

Region V.

		FTS	Commercial
Y.J. Kim		886-6147	312/886-6147
Juan Rojo (Illinois)		886-0990	312/886-0990
Gary Victorine (Indiana)		886-1479	312/886-1479
Lorna Jereza (Michigan)		353-5110	312/353-5110
Wen Haung (Minnesota)		886-6191	312/886-6191
Thelma Codina (Ohio)		886-6181	312/886-6181
Wen Haung (Wisconsin)		886-6191	312/886-6191
Illinois	Robert Watson		217/785-8410
Indiana	Elaine Greg		317/232-8866
Michigan	Steve Buda		517/373-2730

Minnesota	Fred Jenness		612/297-1792
Ohio	Bob Babik		614/644-2949
Wisconsin	Ed Lynch		608/266-3084

Region VI.

Henry Onsgard		655-6785	214/655-6785
Jim Sales (Texas)		655-6785	214/655-6785
Stan Burger (Arkansas, Louisiana Oklahoma, New Mexico)		655-6785	214/655-6785
Arkansas	Mike Bates		501/562-7444
Louisiana	Karen Fisher		504/342-4685
New Mexico	Dr. Elizabeth Gordon		505/827-2934
Oklahoma	Catherine Sharp		405/271-7062
Texas	Wayne Harry		512/463-8173

Region VII.

Joe Galbraith		276-7057	913/551-7057
Luetta Flournoy (Iowa)		276-7058	913/551-7653
Iowa	See Regional contact.		
Kansas	John Ramsey		913/296-1610
Missouri	John Doyle		314/751-3176
Nebraska	Glen Dively		402/471-4176

Region VIII.

Nat Miullo		330-1500	303/330-1500
Colorado	Neal Kolwey		303/331-4830
Montana	See Regional contact.		
N. Dakota	See Regional contact.		
S. Dakota	See Regional contact.		
Utah	Connie Nakahara		801/538-6170
Wyoming	See Regional contact.		

Region IX.

Larry Bowerman		484-1471	415/744-1471
Arizona	Al Roesler		602/257-2249
California	Sangat Kals		916/324-9611
Region 1	Eric Hong		916/855-7726
Region 2	Don F. Murphy		415/540-3969
Region 3	Gautum Guha		818/567-3123
Region 4	Anand Rege		213/590-4896
Hawaii	Les Segunda		808/548-8837
Nevada	Don Gross		702/885-5872

Region X.

Cathy Massimino		399-4153	206/442-4153
Alaska	David Ditraglia		907/465-2671
Idaho	Jay Skabo		208/334-5879
Oregon	Ed Chiong		503/229-5326
Washington	Cindy Gilder		206/438-7019

OTHER TECHNICAL CONTACTS

In addition to the EPA Regional and state technical advisors listed above, the following people can be contacted for specialized information:

Robin Anderson
EPA, Washington
202/398-8739

EPA policies and practices,
    remedial operations

James Cudahy
Focus Environmental
615/694-7517

Full-scale, mobile, thermal
remediation projects

Paul Lemieux
EPA, RTP
919/541-0962

Secondary combustion
chamber/afterburner impact on
toxic air emissions

Kevin Smith
International Technology
615/690-3211

Mobile incinerator markets and
technology

Sonya Sasseville
EPA, Washington
202/382-3132

EPA policies and practices, RCRA
incineration permits

Lionel Vega
EPA, Washington
202/475-8988

EPA policies and practices, RCRA
incinerator permits

Y.J. Kim
EPA Region 5
312/886-6147

National incinerator expert

Dr. Barry Dellinger
University of Dayton
Research Institute
513-229-2846

National incinerator expert

Laurel Staley
RREL
USEPA, Cincinnati
513-569-7863

POHC and PIC thermal stability
technology

ACKNOWLEDGMENTS

This chapter was prepared by the Risk Reduction Engineering Laboratory's (RREL) Engineering and Treatment Technical Support Center, under the technical direction of Laurel Staley (RREL) and Paul Leonard (Region III), with the support of the Superfund Technical Support Project. The first draft of this chapter was prepared by PEI Associates under Work Assignment No. 19-2V of Contract 68-03-3413. The chapter received in-depth technical reviews from the following individuals: Ed Hanlon, Beverly Houston, Sonya Sasserville, Phil Taylor, Paul Leonard, Nat Muillo, Joseph Santoleri, Robert Thurnau, Marta Richards, James Scarborough, Richard Carnes, and Ernest Franke. Their comments are much appreciated and have significantly improved the accuracy and completeness of the final chapter.

REFERENCES

American Society of Mechanical Engineers (ASME). 1988. Hazardous Waste Incineration, A Resource Document. The ASME Research Committee on Industrial and Municipal Waste. New York City.

Brunner, C.R. 1988a. Site Cleanup by Incineration. Hazardous Materials Control Research Institute, Silver Spring, MD.

Brunner, C.R. 1988b. Industrial Waste Incineration. Hazardous Materials Controls 1(4):26+, July/August.

Buonicore, A.J. 1990. Experience with Air Pollution Control Equipment on Hazardous Waste Incinerators. Paper No. 9033.2. Presented at the 83rd Annual Meeting of the Air and Waste Management Association held in Pittsburgh, PA, June 24–29, 1990.

Cudahy, J., S. DeCicco, and W. Troxler. 1987. Thermal Treatment Technologies for Site Remediation. Presented at the International Conference on Hazardous Materials Management, Chattanooga, TN, June 9, 1987.

Cudahy, J.J. and W.L. Troxler. 1990. Thermal Remediation Industry Update-II. Paper presented at the Air & Waste Management Association Symposium on Treatment of Contaminated Soils, Cincinnati, Ohio, February 6, 1990.

Daniels, S.L. 1989. Products of Incomplete Combustion. Journal of Hazardous Materials, 22(2):161–174.

Dellinger, B., P.H. Taylor, and D.A. Tiery. 1989. Pathways of Formation of Chlorinated PICs from the Thermal Degradation of Simple Chlorinated Hydrocarbons. Journal of Hazardous Materials, 22(2):175–186.

Dellinger, B., Torres, J.L., Rubey, W.A., Hall, D.L., Graham, J.L., and Carnes, R.A. Determination of the Thermal Stability of Selected Hazardous Organic Compounds. Hazardous Waste, Vol. 1, pp. 137–157 (1984).

Dellinger, B., Taylor, P.H., and Tirey, D.A., Minimization and Control of Hazardous Combustion Byproducts, Final Report and Project Summary. Prepared for U.S. EPA under cooperative agreement CR-813938-01-0, April 1990.

Evans, G.M. 1990. Estimating Innovative Technology Costs for the Site Program. Journal of the Air & Waste Management Association, 40(7):1047–1051.

McCormick, R.J. and M.L. Duke. 1989. On-Site Incineration as a Remedial Action Alternative. Pollution Engineering, 21(8):68–73.

McCoy & Associates, Inc. 1989. Mobile Treatment Technologies-Regulations, Outlook, and Directory of Commercial Vendors. The Hazardous Waste Consultant, 7(1):4-1+, January/February.

McGraw-Hill, Inc. 1990. Superfund Cleanup Plans. Inside EPA's Superfund Report, 4(5):32, February 28. [Not cited in text.]

Oppelt, E.T. 1987. Incineration of Hazardous Wastes, A Critical Review. Journal of the Air Pollution Control Association, 27(5):558–586.

Santoleri, J.J. 1989a. Design and Operating Problems of Hazardous Waste Incinerators. Environmental Progress, 4(4)246–251.

Santoleri, J.J. 1989b. Liquid-Injection Incinerators. In: Standard Handbook of Hazardous Waste Treatment and Disposal. H.M. Freeman, ed. McGraw-Hill, New York.

Santoleri, J.J. 1989c. Rotary Kiln Incineration Systems: Operating Techniques for Improved Performance. In: Proceedings of the Third International Conference on New Frontiers for Hazardous Waste Management, Pittsburgh, PA, September 10–13, 1989. EPA/600/9-89-072.

Schaefer, C.F. and A.A. Albert. 1989. Rotary Kilns. In: Standard Handbooks of Hazardous Waste Treatment and Disposal. H.M. Freeman, ed. McGraw-Hill, New York.

Stumbar. J.P., et al. 1989. Operating Experiences of the EPA Mobile Incineration System with Various Feed Materials. In: Proceedings of the Third International Conference on New Frontiers for Hazardous Waste Management, Pittsburgh, PA, September 10–13, 1989. EPA/600/9-89/072.

Taylor, P.H. and P. Dellinger. 1988. Thermal Degradation Characteristics of Chlorinated Methane Mixtures. Environmental Science & Technology 22:438–447.

Taylor, P.H. and B. Dellinger. 1990. Development of a Thermal Stability Based Ranking of Hazardous Organic Compound Incinerability. Environmental Science & Technology 24:316–328.

Tillman, D., A. Rossi, and K. Vick. 1990. Rotary Incineration Systems for Solid Hazardous Wastes. Chemical Engineering Progress, 86(7):19–30.

Tirey, D.A., P.H. Taylor, and B. Dellinger. 1990. Products of Incomplete Combustion from the High Temperature Pyrolysis of the Chlorinated Methanes. In: Emissions from Combustion Processes: Origin Measurement and Control, Lewis Publishers, Chelsea, MI, pp. 109–120.

U.S. Environmental Protection Agency (EPA). 1986a. Mobile Treatment Technologies for Superfund Waste. EPA/2-86/003(F) (NTIS PB87-110656).

U.S. Environmental Protection Agency (EPA). 1986b. Handbook Permit Writer's Guide to Test Burn Data, Hazardous Waste Incineration. EPA 625/6-86-012.

U.S. Environmental Protection Agency (EPA). 1988a. CERCLA Compliance with Other Laws Manual: Interim Final. EPA 540/G-89-006.

U.S. Environmental Protection Agency (EPA). 1988b. Experience in Incineration Applicable to Superfund Site Remediation. EPA/625/9-88/008.

U.S. Environmental Protection Agency (EPA). 1989a. CERCLA Compliance with Other Laws Manual: Part II. Clean Air Act and Other Environmental Statutes and State Requirements. EPA 540/G-89-009.

U.S. Environmental Protection Agency (EPA). 1989b. The Superfund Innovative Technology Evaluation Program: Technology Profiles. EPA 540/5-89-013.

U.S. Environmental Protection Agency (EPA). 1990. ROD Annual Report: FY 1989. EPA-540/8-90-006.

Weinberger, L., et al. 1984. Supporting Documentation for the RCRA Incinerator Regulations, 40 CFR 265, Subpart O—Incinerators. U.S. Environmental Protection Agency Contract No. 68-01-6901.

Wilson, R. 1978. Analyzing the Daily Risks of Life. Technology Review, February 1979. [Not cited in text.]

Chapter 26

Thermal Desorption Treatment[1]

Jim Rawe and Eric Saylor, Science Applications International Corporation (SAIC), Cincinnati, OH

ABSTRACT

Thermal desorption is an ex situ means to physically separate volatile and some semivolatile contaminants from soil, sediments, sludges, and filter cakes by heating them at temperatures high enough to volatilize the organic contaminants. For wastes containing up to 10 percent organics or less, thermal desorption can be used in conjunction with off-gas treatment for site remediation. It also may find applications in conjunction with other technologies at a site.

Thermal desorption is applicable to organic wastes and generally is not used for treating metals and other inorganics. The technology thermally heats contaminated media, generally between 300 to 1,000°F, thus driving off the water, volatile contaminants, and some semivolatile contaminants from the contaminated solid stream and transferring them to a gas stream. The organics in the contaminated gas stream are then treated by being burned in an afterburner, condensed in a single- or multistage condenser, or captured by carbon adsorption beds.

The use of this well-established technology is a site-specific determination. Thermal desorption technologies are the selected remedies at 31 Superfund sites [2]. Geophysical investigations and other engineering studies need to be performed to identify the appropriate measure or combination of measures to be implemented based on the site conditions and constituents of concern at the site. Site-specific treatability studies may be necessary to establish the applicability and project the likely performance of a thermal desorption system. The EPA contact indicated at the end of this chapter can assist in the identification of other contacts and sources of information necessary for such treatability studies.

This chapter discusses various aspects of the thermal desorption technology including applicability, limitations of its use, residuals produced, performance data, site requirements, status of the technology, and sources of further information.

TECHNOLOGY APPLICABILITY

Thermal desorption has been proven effective in treating organic-contaminated soils, sediments, sludges, and various filter cakes. Chemical contaminants for which bench-scale through full-scale treatment data exist include primarily volatile organic compounds (VOCs), semivolatile organic compounds (SVOCs), polychlorinated biphenyls (PCBs), pentachlorophenols (PCPs), pesticides, and herbicides [1, 3–7]. The technology is not effective in separating inorganics from the contaminated medium.

Extremely volatile metals may be removed by higher temperature thermal desorption systems. However, the temperature of the medium produced by the process generally does not oxidize the metals present in the contaminated medium [8, p. 85]. The presence of chlorine in the waste can affect the volatilization of some metals, such as lead. Generally, as the chlorine content increases, so will the likelihood of metal volatilization [9].

The technology is also applicable for the separation of organics from refinery wastes, coal tar wastes, wood-treating wastes, creosote-contaminated soils, hydrocarbon-contaminated soils, mixed (radioactive and hazardous) wastes, synthetic rubber processing wastes, and paint wastes [4; 10, p. 2; 11].

---

[1] EPA/540/S-94/501.

Table 26-1. RCRA Codes for Wastes Treated by Thermal Desorption

Wood Treating Wastes	K001
Dissolved Air Flotation	K048
Stop Oil Emulsion Solids	K049
Heat Exchanger Bundles Cleaning Sludge	K050
American Petroleum Institute (API) Separator Sludge	K051
Tank Bottoms (leaded)	K052

Performance data presented in this chapter should not be considered directly applicable to other Superfund sites. A number of variables, such as concentration and distribution and contaminants, soil particle size, and moisture content, can all affect system performance. A thorough characterization of the site and well-designed and conducted treatability studies of all potential treatment technologies are highly recommended.

Table 26-1 lists the codes for the specific Resource Conservation and Recovery Act (RCRA) wastes that have been treated by this technology [4; 10, p. 7; 11]. The indicated codes were derived from vendor data where the objective was to determine thermal desorption effectiveness for these specific industrial wastes.

The effectiveness of thermal desorption on general contaminant groups for various matrices is shown in Table 26-2. Examples of constituents within contaminant groups are provided in Technology Screening Guide for Treatment of CERCLA Soils and Sludges [8, p. 10]. This table has been updated and is based on the current available information or professional judgment where no information was available. The proven effectiveness of the technology for a particular site or waste does not ensure that it will be effective at all sites or that the treatment efficiencies achieved will be acceptable at other sites. For the ratings used for this table, demonstrated effectiveness means that, at some scale, treatability was tested to show the technology was effective for that particular contaminant and medium. The ratings of potential effectiveness or no-expected-effectiveness are both based upon expert judgment. Where potential effectiveness is indicated, the technology is believed capable of successfully treating the contaminant group in a particular medium. When the technology is not applicable or will likely not work for a particular combination of contaminant group and medium, a no-expected-effectiveness rating is given.

Another source of general observations and average removal efficiencies for different treatability groups is contained in the Superfund Land Disposal Restrictions (LDR) Guide #6A, Obtaining a Soil and Debris Treatability Variance for Remedial Actions, (OSWER Directive 9347.3-06FS, September 1990) [12] and Superfund LDR Guide #6B, Obtaining a Soil and Debris Treatability Variance for Removal Actions, (OSWER Directive 9347.3-06BFS, September 1990) [13].

A further source of information is the U.S. EPA's Risk Reduction Engineering Laboratory Treatability Database (accessible via ATTIC).

TECHNOLOGY LIMITATIONS

Inorganic constituents or metals that are not particularly volatile will be unlikely to be effectively removed by thermal desorption. If there is a need to remove a portion of them, a vendor process with a very high bed temperature is recommended, due to the fact that a higher bed temperature will generally result in a greater volatilization of contaminants. If chlorine or another chlorinated compound is present, some volatilization of inorganic constituents in the waste may also occur [14, p. 8].

The contaminated medium must contain at least 20 percent solids to facilitate placement of the waste material into the desorption equipment [3, p. 9]. Some systems specify a minimum of 30 percent solids [15, p. 6].

As the medium is heated and passes through the kiln or desorber, energy is consumed in heating moisture contained in the contaminated soil. A very high moisture content may result in low contaminant volatilization, a need to recycle the soil through the desorber, or a need to dewater the material prior to treatment to reduce the energy required to volatilize the water.

Material handling of soils that are tightly aggregated or largely clay can result in poor processing performance due to caking. Rock fragments or particles greater than 1 to 2 inches may have to be

Table 26-2. Effectiveness of Thermal Desorption on General Contaminant Groups for Soil, Sludge, Sediments, and Filter Cakes

Contaminant Groups	Effectiveness			
	Soil	Sludge	Sediments	Filter Cakes
Organic				
Halogenated volatiles	■	▼	▼	■
Halogenated semivolatiles	■	■	▼	■
Nonhalogenated volatiles	■	▼	▼	■
Nonhalogenated semivolatiles	■	▼	▼	■
PCBs	■	▼	■	▼
Pesticides	■	▼	▼	▼
Dioxins/furans	■	▼	▼	▼
Organic cyanides	▼	▼	▼	▼
Organic corrosives	●	●	●	●
Inorganic				
Volatile metals	■	▼	▼	▼
Nonvolatile metals	●	●	●	●
Asbestos	●	●	●	●
Radioactive materials	●	●	●	●
Inorganic corrosives	●	●	●	●
Inorganic cyanides	●	●	●	●
Reactive				
Oxidizers	●	●	●	●
Reducers	●	●	●	●

■ Demonstrated Effectiveness: Successful treatability test at some scale completed.
▼ Potential Effectiveness: Expert opinion that technology will work.
● No Expected Effectiveness: Expert opinion that technology will not work.

prepared by being crushed, screened, or shredded in order to meet the minimum treatment size. However, one advantage to the soil preparation is that the contaminated medium is mixed and exhibits a more uniform moisture and BTU content.

If a high fraction of the fine silt or clay exists in the matrix, fugitive dusts will be generated [8, p. 83], and a greater dust loading will be placed on the downstream air pollution control equipment [15, p. 6].

The treated medium will typically contain less than 1 percent moisture. Dust can easily form in the transfer of the treated medium from the desorption unit, but can be mitigated by water sprays. Normally, clean water from air pollution control devices can be used for this purpose. Some type of enclosure may be required to control fugitive dust if water sprays are not effective.

Although volatile and semivolatile organics are the primary target of the thermal desorption technology, the total organic loading is limited by some systems to 10 percent or less [16, p. 11-30]. As in most systems that use a reactor or other equipment to process wastes, a medium exhibiting a very high pH (greater than 11) or very low pH (less than 5) may corrode the system components [8, p. 85].

There is evidence with some system configurations that polymers may foul or plug heat transfer surfaces [3, p. 9]. Laboratory/field tests of thermal desorption systems have documented the deposition of insoluble brown tars (presumably phenolic tars) on internal system components [16, p. 76].

Caution should be taken regarding the disposition of the treated material. For example, this material could be susceptible to such destabilizing forces as liquefaction, where pore pressures are able to weaken the material on sloped areas or places where materials must support a load (i.e., roads for vehicles, subsurfaces of structures, etc.). To achieve or increase the required stability of the treated material, it may have to be mixed with other stabilizing materials or compacted in multiple lifts. A thorough geotechnical evaluation of the treated product would first be required [14, p. 8].

There is also a possibility that during the cleanup process at a particular site dioxins and furans may form and be released from the exhaust stack into the environment. The possibility of this occurring should be determined on a case-by-case basis.

TECHNOLOGY DESCRIPTION

Thermal desorption is a process that uses either indirect or direct heat exchange to heat organic contaminants to a temperature high enough to volatilize and separate them from a contaminated solid medium. Air, combustion gas, or an inert gas is used as the transfer medium for the vaporized components. Thermal desorption systems are physical separation processes that transfer contaminants from one phase to another. They are not designed to provide high levels of organic destruction, although the higher temperature of some systems will result in localized oxidation or pyrolysis. Thermal desorption is not incineration, since the destruction of organic contaminants is not the desired result. The bed temperatures achieved and residence times used by thermal desorption systems will volatilize selected contaminants, but usually not oxidize or destroy them. System performance is usually measured by the comparison of untreated solid contaminant levels with those of the processed solids. The contaminated medium is typically heated to 300 to 1,000°F, based on the thermal desorption system selected.

Figure 26-1 is a general schematic of the thermal desorption process.

Material handling (1) requires excavation of the contaminated solids or delivery of filter cake to the system. Typically, large objects (greater than 2 inches in diameter) are screened, crushed, or shredded and, if still too large, rejected. The material to be treated is then delivered by gravity to the desorber inlet or conveyed by augers to a feed hopper [6, p. 1].

Desorption (2) of contaminants can be affected by use of a rotary dryer, thermal screw, vapor extractor (fluidized bed), or distillation chamber [15].

As the waste is heated, the contaminants vaporize, and are then transferred to the gas stream. An inert gas, such as nitrogen, may be injected as a sweep stream to prevent contaminant combustion and to aid in vaporizing and removing the contaminants [4; 10, p. 1]. Other systems simply direct the hot gas stream from the desorption unit [3, p. 5; 5].

The actual bed temperature and residence time are primary factors affecting performance in the desorption stage. These factors are controlled in the desorption unit by using a series of increasing temperature zones [10, p. 1], multiple passes of the medium through the desorber where the operating temperature is sequentially increased, separate compartments where the heat transfer fluid temperature is higher, or sequential processing into higher temperature zones [17; 18]. Heat transfer fluids used include hot combustion gases, hot oil, steam, and molten salts.

Off-gas from desorption is typically processed (3) to remove particulates that were entrained into the gas stream during the desorption step. Volatiles in the off-gas may be burned in an afterburner, collected on activated carbon, or recovered in condensation equipment. The selection of the gas treatment system will depend on the concentrations of the contaminants, air emission standards, and the economics of the off-gas treatment system(s) employed. Some methods commonly used to remove the particulates from the gas stream are cyclones, wet scrubbers, and baghouses. In a cyclone, particulates are removed by centrifugal force. In a wet scrubber, the contaminated gas stream passes upward through water sprays, causing the particulates to be washed out at the bottom of the scrubber. In a baghouse, particulates are caught by bags and discharged out of the system.

PROCESS RESIDUALS

Operation of thermal desorption systems may create up to six process residual streams: treated medium; oversized medium and debris rejects; condensed contaminants and water; spent aqueous and vapor phase activated carbon; particulate dust; and clean off-gas. Treated medium, debris, and oversized rejects may be suitable for return onsite.

The vaporized organic contaminants can be captured by condensation or passing the off-gas through a carbon adsorption bed or other treatment system. Organic compounds may also be destroyed by using an off-gas combustion chamber or a catalytic oxidation unit [14, p. 5].

When off-gas is condensed, the resulting water stream may contain significant contamination, depending on the boiling points and solubility of the contaminants, and may require further treatment (i.e., carbon adsorption). If the condensed water is relatively clean, it may be used to suppress the dust from the treated medium. If carbon adsorption is used to remove contaminants from the off-gas or condensed water, spent carbon will be generated, and is either returned to the supplier for reactivation/incineration or regenerated onsite [14, p. 5].

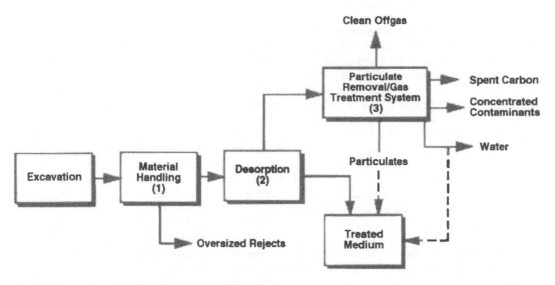

Figure 26-1. Schematic diagram of thermal desorption.

Off-gas from a thermal desorption unit will contain particulates from the medium, vaporized organic contaminants, and water vapor. Particulates are removed by conventional equipment such as cyclones, wet scrubbers, and baghouses. Collected particulates may be recycled through the thermal desorption unit or blended with the treated medium, depending on the concentration of organic contaminants present on the particulates. Very small particles (<1 micron) can cause a visible plume from the stack [14, p. 5].

When off-gas is destroyed by a combustion process, compliance with incineration emission standards may be required. Obtaining the necessary permits and demonstrating compliance may be advantageous, however, since the incineration process would not leave residuals requiring further treatment [14, p. 5]

SITE REQUIREMENTS

Thermal desorption systems typically are transported on specifically adapted flatbed semitrailers. Most systems consist of three components (desorber, particulate control, and gas treatment). Space requirements onsite are typically less than 150 feet by 150 feet, exclusive of materials handling and decontamination areas.

Standard 440V, three-phase electrical service is needed. Water must be available at the site. The quantity of water needed is vendor- and site-specific.

Treatment of contaminated soils of other waste materials require that a site safety plan be developed to provide for personnel protection and special handling measures. Storage should be provided to hold the process product streams until they have been tested to determine their acceptability for disposal or release. Depending upon the site, a method to store waste that has been prepared for treatment may also be necessary. Storage capacity will depend on waste volume. Onsite analytical equipment capable of determining the residual concentration of organic compounds in process residuals makes the operation more efficient and provides better information for process control.

PERFORMANCE DATA

Performance data in this chapter are included as a general guideline to the performance of the thermal desorption technology and may not always be directly transferable to other Superfund sites. Thorough site characterization and treatability studies are essential in determining the potential effectiveness of the technology at a particular site. Most of the data on thermal desorption come from studies conducted for EPA's Risk Reduction Engineering Laboratory under the Superfund Innovative Technology Evaluation (SITE) Program.

Seaview Thermal Systems (formerly T.D.I. Services, Inc.) conducted a pilot-scale test of their HT-5 thermal desorption system at the U.S. DOE's Y-12 plant at Oak Ridge, Tennessee. The test was run to evaluate the capability of the unit to remove and recover mercury from a soil matrix. Initial mercury concentrations in the soil were 1,140 mg/kg. The mercury was removed to concentrations of 0.19 mg/kg with a detection limit of 0.03 mg/kg. A full-scale cleanup (80 tons per day) using the HT-5 system, was conducted for Chevron U.S.A. at their El Segundo Refinery. The primary contaminants and their initial and final concentrations are indicated in Table 26-3 [20].

In September 1992, an EPA SITE demonstration was performed at a confidential Arizona pesticide site using Canonie Environment's Low Temperature Thermal Aeration (LTTA®) system. The unit had a 35-ton-per-hour capacity. Approximately 1,180 tons of pesticide-contaminated soil were treated during the demonstration over three 10-hour replicate runs. The primary pesticides were di(chlorophenyl) trichloroethane (DDT), di(chlorophenyl)dichloroethene (DDE), di(chlorophenyl) dichloroethane (DDD), and toxaphene. Concentrations of pesticides in contaminated soils ranged from 7,080 µg/kg to 1,540,000 µg/kg. The LTTA® system obtained pesticide removal efficiencies ranging from 82.4 percent to greater than 99.9 percent. All pesticides, with the exception of DDE, were removed to near or below method detection limits in the soil. Table 26-4 presents a summary of four case studies involving full-scale applications of the LTTA® process [21].

An EPA SITE demonstration was performed at the Anderson Development Company (ADC) Superfund site in Adrian, Michigan using Roy F. Weston's Low Temperature Thermal Treatment (LT³®) system. The unit had a 2.1-ton-per-hour capacity. Approximately 80 tons of contaminated sludge were treated during the demonstration, which consisted of six 6-hour replicate tests. The lagoon sludge was primarily contaminated with VOCs and SVOCs, including 4,4'-methylenebis(2-chloroaniline)(MBOCA). Initial VOC concentrations ranged from 35 to 25,000 µg/kg. In the treated sludge, VOC concentrations were below method detection limits (less than 30 µg/kg) for most compounds. MBOCA concentrations in the untreated sludge ranged from 43.6 to 960 mg/kg. The treated sludge ranged in concentration from 3 to 9.6 mg/kg. The LT³® system also decreased the concentration of all SVOCs present in the sludge, with two exceptions: chrysene and phenol. The increase of chrysene concentration was likely caused by a minor leak of heat transfer fluid. Chemical transformations during heating likely caused the phenol concentrations to increase. PCDDs and PCDFs were formed in the system, but were removed from the exhaust gas by the unit's vapor-phase carbon column with removal efficiencies, varying with congener, from 20 to 100 percent. Particulate concentrations in the stack gas ranged from less than 8.5×10^{-4} to 6.7×10^{-3} grains per dry standard cubic meter (gr/dscm) and particulate emissions ranged from less than 1.2×10^{-4} to 9.2×10^{-4} pounds per hour. Table 26-5 presents a summary of three case studies involving pilot- and full-scale applications of the LT³® systems [22].

In May 1991, an EPA SITE demonstration was performed at the Wide Beach Development site in Brand, New York using Soil Tech's Anaerobic Thermal Processor (ATP) system. Approximately 104 tons of contaminated soil were treated during three replicate test runs. The soil and sediment at the site were primarily contaminated with PCBs, along with VOCs and SVOCs. The average total PCB concentration was reduced from 28.2 mg/kg in the contaminated soil and sediment to 0.043 mg/kg in the treated soil (a 99.8 percent removal efficiency). The test indicated that an average concentration of 23.1 µg/dscm of PCBs was discharged from the unit's stack to the atmosphere. The high PCB concentrations in the emissions may have been caused by low removal efficiencies in the unit's vapor phase carbon system, high particulate loadings (0.467 gr/dscm) in the stack, or a combination of the two. Low levels of dioxins and furans were present in the feed soil, but none were detected in the treated soils, baghouse fines, or the cyclone's flue gas. The 2,3,7,8-TCDD toxicity equivalents (TEQ) of the stack gas ranged from 0.0106 to 0.0953 ng/dscm [23].

In June 1991, an EPA SITE demonstration test was performed at the Waukegan Harbor Superfund site in Waukegan Harbor, IL. The site was primarily contaminated with PCBs, along with VOCs, SVOCs, and metals. Approximately 253 tons of contaminated soil were treated during four runs using Soil Tech's ATP thermal desorption system. The system used was a combination thermal desorption and dechlorination process. The average PCB concentration in the feed soil was 9,173 mg/kg; the average final concentration was 2 mg/kg, which is a 99.98 percent removal efficiency. The concentration of PCBs in the stack gas was 0.834 µg/dscm (a 99.999987 percent removal efficiency). Tetrachlorinated dibenzofurans were the only dioxins and furans detected in the stack gas at an average concentration of 0.0787 ng/dscm. The total concentration of SVOCs in the feed soil was

Table 26-3. Full-Scale Cleanup Results of the HT-5 System [20]

Contaminant	Feed Soil Concentration (mg/kg)	Treated Soil Concentration (µ/kg)	Removal Efficiency (%)
Toluene	30	<620	<97.93
Benzene	38	<620	<98.36
Ethylbenzene	93	<620	<99.79
Xylenes	290	<620	<99.78
Naphthalene	550	<620	<99.89
2-Methylnaphthalene	1400	<330	<99.98
Acenaphthylene	57	<330	<99.42
Phenanthrene	320	<330	<99.90
Anthracene	320	<330	<99.90
Pyrene	38	<330	<99.13
Benzo(a)Anthracene	36	<330	<99.08
Chrysene	45	<330	<99.27
Styrene	13	<620	<99.23

Table 26-4. Full-Scale Cleanup Results of the LTTA® System [21]

Site	Volume/Mass Treated	Primary Contaminant(s)	Feed Soil Concentration (mg/kg)	Treated Soil Concentration (mg/kg)
South Kearney	16,000 tons	Total VOCs	308.2	0.51
			0.7–15	ND–1.0
McKin	11.500 cubic yards	VOCs	2.7–3,310	<0.05[a]
		SVOCs	0.44–1.2	<0.33–0.51
Ottati and Goss	4,500 cubic yards	1,1,1-TCA	12–270	<0.025
		TCE	6.5–460	<0.025
		Tetrachloroethene	4.9–1200	<0.025
		Toluene	>87–3,000	<0.025–0.11
		Ethylbenzene	>50–440	<0.025
		Total Xylenes	>170->1100	<0.025–0.14
Cannon Bridgewater	11,300 tons	VOCs	5.30[b]	<0.025
Former Spencer	6,500 tons	Total VOCs	5.42	0.45
Kellogg Facility		SVOCs	0.15–4.7	0.042–<0.39

[a] Average concentration.

[b] Maximum concentration.

61.8 mg/kg. In the treated soils SVOC concentrations totaled only 8.52 mg/kg; only two samples were identified below the detection limit. In the contaminated soil, VOC concentrations totaled 17 mg/kg; while in the treated soil the total was only 0.03 mg/kg. Concentrations of metals were approximately the same in both the contaminated and treated soil. This was because the unit does not operate at temperatures high enough to significantly remove metals. The pH of the soil rose from 8.59 in the contaminated soil to 11.35 in the treated soil. This was likely due to the addition of sodium bicarbonate used to reduce PCB emissions [23].

In May 1992, an EPA SITE demonstration was performed at the Re-Solve Superfund site in North Dartmouth, Massachusetts using the Chemical Waste Management X*TRAX™ system. The unit had a capacity of 4.9 tons per hour. Approximately 215 tons of contaminated soil were treated over a period of three duplicate 6-hour tests. The soil is primarily contaminated with PCBs, along with some oil and grease and metals. Initial PCB concentrations ranged from 181 to 515 mg/kg. The PCB concentration in the treated soil was less than 1.0 mg/kg, with an average concentration of 0.25 mg/kg (a 99.9 percent

Table 26-5. Full-Scale Cleanup Results of the LT³® System [22]

Site	Volume/Mass Treated	Primary Contaminant(s)	Feed Soil Concentration	Treated Soil Concentration
Confidential	1,000 cubic feet	Benzene	1 ppm	5.2 ppb
		Toluene	24 ppm	5.2 ppb
		Xylene	110 ppm	<1.0 ppb
		Ethylbenzene	20 ppm	4.8 ppb
		Naphthalene	4.9 ppm	<0.33 ppm
		PAHs	0.890–<6 ppm	<330–590 ppb
Tinker AFB, OK	3,000 cubic yards	Volatiles	18 μg/kg–37,250 μg/kg	0.1 μg/L–2.3 μg/L
		Semivolatiles	90 μg/kg–53,000 μg/kg	6 μg/L–<500 μg/L
Letterkenny Army Depot	7.5 tons	Benzene	590 ppm	0.73 ppm
		Trichloroethene	2,680 ppm	1.8 ppm
		Tetrachloroethene	1,420 ppm	1.4 ppm
		Xylene	27,200 ppm	0.55 ppm
		Other VOCs	39 ppm	BDL

BDL = below detection limit.

removal efficiency). PCDDs and PCDFs were not formed during the demonstration. Concentrations of oil and grease, total recoverable petroleum hydrocarbons, and tetrachloroethane were reduced to below detectable levels. Metal concentrations were not reduced during the test. This was expected because the unit does not operate at temperature high enough to significantly remove metals [24].

RCRA LDRs that require treatment of wastes to best demonstrated available technology (BDAT) levels prior to land disposal may sometimes be determined to be applicable or relevant and appropriate requirements for CERCLA response actions. Thermal desorption often can produce a treated waste that meets treatment levels set by BDAT but may not reach these treatment levels in all cases. The ability to meet required treatment levels is dependent upon the specific waste constituents, the waste matrix, and the thermal desorption system operating parameters. In cases where thermal desorption does not meet these levels, it still may, in certain situations, be selected for use at the site if a treatability variance establishing alternative treatment levels is obtained. Treatability variances are justified for handling complex soil and debris matrices. The following guides describe when and how to seek a treatability variance for soil and debris: Superfund LDR Guide #6A, Obtaining a Soil and Debris Treatability Variance for Remedial Actions (OSWER Directive 9347.3-06FS, September 1990) [12], and Superfund LDR Guide #6B, Obtaining a Soil and Debris Treatability Variance for Removal Actions (OSWER Directive 9347.3-06BFS, September 1990) [13].

TECHNOLOGY STATUS

Several firms have experience in implementing this technology. Therefore, there should not be significant problems of availability. The engineering and configuration of the systems are similarly refined, so that once a system is designed full-scale, little or no prototyping or redesign is generally required.

An EPA SITE demonstration took place at the end of 1993 at the Niagara Mohawk Power Corporation site in Utica, New York. The facility is a former gas manufacturing plant. Approximately 800 tons of contaminated soils were treated during the demonstration. The soil is primarily contaminated with polyaromatic hydrocarbons (PAHs); benzene, toluene, ethylbenzene, and xylenes (BTEXs); lead; arsenic; and cyanide. An EPA Innovation Technology Evaluation Report will be developed to evaluate the performance of and the cost to implement the system.[2]

Thermal desorption technologies are the selected remedies at 31 Superfund sites. Table 26-6 presents the status of selected Superfund sites employing the thermal desorption technology [2].

Several vendors have experience in the operation of this technology and have documented processing costs per ton of feed processed. The overall range varies from approximately $100 to $400 (1993 dollars) per ton processed. Caution is recommended in using costs out of context because the base year of the estimates varies. Costs also are highly variable due to the quantity of waste to be processed, terms of the remediation contract, moisture content, organic constituency of the contaminated medium, and cleanup standards to be achieved. Similarly, cost estimates should include such items as preparation of Work Plans, permitting, excavation, processing, QA/QC verification of treatment performance, and reporting data.

EPA CONTACTS

Technology-specific questions regarding thermal desorption may be directed to Paul dePercin (513) 569-7797 or Marta Richards (513) 569-7962, RREL-Cincinnati (see Preface for mailing address).

ACKNOWLEDGMENTS

This updated chapter was prepared for the U.S. Environmental Protection Agency, Office of Research and Development (ORD), Risk Reduction Engineering Laboratory (RREL), Cincinnati, Ohio, by

---

[2] Editor's Note: The report on the Maxymillian Technologies (formerly Clean Berkshires, Inc.) Mobile Thermal Desorption System was not yet available as of late 1995.

Table 26-6. Selected Superfund Sites Specifying Thermal Desorption as the Remediation Technology [2]

Site	Location (Region)	Primary Contaminants	Status
Cannons/Bridgewater	Bridgewater, MA (1)	VOCs (Benzene, TCE, Toluene, Vinyl Chloride)	Site remediated 10/90
McKin	McKin, ME (1)	VOCs (TCE, BTX)	Site remediated 2/87
Ottati & Goss	New Hampshire (1)	VOCs (TCE, PCE, 1,2-DCE, Benzene)	Site remediated 9/89
Wide Beach Development	Brandt, NY (2)	PCBs	Site remediated 6/92
Metaltec/Aerosystems	Franklin Borough, NJ (2)	VOCs (TCE)	Design completed
Caldwell Trucking	Fairfield, NJ (2)	VOCs (TCE, PCE, TCA)	Design completed
Outboard Marine/Waukegan Harbor	Waukegan Harbor, IL (5)	PCBs	Site remediated 6/92
Reich Farms	Dover Township, NJ (2)	VOCs (TCE, PCE, TCA), SVOCs	Pre-design
Re-Solve	North Dartmouth, MA (2)	PCBs	Pilot study completed 5/92
Waldick Aerospace Devices	New Jersey (2)	VOCs (TCE, PCE), Metals (Cadmium, Chromium)	Design completed
Wamchem	Burton, SC (4)	VOCs, BTX	In design
Fulton Terminals	Fulton, NJ (2)	VOCs (Xylenes, TCE, Benzene, DCE)	In design
Anderson Development Company	Adrian, MI (5)	VOCs, SVOCs	Site remediated 12/92

Science Applications International Corporation (SAIC) under Contract No. 68-C0-0048. Mr. Eugene Harris served as the EPA Technical Project Monitor. Mr. Jim Rawe (SAIC) was the Work Assignment Manager. He and Mr. Eric Saylor (SAIC) co-authored the revised chapter. The authors are especially grateful to Mr. Paul dePercin of EPA-RREL, who contributed significantly by serving as a technical consultant during the development of this chapter. The authors also want to acknowledge the contributions of those who participated in the development of and are listed in the original bulletin.

The following other contractor personnel have contributed their time and comments by participating in the expert review of the chapter:

Mr. William Troxler, Focus Environmental, Inc.
Dr. Steve Lanier, Energy and Environmental Research Corp.

REFERENCES

1. Thermal Desorption Treatment. Engineering Bulletin. U.S. Environmental Protection Agency, EPA/540/2-91/008, May 1991.
2. Innovative Treatment Technologies. Semi-Annual Status Report (Fourth Edition), U.S. Environmental Protection Agency, EPA/542/R-92/011, October 1992. [Annual Status Report (6th Edition) was published in September 1994 as EPA 542-R-94-005.]
3. Abrishamian, R. Thermal Treatment of Refinery Sludges and Contaminated Soils. Presented at American Petroleum Institute, Orlando, Florida, 1990.
4. Swanstrom, C. and C. Palmer. X*TRAX™ Transportable Thermal Separator for Organic Contaminated Solids. Presented at Second Forum on Innovative Hazardous Waste Treatment Technologies: Domestic and International, Philadelphia, Pennsylvania, 1990.
5. Canonie Environmental Services Corp. Low Temperature Thermal Aeration (LTTA®) Marketing Brochures, circa 1990.
6. VISITT Database, U.S. Environmental Protection Agency, 1993.
7. Nielson, R. and M. Cosmos. Low Temperature Thermal Treatment (LT³®) of Volatile Organic Compounds from Soil: A Technology Demonstrated. Presented at the American Institute of Chemical Engineers Meeting, Denver, Colorado, 1988.
8. Technology Screening Guide for Treatment of CERCLA Soils and Sludges. EPA/540/2-88/004 (NTIS PB89-132674), 1988.
9. Considerations for Evaluating the Impact of Metals Partitioning During the Incineration of Contaminated Soils from Superfund Sites. Superfund Engineering Issue. EPA/540/S-92/014, September 1992.
10. T.D.I. Services. Marketing brochures, circa 1990.
11. Cudahy, J. and W. Troxler. 1990. Thermal Remediation Industry Update - II. Presented at Air and Waste Management Association Symposium on Treatment of Contaminated Soils, Cincinnati, Ohio, 1990.
12. Superfund LDR Guide #6A: (2nd Edition) Obtaining a Soil and Debris Treatability Variance for Remedial Actions. OSWER Directive 9347.3-06FS (NTIS PB91-921327), 1990.
13. Superfund LDR Guide #6B: Obtaining a Soil and Debris Treatability Variance for Removal Actions. OSWER Directive 9347.3-06BFS (NTIS PB91-921310), 1990.
14. Guide for Conducting Treatability Studies Under CERCLA: Thermal Desorption Remedy Selection, Interim Guidance. U.S. Environmental Protection Agency, EPA/540/R-92/074A, September 1992.
15. Recycling Sciences International, Inc., DAVES marketing brochures, circa 1990.
16. The Superfund Innovative Technology Evaluation Program-Progress and Accomplishments Fiscal Year 1989, A Third Report to Congress. U.S. Environmental Protection Agency, EPA/540/5-90/001, 1990.
17. Superfund Treatability Clearinghouse Abstracts. EPA/540/2-89/001 (NTIS PB90-119751), 1989.
18. Soil Tech, Inc. AOSTRA—Taciuk Processor marketing brochure, circa 1990.
19. Ritcey, R. and F. Schwartz. Anaerobic Pyrolysis of Waste Solids and Sludges - The AOSTRA Taciuk Process System. Presented at Environmental Hazards Conference and Exposition, Seattle, Washington, 1990.
20. Seaview Thermal Systems. Marketing brochures, circa 1993.
21. Low Temperature Thermal Treatment Aeration (LTTA®) Technology, Canonie Environmental Services Corporation, Applications Analysis Report. EPA/540/AR-93/504, July, 1995. [See also Demonstration Bulletin EPA/540/MR-93/504.]*

22.   Low Temperature Thermal Treatment (LT³) System, Roy F. Weston, Inc., Applications Analysis Report. EPA/540/AR-92/019, December 1992. [See also Demonstration Bulletin EPA/540/MR-92/019.]*

23.   Soil Tech ATP Systems, Inc. Anaerobic Thermal Processor. Applications Analysis Report. Wide Beach Development Site and Outboard Marine Corporation Site. U.S. Environmental Protection Agency (Preliminary Draft-February 1993). [Editor's Note: final report had not been published as of late 1995; see Demonstration Bulletins EPA/540/MR-92/008 and EPA/MR-92/078 for additional information.]*

24.   X*TRAX™ Model 200 Thermal Desorption System. Chemical Waste Management, Inc. Demonstration Bulletin, EPA/540/MR-93/502, February 1993.*

ADDITIONAL SITE PROGRAM REFERENCES*

Clean Berkshires, Inc. Thermal Desorption System:

Demonstration Bulletin (EPA/540/MR-94/507), April 1994
SITE Technology Capsule (EPA/540/R-94/507a) August 1994

Eco Logic International, Inc. Thermal Desorption Unit:

Demonstration Bulletin (EPA/540/MR-94/504)
Applications Analysis Report (EPA/540/AR-94/504), September 1994

* See Appendix A for information on how to obtain SITE program reports.

Chapter 27

Pyrolysis Treatment[1]

Sharon Krietmeyer and Richard Gardner, Science Applications International Corporation (SAIC), Cincinnati, OH

ABSTRACT

Pyrolysis is formally defined as chemical decomposition induced in organic materials by heat in the absence of oxygen. In practice, it is not possible to achieve a completely oxygen-free atmosphere; actual pyrolytic systems are operated with less than stoichiometric quantities of oxygen. Because some oxygen will be present in any pyrolytic system, nominal oxidation will occur. If volatile or semivolatile materials are present in the waste, thermal desorption will also occur.

Pyrolysis is a thermal process that transforms hazardous organic materials into gaseous components and a solid residue (coke) containing fixed carbon and ash. Upon cooling, the gaseous components condense, leaving an oil/tar residue. Pyrolysis typically occurs at operating temperatures above 800°F [1, pp. 165, 167; 2, p. 5].[2] This chapter does not address other thermal processes that operate at lower temperatures or those that operate at very high temperatures, such as a plasma arc. Pyrolysis is applicable to a wide range of organic wastes and is generally not used in treating wastes consisting primarily of inorganics and metals.

Pyrolysis should be considered an emerging technology. (An emerging technology is a technology for which performance data have not been evaluated according to methods approved by EPA and adhering to the EPA quality assurance/quality control standards, although the basic concepts of the process have been validated [3, pp. 1–2].) Performance data are currently available only from vendors. In addition, existing data are limited in scope and quantity, and are frequently of a proprietary nature.

This chapter provides information on the technology applicability, the types of residuals resulting from the use of the technology, the latest performance data, site requirements, the status of the technology, and where to go for further information.

TECHNOLOGY APPLICABILITY

Pyrolysis systems may be applicable to a number of organic materials that "crack," or undergo a chemical decomposition in the presence of heat. Pyrolysis has shown promise in treating organic contaminants in soils and oily sludges. Chemical contaminants for which treatment data exist include polychlorinated biphenyls (PCBs), dioxins, polycyclic aromatic hydrocarbons, and many other organics. Treatment data discussed in this chapter were taken from treatability studies conducted by three vendors.

Pyrolysis is not effective in either destroying or physically separating inorganics from the contaminated medium. Volatile metals may be desorbed as a result of the higher temperatures associated with the process, but are similarly not destroyed.

The probable effectiveness of pyrolysis on general contaminant groups for various matrices is shown in Table 27-1. Examples of constituents within contaminant groups are provided in Technology Screening Guide for Treatment of CERCLA Soils and Sludges [4, pp. 10–12]. Table 27-1 is based on current available information or professional judgment where no information was avail-

---

[1] EPA/540/S-92/010.
[2] [Reference number, page number.]

Table 27-1. Effectiveness of Pyrolysis on General Contaminant Groups for Soil and Sediment/Sludge

Contaminant Groups	Effectiveness	
	Soil	Sediment/Sludge
Organic		
Halogenated volatiles	▼	▼
Halogenated semivolatiles	▼	▼
Nonhalogenated volatiles	▼	■
Nonhalogenated semivolatiles	■	■
PCBs	■	■
Pesticides (halogenated)	▼	▼
Dioxins/furans	▼	■
Organic cyanides	▼	▼
Organic corrosives	●	●
Inorganic		
Volatile metals	●	●
Nonvolatile metals	●	●
Asbestos	●	●
Radioactive materials	●	●
Inorganic corrosives	●	●
Inorganic cyanides	●	●
Reactive		
Oxidizers	●	●
Reducers	●	●

■ Demonstrated Effectiveness: Successful treatability test at some scale completed.
▼ Potential Effectiveness: Expert opinion that technology will work.
● Not Expected Effectiveness: Expert opinion that technology will not work.

able [1, pp. 165, 168; 2, pp. 9–14; 5, pp. 10–15; 6, p. 9]. The proven effectiveness of the technology for a particular site or waste does not ensure that it will be effective at all sites or that the treatment efficiencies achieved will be acceptable at other sites. For the ratings used in this table, demonstrated effectiveness means that, at some scale, treatment results indicated that the technology was effective for that particular contaminant and medium. The ratings of potential effectiveness or no expected effectiveness are both based upon expert judgment. Where potential effectiveness is indicated, the technology is believed capable of successfully treating the contaminant group in a particular medium. When the technology is not applicable or will probably not work for a particular combination of contaminant group and medium, a no-expected-effectiveness rating is given.

LIMITATIONS

The primary technical factors affecting pyrolytic performance are the temperature, residence time, and heat transfer rate to the material. There are also several practical limitations which should be considered.

As the medium is heated and passes through a pyrolytic system, energy is consumed in heating moisture contained in the contaminated medium. A very high moisture content would result in lower throughput. High moisture content, therefore, causes increased treatment costs. For some wastes, dewatering prior to pyrolysis may be desirable.

The treated medium will typically contain less than one percent moisture. Dust can easily form in the transfer of the treated medium from the treatment unit, but this problem can be mitigated by water sprays.

A very high pH (greater than 11) or very low pH (less than 5) may corrode the system components. The pyrolysis of halogenated organics will yield hydrogen halides; the pyrolysis of sulfur-containing organics will yield various sulfur compounds including hydrogen sulfide (H_2S). Because hydrogen halides and hydrogen sulfide are corrosive chemicals, corrosion control measures should be taken for any pyrolytic system which will be processing wastes with high concentrations of halogenated or sulfur-containing organics.

TECHNOLOGY DESCRIPTION

Pyrolysis is formally defined as chemical decomposition induced in organic materials by heat in the absence of oxygen. Pyrolysis is a thermal process that transforms organic materials into gaseous components and a solid residue (coke) containing fixed carbon and ash. The pyrolysis of organics yields combustible gases including carbon monoxide, hydrogen, methane, and other low molecular weight hydrocarbons [7, pp. 252–253]. Pyrolysis occurs to some degree whenever heat is applied to an organic material. The rate at which pyrolysis occurs increases with temperature. At low temperatures and in the presence of oxygen, the rates are typically negligible. In addition, the final percent weight loss for the treated material is directly proportional to the operating temperature. Similarly, the hydrogen fraction in the treated material is inversely proportional to the temperature.

The primary cleanup mechanisms in pyrolytic systems are destruction and removal. Destruction occurs when organics are broken down into lower molecular weight compounds. Removal occurs when pollutants are desorbed from the contaminated material and leave the pyrolysis portion of the system without being destroyed.

Pyrolysis systems typically generate solid, liquid, and gaseous products. Solid products include the treated (and dried) medium and the carbon residue (coke) formed from hydrocarbon decomposition. Various gases are produced during pyrolysis, and certain low-boiling compounds may volatilize rather than decompose. This is not typically a problem. Gases may be condensed, treated, incinerated in an afterburner, flared, or a combination of the above. Depending on the specific components, organic condensate may be reusable. Other liquid streams will include process water used throughout the system. A general schematic of a pyrolytic process is shown in Figure 27-1.

As shown in Figure 27-1, the first step in the treatment process is the excavation of the contaminated soil, sludge, or sediment. Oversized rejects such as large rocks or branches are removed and the material is transferred to the pyrolysis unit. The treatment system may include a desorption stage prior to pyrolysis. If so, the desorbed gases flow to the gas treatment system for treatment and/or recovery, and the contaminated matrix (minus any desorbed chemicals) is transferred to the pyrolysis chamber [1, p. 166; 2, pp. 3–6].

The temperature in the pyrolysis chamber is typically between 800 and 2,100°F, and the quantity of the oxygen present is not sufficient for the complete oxidation of all contaminants. In pyrolysis, organic materials are transformed into coke and gaseous components. Gas treatment options include: (1) condensation plus gas cleaning and (2) incineration plus gas cleaning.

Pyrolysis forms new compounds whose presence could impact the design of the off-gas management system. For example, compounds such as hydrogen halides and sulfur-containing compounds may be formed. These must be accounted for within the design of the Air Pollution Control (APC) system.

There are three pyrolytic systems which will be discussed in this chapter. These systems are: the HT-5 system marketed by TDI Thermal Dynamics (formerly Southdown Thermal Dynamics), a process developed by Deutsche Babcock Anlagen AG, and an "anaerobic thermal processor" (ATP) marketed by Soil Tech, Inc.

The HT-5 Thermal Distillation System is a mobile thermal desorption system which may be operated in a pyrolytic mode. The Thermal Distillation System processes waste by applying heat in a nitrogen atmosphere. Gravity and a system of annular augers are used to transfer waste through a series of three electrically heated distillation chambers. The temperature is ambient at the entrance to the distillation chambers and increases to full operating temperature (up to 2,100°F) as the waste progresses through the chambers. The continuous introduction of a nitrogen sweep gas removes and separates the volatile contaminants [8, p. 3]. The sweep gas must be periodically sent to a flare to reduce the noncondensible combustible portion.

TDI is currently conducting bench-scale tests on the Thermal Degradation System, which was developed for use in conjunction with the Thermal Distillation System. The full-scale design of the system is currently theoretical, but TDI envisions that Thermal Degradation will follow Thermal Distillation and will be used primarily for pyrolysis. In recent bench-scale tests, the Thermal Degrada-

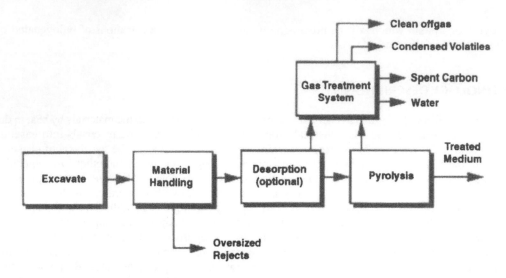

Figure 27-1. Schematic diagram of pyrolysis.

tion System was operated at approximately 2,000°F and a copper catalyst was used to enhance the pyrolysis of halogenated organics [2, pp. 3–6; 5, pp. 3–7].

A German company, Deutsche Babcock Anlagen AG, developed a pyrolytic process which utilizes an indirectly heated rotary kiln. In the first step of the Deutsche Babcock system, pyrolysis occurs at a temperature of 1,100 to 1,200°F. If volatile or semivolatile organics are present, they will be desorbed in this step. In the second step, the gases produced by pyrolysis (as well as other volatilized organics) are combusted in an afterburner at a high temperature (1,800 to 2,400°F). Heat produced during the second step may provide at least a portion of the energy for the first step, which is endothermic. Prior to discharge, effluent gases from the second step are scrubbed to remove various pollutants, including hydrogen halides and sulfur oxides [1, p. 166].

The pyrolysis systems marketed by Deutsche Babcock are not currently available in mobile or transportable configurations and are therefore not directly applicable to onsite remediation of Superfund sites. These systems were included in this discussion to provide additional data and to indicate the potential viability of pyrolysis. In addition, full-scale applications and testing of the Deutsche Babcock system have included the cleanup of contaminated soils [1, pp. 165–168].

Finally, Soil Tech, Inc. (Canonie Environmental) markets an anaerobic thermal processor (ATP) which may be operated in a pyrolytic mode. The ATP is also known as the AOSTRA-Taciuk process, and is essentially an indirectly-heated rotary kiln. A transportable ATP with a nominal processing rate of 10 tons per hour is available for onsite demonstrations and remediation [9, p. 3].

The ATP unit includes four chambers: preheat, reaction, combustion, and cooling. In the preheat chamber, volatile materials are desorbed at temperatures up to 500°F. Pyrolytic conditions and temperatures between 700 and 1,150°F are maintained in the reaction chamber. The desorption and/or pyrolysis of heavier organics will occur in this chamber. Coke and noncombustible hydrocarbons produced by pyrolysis are transferred to the combustion chamber and burned [9, pp. A-1 to A-2]. Additional fuels such as gas or oil must be available for startup, for control, and to supplement the pyrolysis products when they do not provide adequate fuel. Solids and gases from the combustion chamber proceed into the cooling zone. The cooling zone and the preheat zone function as a heat exchanger in which heat is transferred from the combustion residuals to the feed [10, p. 3].

PROCESS RESIDUALS

The effluents generated by pyrolytic systems typically include solid, liquid, and gaseous residuals. Solid products include debris, oversized rejects, dust, ash, and the treated medium. Dust collected from particulate control devices may be combined with the treated medium or, depending on analyses for carryover contamination, recycled through the treatment unit.

Depending on the individual system, the flue gases from the pyrolysis unit will generally be treated by wet or dry APC systems before discharge through a stack. In the Deutsche Babcock System, off-gases are treated by incineration [1, p. 166].

Ash and treated soil/solids from pyrolysis may be contaminated with heavy metals. APC system solids, such as fly ash, may contain high concentrations of volatile metals. If these residues fail required leachate toxicity tests, they can be treated by a process such as solidification/stabilization and disposed of onsite or in an approved landfill [11, p. 8.97]. If the treated medium and ash pass all required tests, they may be disposed of onsite without further treatment.

Depending on the specific pyrolysis system, liquid streams may include condensed organics or water from the APC system. After organics are removed, condensed water may be used as a dust suppressant for the treated medium. Scrubber purge water can be purified and returned to the site wastewater treatment facility (if available), discharged to the sewer, or used for rehumidification and cooling of the hot, dusty media.

Liquid waste from the APC system may contain excess alkali, high chlorides, volatile metals, organics, metals particulates, and inorganic particulates. Treatment may require neutralization, chemical precipitation, settling, filtration, or carbon adsorption before discharge.

SITE REQUIREMENTS

Pyrolytic treatment processes are not expected to have significantly different site requirements than those for thermal desorption or incineration processes.

Note that the pyrolytic systems marketed by Deutsche Babcock are not currently available in mobile or transportable configurations. The HT-5 system and the ATP are transportable, and vendors claim that they can be set up in a matter of days.

Standard site requirements include electric power (440 or 480 V, 3-phase) and water. The quantity of water required is design- and site-specific.

Treatment of contaminated soils or other waste materials require that a site safety plan be developed to provide for personnel protection and special handling measures. Storage should be provided to hold the process product streams until they have been tested to determine their acceptability for disposal or release. Depending upon the site, a method to store waste that has been prepared for treatment may be necessary. Storage capacity will depend on waste volume.

Onsite analytical equipment capable of monitoring site-specific organic compounds for performance assessment make the operation more efficient and provide better information for process control.

PERFORMANCE DATA

Limited performance data are available for pyrolytic systems treating hazardous wastes containing PCBs, dioxins, and other organics [1, pp. 165, 168; 2, pp. 9–14; 5, pp. 10–15; 6, p. 9]. The quality of this information has not been determined. These data are included as a general indication of the performance of pyrolysis equipment and may not be directly transferable to a specific Superfund site. Good site characterization and treatability studies are essential in further refining and screening the pyrolysis technology.

The HT-5 system's performance on oily sludges contaminated with dioxins and PCBs was evaluated in bench-scale treatability tests conducted by Law Environmental on April 25, 1991 [2, pp. 9–14; 5, pp. 10–15]. The simulated waste used in the dioxin test was contaminated with 2,3,7,8-tetrachlorodibenzo-p-dioxin (TCDD). A decontamination efficiency of over 99.99% was calculated, as no 2,3,7,8-TCDD was detected in the treated residue, off-gases, or condensate. In addition, the test report claims that no significant quantities of new toxic compounds were synthesized by the process [2, pp. 9–14].

A second bench-scale treatability study was conducted on a mixture of PCB-contaminated soil, PCB-contaminated oil, and water. All process streams were sampled and analyses indicated a decontamination efficiency of over 99.99%. PCB levels were below the detection limits in all effluent streams and the test report claims that no significant quantities of new toxic compounds were synthesized by the process [5, pp. 10–15]. Although these results appear promising, complete closures of mass balances are not possible with the information collected during the HT-5 treatability tests.

The Deutsche Babcock system was tested in an industrial-scale demonstration in May and June 1988. Prior to this demonstration, the same system was used to treat 35,000 tons of soil. The plant is located in Unna-Bonen, Germany, at a former coke oven site. The unit had a design rate of 7 tons/hour with a soil moisture content of 21 percent and 5 percent volatile compounds. The destruction of 17 polycyclic aromatics was measured. A system decontamination efficiency of 99.77 percent was achieved. The results are summarized in Table 27-2 [1, p. 168]. Note that this test was conducted in Germany and that the majority of the applications of the Deutsche Babcock system have been in Germany. German requirements regarding incineration were not researched and may differ significantly from U.S. requirements.

The Soil Tech ATP is being used in conjunction with chemical dehalogenation to remediate the Wide Beach Superfund site. Much of the soil in the small community of Wide Beach, New York is contaminated with PCBs from road oils. PCB levels range from approximately 10 ppm to over 5,000 ppm; the primary cleanup requirement is to reduce PCB concentrations to less than 2 ppm [6, pp. 2–3].

The system used at Wide Beach is similar to the ATP described previously, but also includes a reagent mix system. The reagent mix system adds dechlorination chemicals (potassium hydroxide and polyethylene glycol) to a stream of oils recycled from the system effluent [6, p. 4; 12, p. 45].

PCB concentrations in the treated soil were below the reporting limit of 70 ppb, which is significantly below the required level. In addition, the process water contained no more than 1 ppb PCBs, stack gas PCB levels were less than 33 percent of the New York State Department of Environmental Conservation (NYDEC) limits, fugitive emissions were within NYDEC limits, and treated soils passed the toxicity characteristic leaching procedure (TCLP) [6, pp. 2, 9]. At the beginning of the cleanup effort, treated soil was returned to local sites. The treated soil, however, does not have the same consistency as untreated soil, and current plans are to landfill the soil rather than returning it to the original sites [12, p. 45].

TECHNOLOGY STATUS

Pyrolysis has been used to treat various hazardous wastes as documented in the Performance Data section of this chapter. In particular, pyrolysis has been applied to the remediation of the Wide Beach Superfund site (in conjunction with chemical dehalogenation) [6, pp. 1–2] and to the cleanup of contaminated soils in Germany [1, pp. 165–168].

EPA CONTACT

Technology-specific questions regarding pyrolysis may be directed to Marta Richards, RREL-Cincinnati, (513) 569-7692 (see Preface for mailing address).

ACKNOWLEDGMENTS

This chapter was prepared for the U.S. Environmental Protection Agency, Office of Research and Development (ORD), Risk Reduction Engineering Laboratory (RREL), Cincinnati, Ohio, by Science Applications International Corporation (SAIC) under Contract No. 68-C8-0062. Mr. Eugene Harris served as the EPA Technical Project Monitor. Mr. Gary Baker (SAIC) was the Work Assignment Manager, and Ms. Sharon Krietemeyer and Mr. Richard Gardner (SAIC) were co-authors of this chapter. The authors are especially grateful to Mr. Donald Oberacker and Mr. Paul de Percin of EPA, RREL, who have contributed significantly by serving as technical consultants during the development of this chapter.

The following other contractor personnel have contributed their time and comments by participating in the expert review meetings and/or peer-reviewing the chapter:

Mr. James Cudahy, Focus Environmental, Inc.
Dr. Steve Lanier, Energy and Environmental Research Corp.

Table 27-2. Deutsche Babcock Pyrolytic Rotary Kiln Contaminated Soil Results

Pollutant	March 8, 1989 Input mg/kg	Output mg/kg	January 27, 1989 Input mg/kg	Output mg/kg
Naphthalene	101.00	1.7	161.60	0.5
2-Methylnaphthalene	40.20	0.5	73.80	0.1
1-Methylnaphthalene	23.40	0.3	42.90	0.1
Dimethylnaphthalene	ND	ND	93.20	0.3
Acenaphthylene	ND	ND	68.20	0.1
Acenaphthene	ND	ND	42.30	0.1
Fluorene	156.00	0.1	238.00	0.1
Phenanthrene	686.00	0.6	1055.30	1.4
Anthracene	281.00	0.1	226.00	0.3
Fluoranthene	ND	ND	688.60	1.3
Pyrene	236.00	0.1	398.20	0.6
Benz(a)anthracene	155.00	0.2	2259.20	0.3
Chrysene	214.00	0.5	134.60	0.9
Benzo(e)pyrene	66.60	0.4	111.50	1.1
Benzo(b)fluoranthene	112.00	0.1	168.50	5.2
Benzo(k)fluoranthene	43.70	0.1	81.90	0.3
Benzo(a)pyrene	86.60	0.2	138.10	0.4
Dibenz(a,h)anthracene	16.80	0.1	23.20	0.1
Benzo(g,h,i)perylene	14.00	0.1	60.20	0.1
Indeno(1,2,3-cd)pyrene	33.80	0.1	69.50	0.1
Sum	2266.10	5.2	6134.80	13.4
Decontamination efficiency in %		99.77		99.78

ND = not detectable

REFERENCES

1. Schneider, D. and B.D. Beckstrom. Cleanup of Contaminated Soils by Pyrolysis in an Indirectly Heated Rotary Kiln. Environmental Progress, Volume 9, No. 3, pp. 165–168. August 1990.
2. Test Report of Bench Scale Unit (BSU) Treatability Test for Dioxin Contaminated Oily Sludge. Test Date: April 25, 1991. Prepared by Law Environmental, Inc. for Southdown Thermal Dynamics. June 1991.
3. The Superfund Innovative Technology Evaluation Program: Technology Profiles. U.S. Environmental Protection Agency, Office of Solid Waste and Emergency Response and Office of Research and Development, Washington, D.C. EPA/540/5-90/006. November 1990.
4. Technology Screening Guide for Treatment of CERCLA Soils and Sludges. EPA/540/2-88/004 (NTIS PB89-132674), 1988.
5. Test Report of Bench Scale Unit (BSU) Treatability Test for PCB Contaminated Oily Sludge. Test Date: April 25, 1991. Prepared by Law Environmental, Inc. for Southdown Thermal Dynamics. June 1991.
6. Vorum, M. PCB-Soil Dechlorination at the Wide Beach Superfund Site: The Commercial Experience of Soil Tech, Inc. May 1991.
7. Incinerating Hazardous Wastes, H.M. Freeman, Editor. Technomic Publishing Co., Lancaster, PA 1988.
8. Southdown Thermal Dynamics, marketing brochures, circa 1990.
9. The Taciuk Process Technology: Thermal Remediation of Solid Wastes and Sludges. Technical Information. Submitted by Soil Tech, Inc. [See also SITE report listed below.]
10. Ritcey, R. and F. Schwartz. Anaerobic Pyrolysis of Waste Solids and Sludges: The AOSTRA Taciuk Process System. Presented to the Environmental Hazards Conference & Exposition, Environmental Hazards Management Institute, Seattle, WA. May 1990.
11. Standard Handbook of Hazardous Waste Treatment and Disposal. H.M. Freeman, Editor. McGraw-Hill Book Company, New York, pp. 8.91 to 8.104.
12. Turning "Dirty" Soil into "Clean" Mush. Soils. September–October 1991.

ADDITIONAL SITE REPORTS (see Appendix A for information on how to obtain SITE program reports)

AOSTRA-Soiltech Anaerobic Thermal Process, Soil Tech ATP Systems:

Demonstration Bulletin (EPA/540/MR-92/008)

Eco Logic International Gas-Phase Chemical Reduction Process:

Demonstration Bulletin (EPA/540/MR-93/522), September 1993
Applications Analysis Report (EPA/540/AR-93/522), September 1994

Chapter 28

Supercritical Water Oxidation[1]

Sharon Krietmeyer, Science Applications International (SAIC), Cincinnati, OH

TECHNOLOGY STATUS

Supercritical water oxidation (SCWO) has existed as an emerging waste treatment technology for approximately 10 years [1].[2] There are currently no full-scale SCWO systems in operation, but considerable bench- and pilot-scale data are available. The largest existing SCWO system can process waste at a rate of approximately 4 gallons per minute (gpm) [2].

Several universities and research institutes are studying SCWO. The U.S. Air Force is investigating SCWO for destruction of rocket fuels and explosives. The U.S. Department of Energy is considering SCWO for treatment of wastes generated at its nuclear plants [3]. SCWO is also being considered by National Aeronautics and Space Administration (NASA) for waste treatment during extended space missions [4; 5].

The Defense Advanced Research Projects Agency (DARPA) is also investigating SCWO. Ongoing work under a DARPA contract includes the design and construction of a mobile SCWO unit for the destruction of military wastes. General Atomics is the prime contractor for this project, and the University of Texas (UT) Balcones Research Center and Eco Waste Technologies (EWT) are subcontractors [6].

EWT is currently developing a proprietary SCWO system which operates above ground (surface SCWO). Besides its involvement in the DARPA project, EWT is also designing a 5-gpm commercial demonstration unit for a small chemical manufacturing facility [6].[3]

Modell Development Corporation (MODEC) is also developing a proprietary surface SCWO system. MODEC hopes to have a 5 dry ton/day pilot plant completed in 1992 and small commercial units available in 1993 [7].[4]

MODAR, Inc. owns and operates the 4-gpm SCWO system mentioned previously [2]. MODAR conducts surface SCWO research and development in conjunction with its licensor, ABB Lummus Crest [8; 9].

GeneSyst International is developing a proprietary SCWO system called a "Gravity Pressure Vessel" which is designed to operate below ground (subsurface SCWO) [10].

Vertech was involved in the development of subsurface SCWO reactors, but it was purchased by Wijnanin N.V., which has Air Products and Chemicals as its U.S. licensee. It is not clear whether Wijnanin N.V. or Air Products and Chemicals plans to pursue SCWO development.

Oxidyne (previously Vertox) was also involved in subsurface SCWO development. Oxidyne developed plans for a full-scale, subsurface subcritical water oxidation reactor in Houston, Texas at Sims Bayou Sewage Treatment Plant. Construction of the reactor was initiated but was not completed due to insufficient funding [11–13]. Oxidyne is no longer involved in SCWO research and therefore sold a number of its patents and designs to City Management Corporation (CMC). CMC has no immediate plans to continue SCWO research [14]. The Oxidyne work in Houston is important because the design

---

[1] EPA/540/S-92/006.

[2] [Reference number.]

[3] Editor's Note: A commercial unit began operating in Austin, Texas in late 1994 (personal communication, Dr. Earnest Gloyna, January 1996).

[4] Editor's Note: Commercial units were not available as of early 1996 (personal communication, Dr. Earnest Gloyna, January 1996).

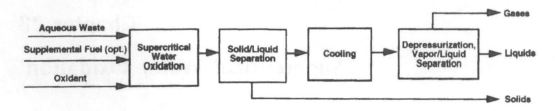

Figure 28-1. SCWO schematic.

of that subcritical system may serve as a basis for the design of subsurface systems which operate at supercritical conditions.

Research currently being conducted by various firms and universities focuses on a better understanding of the SCWO process and will be used in the design of full-scale systems. Specific research topics include kinetics, the mechanisms of SCWO, and fluid flow characteristics [15; 16].[5]

TECHNOLOGY DESCRIPTION

In SCWO, decomposition occurs in the aqueous phase above the critical point of water (374°C/ 221 atmospheres or atm). A schematic of a generic SCWO process is provided in Figure 28-1. As shown in this figure, the feed stream is typically an aqueous waste. An oxidant such as air, oxygen, or hydrogen peroxide must be provided unless the waste itself is an oxidant.

A supplemental fuel source should also be available. Because oxidation is exothermic, SCWO is self-sustaining for a waste stream with an adequate chemical oxygen demand (COD). According to developers, SCWO is self-sustaining, provided the waste stream has a COD of approximately 15,000 mg/L or higher [15]. Theoretically, SCWO may be self-sustaining for CODs as low as 5,000 mg/L [10]. At startup and for dilute wastes that will not autogenically sustain combustion, a supplemental fuel such as waste oil is added [17]. Alternatively, some dilute wastes can be dewatered until they are concentrated enough to sustain SCWO without supplemental fuel [18]. Concentrated wastes, on the other hand, must be diluted if the oxidation of the waste will generate more heat than can be readily removed from the SCWO processing vessel [18].

The streams entering an SCWO reactor must be heated and pressurized to supercritical conditions. Influent streams are frequently heated by thermal contact with the hot effluent. Both influent pressure and backpressure (often a restriction of the outlet) must be provided. The influent streams are then combined at supercritical conditions, and oxidation occurs.

Certain properties of supercritical water make it an excellent medium for oxidation. Many of the properties of water change drastically near its critical point: the hydrogen bonds disappear and water becomes similar to a moderately polar solvent, oxygen and almost all hydrocarbons become completely miscible in water; mass transfer occurs almost instantaneously; and the solubility of inorganic salts drops to the parts per million (ppm) range [19]. Because inorganic salts (as well as certain other solids) are nearly insoluble in supercritical water, solids removal must be considered in the design of a SCWO reactor [7; 20; 21].

The liquid effluent from SCWO is cooled (often by heat exchange with the influent) and returned to ambient pressure. As the effluent is cooled and depressurized, compounds such as carbon dioxide and oxygen will vaporize. According to SCWO developers, the effluent contains relatively innocuous products. Organic materials produce carbon dioxide and water; additional products depend upon the components of the waste. Nitrogen compounds principally produce ammonia and nitrogen as well as small amounts of nitrogen oxides (NO_x); halogens produce the corresponding halogen acids; phosphorus produces phosphoric acid; and sulfur produces sulfuric acid [18].

Vendors are currently developing both surface and subsurface SCWO systems. Figure 28-2 is a schematic of a subsurface SCWO reactor. As shown in Figure 28-2, subsurface SCWO reactors will

---

[5]   Editor's Note: As of early 1996, the U.S. EPA was not actively pursuing development of SCWO technology (personal communication, Ronald Turner, RREL-Cincinnati, January 1996).

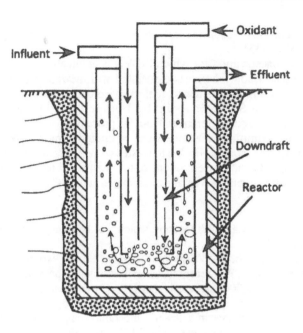

Figure 28-2. Subsurface SCWO reactor.

consist of columns of aqueous waste which are deep enough that the material near the bottom is subjected to a pressure of at least 221 atm [22]. To achieve this pressure solely through hydrostatic head, a water column depth of approximately 12,000 feet will be required [10]. The influent and effluent will flow in opposite directions in concentric vertical tubes [13]. In surface SCWO systems, the majority of the pressure is provided by a source other than gravity, and the reactor is on or above the earth's surface.

APPLICABILITY

Surface and subsurface SCWO systems may have slightly different applications. Because subsurface SCWO systems are below ground, developers claim that the earth will provide protection in the event of a catastrophic reactor failure. Subsurface designs have additional advantages over surface SCWO systems, including fewer mechanical parts (which should lead to lower maintenance) and pressure provided by hydrostatic head [13].

Surface SCWO systems, however, have several advantages over subsurface systems. Surface systems are much more accessible (and therefore easier to monitor) than subsurface reactors [13]. Developers project that it will not be cost-effective to construct subsurface reactors for small waste streams, as the drilling cost for the well is significant [10].

In general, applications of SCWO processes may include liquid wastes, sludges [13], and slurried solid wastes [18]. Potentially treatable compounds include halogenated and nonhalogenated aliphatic and aromatic hydrocarbons; aldehydes; ketones; esters; carbohydrates; organic nitrogen compounds; polychlorinated biphenyls (PCBs), phenols, and benzenes; aliphatic and aromatic alcohols; pathogens and viruses; mercaptans, sulfides, and other sulfur-containing compounds; dioxins and furans; leachable metals; and propellant components [12; 13; 18; 22; 23]. SCWO has been applied to municipal and industrial sludges. Tests performed on pulp mill sludges, for example, showed that SCWO can effectively treat these wastes (a total organic carbon destruction efficiency of 99.3 percent was achieved). Further analysis indicated that treatment of pulp mill sludges by SCWO should be able to compete economically with incineration and, in some regions, with landfilling [7].

SCWO also compares favorably with wet air oxidation (WAO), a commercially available technology which is similar to SCWO. In WAO, thermal decomposition and hydrolysis occur as well as oxidation. WAO is conducted in the aqueous phase and typically utilizes temperatures ranging from 150 to 300°C and pressures up to 200 atm. SCWO provides a number of advantages over WAO, including higher destruction efficiencies (DEs) and lower reaction times [24]. SCWO is also more energy-efficient than WAO [25].

The minimum waste concentration for which SCWO is applicable is waste-specific and can be determined by a cost comparison. The costs associated with dewatering the waste, operating the SCWO system, and purchasing supplemental fuel must be considered. There is also a maximum waste concentration for which SCWO is applicable because the oxidation of the waste must not generate more heat than can be readily removed from the processing vessel [18]. Note, however, that wastes which exceed the maximum concentration can be diluted prior to SCWO. MODAR literature states that its SCWO process is most applicable to wastes with hydrocarbon concentrations of 1 to 30 percent, but it does not specify the concentrations of the wastes fed to the SCWO reactor [21].

SCWO developers claim several advantages associated with SCWO as a means of destroying wastes:

- One vendor plans to design a SCWO system which will be transportable and thus applicable to Superfund sites [6].
- One developer claims that the SCWO process is odor-free and extremely quiet [11].
- According to developers, SCWO reactions are self-sustaining, provided the waste stream has a COD of approximately 5,000 mg/L or higher [10]. By contrast, self-sustaining incineration requires a minimum COD of approximately 300,000 mg/L [15].
- Because SCWO systems operate in a lower temperature range (400 to 600°C) than typical incineration systems, researchers believe that SCWO will produce lower quantities of NO_x [26].

Developers claim that SCWO is relatively safe because the reaction temperature can be controlled through adjustment of the degree of preheating and/or the concentration of the waste [7]. The high temperatures and pressures necessary for SCWO are potentially dangerous, but designing SCWO reactors with large safety factors should reduce the risk. One developer indicates the failure of a heater tube at approximately 3700 psi and 1400°F produced a loud pop and damage to local insulation, but no injuries and no damage to adjacent equipment or instrumentation. The developer further states that fluid loss from the rupture was minimal [6].

A second danger involves the possibility that the process could be interrupted, causing an incomplete reaction which could produce dangerous off-gases. SCWO systems can be designed to provide an emergency shutdown option and it is known that at least one pilot-scale system includes such a provision [6]. Note that the above are only potential dangers, as no safety problems were documented in the literature reviewed.

LIMITATIONS

The density of water drops rapidly between 300 and 400°C, and SCWO systems typically operate at or above 400°C. The low densities associated with the supercritical temperatures can result in the deposition of salts and pyrolytic chars. Deposition may result in plugging problems or added cleaning requirements. Some researchers prefer near-critical water oxidation at approximately 300°C, as the density of water is higher and salts and chars are more likely to remain dissolved [27]. Other developers prefer SCWO and are researching solutions to the deposition problem.

Possible problems due to corrosion must be examined when SCWO is considered. Several studies have been conducted regarding the minimization of corrosion in SCWO systems. Titanium, stainless steel 316, Hastelloy C-276, and Monel 400 were considered as alternative materials of construction for SCWO reactors. The results of these studies indicated that titanium had excellent corrosion resistance but its structural properties were unsatisfactory. Stainless steel 316 exhibited adequate corrosion resistance for use at low supercritical temperatures and moderate pH levels and chloride concentrations. A Hastelloy (or another nickel-chrome alloy) is recommended for more corrosive conditions (low pH levels or high chloride concentrations). The monel had poor corrosion resistance and is therefore not recommended for SCWO reactor construction [25]. The use of ceramics and ceramic coatings in conjunction with the above metals has also been proposed [10].

High-temperature flames which have been observed during SCWO may present an additional equipment problem in both surface and subsurface SCWO systems. Research is being conducted to determine what factors influence these "hydrothermal" flames because there is some concern that these flames will cause 'hot spots" which could weaken SCWO vessels [1].

Other drawbacks associated with SCWO (as well as other oxidation technologies) include the slow oxidation rate of many polyhalogenated hydrocarbons and the production of dioxins from the oxidation of certain halogenated organics [27]. The production of dioxins may not present a significant problem, however, as the destruction of dioxins by SCWO has been documented [7].

Acetic acid is generally considered one of the most refractory by-products of the SCWO of industrial wastes [28]. The acetic acid DEs shown in Table 28-1 reflect a portion of the performance data collected on this compound.

Ammonia, a second refractory compound, is produced by water oxidation of nitrogen-containing wastes at temperatures of 300 to 400°C [19]. Water oxidation does not degrade ammonia at any significant rate at these temperatures. If a water oxidation system is to be operated at or below 400°C, the ammonia may be removed by steam stripping or some other method. Above 425°C, organic nitrogen and ammonia in an SCWO system will decompose at a significant rate [19]. The primary products of this decomposition (below 650°C) are N_2 and N_2O, which further decompose to form N_2 and O_2 [12].

PERFORMANCE DATA

Significant bench- and pilot-scale SCWO performance data are available. Typical DEs for a number of compounds are summarized in Table 28-1. Although several low DEs are included in this table to illustrate the fact that DE is proportional to both temperature and residence time, DEs in excess of 99 percent can be achieved for nearly all the pollutants studied.

Studies have been conducted to examine the effects of various parameters on SCWO DEs. The operating parameters studied include temperature, residence time, pressure, feed concentration, amount of oxidant (as a multiple of stoichiometric requirements), and type of oxidant [13; 16; 28].

As noted above, DE was found to increase with operating temperature and residence time. DE also increases with operating pressure, but only slightly [28]. Recent studies also indicate that the addition of catalysts such as potassium permanganate, manganous sulfate, copper, and iron can enhance DEs [13].

In at least one study, DE was found to increase slightly with feed concentration. The relationship between DE and amount of excess oxidant provided has also been examined. DE increases with increasing amounts of oxidant from 100 to 300 percent of the stoichiometric requirements; adding over 300 percent of the stoichiometric amount of oxidant does not significantly affect DEs [16; 28].

Early SCWO systems used either oxygen or air as oxidants. Bench-scale studies were conducted to compare the DEs resulting from the use of air and oxygen, but no statistical difference was found [13]. In 1987, Welch and Siegwarth developed and patented a variation of SCWO which uses hydrogen peroxide as the oxidant. In Welch and Siegwarth's system, liquid hydrogen peroxide is mixed with the influent wastewater or slurry [13].

Welch, Siegwarth, and other researchers have shown that the use of hydrogen peroxide as an oxidant in SCWO systems produced DEs which were significantly higher than those obtained from the use of air or oxygen for the compounds tested [13; 28]. Oxidation with hydrogen peroxide and oxidation with oxygen or air proceed by different mechanisms. This difference may result in higher DEs for either hydrogen peroxide or oxygen, depending on the particular organic compounds being degraded [28]. Several other factors may influence the choice between oxidants. Hydrogen peroxide is significantly more expensive than oxygen, but aqueous hydrogen peroxide is easier to pump, requires a less expensive feed system, and may be combined with the influent more readily than oxygen [10; 28].

PROCESS RESIDUALS

In general, residuals from SCWO processes include gases, liquids, and solids. The gaseous effluent from the bench-scale treatment of pulp mill sludges was found to primarily consist of oxygen and carbon dioxide, with small concentrations of nitrogen [7]. Gaseous effluent from the bench-scale treatment of propellant components was also analyzed and found to contain nitrous oxide (N_2O) and

Table 28-1. SCWO Performance Data

Pollutant	Temp. (°C)	Pressure (atm.)	DE (%)	React Time (min.)	Oxidant	Ref.	Feed Conc. (mg/L)
1,1,1-Trichloroethane	495		99.99	4	Oxygen	13	
1,1,2,2-Tetrachloroethylene	495		99.99	4	Oxygen	13	
1,2-Ethylene dichloride	495		99.99	4	Oxygen	13	
2,4-Dichlorophenol	400		33.7	2	Oxygen	13	2,000
2,4-Dichlorophenol	400		99.440	1	H_2O_2	13	2,000
2,4-Dichlorophenol	450		63.3	2	Oxygen	28	2,000
2,4-Dichlorophenol	450		99.950	1	H_2O_2	28	2,000
2,4-Dichlorophenol	500		78.2	2	Oxygen	28	2,000
2,4-Dichlorophenol	500		>99.995	1	H_2O_2	28	2,000
2,4-Dimethylphenol	580	443	>99	10	$H_2O_2+O_2$	29	135
2,4-Dinitrotoluene	410	443	83	3	Oxygen	29	84
2,4-Dinitrotoluene	528	287	>99	3	Oxygen	29	180
2-Nitrophenol	515	443	90	10	Oxygen	29	104
2-Nitrophenol	530	430	>99	15	$H_2O_2+O_2$	29	104
Acetic acid	400		3.10	5	Oxygen	13	2,000
Acetic acid	400		61.8	5	H_2O_2	13	2,000
Acetic acid	450		34.3	5	Oxygen	28	2,000
Acetic acid	450		92.0	5	H_2O_2	28	2,000
Acetic acid	500		47.4	5	Oxygen	28	2,000
Acetic acid	500		90.9	5	H_2O_2	28	2,000
Activated sludge (COD)	400	272	90.1	2		30	62,000
Activated sludge (COD)	400	306	94.1	15		30	62,000
Ammonium perchlorate	500	374	99.85	0.2	None	18	12,000

Cyclohexane	445		99.97	7	Oxygen	13	
DDT	505		99.997	4	Oxygen	13	
Dextrose	440		99.6	7	Oxygen	13	
Industrial sludge (TCOD)	425		>99.8	20	Oxygen	19	
Methyl ethyl ketone	505		99.993	4	Oxygen	13	
Nitromethane	400	374	84	3	None	18	10,000
Nitromethane	500	374	>99	0.5	None	18	10,000
Nitromethane	580	374	>99	0.2	None	18	10,000
o-Chlorotoluene	495		99.99	4	Oxygen	13	
o-Xylene	495		99.93	4	Oxygen	13	
PCB 1234	510		99.99	4	Oxygen	13	
PCB 1254	510		99.99	4	Oxygen	13	
Phenol	490	389	92	1	Oxygen	29	1,650
Phenol	535	416	>99	10	Oxygen	29	150

oxygen. Analysis by mass spectroscopy did not detect the presence of chlorine (Cl_2), nitrosyl chloride (NOCl), or nitrogen dioxide (NO_2). These are positive results because they indicate that SCWO avoided the hazardous products such as Cl_2 and NOCl formed in typical thermal decomposition. In addition, SCWO appears to produce relatively little NO_x [18].

The aqueous effluent from the SCWO of pulp mill sludge had a total organic concentration (TOC) of only 27 ppm. The major inorganics present were calcium, chlorine (as chloride ion), nitrogen (as ammonia), sodium, and sulfur (as sulfate). The minor elements identified were all present at concentrations below Environmental Protection Agency (EPA) ground-water pollution criteria [7]. Liquid effluent from the SCWO of propellant components contained sodium chloride (NaCl), nitrite, and nitrate. The developer believes that the majority of the chlorine from the propellant exists as NaCl, but a chlorine mass balance has not yet been attempted [18].

Limited data describing solid residue from SCWO are available. When a bench-scale SCWO system was used to treat pulp mill sludges, benzene and lead were the only pollutants which the toxicity characteristic leaching procedure (TCLP) detected at concentrations above EPA ground-water limits. Benzo(a)pyrene and PCB, however, had detection limits above the ground-water limit. Based on these results, the developer believes that the solid residue from SCWO should easily qualify for disposal in any sanitary landfill [7]. Before disposal in a sanitary landfill will be allowed, however, the residue must be delisted.

TECHNICAL CONTACT

Technology-specific questions regarding SCWO may be directed to:

Dr. Earnest F. Gloyna
University of Texas at Austin
Balcones Research Center
10100 Burnet Road
Austin, TX 78758
(512) 471-7053 or 471-5928

EPA CONTACT

Technology-specific questions regarding SCWO may be directed to Ronald Turner, RREL-Cincinnati, (513) 569-7775 (see Preface for mailing address).

ACKNOWLEDGMENTS

This chapter was prepared for the U.S. EPA, Office of Research and Development (ORD), Risk Reduction Engineering Laboratory (RREL), Cincinnati, Ohio, by Science Applications International Corporation (SAIC) under EPA Contract No. 68-C8-4062. Mr. Ronald Turner served as the EPA Technical Project Monitor. Mr. Thomas Wagner was SAIC's Work Assignment Manager. This chapter was written by Ms. Sharon Krietemeyer of SAIC.

The following Agency, contractor, and vendor personnel have contributed their time and comments by peer reviewing the chapter:

Mr. Thomas Wagner, SAIC
Mr. Michael Carolan, City Management Corporation
Mr. L. Jack Davis, Eco Waste Technologies
Dr. Earnest F. Gloyna, University of Texas at Austin
Mr. Glenn T. Hong, MODAR, Inc.
Mr. James Titmas, GeneSyst International

REFERENCES

1. Supercritical Oxidation Destroys Aqueous Toxic Wastes. NTIS Technical Note prepared by the U.S. Department of Energy, Washington, DC. February 1991.
2. Letter from Glenn T. Hong of MODAR, Inc. March 13, 1992.
3. New Process Purifies Waste Simply, Safely, Experts Say. Associated Press.
4. Oleson, M., T. Slavin, F. Liening, and R.L. Olson. Controlled Ecological Life Support Systems (CELSS) Physiochemical Waste Management Systems Evaluation. Prepared by Boeing Aerospace Company for the National Aeronautics and Space Administration, Washington, DC. June 1986.
5. Tester, J.W., G.A. Huff, R.K. Helling, T.B. Thomasson, and K.C. Swallow. Prepared by Massachusetts Institute of Technology for the National Aeronautics and Space Administration, Washington, DC. 1986.
6. Letter from L. Jack Davis of Eco Waste Technologies. March 27, 1992.
7. Modell, M. Treatment of Pulp Mill Sludges by Supercritical Water Oxidation: Final Report. Prepared for the U.S. Department of Energy, Office of Industrial Programs, Washington, DC. July 1990.
8. Chemical & Engineering News. Letter to the Editor from Herbert E. Barner of ABB Lummus Crest, Inc. March 2, 1992.
9. Chemical & Engineering News. Letter to the Editor from William R. Killilea of MODAR, Inc. March 2, 1992.
10. Letter from James Titmas of GeneSyst International, Inc. March 14, 1992.
11. Scarlett, H. Hot Water, Pressure Process Can Destroy Toxic Waste. Houston Post. April 8, 1990.
12. Gloyna, E.F. and K. Johnston. Supercritical Water and Solvent Oxidation. Presented at the 11th Industrial Symposium on Wastewater Treatment, Montreal, Quebec, Canada, November 21–22, 1988.
13. Adrian, M.A. Partial Literature Survey on Supercritical Water Oxidation. The University of Texas at Austin. May 1991.
14. Letter from Michael Carolan of City Management Corporation. March 18, 1992.
15. Gloyna, E.F. Supercritical Water Oxidation, Deep-Well Reactor Model Development. Second Year Proposal Prepared for the U.S. Environmental Protection Agency, Grants Administration Division, Washington, DC. May 1991.
16. Wilmanns, E., L. Li, and E.F. Gloyna. Supercritical Water Oxidation of Volatile Acids. Presented at the AIChE August 1989 Summer Meeting, Philadelphia, Pennsylvania. August 1989.
17. Staszak, C., K. Malinowski, and W. Killilea. The Pilot-Scale Demonstration of the MODAR Oxidation Process for the Destruction of Hazardous Organic Waste Materials. Environmental Progress. 6(1):39, 1987.
18. Buelow, S.J., R.B. Dyer, C.K. Rofer, J.H. Atencio, and J.D. Wander. Destruction of Propellant Components in Supercritical Water. Submitted to the Workshop on the Alternatives to Open Burning/Open Detonation of Propellants and Explosives. Prepared by the Los Alamos National Laboratory for the U.S. Department of Energy. May 1990.
19. Shanableh, A. and E.F. Gloyna. Supercritical Water Oxidation—Wastewaters and Sludges. Presented at International Association for Water Pollution Research and Control Conference, Kyoto, Japan. August 1990.
20. MODAR marketing brochures. Circa 1987.
21. Lawson, M. New Technology Tackles Dilute Wastes. Chemical Week, October 1986.
22. GeneSyst International, Inc. The Gravity Pressure Vessel. June 1990.
23. Technology Screening Guide for Treatment of CERCLA Soils and Sludges. Office of Solid Waste and Emergency Response, Office of Emergency and Remedial Response, EPA/540/2-88/004, U.S. Environmental Protection Agency, Washington, DC. September 1988.
24. Lee, D.S., A. Kanthasamy, and E.F. Gloyna. Supercritical Water Oxidation of Hazardous Organic Compounds. Prepared for Presentation at AIChE Annual Meeting, November 20–24, 1991.
25. Matthews, C.F. and E.F. Gloyna. Corrosion Behavior of Three High-Grade Alloys in Supercritical Water Oxidation Environments. July 1991.
26. Discussion of Waste Destruction Results (from MODAR marketing literature).
27. Mill, T. and D. Ross. Effective Treatment of Hazardous Waste Under Hydrothermal Conditions. 1991.
28. Lee, D., L. Li, and E.F. Gloyna. Efficiency of H_2O_2 and O_2 in Supercritical Water Oxidation of 2,4-Dichlorophenol and Acetic Acid. Submitted for presentation at AIChE Spring National Meeting, Orlando, Florida, March 18–22, 1990.
29. Lee, D. and E.F. Gloyna. Supercritical Water Oxidation a Microreactor System. Presented at WPCF Specialty Conference, New Orleans, Louisiana. April 1989.
30. Hartmann, G. et al. Water Oxidation of Sludges and Toxic Wastes. Presented at ASCE Conference, Austin, Texas. July 1989.

Appendix A

Sources of Additional Information

J. Russell Boulding, Boulding Soil-Water Consulting, Bloomington, IN

HOW TO OBTAIN DOCUMENTS FROM EPA

EPA publications can be obtained at no charge (while supplies are available) from the following sources:

EPA/625-series documents: ORD Publications, P.O. Box 19968, Cincinnati, OH 45219-0968; phone 513 569-7562, fax 513 569-7562.

Other EPA documents: National Center for Environmental Publications and Information (NCEPI), 11029 Kenwood Road, Cincinnati, OH 45242; fax 513 891-6685.

Other documents, for which an NTIS acquisition number is shown can be obtained from the National Technical Information Service (NTIS), Springfield, VA 22161; 800 336-4700, fax 703/321-8547.

Resource Guide Series (Available from NCEPI):

Bioremediation Resource Guide (EPA/542-B-93-004), September 1993, 28 pp.

Soil Vapor Extraction (SVE) Treatment Technology Resource Guide (EPA/542-B-94-007), September 1994, 31 pp.

Physical/Chemical Treatment Technology Resource Guide (EPA/542-B-94-006), September 1994, 43 pp.

Ground-Water Treatment Technology Resource Guide (EPA/542-B-94-009), September 1994, 29 pp.

Other Major Recent References on Remediation:

Evaluation of Technologies for In-Situ Cleanup of DNAPL Contaminated Sites (EPA/600/R-94/120; NTIS PB94-195039), August 1994, 173 pp.

Contaminants and Remedial Options at Solvent Contaminated Sites (EPA/600/R-94/203), 171 pp.

Remediation Technologies Screening Matrix and Reference Guide (EPA/542/B-94/013; NTIS PB95-104782), October 1994. [Prepared by DOD Environmental Technology Transfer Committee and incorporates remediation technology reports prepared by U.S. Army Environmental Center, Federal Remediation Technologies Roundtable, U.S. EPA, U.S. Department of Energy, U.S. Air Force, California Base Closure Committee and U.S. Navy; single most comprehensive reference source available]

Superfund Innovative Technology Evaluation (SITE) Program Reports:

The annually revised *Superfund Innovative Technology Program Technology Profiles* includes summaries of all demonstration projects and emerging technology projects and identifies all published reports related to specific projects. The 7th edition (EPA/540/R-94/526) was published November 1994. The 8th edition was not available at the time this volume went to press. Requests to be placed on the mailing list to receive information about the SITE program should be addressed to U.S. EPA/ORD Publications Unit (MS-G72), 26 West Martin Luther King Drive, Cincinnati, OH 45268.

Guides for Conducting Treatability Studies Under CERCLA Series *(First volume is the full report and the second volume is a Quick Reference Fact Sheet; most are still available from NCEPI):*

    Aerobic Biodegradation Remedy Screening (EPA/540/2-91/013A&B)
    Biodegradation Remedy Selection (EPA/540/R-93/519A&B)
    Soil Vapor Extraction (EPA/540/2-91/019A&B)
    Soil Washing (EPA/540/2-91/020A&B)
    Chemical Dehalogenation (EPA/540/R-92/013A&B)
    Solvent Extraction (EPA/540/R-92/016A&B)
    Thermal Desorption (EPA/540/R-92/074A&B)
    Guide for Conducting Treatability Studies Under CERCLA - Update (EPA/540/R-92/071A)

In Situ Remediation Technology Status Reports *(available from NCEPI):*

    Surfactant Enhancement (EPA 542-K-94-003), April 1995, 22 pp.
    Treatment Walls (EPA 542-K-94-004), April 1995, 26 pp.
    Hydraulic and Pneumatic Fracturing (EPA 542-K-94-005), April 1995, 15 pp.
    Cosolvents (EPA 542-K-94-006) April, 1995, 6 pp.
    Electrokinetics (EPA 542-K-94-007) April 1995, 20 pp.
    Thermal Enhancements (EPA 542-K-94-009) April 1995, 22 pp.

PC-Based Database Software:

    RREL Treatability Database: Version 5.0 available Glenn Shaul, RREL-Cincinnati, (513) 569-7408 (See Preface for address).
    VISITT (Vendor Information System for Innovative Treatment Technologies): Software and Version 4.0 software available from NCEPI.

Index

Printed and bound by CPI Group (UK) Ltd, Croydon, CR0 4YY

23/10/2024

01778245-0019